LIST OF EXAMPLES

D0215279

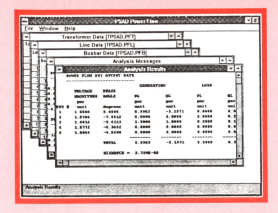

POWER SYSTEM ANALYSIS AND DESIGN Software
to accompany
Glover/Sarma
POWER SYSTEM ANALYSIS AND DESIGN
Second Edition

PWS Publishing Company
*Look to PWS for innovative products
for education and industry.*

This set of nine DOS and Windows programs (for IBM PCs or compatibles with 4 MB RAM, hard drive, and Microsoft Windows 3.1) enables students and professional engineers to solve design-related problems in the text and to conduct analysis and design studies. The nine programs cover the following topics: **Symmetrical Components; Line Constants; Transmission Lines: Steady-State Operation; Symmetrical Short Circuits; Short Circuits; Transmission Line Transients; and Transient Stability** (all in DOS format); **Matrix Operations; and Power Flow** (in Windows).

- -

Detach and mail or fax to:
PWS PUBLISHING COMPANY 20 Park Plaza, Boston, MA 02116 Fax: (617) 338-6134

_____**Yes**, I would like to order the POWER SYSTEM ANALYSIS AND DESIGN SOFTWARE User's Guide and IBM PC disk package (ISBN: 0-534-93962-7) for my own use.

Please send me _____copies @ 24.95 each, plus $2.00 handling charges*
*Residents of AL, AZ, CA, CO, CT, FL, GA, IL, KY, LA, MA, MD, MI, MN, MO, NC, NJ, OH, PA, RI, SC, TN, TX, UT, VA, WA, WI must add sales tax.

_____**Payment enclosed** (check, money order, purchase order) OR Please charge my:
_____**MasterCard**, _____**Visa**, _____**Amex**. For individual telephone credit card orders, call 1-800-842-3636.

Account number_____Expiration Date_____

Signature _____Name_____

School_____Dept_____

Address_____

City_____State_____Zip_____Phone_____

_____**Yes**, I would like to receive the POWER SYSTEM ANALYSIS AND DESIGN SOFTWARE User's Guide and IBM PC disk package. For qualified instructors only; please make requests on department letterhead. (ISBN: 0-534-93962-7)
_____**Yes**, I would like to receive the Instructor's Manual for the Glover/Sarma text. (ISBN: 0-534-93961-9). For qualified instructors only.

9402235

POWER SYSTEM ANALYSIS
AND DESIGN

NASA LIBRARY
AMES RESEARCH CENTER
MOFFETT FIELD, CALIF.

THE PWS SERIES IN ENGINEERING

Anderson, *Thermodynamics*

Askeland, *The Science and Engineering of Materials*, Third Edition

Bolluyt, Stewart and Oladipupo, *Modeling For Design Using SilverScreen*

Borse, *FORTRAN 77 and Numerical Methods for Engineers*, Second Edition

Clements, *68000 Family Assembly Language*

Clements, *Microprocessor Systems Design*, Second Edition

Clements, *Principles of Computer Hardware*, Second Edition

Fleischer, *Introduction to Engineering Economy*

Gere and Timoshenko, *Mechanics of Materials*, Third Edition

Glover and Sarma, *Power System Analysis and Design*, Second Edition

Keedy, *Introduction to CAD using CADKEY*, Second Edition

Knight, *The Finite Element Method in Mechanical Design*

Logan, *A First Course in the Finite Element Method*, Second Edition

McGill and King, *Engineering Mechanics: Statics*, Second Edition

McGill and King, *Engineering Mechanics: An Introduction to Dynamics*, Second Edition

McGill and King, *Engineering Mechanics: Statics and An Introduction to Dynamics*, Second Edition

Raines, *Software for Mechanics of Materials*

Shen and Kong, *Applied Electromagnetism*, Second Edition

Strum and Kirk, *Contemporary Linear Systems using MATLAB*

Sule, *Manufacturing Facilities*, Second Edition

Vardeman, *Statistics for Engineering Problem Solving*

Weinman, *VAX FORTRAN*, Second Edition

Weinman, *FORTRAN for Scientists and Engineers*

SECOND EDITION

POWER SYSTEM ANALYSIS AND DESIGN

J. Duncan Glover
Failure Analysis Associates, Inc.
Electrical Division

Mulukutla Sarma
Northeastern University

PWS Publishing Company

Boston

PWS PUBLISHING COMPANY
20 Park Plaza, Boston, MA 02116-4324

Copyright © 1994, 1987 by PWS Publishing Company.

All rights reserved. No part of this book may be reproduced, stored in a retrieval system, or transcribed in any form or by any means—electronic, mechanical, photocopying, recording, or otherwise—without the prior written permission of PWS Publishing Company.

PWS Publishing Company is a division of Wadsworth, Inc.

I(T)P™
International Thomson Publishing
The trademark ITP is used under license.

Library of Congress Cataloging-in-Publication Data

Glover, J. Duncan.
    Power system analysis and design / J. Duncan Glover, Mulukutla Sarma.—2nd ed.
      p. cm.
    Includes index.
    ISBN: 0-53493-960-0
  1. Electric power systems—Data processing.   I. Sarma, Mulukutla S.
II. Title.
  TK1005.G57 1993   93-34243
  621.319—dc20   CIP

Sponsoring Editor: *Jonathan Plant*
Assistant Editor: *Mary Thomas*
Editorial Assistant: *Cynthia Harris*
Production Coordinator: *Elise S. Kaiser*
Production: *The Book Company*
Interior Designer: *Catherine Griffin*
Cover Designer: *Monique Calello*
Cover Image: *Copyright © Steven Hunt/The Image Bank*
Marketing Manager: *Nathan Wilbur*
Manufacturing Coordinator: *Marcia A. Locke*
Compositor: *Doyle Graphics*
Cover Printer: *Henry N. Sawyer Co.*
Text Printer and Binder: *Arcata Graphics/Martinsburg*

Printed and bound in the United States of America
93 94 95 96 97 98 — 10 9 8 7 6 5 4 3 2 1

To Tatiana and John

CONTENTS

CHAPTER 12 | Transmission Lines: Transient Operation   461

CHAPTER 13 | Transient Stability   529

PREFACE

The objective of this book is to present methods of power system analysis and design, particularly with the aid of a personal computer, in sufficient depth to give the student the basic theory at the undergraduate level. The approach is designed to develop students' thinking process, enabling them to reach a sound understanding of a broad range of topics related to power-system engineering, while motivating their interest in the electrical power industry. Believing that fundamental physical concepts underlie creative engineering and form the most valuable and permanent part of an engineering education, physical concepts are highlighted while also giving due attention to mathematical techniques. Both theory and modeling are developed from simple beginnings so that they can be readily extended to new and complex situations.

This edition of the text has four new features: a chapter on system protection; case studies in Chapters 4 through 13 describing present-day, practical applications; an updated, user-friendly software package that includes more graphics for output displays and student design projects that have been classroom-tested; and additional problems at the end of each chapter. These features are in response to a questionnaire sent to a number of power-engineering educators, whose suggestions and valuable comments are greatly appreciated.

The text is intended to be fully covered in a two-semester or three-quarter course offered to seniors and first-year graduate students. The organization of chapters and individual sections is flexible enough to give the instructor sufficient latitude in choosing topics to cover, especially in a one-semester course. The text is supported by an ample number of worked examples covering most of the theoretical points raised. The numerous problems to be worked with a calculator as well as problems to be worked using a personal computer have been expanded in this edition.

As background for this course, it is assumed that students have had courses in electric network theory (including transient analysis) and ordinary differential equations, and have been exposed to linear systems, matrix algebra, and computer programming. In addition, it would be helpful, but not necessary, to have had an electric machines course.

After an introduction to the history of electric power systems along with present and future trends, Chapter 2 on fundamentals orients the students to the terminology and serves as a brief review. The chapter reviews phasor concepts, power, and single-phase as well as three-phase circuits. Chapter 3, "Symmetrical Components," discusses the advantages of linear transformations and includes some applications.

Chapters 4 through 7 examine power transformers, transmission-line parameters, steady-state operation of transmission lines, and power flows including the Newton–Raphson method. These chapters provide a basic understanding of power systems under balanced three-phase, steady-state, normal operating conditions.

Chapters 8 through 10, which cover symmetrical faults, unsymmetrical faults, and system protection, come under the general heading of power system short-circuit protection. Chapter 11 is a self-contained chapter on power system controls, including turbine-generator controls, load-frequency control, and economic dispatch.

The last two chapters are on transient operation of transmission lines, including surge protection; and transient stability, which includes the swing equation, the equal-area criterion, and multimachine stability. These self-contained chapters come under the general heading of power-system transients.

A novel feature of this text is a personal computer (PC) software package that includes a manual and a 3.5″ disk with computer programs for nine of the chapters. The programs may be taught as an integral part of the course or may be omitted without loss of continuity. The programs are written for IBM and compatible PCs that can run Microsoft Windows 3.1 or higher. Specific problems with practical engineering design orientation to be solved with the help of the PC package are included at the end of chapters. Operating instructions with sample runs, flowcharts, and student design projects that have been tested at three universities are given in the second edition of the software manual. The data-handling and number-crunching capabilities of today's PC allow students to work on more difficult and realistic problems and make it an innovative tool in the learning process. The computer programs that accompany this text can also be used by students later in their careers and by professionals.

Acknowledgments

The material in this text was gradually developed to meet the needs of classes taught at universities in the United States and abroad over the past 25 years. The 13 chapters were written by the first author, J. D. Glover, who is indebted to many people who helped during the planning and writing of this book. The profound influence of earlier texts written on Power Systems, particularly by W. D. Stevenson, Jr., and the developments made by various outstanding engineers are gratefully acknowledged. Details of sources can only be made through references at the end of each chapter, as they are otherwise too numerous to mention.

Jonathan Plant, Mary Thomas, and Elise Kaiser of PWS are commended for their broad knowledge, skills, and diligence in publishing this edition. The following reviewers have made substantial contributions to this edition and provided prerevision survey comments that were extremely helpful in preparation of the second edition:

Max D. Anderson
University of Missouri—Rolla

Sohrab Asgarpoor
University of Nebraska—Lincoln

Kaveh Ashenayi
University of Tulsa

Richard D. Christie, Jr.
University of Washington

Mariesa L. Crow
University of Missouri—Rolla

Richard G. Farmer
Arizona State University

Saul Goldberg
*California Polytechnic State
    University*

Clifford H. Grigg
*Rose–Hulman Institute
    of Technology*

Howard B. Hamilton
University of Pittsburgh

Leo Holzenthal, Jr.
University of New Orleans

Walid Hubbi
New Jersey Institute of Technology

Charles W. Isherwood
*University of Massachusetts—
    Dartmouth*

W. H. Kersting
New Mexico State University

Wayne E. Knabach
South Dakota State University

Pierre-Jean Lagacé
*IREQ Institut de Recherche
    d'Hydro-Québec*

James T. Lancaster
Alfred University

Kwang Y. Lee
Pennsylvania State University

Mohsen Lotfalian
University of Evansville

René B. Marxheimer
San Francisco State University

Lamine Mili
*Virginia Polytechnic Institute and
    State University*

Osama A. Mohammed
Florida International University

Clifford C. Mosher
Washington State University

Anil Pahwa
Kansas State University

M. A. Pai
*University of Illinois at
    Urbana-Champaign*

R. Ramakumar
Oklahoma State University

Teodoro C. Robles
Milwaukee School of Engineering

Ronald G. Schultz
Cleveland State University

Stephen A. Sebo
Ohio State University

Raymond Shoults
University of Texas at Arlington

Richard D. Shultz
*University of Wisconsin—
    Platteville*

Charles Slivinsky
University of Missouri—Columbia

John P. Stahl
Ohio Northern University

E. K. Stanek
University of Missouri—Rolla

Robert D. Strattan
University of Tulsa

Tian-Shen Tang
Texas A&M University—Kingsville

S. S. Venkata
University of Washington

Francis M. Wells
Vanderbilt University

Bill Wieserman
*University of Pennsylvania—
    Johnstown*

Stephen Williams   Salah M. Yousif
U.S. Naval Postgraduate School   *California State University—*
                                          Sacramento

In addition, the following reviewers made many contributions to the first edition: Frederick C. Brockhurst, *Rose-Hulman Institute of Technology*; Bell A. Cogbill, *Northeastern University*; Saul Goldberg, *California Polytechnic State University*; Mack Grady, *University of Texas at Austin*; Leonard F. Grigsby, *Auburn University*; Howard Hamilton, *University of Pittsburgh*; William F. Horton, *California Polytechnic State University*; W. H. Kersting, *New Mexico State University*; John Pavlat, *Iowa State University*; R. Ramakumar, *Oklahoma State University*; B. Don Russell, *Texas A & M*; Sheppard Salon, *Rensselaer Polytechnic Institute*; Stephen A. Sebo, *Ohio State University*; and Dennis O. Wiitanen, *Michigan Technological University*.

In conclusion, may we add that our objective in writing this text and the accompanying software package will have been fulfilled if the book is considered to be student-oriented, comprehensive, and up to date, with consistent notation and necesary detailed explanation at the level for which it is intended.

LIST OF SYMBOLS, UNITS, AND NOTATION

SYMBOL	DESCRIPTION
a	operator $1\underline{/120°}$
a_t	transformer turns ratio
A	area
A	transmission line parameter
A	symmetrical components transformation matrix
B	loss coefficient
B	frequency bias constant
B	phasor magnetic flux density
B	transmission line parameter
C	capacitance
C	transmission line parameter
D	distance
D	transmission line parameter
E	phasor source voltage
E	phasor electric field strength
f	frequency
G	conductance
G	conductance matrix
H	normalized inertia constant
H	phasor magnetic field intensity
$i(t)$	instantaneous current
I	current magnitude (rms unless otherwise indicated)
I	phasor current
I	vector of phasor currents
j	operator $1\underline{/90°}$
J	moment of inertia
l	length
l	length
L	inductance
L	inductance matrix
N	number (of buses, lines, turns, etc.)
p.f.	power factor
$p(t)$	instantaneous power
P	real power
q	charge
Q	reactive power
r	radius
R	resistance
R	turbine-governor regulation constant

SYMBOL	DESCRIPTION
$\mathbf{R}$	resistance matrix
s	Laplace operator
S	apparent power
S	complex power
t	time
T	period
T	temperature
T	torque
$v(t)$	instantaneous voltage
V	voltage magnitude (rms unless otherwise indicated)
V	phasor voltage
$\mathbf{V}$	vector of phasor voltages
X	reactance
$\mathbf{X}$	reactance matrix
Y	phasor admittance
$\mathbf{Y}$	admittance matrix
Z	phasor impedance
$\mathbf{Z}$	impedance matrix
α	angular acceleration
α	transformer phase shift angle
β	current angle
β	area frequency response characteristic
δ	voltage angle
δ	torque angle
ε	permittivity
Γ	reflection or refraction coefficient
λ	magnetic flux linkage
λ	penalty factor
	magnetic flux
ρ	resistivity
τ	time in cycles
τ	transmission line transit time
θ	impedance angle
θ	angular position
μ	permeability
ν	velocity of propagation
ω	radian frequency

SI UNITS

A	ampere
C	coulomb
F	farad
H	henry
Hz	hertz
J	joule
kg	kilogram
m	meter
N	newton
rad	radian
s	second
S	siemen
VA	voltampere
var	voltampere reactive
W	watt
Wb	weber
Ω	ohm

ENGLISH UNITS

BTU	British thermal unit
cmil	circular mil
ft	foot
hp	horsepower
in	inch
mi	mile

NOTATION

Lowercase letters such as $v(t)$ and $i(t)$ indicate instantaneous values.

Uppercase letters such as V and I indicate rms values.

Uppercase letters in italic such as V and I indicate rms phasors.

Matrices and vectors with real components such as **R** and **I** are indicated by boldface type.

Matrices and vectors with complex components such as $\boldsymbol{Z}$ and $\boldsymbol{I}$ are indicated by boldface italic type.

Superscript T denotes vector or matrix transpose.

Asterisk (*) denotes complex conjugate.

■ indicates the end of an example and continuation of text

▢ highlights sections and problems that utilize personal computer programs.

CHAPTER 1

INTRODUCTION

1300 MW coal-fired power plant
(Courtesy of American Electric Power)

Electrical engineers are concerned with every step in the process of generation, transmission, distribution, and utilization of electrical energy. The electric utility industry is probably the largest and most complex industry in the world. The electrical engineer who works in that industry will encounter challenging problems in designing future power systems to deliver increasing amounts of electrical energy in a safe, clean, and economical manner.

The objectives of this chapter are to review briefly the history of the electric utility industry, to discuss present and future trends in electric power systems, and to introduce a set of digital computer programs employed in power-system engineering that will be studied in later chapters.

SECTION 1.1

HISTORY OF ELECTRIC POWER SYSTEMS

In 1878 Thomas A. Edison began work on the electric light and formulated the concept of a centrally located power station with distributed lighting serving a surrounding area. He perfected his light by October 1879, and the opening of his historic Pearl Street Station in New York City on September 4, 1882, marked the beginning of the electric utility industry (see Figure 1.1).

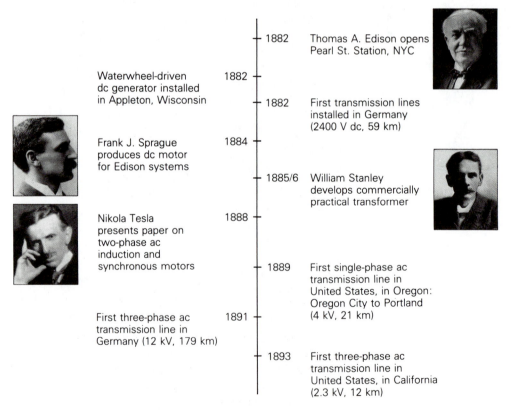

Figure 1.1   Milestones of the early electric utility industry [1]

At Pearl Street, dc generators, then called dynamos, were driven by steam engines to supply an initial load of 30 kW for 110-V incandescent lighting to 59 customers in a 1-square-mile area. From this beginning in 1882 through 1972 the electric utility industry grew at a remarkable pace—a growth based on continuous reductions in the price of electricity due primarily to technological acomplishment and creative engineering.

The introduction of the practical dc motor by Sprague Electric, as well as the growth of incandescent lighting, promoted the expansion of Edison's dc systems. The development of three-wire 220-V dc systems allowed load to increase somewhat, but as transmission distances and loads continued to increase, voltage problems were encountered. These limitations of maximum distance and load were overcome in 1885 by William Stanley's development of a commercially practical transformer. Stanley installed an ac distribution system in Great Barrington, Massachusetts, to supply 150 lamps. With the transformer, the ability to transmit power at high voltage with corresponding lower current and lower line-voltage drops made ac more attractive than dc. The first single-phase ac line in the United States operated in 1889 in Oregon, between Oregon City and Portland—21 km at 4 kV.

The growth of ac systems was further encouraged in 1888 when Nikola Tesla presented a paper at a meeting of the American Institute of Electrical Engineers describing two-phase induction and synchronous motors, which made evident the advantages of polyphase versus single-phase systems. The first three-phase line in Germany became operational in 1891, transmitting power 179 km at 12 kV. The first three-phase line in the United States (in California) became operational in 1893, transmitting power 12 km at 2.3 kV. The three-phase induction motor conceived by Tesla went on to become the workhorse of the industry.

In the same year that Edison's steam-driven generators were inaugurated, a waterwheel-driven generator was installed in Appleton, Wisconsin. Since then, most electric energy has been generated in steam-powered and in water-powered (called hydro) turbine plants. Today, steam turbines account for more than 85% of U.S. electric energy generation, whereas hydro turbines account for about 10%. Gas turbines are used in some cases during short periods to meet peak loads.

Steam plants are fueled primarily by coal, gas, oil, and uranium. Of these, coal is the most widely used fuel in the United States due to its abundance in the country. Although many of these coal-fueled power plants were converted to oil during the early 1970s, that trend has been reversed back to coal since the 1973/74 oil embargo, which caused an oil shortage and created a national desire to reduce dependency on foreign oil. In 1957 nuclear units with 90 MW steam-turbine capacity, fueled by uranium, were installed, and today nuclear units with 1280 MW steam-turbine capacity are in service. But the recent growth of nuclear capacity has been slowed by rising construction costs, licensing delays, and public opinion.

Other types of electric power generation are also being used, including wind-turbine generators; geothermal power plants, wherein energy in the form of steam or hot water is extracted from the earth's upper crust; solar cell arrays; and tidal power plants. These sources of energy cannot be ignored, but they are not expected to supply a large percentage of the world's future energy needs. On the other hand, nuclear fusion energy just may. Substantial research efforts now underway show nuclear fusion energy to be the most promising technology for producing safe, pollution-free, and economical electric energy in the next century and beyond. The fuel consumed in a nuclear fusion reaction is deuterium, of which a virtually inexhaustible supply is present in seawater.

The early ac systems operated at various frequencies including 25, 50, 60 and 133 Hz. In 1891, it was proposed that 60 Hz be the standard frequency in the United States. In 1893, 25-Hz systems were introduced with the synchronous converter. However, these systems were used primarily for railroad electrification (and many are now retired) since they had the disadvantage of causing incandescent lights to flicker. In California, the Los Angeles Department of Power and Water operated at 50 Hz, but converted to 60 Hz when power from the Hoover Dam became operational in 1937. In 1949 Southern California Edison also converted from 50 to 60 Hz. Today, the two standard frequencies for generation, transmission, and distribution of electric power in the world are 60 Hz (in the United States, Canada, Japan, Brazil) and 50 Hz (in Europe, the former Soviet republics, South America except Brazil, India, also Japan). The advantage of 60-Hz systems is that generators, motors, and transformers in these systems are generally smaller than 50-Hz equipment with the same ratings. The advantage of 50-Hz systems is that transmission lines and transformers have smaller reactances at 50 Hz than at 60 Hz.

As shown in Figure 1.2, the rate of growth of electric energy in the United States was approximately 7% per year from 1902 to 1972. This corresponds to a doubling of electric energy consumption every 10 years over the 70-year period. In other words, every 10 years the industry installed a new electric system equal in energy-producing capacity to the total of what it had built since the industry began. The annual growth rate slowed after the oil embargo of 1973/74. Kilowatt-hour consumption in the United States increased by 3.4% per year from 1972 to 1980, and by 2.1% per year from 1980 to 1990.

Along with increases in load growth, there have been continuing increases in the size of generating units (Table 1.1). The principal incentive to build larger units has been economy of scale—that is, a reduction in installed cost per kilowatt of capacity for larger units. However, there have also been steady improvements in generation efficiency. For example, in 1934 the average heat rate for steam generation in the U.S. electric industry was 17,950 BTU/kWh, which corresponds to 19% efficiency. By 1991 the average heat rate was 10,367 BTU/kWh, which corresponds to 33% efficiency. These improvements in thermal efficiency due to increases in unit size and in steam temperature and pressure, as well as to the use of steam reheat, have resulted in savings in fuel costs and overall operating costs.

There have been continuing increases, too, in transmission voltages (Table 1.2). From Edison's 220-V three-wire dc grid to 4-kV single-phase and

Figure 1.2

Growth of U.S. electric energy consumption [1, 2, 3]

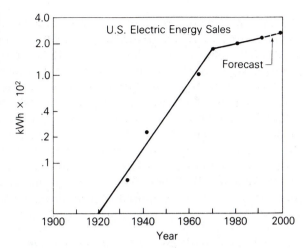

Table 1.1

Growth of generator sizes in the United States [1]

HYDROELECTRIC GENERATORS		GENERATORS DRIVEN BY SINGLE-SHAFT, 3600 r/min FOSSIL-FUELED STEAM TURBINES	
SIZE	YEAR OF INSTALLATION	SIZE	YEAR OF INSTALLATION
MVA		MVA	
4	1895	5	1914
108	1941	50	1937
158	1966	216	1953
232	1973	506	1963
615	1975	907	1969
718	1978	1120	1974

Table 1.2

History of increases in three-phase transmission voltages in the United States [1]

VOLTAGE	YEAR OF INSTALLATION
kV	
2.3	1893
44	1897
150	1913
165	1922
230	1923
287	1935
345	1953
500	1965
765	1969

2.3-kV three-phase transmission, ac transmission voltages in the United States have risen progressively to 150, 230, 345, 500, and now 765 kV. And ultra-high voltages (UHV) above 1000 kV are now being studied. The incentives for increasing transmission voltages have been: (1) increases in transmission distance and transmission capacity, (2) smaller line-voltage drops, (3) reduced line losses, (4) reduced right-of-way requirements per MW transfer, and (5) lower capital and operating costs of transmission. Today, one 765-kV three-phase line can transmit thousands of megawatts over hundreds of kilometers.

The technological developments that have occurred in conjunction with ac transmission, including developments in insulation, protection, and control, are in themselves important. The following examples are noteworthy:

1. The suspension insulator

2. The high-speed relay system, currently capable of detecting short-circuit currents within one cycle (0.017 s)

3. High-speed, extra-high-voltage (EHV) circuit breakers, capable of interrupting up to 63-kA three-phase short-circuit currents within two cycles (0.033s)

4. High-speed reclosure of EHV lines, which enables automatic return to service within a fraction of a second after a fault has been cleared

5. The EHV surge arrester, which provides protection against transient overvoltages due to lightning strikes and line-switching operations

6. Power-line carrier, microwave, and fiber optics as communication mechanisms for protecting, controlling, and metering transmission lines

7. The principle of insulation coordination applied to the design of an entire transmission system

8. Energy control centers with supervisory control and data acquisition (SCADA) and with automatic generation control (AGC) for centralized computer monitoring and control of generation, transmission, and distribution

9. Automated distribution features including reclosers and remotely controlled sectionalizing switches with fault-indicating capability, along with automated mapping/facilities management (AM/FM) for quick isolation and identification of outages and for rapid restoration of customer services

10. Microprocessor-based protective relays capable of circuit breaker control, data logging, fault locating, self-checking, fault analysis, remote query, and control

In 1954 the first modern high-voltage dc (HVDC) transmission line was put into operation in Sweden between Vastervik and the island of Gotland in the Baltic sea; it operated at 100 kV for a distance of 100 km. The first

HVDC line in the United States was the ± 400-kV, 1360-km Pacific Intertie line installed between Oregon and California in 1970. As of 1991, four other HVDC lines up to 400 kV and five back-to-back ac-dc links had been installed in the United States, and a total of 30 HVDC lines up to 533 kV had been installed worldwide.

For an HVDC line embedded in an ac system, solid-state converters at both ends of the dc line operate as rectifiers and inverters. Since the cost of an HVDC transmission line is less than that of an ac line with the same capacity, the additional cost of converters for dc transmission is offset when the line is long enough. Studies have shown that overhead HVDC transmission is economical in the United States for transmission distances longer than about 600 km.

In the United States electric utilities grew first as isolated systems, with new ones continuously starting up throughout the country. Gradually, however, neighboring electric utilities began to interconnect, to operate in parallel. This improved both reliability and economy. Figure 1.3 shows major 230-kV and higher-voltage, interconnected transmission in the United States in 1991. An interconnected system has many advantages. An interconnected company can draw upon another's rotating generator reserves during a time

Figure 1.3   Major transmission in the United States—1991 [8]

of need (such as a sudden generator outage or load increase), thereby maintaining continuity of service, increasing reliability, and reducing the total number of generators that need to be kept running under no-load conditions. Also, interconnected companies can schedule power transfers during normal periods to take advantage of energy-cost differences in respective areas, load diversity, time zone differences, and seasonal conditions. For example, utilities whose generation is primarily hydro can supply low-cost power during high-water periods in spring/summer, and can receive power from the interconnection during low-water periods in fall/winter. Interconnections also allow shared ownership of larger, more efficient generating units.

While sharing the benefits of interconnected operation, each utility is obligated to help neighbors who are in trouble, to maintain scheduled intertie transfers during normal periods, and to participate in system frequency regulation.

In addition to the benefits/obligations of interconnected operation, there are disadvantages. Interconnections, for example, have increased fault currents that occur during short circuits, thus requiring the use of circuit breakers with higher interrupting capability. Furthermore, although overall system reliability and economy have improved dramatically through interconnection, there is a remote possibility that an initial disturbance may lead to a regional blackout, such as the one that occurred in 1966 in the northeastern United States.

SECTION 1.2

PRESENT AND FUTURE TRENDS

Present trends indicate that the United States is becoming more electrified as it shifts away from a dependence on the direct use of fossil fuels. According to the Edison Electric Institute [2], electricity's share of U.S. primary energy use grew from 27% in 1974 to almost 36% in 1989, and is expected to reach 46% by the year 2010.

As shown in Figure 1.2, the growth rate in the use of electric energy in the United States is projected to increase by about 2.4% per year for the rest of this decade [3]. Although energy forecasts for the next 10 years are uncertain and based on economic and social factors that are subject to change, the predicted 2.4% annual growth rate is considered necessary to generate the GNP anticipated over that period. Longer-term forecasts of 2% to 3% annual growth are also based on predictions that the population growth will slacken, that conservation practices will continue, and that electrical energy will be used more efficiently. The cost of electricity, which is expected to rise somewhat higher than the inflation rate, is the major impediment to growth.

Figure 1.4 shows the percentages of various fuels used to meet U.S. electric energy requirements over the recent past and those projected for the

Figure 1.4

Electric energy generation in the United States, by principal fuel types [2, 3]

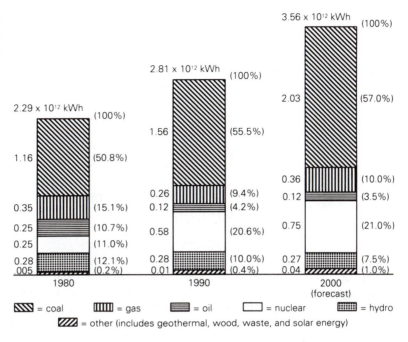

= coal   = gas   = oil   = nuclear   = hydro

= other (includes geothermal, wood, waste, and solar energy)

near future. Several trends are apparent in the chart. One is the continuing shift away from the use of gas and oil and toward the increasing use of coal. This trend is due primarily to the large amount of U.S. coal reserves, which, according to some estimates, is sufficient to meet U.S. energy needs for the next 500 years. An increase in nuclear fuel consumption is also evident; however, the increases seen in the figure are based only on the use of existing nuclear units and those presently under construction. New orders for nuclear units are not foreseen unless the 12- to 15-year construction time of previous units is cut to 4 or 5 years, via standardized plants that cost $500/kW instead of the previous $2000/kW charges. Safety concerns will also require passive or inherently safe reactor designs with standardized, modular construction of nuclear units. Another trend in Figure 1.4 is the continuing percentage decline in hydroelectric energy consumption, which is based on the fact that major U.S. hydroelectric sites (except in Alaska) have been almost fully developed.

Figure 1.5 shows the recent past and projected U.S. generating capability by the type of prime mover driving the generator. As shown, total U.S. generating capacity is projected to reach 817 GW (1 GW = 1000 MW) by the year 2000. Due to high reserve margins, capacity problems were not encountered during the 1980s. But adequate capacity through the 1990s is based on the forecast that a substantial number of large units in the planning or construction phase will be completed on schedule. Current lead times of about 10 years for licensing and construction of large, coal-fired units may create insufficient reserve margins in some regions of the United States [3, 4].

Growth in transmission-line construction roughly correlates with generation additions. As of December 31, 1989, U.S. transmission systems comprised about 146,600 circuit-miles of high-voltage transmission [3, 5, 6].

Figure 1.5

Installed generating capability in the United States by type of prime mover driving the generator

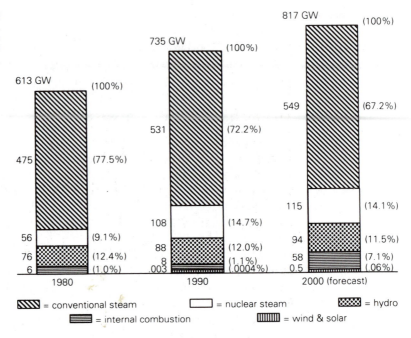

During the 1990s, planned U.S. transmission additions are expected to total 13,353 circuit-miles, including 4883 miles of 230-kV, 3992 miles of 345-kV, and 4245 miles of 500-kV lines. Although not considered a boon, this growth is expected to strengthen the presently aging transmission system, keep pace with new generation, and improve interconnections among utilities.

Growth in distribution construction roughly correlates with growth in electric energy consumption. In terms of projections, U.S. distribution construction is expected to increase steadily over the next 10 years [6, 7]. Many utilities have been converting older 2.4-, 4.1-, and 5-kV distribution systems to 12 or 15 kV. The 15-kV voltage class is widely preferred by large U.S. utilities for new installations; 25 kV, 34.5 kV, and higher distribution voltages are also utilized [7].

SECTION 1.3

COMPUTERS IN POWER-SYSTEM ENGINEERING

As electric utilities have grown in size, and the number of interconnections has increased, planning for future expansion has become increasingly complex. The increasing cost of additions and modifications has made it imperative that utilities consider a range of design options, and perform detailed studies of the effects on the system of each option, based on a number of assumptions: normal and abnormal operating conditions, peak and off-peak loadings, and present and future years of operation. A large volume of network data must also be collected and accurately handled. To assist the engineer in this power-system planning, digital computers and highly developed computer programs are used. Such programs include power-flow, stability, short-circuit, and transients programs.

Power-flow programs compute the voltage magnitudes, phase angles, and transmission-line power flows for a network under steady-state operating conditions. Other results, including transformer tap settings and generator reactive power outputs, are also computed. Today's computers have sufficient storage and speed to compute in less than 1 minute power-flow solutions for networks with more than 2000 buses and 2500 transmission lines. High-speed printers then print out the complete solution in tabular form for analysis by the planning engineer. Also available are interactive power-flow programs, whereby power-flow results are displayed on CRTs in the form of single-line diagrams; the engineer uses these to modify the network with a mouse or from a keyboard and can readily visualize the results. The computer's large storage and high-speed capabilities allow the engineer to run the many different cases necessary to analyze and design transmission and generation-expansion options.

Stability programs are used to study power systems under disturbance conditions to determine whether synchronous generators and motors remain in synchronism. System disturbances can be caused by the sudden loss of a generator or transmission line, by sudden load increases or decreases, and by short circuits and switching operations. The stability program combines power-flow equations and machine-dynamic equations to compute the angular swings of machines during disturbances. The program also computes critical clearing times for network faults, and allows the engineer to investigate the effects of various machine parameters, network modifications, disturbance types, and control schemes.

Short-circuits programs are used to compute three-phase and line-to-ground faults in power-system networks in order to select circuit breakers for fault interruption, select relays that detect faults and control circuit breakers, and determine relay settings. Short-circuit currents are computed for each relay and circuit-breaker location, and for various system-operating conditions such as lines or generating units out of service, in order to determine minimum and maximum fault currents.

Transients programs compute the magnitudes and shapes of transient overvoltages and currents that result from lightning strikes and line-switching operations. The planning engineer uses the results of a transients program to determine insulation requirements for lines, transformers, and other equipment, and to select surge arresters that protect equipment against transient overvoltages.

Other computer programs for power-system planning include relay-coordination programs and distribution-circuits programs. Computer programs for generation-expansion planning include reliability analysis and loss-of-load probability (LOLP) programs, production-cost programs, and investment-cost programs.

A computer software package, including a set of programs stored on floppy disk and a user's manual, accompanies this text. All the programs are written in BASIC to run on IBM and similar personal computers. Operating instructions, flow charts, and sample runs are given in the user's manual. The following sections in the text discuss the programs:

Section 2.5   Matrix Operations

Section 3.6   Symmetrical Components

Section 5.14 Line Constants

Section 6.8   Transmission Lines—Steady-State Operation

Section 7.8   Power Flow

Section 8.6   Symmetrical Short Circuits

Section 9.6   Short Circuits

Section 12.9 Transmission-Line Transients

Section 13.6 Transient Stability

These programs are intended to help users understand the process of power-system analysis and design. Specific problems with practical engineering design orientation that are to be solved with the programs are included in the text, along with conventional problems to be solved with only a calculator.

References

1.   H. M. Rustebakke et al., *Electric Utility Systems Practice*, 4th ed. (New York: Wiley, 1983). Photos courtesy of Westinghouse Historical Collection.

2.   W. M. McManus et al., *Statistical Yearbook of the Electric Utility Industy/1990* (Washington, DC: Edison Electric Institute, October, 1991).

3.   "1991 Annual Statistical Report," *Electrical World*, *205*, 4 (New York: McGraw-Hill, April 1991), pp. 9–14.

4.   "1991 New Plant Construction Survey," *Electrical World*, *205*, 1 (New York: McGraw-Hill, January 1991), pp. 47–53.

5.   "T&D 1991 Forecast," *Transmission & Distribution*, *43*, 1 (January 1991), pp. 28–40.

6.   "Planned Transmission Construction Compared to Recent In-Service Lines," *Transmission & Distribution*, *43*, 1 (January 1991), pp. 40, 41.

7.   "15 kV Is Still Preferred Distribution Voltage for Larger Utilities," *Transmission & Distribution*, *43*, 1 (January 1991), pp. 42, 43.

8.   North American Electric Reliability Council (NERC), *Facilities Existing as of January 1, 1991* (Princeton, NJ: NERC, 1991).

CHAPTER 2

FUNDAMENTALS

Fossil-fuel (oil/gas) power plant with two 850-MVA generating units
(Courtesy of Florida Power and Light Company)

The objective of this chapter is to review basic concepts and establish terminology and notation. In particular, the following are reviewed: phasors, instantaneous power, complex power, and network equations. A set of matrix operations from the computer software package included with this text is described, and elementary aspects of balanced three-phase circuits are introduced. Students who have already had courses in electric network theory and basic electric machines should find this chapter primarily refresher material.

SECTION 2.1

PHASORS

A sinusoidal voltage or current at constant frequency is characterized by two parameters: a maximum value and a phase angle. A voltage

$$v(t) = V_{max}\cos(\omega t + \delta) \qquad (2.1.1)$$

has a maximum value V_{max} and a phase angle δ when referenced to $\cos(\omega t)$. The root-mean-square (rms) value, also called *effective value*, of the sinusoidal voltage is

$$V = \frac{V_{max}}{\sqrt{2}} \qquad (2.1.2)$$

Euler's identity, $e^{j\phi} = \cos\phi + j\sin\phi$, can be used to express a sinusoid in terms of a phasor. For the above voltage,

$$v(t) = \text{Re}[V_{max}e^{j(\omega t + \delta)}]$$
$$= \text{Re}[\sqrt{2}(Ve^{j\delta})e^{j\omega t}] \qquad (2.1.3)$$

where $j = \sqrt{-1}$ and Re denotes "real part of." The rms phasor representation of the voltage is given in three forms—exponential, polar, and rectangular:

$$V = \underbrace{Ve^{j\delta}}_{\text{exponential}} = \underbrace{V\underline{/\delta}}_{\text{polar}} = \underbrace{V\cos\delta + jV\sin\delta}_{\text{rectangular}} \qquad (2.1.4)$$

A phasor can be easily converted from one form to another. Conversion from polar to rectangular is shown in the phasor diagram of Figure 2.1. Euler's identity can be used to convert from exponential to rectangular form. As an example, the voltage

$$v(t) = 169.7\cos(\omega t + 60°) \quad \text{volts} \qquad (2.1.5)$$

has a maximum value $V_{max} = 169.7$ volts, a phase angle $\delta = 60°$ when referenced to $\cos(\omega t)$, and an rms phasor representation in polar form of

$$V = 120\underline{/60°} \quad \text{volts} \qquad (2.1.6)$$

Also, the current

$$i(t) = 100\cos(\omega t + 45°) \quad \text{A} \qquad (2.1.7)$$

Figure 2.1

Phasor diagram for converting from polar to rectangular form

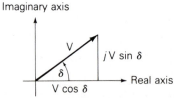

Figure 2.2

Summary of relationships
between phasors *V* and *I*
for constant R, L, and C
elements with sinusoidal-
steady-state excitation

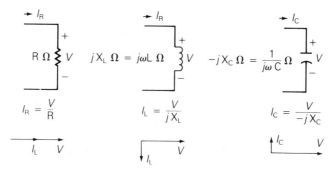

has a maximum value $I_{max} = 100$ A, an rms value $I = 100/\sqrt{2} = 70.7$ A, a phase angle of 45°, and a phasor representation

$$I = 70.7\underline{/45°} = 70.7e^{j45} = 50 + j50 \quad A \tag{2.1.8}$$

The relationships between the voltage and current phasors for the three passive elements — resistor, inductor, and capacitor — are summarized in Figure 2.2, where sinusoidal-steady-state excitation and constant values of R, L, and C are assumed.

When voltages and currents are discussed in this text, lowercase letters such as $v(t)$ and $i(t)$ indicate instantaneous values, uppercase letters such as V and I indicate rms values, and uppercase letters in italics such as V and I indicate rms phasors. When voltage or current values are specified, they shall be rms values unless otherwise indicated.

<hr/>

SECTION 2.2

INSTANTANEOUS POWER IN SINGLE-PHASE AC CIRCUITS

Power is the rate of change of energy with respect to time. The unit of power is a watt, which is a joule per second. Instead of saying that a load absorbs energy at a rate given by the power, it is common practice to say that a load absorbs power. The instantaneous power in watts absorbed by an electrical load is the product of the instantaneous voltage across the load in volts and the instantaneous current into the load in amperes. Assume that the load voltage is

$$v(t) = V_{max} \cos(\omega t + \delta) \quad \text{volts} \tag{2.2.1}$$

We now investigate the instantaneous power absorbed by purely resistive, purely inductive, purely capacitive, and general RLC loads. We also introduce the concepts of real power, power factor, and reactive power. The physical significance of real and reactive power is also discussed.

Purely Resistive Load

For a purely resistive load, the current into the load is in phase with the load voltage, $I = V/R$, and the current into the resistive load is

$$i_R(t) = I_{Rmax} \cos(\omega t + \delta) \quad \text{A} \tag{2.2.2}$$

where $I_{Rmax} = V_{max}/R$. The instantaneous power absorbed by the resistor is

$$
\begin{aligned}
p_R(t) = v(t)i_R(t) &= V_{max} I_{Rmax} \cos^2(\omega t + \delta) \\
&= \tfrac{1}{2} V_{max} I_{Rmax} \{1 + \cos[2(\omega t + \delta)]\} \\
&= VI_R \{1 + \cos[2(\omega t + \delta)]\} \quad \text{W}
\end{aligned}
\tag{2.2.3}
$$

As indicated by (2.2.3), the instantaneous power absorbed by the resistor has an average value

$$P_R = VI_R = \frac{V^2}{R} = I_R^2 R \quad \text{W} \tag{2.2.4}$$

plus a double-frequency term $VI_R \cos[2(\omega t + \delta)]$.

Purely Inductive Load

For a purely inductive load, the current lags the voltage by 90°, $I_L = V/(jX_L)$, and

$$i_L(t) = I_{Lmax} \cos(\omega t + \delta - 90°) \quad \text{A} \tag{2.2.5}$$

where $I_{Lmax} = V_{max}/X_L$, and $X_L = \omega L$ is the inductive reactance. The instantaneous power absorbed by the inductor is*

$$
\begin{aligned}
p_L(t) = v(t)i_L(t) &= V_{max} I_{Lmax} \cos(\omega t + \delta) \cos(\omega t + \delta - 90°) \\
&= \tfrac{1}{2} V_{max} I_{Lmax} \cos[2(\omega t + \delta) - 90°] \\
&= VI_L \sin[2(\omega t + \delta)] \quad \text{W}
\end{aligned}
\tag{2.2.6}
$$

As indicated by (2.2.6), the instantaneous power absorbed by the inductor is a double-frequency sinusoid with *zero* average value.

Purely Capacitive Load

For a purely capacitive load, the current leads the voltage by 90°, $I_C = V/(-jX_C)$, and

$$i_C(t) = I_{Cmax} \cos(\omega t + \delta + 90°) \quad \text{A} \tag{2.2.7}$$

where $I_{Cmax} = V_{max}/X_C$, and $X_C = 1/(\omega C)$ is the capacitive reactance. The

---

*Use the identity: $\cos A \cos B = \tfrac{1}{2}[\cos(A - B) + \cos(A + B)]$.

instantaneous power absorbed by the capacitor is

$$p_C(t) = v(t)i_C(t) = V_{max}I_{Cmax}\cos(\omega t + \delta)\cos(\omega t + \delta + 90°)$$

$$= \tfrac{1}{2}V_{max}I_{Cmax}\cos[2(\omega t + \delta) + 90°)]$$

$$= -VI_C\sin[2(\omega t + \delta)] \quad W \tag{2.2.8}$$

The instantaneous power absorbed by a capacitor is also a double-frequency sinusoid with *zero* average value.

General RLC Load

For a general load composed of RLC elements under sinusoidal-steady-state excitation, the load current is of the form

$$i(t) = I_{max}\cos(\omega t + \beta) \quad A \tag{2.2.9}$$

The instantaneous power absorbed by the load is then*

$$p(t) = v(t)i(t) = V_{max}I_{max}\cos(\omega t + \delta)\cos(\omega t + \beta)$$

$$= \tfrac{1}{2}V_{max}I_{max}\{\cos(\delta - \beta) + \cos[2(\omega t + \delta) - (\delta - \beta)]\}$$

$$= VI\cos(\delta - \beta) + VI\cos(\delta - \beta)\cos[2(\omega t + \delta)]$$

$$+ VI\sin(\delta - \beta)\sin[2(\omega t + \delta)]$$

$$p(t) = VI\cos(\delta - \beta)\{1 + \cos[2(\omega t + \delta)]\} + VI\sin(\delta - \beta)\sin[2(\omega t + \delta)]$$

Letting $I\cos(\delta - \beta) = I_R$ and $I\sin(\delta - \beta) = I_X$ gives

$$p(t) = \underbrace{VI_R\{1 + \cos[2(\omega t + \delta)]\}}_{p_R(t)} + \underbrace{VI_X\sin[2(\omega t + \delta)]}_{p_X(t)} \tag{2.2.10}$$

As indicated by (2.2.10), the instantaneous power absorbed by the load has two components: One can be associated with the power $p_R(t)$ absorbed by the resistive component of the load, and the other can be associated with the power $p_X(t)$ absorbed by the reactive (inductive or capacitive) component of the load. The first component $p_R(t)$ in (2.2.10) is identical to (2.2.3), where $I_R = I\cos(\delta - \beta)$ is the component of the load current in phase with the load voltage. The phase angle $(\delta - \beta)$ represents the angle between the voltage and current. The second component $p_X(t)$ in (2.2.10) is identical to (2.2.6) or (2.2.8), where $I_X = I\sin(\delta - \beta)$ is the component of load current 90° out of phase with the voltage.

Real Power

Equation (2.2.10) shows that the instantaneous power $p_R(t)$ absorbed by the resistive component of the load is a double-frequency sinusoid with average value P given by

$$P = VI_R = VI\cos(\delta - \beta) \quad W \tag{2.2.11}$$

---

*Use the identity: $\cos A \cos B = \tfrac{1}{2}[\cos(A - B) + \cos(A + B)]$.

The *average power* P is also called *real power* or *active power*. All three terms indicate the same quantity P given by (2.2.11).

Power Factor

The term $\cos(\delta - \beta)$ in (2.2.11) is called the *power factor*. The phase angle $(\delta - \beta)$, which is the angle between the voltage and current, is called the *power factor angle*. For dc circuits, the power absorbed by a load is the product of the dc load voltage and the dc load current; for ac circuits the average power absorbed by a load is the product of the rms load voltage V, rms load current I, and the power factor $\cos(\delta - \beta)$, as shown by (2.2.11). For inductive loads, the current lags the voltage, which means β is less than δ, and the power factor is said to be *lagging*. For capacitive loads, the current leads the voltage, which means β is greater than δ, and the power factor is said to be *leading*. By convention the power factor $\cos(\delta - \beta)$ is positive. If $|\delta - \beta|$ is greater than $90°$, then the reference direction for current may be reversed, resulting in a positive value of $\cos(\delta - \beta)$.

Reactive Power

The instantaneous power absorbed by the reactive part of the load, given by the component $p_X(t)$ in (2.2.10), is a double-frequency sinusoid with zero average value and with amplitude Q given by

$$Q = VI_X = VI \sin(\delta - \beta) \quad \text{var} \tag{2.2.12}$$

The term Q is given the name *reactive power*. Although it has the same units as real power, the usual practice is to define units of reactive power as volt-amperes reactive, or var.

| EXAMPLE 2.1 | **Instantaneous, real, and reactive power; power factor** |

The voltage $v(t) = 141.4 \cos(\omega t)$ is applied to a load consisting of a 10-Ω resistor in parallel with an inductive reactance $X_L = \omega L = 3.77\,\Omega$. Calculate the instantaneous power absorbed by the resistor and by the inductor. Also calculate the real and reactive power absorbed by the load, and the power factor.

Solution   The circuit and phasor diagram are shown in Figure 2.3(a). The load voltage is

$$V = \frac{141.4}{\sqrt{2}} \underline{/0°} = 100\underline{/0°} \quad \text{volts}$$

The resistor current is

$$I_R = \frac{V}{R} = \frac{100}{10} \underline{/0°} = 10\underline{/0°} \quad \text{A}$$

Figure 2.3

Circuit and phasor diagram for
Example 2.1

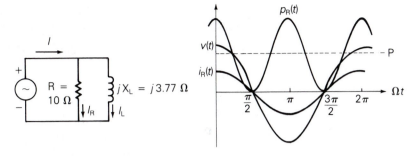

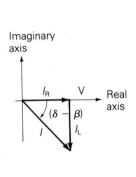

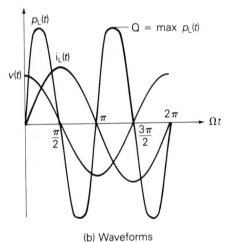

(a) Circuit and phasor diagram   (b) Waveforms

The inductor current is

$$I_L = \frac{V}{jX_L} = \frac{100}{(j\,3.77)}\underline{/0°} = 26.53\underline{/-90°} \quad A$$

The total load current is

$$I = I_R + I_L = 10 - j\,26.53 = 28.35\underline{/-69.34°} \quad A$$

The instantaneous power absorbed by the resistor is, from (2.2.3),

$$p_R(t) = (100)(10)[1 + \cos(2\omega t)]$$

$$= 1000[1 + \cos(2\omega t)] \quad W$$

The instantaneous power absorbed by the inductor is, from (2.2.6),

$$p_L(t) = (100)(26.53)\sin(2\omega t) \qquad 90° \text{ out of phase from } \cos 2\omega t$$

$$= 2653\sin(2\omega t) \quad W$$

The real power absorbed by the load is, from (2.2.11),

$$P = VI\cos(\delta - \beta) = (100)(28.53)\cos(0° + 69.34°)$$

$$= 1000 \quad W$$

These are phasors

(*Note*: P is also equal to $VI_R = V^2/R$.)

The reactive power absorbed by the load is, from (2.2.12),

$$Q = VI \sin(\delta - \beta) = (100)(28.53) \sin(0° + 69.34°)$$

$$= 2653 \quad \text{var}$$

(*Note:* Q is also equal to $VI_L = V^2/X_L$.)

The power factor is

$$\text{p.f.} = \cos(\delta - \beta) = \cos(69.34°) = 0.3528 \quad \text{lagging}$$

Voltage, current, and power waveforms are shown in Figure 2.3(b).

Note that $p_R(t)$ and $p_X(t)$, given by (2.2.10), are strictly valid only for a parallel R-X load. For a general RLC load, the voltages across the resistive and reactive components may not be in phase with the source voltage $v(t)$, resulting in additional phase shifts in $p_R(t)$ and $p_X(t)$ (see Problem 2.9). However, (2.2.11) and (2.2.12) for P and Q are valid for a general RLC load.

∎

Physical Significance of Real and Reactive Power

The physical significance of real power P is easily understood. The total energy absorbed by a load during a time interval T, consisting of one cycle of the sinusoidal voltage, is PT watt-seconds (Ws). During a time interval of n cycles, the energy absorbed is $P(nT)$ watt-seconds, all of which is absorbed by the resistive component of the load. A kilowatt-hour meter is designed to measure the energy absorbed by a load during a time interval $(t_2 - t_1)$, consisting of an integral number of cycles, by integrating the real power P over the time interval $(t_2 - t_1)$.

The physical significance of reactive power Q is not as easily understood. Q refers to the maximum value of the instantaneous power absorbed by the reactive component of the load. The instantaneous reactive power, given by the second term $p_X(t)$ in (2.2.10), is alternately positive and negative, and it expresses the reversible flow of energy to and from the reactive component of the load. Q may be positive or negative, depending on the sign of $(\delta - \beta)$ in (2.2.12). Reactive power Q is a useful quantity when describing the operation of power systems (this will become evident in later chapters). As one example, shunt capacitors can be used in transmission systems to deliver reactive power and thereby increase voltage magnitudes during heavy load periods (see Chapter 6).

SECTION 2.3

COMPLEX POWER

For circuits operating in sinusoidal-steady-state, real and reactive power are conveniently calculated from complex power, defined below. Let the voltage across a circuit element be $V = V\underline{/\delta}$, and the current into the element be $I = I\underline{/\beta}$. Then the complex power S is the product of the voltage and the conjugate of the current:

$$S = VI^* = [V\underline{/\delta}][I\underline{/\beta}]^* = VI\underline{/\delta - \beta}$$
$$= VI\cos(\delta - \beta) + j\,VI\sin(\delta - \beta) \tag{2.3.1}$$

where $(\delta - \beta)$ is the angle between the voltage and current. Comparing (2.3.1) with (2.2.11) and (2.2.12), S is recognized as

$$S = P + jQ \tag{2.3.2}$$

The magnitude $S = VI$ of the complex power S is called the *apparent power*. Although it has the same units as P and Q, it is common practice to define the units of apparent power S as voltamperes or VA. The real power P is obtained by multiplying the apparent power $S = VI$ by the power factor p.f. $= \cos(\delta - \beta)$.

The procedure for determining whether a circuit element absorbs or delivers power is summarized in Figure 2.4. Figure 2.4(a) shows the *load convention*, where the current *enters* the positive terminal of the circuit element, and the complex power *absorbed* by the circuit element is calculated from (2.3.1). This equation shows that, depending on the value of $(\delta - \beta)$, P may have either a positive or negative value. If P is positive, then the circuit element absorbs positive real power. However, if P is negative, the circuit element absorbs negative real power, or alternatively, it delivers positive real power. Similarly, if Q is positive, the circuit element in Figure 2.4(a) absorbs positive reactive power. However, if Q is negative, the circuit element absorbs negative reactive power, or it delivers positive reactive power.

Figure 2.4(b) shows the *generator convention*, where the current *leaves* the positive terminal of the circuit element, and the complex power *delivered* is calculated from (2.3.1). When P is positive (negative) the circuit element *delivers* positive (negative) real power. Similarly, when Q is positive (negative), the circuit element *delivers* positive (negative) reactive power.

Figure 2.4

Load and generator conventions

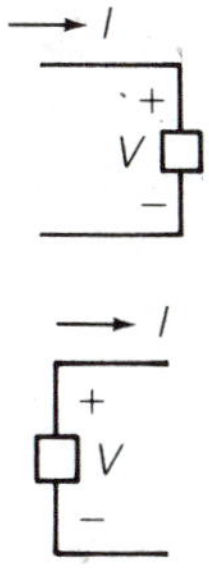

(a) *Load convention.* Current *enters* positive terminal of circuit element. If P is positive, then positive real power is *absorbed*. If Q is positive, the postive reactive power is *absorbed*. If P (Q) is negative, then postive real (reactive) power is *delivered*.

(b) *Generator convention.* Current *leaves* positive terminal of the circuit element. If P is positive, then positive real power is *delivered*. If Q is positive, then postive reactive power is *delivered*. If P (Q) is negative, then positive real (reactive) power is *absorbed*.

EXAMPLE 2.2

Real and reactive power, delivered or absorbed

A single-phase voltage source with $V = 100\underline{/130°}$ volts delivers a current $I = 10\underline{/10°}$ A, which leaves the positive terminal of the source. Calculate the source real and reactive power and state whether the source delivers or absorbs each of these.

Solution   Since *I* leaves the positive terminal of the source, the generator convention is assumed, and the complex power delivered is, from (2.3.1),

$$S = VI^* = [100\underline{/130°}][10\underline{/10°}]^*$$

$$S = 1000\underline{/120°} = -500 + j866$$

$$P = \text{Re}[S] = -500 \quad \text{W}$$

$$Q = \text{Im}[S] = +866 \quad \text{var}$$

where Im denotes "imaginary part of." The source absorbs 500 W and delivers 866 var. Readers familiar with electric machines will recognize that one example of this source is a synchronous motor. When a synchronous motor operates at a leading power factor, it absorbs real power and delivers reactive power. ∎

The *load convention* is used for the RLC elements shown in Figure 2.2. Therefore, the complex power *absorbed* by any of these three elements can be calculated as follows. Assume a load voltage $V = V\underline{/\delta}$. Then, from (2.3.1),

$$\text{resistor: } S_R = VI_R^* = [V\underline{/\delta}]\left[\frac{V}{R}\underline{/-\delta}\right] = \frac{V^2}{R} \tag{2.3.3}$$

$$\text{inductor: } S_L = VI_L^* = [V\underline{/\delta}]\left[\frac{V}{-jX_L}\underline{/-\delta}\right] = +j\frac{V^2}{X_L} \tag{2.3.4}$$

$$\text{capacitor: } S_C = VI_C^* = [V\underline{/\delta}]\left[\frac{V}{jX_C}\underline{/-\delta}\right] = -j\frac{V^2}{X_C} \tag{2.3.5}$$

From these complex power expressions, the following can be stated:

A (positive-valued) resistor absorbs (positive) real power, $P_R = V^2/R$ W, and zero reactive power, $Q_R = 0$ var.

An inductor absorbs zero real power, $P_L = 0$ W, and positive reactive power, $Q_L = V^2/X_L$ var.

A capacitor absorbs zero real power, $P_C = 0$ W, and *negative* reactive power, $Q_C = -V^2/X_C$ var. Alternatively, a capacitor *delivers positive* reactive power, $+V^2/X_C$.

For a general load composed of RLC elements, complex power S is also calculated from (2.3.1). The real power $P = \text{Re}(S)$ absorbed by a passive load is always positive. The reactive power $Q = \text{Im}(S)$ absorbed by a load may be either positive or negative. When the load is inductive, the current lags the voltage, which means β is less than δ in (2.3.1), and the reactive power absorbed is positive. When the load is capacitive, the current leads the voltage, which means β is greater than δ, and the reactive power absorbed is negative; or, alternatively, the capacitive load delivers positive reactive power.

Figure 2.5

Power triangle

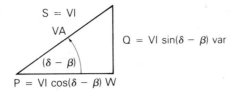

$$S = \sqrt{P^2 + Q^2} \tag{2.3.6}$$

Complex power can be summarized graphically by use of the power triangle shown in Figure 2.5. As shown, the apparent power S, real power P, and reactive power Q form the three sides of the power triangle. The power factor angle $(\delta - \beta)$ is also shown, and the following expressions can be obtained:

$$S = \sqrt{P^2 + Q^2} \tag{2.3.6}$$

$$(\delta - \beta) = \tan^{-1}(Q/P) \tag{2.3.7}$$

$$Q = P\tan(\delta - \beta) \tag{2.3.8}$$

$$\text{p.f.} = \cos(\delta - \beta) = \frac{P}{S} = \frac{P}{\sqrt{P^2 + Q^2}} \tag{2.3.9}$$

EXAMPLE 2.3   **Power triangle and power factor correction**

A single-phase source delivers 100 kW to a load operating at a power factor of 0.8 lagging. Calculate the reactive power to be delivered by a capacitor connected in parallel with the load in order to raise the source power factor to 0.95 lagging. Also draw the power triangle for the source and load. Assume that the source voltage is constant, and neglect the line impedance between the source and load.

Solution   The circuit and power triangle are shown in Figure 2.6. The real power $P = P_S = P_R$ delivered by the source and absorbed by the load is not changed when the capacitor is connected in parallel with the load, since the capacitor delivers only reactive power Q_C. For the load, the power factor angle, reactive power absorbed, and apparent power are

$$\theta_L = (\delta - \beta_L) = \cos^{-1}(0.8) = 36.87°$$

$$Q_L = P\tan\theta_L = 100\tan(36.87°) = 75 \quad \text{kvar}$$

$$S_L = \frac{P}{\cos\theta_L} = 125 \quad \text{kVA}$$

After the capacitor is connected, the power factor angle, reactive power delivered, and apparent power of the source are

$$\theta_S = (\delta - \beta_S) = \cos^{-1}(0.95) = 18.19°$$

$$Q_S = P\tan\theta_S = 100\tan(18.19°) = 32.87 \quad \text{kvar}$$

$$S_S = \frac{P}{\cos\theta_S} = \frac{100}{0.95} = 105.3 \quad \text{kVA}$$

Figure 2.6

Circuit and power triangle for Example 2.3

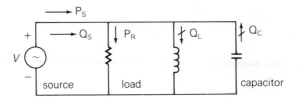

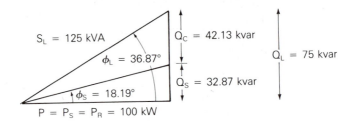

The capacitor delivers

$$Q_C = Q_L - Q_S = 75 - 32.87 = 42.13 \quad \text{kvar}$$

The method of connecting a capacitor in parallel with an inductive load is known as *power factor correction.* The effect of the capacitor is to increase the power factor of the source that delivers power to the load. Also, the source apparent power S_S decreases. As shown in Figure 2.6, the source apparent power for this example decreases from 125 kVA without the capacitor to 105.3 kVA with the capacitor. The source current $I_S = S_S/V$ also decreases. When line impedance between the source and load is included, the decrease in source current results in lower line losses and lower line-voltage drops. The end result of power factor correction is improved efficiency and improved voltage regulation. ∎

SECTION 2.4

NETWORK EQUATIONS

For circuits operating in sinusoidal-steady-state, Kirchhoff's current law (KCL) and voltage law (KVL) apply to phasor currents and voltages. Thus the sum of all phasor currents entering any node is zero and the sum of the phasor-voltage drops around any closed path is zero. Network analysis techniques based on Kirchhoff's laws, including nodal analysis, mesh or loop analysis, superposition, source transformations, and Thévenin's theorem or Norton's theorem, are useful for analyzing such circuits.

Various computer solutions of power-system problems are formulated from nodal equations, which can be systematically applied to circuits. The

Figure 2.7

Circuit diagram for reviewing nodal analysis

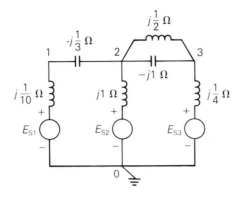

circuit shown in Figure 2.7, which is used here to review nodal analysis, is assumed to be operating in sinusoidal-steady-state; source voltages are represented by phasors E_{S1}, E_{S2}, and E_{S3}; circuit impedances are specified in ohms. Nodal equations are written in the following three steps:

Step 1 For a circuit with $(N + 1)$ nodes (also called buses), select one bus as the reference bus and define the voltages at the remaining buses with respect to the reference bus.

The circuit in Figure 2.7 has four buses — that is, $N + 1 = 4$ or $N = 3$. Bus 0 is selected as the reference bus, and bus voltages V_{10}, V_{20}, and V_{30} are then defined with respect to bus 0.

Step 2 Transform each voltage source in series with an impedance to an equivalent current source in parallel with that impedance. Also, show admittance values instead of impedance values on the circuit diagram. Each current source is equal to the voltage source divided by the source impedance.

In Figure 2.8 equivalent current sources I_1, I_2, and I_3 are shown, and all impedances are converted to corresponding admittances.

Step 3 Write nodal equations in matrix format as follows:

$$\begin{bmatrix} Y_{11} & Y_{12} & Y_{13} & \cdots & Y_{1N} \\ Y_{21} & Y_{22} & Y_{23} & \cdots & Y_{2N} \\ Y_{31} & Y_{32} & Y_{33} & \cdots & Y_{3N} \\ \vdots & \vdots & \vdots & & \vdots \\ Y_{N1} & Y_{N2} & Y_{N3} & \cdots & Y_{NN} \end{bmatrix} \begin{bmatrix} V_{10} \\ V_{20} \\ V_{30} \\ \vdots \\ V_{NO} \end{bmatrix} = \begin{bmatrix} I_1 \\ I_2 \\ I_3 \\ \vdots \\ I_N \end{bmatrix} \qquad (2.4.1)$$

Using matrix notation, (2.4.1) becomes

$$Y V = I \qquad (2.4.2)$$

where Y is the $N \times N$ bus admittance matrix, V is the column vector of N bus voltages, and I is the column vector of N current

Figure 2.8

Circuit of Figure 2.7 with equivalent current sources replacing voltage sources. Admittance values are also shown

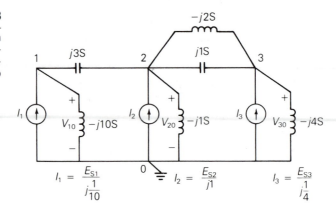

$$I_1 = \frac{E_{S1}}{j\frac{1}{10}} \qquad I_2 = \frac{E_{S2}}{j1} \qquad I_3 = \frac{E_{S3}}{j\frac{1}{4}}$$

sources. The elements Y_{kn} of the bus admittance matrix Y are formed as follows:

diagonal elements: Y_{kk} = sum of admittances connected to bus k $(k = 1, 2, \ldots, N)$ (2.4.3)

off-diagonal elements: Y_{kn} = $-$(sum of admittances connected between buses k and n) $(k \neq n)$ (2.4.4)

The diagonal element Y_{kk} is called the *self-admittance* or the *driving-point admittance* of bus k, and the off-diagonal element Y_{kn} for $k \neq n$ is called the *mutual admittance* or the *transfer admittance* between buses k and n. Since $Y_{kn} = Y_{nk}$, the matrix Y is symmetric.

For the circuit of Figure 2.8, (2.4.1) becomes

$$\begin{bmatrix} (j3 - j10) & -(j3) & 0 \\ -(j3) & (j3 - j1 + j1 - j2) & -(j1 - j2) \\ 0 & -(j1 - j2) & (j1 - j2 - j4) \end{bmatrix} \begin{bmatrix} V_{10} \\ V_{20} \\ V_{30} \end{bmatrix} = \begin{bmatrix} I_1 \\ I_2 \\ I_3 \end{bmatrix}$$

$$j \begin{bmatrix} -7 & -3 & 0 \\ -3 & 1 & 1 \\ 0 & 1 & -5 \end{bmatrix} \begin{bmatrix} V_{10} \\ V_{20} \\ V_{30} \end{bmatrix} = \begin{bmatrix} I_1 \\ I_2 \\ I_3 \end{bmatrix} \qquad (2.4.5)$$

The advantage of this method of writing nodal equations is that a digital computer can be used both to generate the admittance matrix Y and to solve (2.4.2) for the unknown bus voltage vector V. Once a circuit is specified with the reference bus and other buses identified, the circuit admittances and their bus connections become computer input data for calculating the elements Y_{kn} via (2.4.3) and (2.4.4). After Y is calculated and the current source vector I is given as input, standard computer programs for solving simultaneous linear equations can then be used to determine the bus voltage vector V. See Example 2.4 for an illustration.

When double subscripts are used to denote a voltage in this text, the voltage shall be that at the node identified by the first subscript with respect to the node identified by the second subscript. For example, the voltage V_{10} in Figure 2.8 is the voltage at node 1 with respect to node 0. Also, a current I_{ab} shall indicate the current from node a to node b. Voltage polarity marks $(+/-)$ and current reference arrows $(\rightarrow)$ are not required when double subscript notation is employed. The polarity marks in Figure 2.8 for V_{10}, V_{20}, and V_{30}, although not required, are shown for clarity. The reference arrows for sources I_1, I_2, and I_3 in Figure 2.8 are required, however, since single subscripts are used for these currents. Matrices and vectors shall be indicated in this text by boldface type (for example, Y or V).

SECTION 2.5

PERSONAL COMPUTER PROGRAM: MATRIX OPERATIONS

The following matrix operation subroutines are included in the software package that accompanies this text:

1. RMA(**A,B,C**,N,M) computes the matrix sum $C = A + B$ of the two $N \times M$ real matrices **A** and **B**.
2. CMA(A,B,C,N,M) computes the matrix sum $C = A + B$ of the two $N \times M$ complex matrices A and B.
3. RMM(**A,B,C**,N,M,P) computes the matrix product $C = AB$ of the $N \times M$ real matrix **A** and the $M \times P$ real matrix **B**.
4. CMM(A,B,C,N,M,P) computes the matrix product $C = AB$ of the $N \times M$ complex matrix A and the $M \times P$ complex matrix B.
5. RMI(**A**,N) computes the matrix inverse A^{-1} of the $N \times N$ real matrix **A**, whose determinant is assumed to be nonzero.
6. CMI(A,N) computes the matrix inverse A^{-1} of the $N \times N$ complex matrix A, whose determinant is assumed to be nonzero.
7. RMT(**A,AT**,N,M) computes the matrix transpose $AT = A^{T}$ of the $N \times M$ matrix **A** by interchanging its rows and columns.
8. CMT(A,AT,N,M) computes the matrix transpose $AT = A^{T}$ of the $N \times M$ complex matrix A.
9. CMC(A,AC,N,M) computes the complex conjugate $AC = A^{*}$ of the $N \times M$ complex matrix A.

In this text, the matrix inverse, transpose, and complex conjugate operations shall be denoted by the superscripts -1, T, and *, respectively.

EXAMPLE 2.4

Computer solution of nodal equations

The voltage sources shown in the circuit of Figure 2.7 are $E_{S1} = 1\underline{/0°}$, $E_{S2} = 1.5\underline{/30°}$, and $E_{S3} = 2.0\underline{/-30°}$ volts. Using these sources as well as the bus connections and impedances shown in Figure 2.7 as input data, write a computer program to:

1. Convert the voltage sources to current sources and the impedances to corresponding admittances.
2. Compute the 3×3 admittance matrix Y.
3. Compute Y^{-1} using the CMI subroutine.
4. Compute the bus voltage vector V from (2.4.2) using the CMM subroutine.

Solution   Table 2.1 shows the solution, including the program and input data and output data listings.

Table 2.1(a)   Example 2.4 computer program listing

```
1    '***** EXAMPLE 2.4 *****
5    '***** BUS DATA *****
10   DATA 1,1.0,0,0,0.1
20   DATA 2, 1.29904,0.75,0,1.0
30   DATA 3,1.73206, −1.0,0,0.25
40   DIM NBUS(3), ESR(3), ESI(3), RSRC(3),
        XSRC(3)
50   FOR I = 1 TO 3
60   READ NBUS(I), ESR(I), ESI(I), RSRC(I),
        XSRC(I)
70   NEXT I
100  '***** LINE DATA *****
110  DATA 1,2,0, −0.3333333333333
120  DATA 2,3,0, −1.0
130  DATA 2,3,0,0.5
140  DIM NBUS1(3), NBUS2(3), RLINE(3),
        XLINE(3)
150  FOR L = 1 TO 3
160  READ NBUS1(L), NBUS2(L), RLINE(L),
        XLINE(L)
170  NEXT L
200  '***** STEP 1 *****
210  'CONVERT VOLTAGE SOURCES TO
        CURRENT SOURCES
220  DIM ISR(3,1), ISI(3,1)
230  FOR I = 1 TO 3
235  ZSQ = RSRC(I) ^2 + XSRC(I) ^2
240  ISR(I,1) = (ESR(I) * RSRC(I)
        +ESI(I) * XSRC(I))/ZSQ
250  ISI(I,1) = (−ESR(I) * XSRC(I)
        +ESI(I) * RSRC(I))/ZSQ
260  NEXT I
300  '**STEP 2: COMPUTE Y = YR + jYI**
350  '***** OFF-DIAGONAL ELEMENTS *****
360  DIM YR(3,3), YI(3,3)
370  FOR I = 1 TO 3
380  FOR J = 1 TO 3
385  YR(I,J) = 0
390  YI(I,J) = 0
400  NEXT J
410  NEXT I
420  FOR I = 1 TO 3
430  FOR J = 1 TO 3
435  IF I = J THEN GOTO 480
440  FOR L = 1 TO 3
450  IF (NBUS1(L) = I) AND (NBUS2(L) = J)
        THEN 455 ELSE 470
455  ZLSQ = RLINE(L) ^2 +XLINE(L) ^2
457  YR(I,J) = YR(I,J) − RLINE(L)/ZLSQ
460  YI(I,J) = YI(I,J) + XLINE(L)/ZLSQ
462  YR(J,I) = YR(I,J)
465  YI(J,I) = YI(I,J)
470  NEXT L
480  NEXT J
```

Table 2.1(a)   *Continued*

```
490 NEXT I
500 '***** DIAGONAL ELEMENTS *****
540 FOR I = 1 TO 3
550 FOR J = 1 TO 3
560 IF J = I THEN 580
565 YR(I,I) = YR(I,I) - YR(I,J)
570 YI(I,I) = YI(I,I) - YI(I,J)
580 NEXT J
590 NEXT I
600 FOR I = 1 TO 3
605 ZSQ = RSRC(I) ^2 + XSRC(I) ^2
608 YR(I,I) = YR(I,I) + RSRC(I)/ZSQ
610 YI(I,I) = YI(I,I) - XSRC(I)/ZSQ
615 NEXT I
700 '*** STEP 3: INVERT Y ***
720 N = 3
730 DIM AR(3,3), AI(3,3), YINVR(3,3), YINVI(3,3)
740 FOR I = 1 TO 3
750 FOR J = 1 TO 3
755 AR(I,J) = YR(I,J)
760 AI(I,J) = YI(I,J)
770 NEXT J
780 NEXT I
CMI subroutine
790 GOSUB 6000
800 FOR I = 1 TO 3
810 FOR J = 1 TO 3
820 YINVR(I,J) = AR(I,J)
830 YINVI(I,J) = AI(I,J)
840 NEXT J
850 NEXT I
900 '** STEP 4: COMPUTE COMPLEX VOLTAGE
        VECTOR**
920 DIM VR(3,1), VI(3,1), BR(3,3)
925 DIM BI(3,3), CR(3,1), CI(3,1)
990 FOR I = 1 TO 3
1000 BR(I,1) = ISR(I,1)
1010 BI(I,1) = ISI(I,1)
1020 NEXT I
1030 N = 3
1040 M = 3
1050 P = 1
CMM subroutine
1055 GOSUB 4000
1060 FOR I = 1 TO 3
1070 VR(I,1) = CR(I,1)
1080 VI(I,1) = CI(I,1)
1090 NEXT I
2000 '***** PRINT INPUT/OUTPUT DATA *****
2010 PRINT"***** BUS DATA *****"
2020 PRINT
2030 PRINT TAB(2); "BUS #"; TAB(17); "SOURCE
        VOLTAGE";
2035 PRINT TAB(40); "SOURCE IMPEDANCE"
2040 PRINT TAB(14); "REAL PART"; TAB(24);
        "IMAG PART";
2042 PRINT TAB(37); "REAL PART"; TAB(48);
        "IMAG PART"
2045 PRINT TAB(14);" VOLTS ";
2047 PRINT TAB(24);" VOLTS"; TAB(45); "OHMS"
2050 PRINT "————————————————"
2065 FOR I = 1 TO 3
2070 PRINT TAB(2); NBUS(I); TAB(14); ESR(I);
2072 PRINT TAB(27);"j"; ESI(I); TAB(40); RSRC(I)
2074 PRINT TAB(52); "j"; XSRC(I)
2075 NEXT I
2090 PRINT
2095 PRINT
2110 PRINT "***** "LINE DATA" *****"
2130 PRINT TAB(2); "FROM"; TAB(12); " TO ";
2135 PRINT TAB(25); "LINE IMPEDANCE"
2140 PRINT TAB(3); "BUS"; TAB(11); "BUS";
2142 PRINT TAB(22); "REAL PART"; TAB(34);
        "IMAG PART"
2144 PRINT TAB(30); "OHMS"
2150 PRINT "————————————————"
2160 PRINT
2165 FOR L = 1 TO 3
2170 PRINT TAB(2); NBUS1(L); TAB(12);
        NBUS2(L);
2172 PRINT TAB(25); RLINE(L); TAB(37); "j";
        XLINE(L)
2175 NEXT L
2210 PRINT "***** BUS VOLTAGES *****"
2220 PRINT
2230 PRINT TAB(2); "BUS #"; TAB(22); "BUS
        VOLTAGE"
2240 PRINT TAB(14); "REAL PART"; TAB(30);
        "IMAG PART"
2245 PRINT TAB(14); " VOLTS "; TAB(30);
        " VOLTS "
2250 PRINT "————————————————"
2265 FOR I = 1 TO 3
2270 PRINT TAB(2); NBUS(I); TAB(10); VR(I,1);
2272 PRINT TAB(27); "j"; VI(I,1)
2275 NEXT I
2297 PRINT "*****************************************"
3000 END
```

Table 2.1(b)

Example 2.4 input/output data files

BUS DATA				
BUS #	SOURCE VOLTAGE		SOURCE IMPEDANCE	
	REAL PART	IMAG PART	REAL PART	IMAG PART
	VOLTS	VOLTS	OHMS	OHMS
1	1	j 0	0	j .1
2	1.29904	j .75	0	j 1
3	1.73206	$j-1$	0	j .25

LINE DATA			
FROM BUS	TO BUS	LINE IMPEDANCE	
		REAL PART	IMAG PART
		OHMS	OHMS
1	2	0	$j-$ .3333333333333
2	3	0	$j-1$
2	3	0	j .5

BUS VOLTAGES		
BUS #	BUS VOLTAGE	
	REAL PART	IMAG PART
	VOLTS	VOLTS
1	1.1525324137931	$j-$ .00862068965517
2	.64409103448291	j .02011494252872
3	1.5144662068966	$j-$ .79597701149423

■

SECTION 2.6

BALANCED THREE-PHASE CIRCUITS

In this section we introduce the following topics for balanced three-phase circuits: Y connections, line-to-neutral voltages, line-to-line voltages, line currents, Δ loads, $\Delta - Y$ conversions, and equivalent line-to-neutral diagrams.

Balanced-Y Connections

Figure 2.9 shows a three-phase Y-connected (or "wye-connected") voltage source feeding a balanced-Y-connected load. For a Y connection, the neutrals of each phase are connected. In Figure 2.9 the source neutral connection is

Figure 2.9

Circuit diagram of a three-phase Y-connected source feeding a balanced-Y load

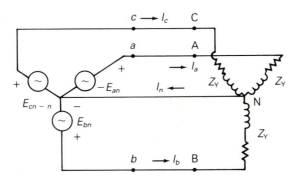

labeled bus *n* and the load neutral connection is labeled bus *N*. The three-phase source is assumed to be ideal since source impedances are neglected. Also neglected are the line impedances between the source and load terminals and the neutral impedance between buses *n* and *N*. The three-phase load is *balanced,* which means the load impedances in all three phases are identical.

Balanced Line-to-Neutral Voltages

In Figure 2.9, the terminal buses of the three-phase source are labeled *a*, *b*, and *c*, and the source line-to-neutral voltages are labeled E_{an}, E_{bn}, and E_{cn}. The source is *balanced* when these voltages have equal magnitudes and an equal 120°-phase difference between any two phases. An example of balanced three-phase line-to-neutral voltages is

$$E_{an} = 10\underline{/0°}$$

$$E_{bn} = 10\underline{/-120°} = 10\underline{/+240°}$$ *pos seq*

$$E_{cn} = 10\underline{/+120°} = 10\underline{/-240°} \quad \text{volts} \tag{2.6.1}$$

where the line-to-neutral voltage magnitude is 10 volts and E_{an} is the reference phasor. The phase sequence is called *positive sequence* or *abc* sequence when E_{an} leads E_{bn} by 120° and E_{bn} leads E_{cn} by 120°. The phase sequence is called *negative sequence* or *acb* sequence when E_{an} leads E_{cn} by 120° and E_{cn} leads E_{bn} by 120°. The voltages in (2.6.1) are positive-sequence voltages, since E_{an} leads E_{bn} by 120°. The corresponding phasor diagram is shown in Figure 2.10.

Figure 2.10

Phasor diagram of balanced positive-sequence line-to-neutral voltages with E_{an} as the reference

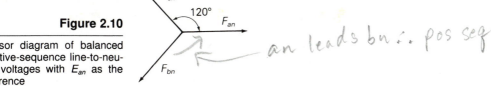

an leads bn ∴ pos seq

Balanced Line-to-Line Voltages

The voltages E_{ab}, E_{bc}, and E_{ca} between phases are called line-to-line voltages. Writing a KVL equation for a closed path around buses a, b, and n in Figure 2.9,

$$E_{ab} = E_{an} - E_{bn} \qquad (2.6.2)$$

For the line-to-neutral voltages of (2.6.1),

$$E_{ab} = 10\underline{/0°} - 10\underline{/-120°} = 10 - 10\left[\frac{-1-j\sqrt{3}}{2}\right]$$

$$E_{ab} = \sqrt{3}(10)\left(\frac{\sqrt{3}+j1}{2}\right) = \sqrt{3}(10\underline{/30°}) \quad \text{volts} \qquad (2.6.3)$$

Similarly, the line-to-line voltages E_{bc} and E_{ca} are

$$E_{bc} = E_{bn} - E_{cn} = 10\underline{/-120°} - 10\underline{/+120°}$$

$$= \sqrt{3}(10\underline{/-90°}) \quad \text{volts} \qquad (2.6.4)$$

$$E_{ca} = E_{cn} - E_{an} = 10\underline{/+120°} - 10\underline{/0°}$$

$$= \sqrt{3}(10\underline{/150°}) \quad \text{volts} \qquad (2.6.5)$$

The line-to-line voltages of (2.6.3)–(2.6.5) are also balanced, since they have equal magnitudes of $\sqrt{3}(10)$ volts and $120°$ displacement between any two phases. Comparison of these line-to-line voltages with the line-to-neutral voltages of (2.6.1) leads to the following conclusion:

In a balanced three-phase Y-connected system with positive-sequence sources, the line-to-line voltages are $\sqrt{3}$ times the line-to-neutral voltages and lead by 30°. That is,

$$E_{ab} = \sqrt{3}\,E_{an}\underline{/+30°}$$

$$E_{bc} = \sqrt{3}\,E_{bn}\underline{/+30°}$$

$$E_{ca} = \sqrt{3}\,E_{cn}\underline{/+30°} \qquad (2.6.6)$$

This very important result is summarized in Figure 2.11. In Figure 2.11(a) each phasor begins at the origin of the phasor diagram. In Figure 2.11(b) the line-to-line voltages form an equilateral triangle with vertices labeled a, b, c corresponding to buses a, b, and c of the system; the line-to-neutral voltages begin at the vertices and end at the center of the triangle, which is labeled n for neutral bus n. Also, the clockwise sequence of the vertices *abc* in Figure 2.11(b) indicates positive-sequence voltages. In both diagrams, E_{an} is the reference. However, the diagrams could be rotated to align with any other reference.

Since the balanced line-to-line voltages form a closed triangle in Figure 2.11, their sum is zero. In fact, the sum of line-to-line voltages

Figure 2.11

Positive-sequence line-to-neutral and line-to-line voltages in a balanced three-phase Y-connected system

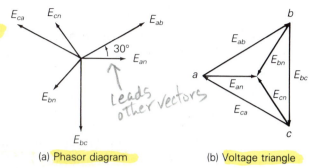

(a) Phasor diagram   (b) Voltage triangle

$(E_{ab} + E_{bc} + E_{ca})$ is *always* zero, even if the system is unbalanced, since these voltages form a closed path around buses a, b, and c. Also, in a balanced system the sum of the line-to-neutral voltages $(E_{an} + E_{bn} + E_{cn})$ equals zero.

Balanced Line Currents

Since the impedance between the source and load neutrals in Figure 2.9 is neglected, buses n and N are at the same potential, $E_{nN} = 0$. Accordingly, a separate KVL equation can be written for each phase, and the line currents can be written by inspection:

$$I_a = E_{an}/Z_Y$$
$$I_b = E_{bn}/Z_Y$$
$$I_c = E_{cn}/Z_Y \tag{2.6.7}$$

For example, if each phase of the Y-connected load has an impedance $Z_Y = 2\underline{/30°}\ \Omega$, then

$$I_a = \frac{10\underline{/0°}}{2\underline{/30°}} = 5\underline{/-30°}\quad \text{A}$$

$$I_b = \frac{10\underline{/-120°}}{2\underline{/30°}} = 5\underline{/-150°}\quad \text{A}$$

$$I_c = \frac{10\underline{/+120°}}{2\underline{/30°}} = 5\underline{/90°}\quad \text{A} \tag{2.6.8}$$

The line currents are also balanced, since they have equal magnitudes of 5 A and 120° displacement between any two phases. The neutral current I_n is determined by writing a KCL equation at bus N in Figure 2.9.

$$I_n = I_a + I_b + I_c \tag{2.6.9}$$

Using the line currents of (2.6.8),

$$I_n = 5\underline{/-30°} + 5\underline{/-150°} + 5\underline{/90°}$$

$$I_n = 5\left(\frac{\sqrt{3} - j1}{2}\right) + 5\left(\frac{-\sqrt{3} - j1}{2}\right) + j5 = 0 \tag{2.6.10}$$

Figure 2.12

Phasor diagram of line currents in a balanced three-phase system

equal sides,∴ balanced

The phasor diagram of the line currents is shown in Figure 2.12. Since these line currents form a closed triangle, their sum, which is the neutral current I_n, is zero. In general, the sum of any balanced three-phase set of phasors is zero, since balanced phasors form a closed triangle. Thus, although the impedance between neutrals n and N in Figure 2.9 is assumed to be zero, the neutral current will be zero for *any* neutral impedance ranging from short circuit ($0\,\Omega$) to open circuit ($\infty\,\Omega$), as long as the system is balanced. If the system is not balanced — which could occur if the source voltages, load impedances, or line impedances were unbalanced — then the line currents will not be balanced and a neutral current I_n may flow between buses n and N.

neutral current note!

Balanced Δ Loads

Figure 2.13 shows a three-phase Y-connected source feeding a balanced-Δ-connected (or "delta-connected") load. For a balanced-Δ connection, equal load impedances Z_Δ are connected in a triangle whose vertices form the buses, labeled A, B, and C in Figure 2.13. The Δ connection does not have a neutral bus.

Since the line impedances are neglected in Figure 2.13, the source line-to-line voltages are equal to the load line-to-line voltages, and the Δ-load currents I_{AB}, I_{BC}, and I_{CA} are

$$I_{AB} = E_{ab}/Z_\Delta$$

$$I_{BC} = E_{bc}/Z_\Delta$$

$$I_{CA} = E_{ca}/Z_\Delta \tag{2.6.11}$$

For example, if the line-to-line voltages are given by (2.6.3)–(2.6.5) and

Figure 2.13

Circuit diagram of a Y-connected source feeding a balanced-Δ load

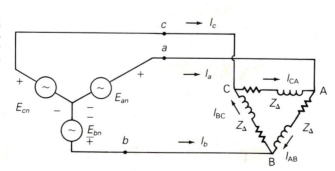

if $Z_\Delta = 5\underline{/30°}\,\Omega$, then the Δ-load currents are

(handwritten: load current →)

$$I_{AB} = \sqrt{3}\left(\frac{10\underline{/30°}}{5\underline{/30°}}\right) = 3.464\underline{/0°} \text{ A} \quad = \frac{\sqrt{3}\,E_{an}}{Z_\Delta}$$

$$I_{BC} = \sqrt{3}\left(\frac{10\underline{/-90°}}{5\underline{/30°}}\right) = 3.464\underline{/-120°} \text{ A} \quad \stackrel{?}{=} \frac{\sqrt{3}\,E_{bn}}{Z_\Delta}$$

(handwritten: load)

$$I_{CA} = \sqrt{3}\left(\frac{10\underline{/150°}}{5\underline{/30°}}\right) = 3.464\underline{/+120°} \text{ A} \quad \stackrel{?}{=} \frac{\sqrt{3}\,E_{cn}}{Z_\Delta} \qquad (2.6.12)$$

Also, the line currents can be determined by writing a KCL equation at each bus of the Δ load, as follows:

(handwritten: line current, $I_a = \sqrt{3}\,I_{AB}$)

$$I_a = I_{AB} - I_{CA} = 3.464\underline{/0°} - 3.464\underline{/120°} = \sqrt{3}(3.464\underline{/-30°})$$

$$I_b = I_{BC} - I_{AB} = 3.464\underline{/-120°} - 3.464\underline{/0°} = \sqrt{3}(3.464\underline{/-150°})$$

$$I_c = I_{CA} - I_{BC} = 3.464\underline{/120°} - 3.464\underline{/-120°} = \sqrt{3}(3.464\underline{/+90°})$$

$$(2.6.13)$$

Both the Δ-load currents given by (2.6.12) and the line currents given by (2.6.13) are balanced. Thus the sum of balanced Δ-load currents $(I_{AB} + I_{BC} + I_{CA})$ equals zero. The sum of line currents $(I_a + I_b + I_c)$ is always zero for a Δ-connected load even if the system is unbalanced, since there is no neutral wire. Comparison of (2.6.12) and (2.6.13) leads to the following conclusion:

(handwritten margin: if un-bal load)

For a balanced-Δ load supplied by a balanced positive-sequence source, the line currents into the load are $\sqrt{3}$ times the Δ-load currents and lag by 30°. That is,

(handwritten: line currents lag load current, -30°; phase)

$$I_a = \sqrt{3}\,I_{AB}\underline{/-30°}$$

$$I_b = \sqrt{3}\,I_{BC}\underline{/-30°}$$

$$I_c = \sqrt{3}\,I_{CA}\underline{/-30°} \qquad (2.6.14)$$

This result is summarized in Figure 2.14.

Figure 2.14

Phasor diagram of line currents and load currents for a balanced-Δ load

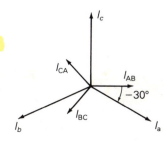

Figure 2.15

Δ–Y conversion for balanced loads

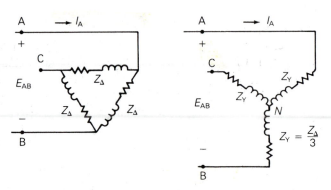

(a) Balanced-Δ load   (b) Equivalent balanced-Y load

Δ–Y Conversion for Balanced Loads

Figure 2.15 shows the conversion of a balanced-Δ load to a balanced-Y load. If balanced voltages are applied, then these loads will be equivalent as viewed from their terminal buses A, B, and C when the line currents into the Δ load are the same as the line currents into the Y load. For the Δ load,

$$I_A = \sqrt{3} I_{AB} \underline{/-30°} = \frac{\sqrt{3} E_{AB} \underline{/-30°}}{Z_\Delta} \qquad (2.6.15)$$

and for the Y load,

$$I_A = \frac{E_{AN}}{Z_Y} = \frac{E_{AB} \underline{/-30°}}{\sqrt{3} Z_Y} \qquad (2.6.16)$$

Comparison of (2.6.15) and (2.6.16) indicates that I_A will be the same for both the Δ and Y loads when

Note! 

$$Z_Y = \frac{Z_\Delta}{3} \quad \text{balanced loads} \qquad (2.6.17)$$

Also, the other line currents I_B and I_C into the Y load will equal those into the Δ load when $Z_Y = Z_\Delta/3$, since these loads are balanced. Thus a balanced-Δ load can be converted to an equivalent balanced-Y load by dividing the Δ-load impedance by 3. The angles of these Δ- and equivalent Y-load impedances are the same. Similarly, a balanced-Y load can be converted to an equivalent balanced-Δ load using $Z_\Delta = 3Z_Y$.

EXAMPLE 2.5	**Balanced Δ and Y loads**

A balanced, positive-sequence, Y-connected voltage source with $E_{ab} = 480 \underline{/0°}$ volts is applied to a balanced-Δ load with $Z_\Delta = 30 \underline{/40°}\ \Omega$. The line impedance between the source and load is $Z_L = 1 \underline{/85°}\ \Omega$ for each phase. Calculate the line currents, the Δ-load currents, and the voltages at the load terminals.

L–L ↑
not to neutral?

Solution   The solution is most easily obtained as follows. First convert the Δ load to an equivalent Y. Then connect the source and Y-load neutrals with a zero-ohm neutral wire. The connection of the neutral wire has no effect on the circuit, since the neutral current $I_n = 0$ in a balanced system. The resulting circuit is shown in Figure 2.16. The line currents are

$$I_A = \frac{E_{an}}{Z_L + Z_Y} = \frac{\frac{480}{\sqrt{3}}\underline{/-30°}}{1\underline{/85°} + \frac{30}{3}\underline{/40°}}$$

Y load current   line–neutral   → current flows from line to neut   not from line to line

line Z

$$= \frac{277.1\underline{/-30°}}{(0.0872 + j0.9962) + (7.660 + j6.428)}$$

$$I_A = \frac{277.1\underline{/-30°}}{(7.748 + j7.424)} = \frac{277.1\underline{/-30°}}{10.73\underline{/43.78°}} = 25.83\underline{/-73.78°}\quad A$$

$$I_B = 25.83\underline{/166.22°}\quad A$$

$$I_C = 25.83\underline{/46.22°}\quad A \tag{2.6.18}$$

The Δ-load currents are, from (2.6.14),

$$I_{AB} = \frac{I_a}{\sqrt{3}}\underline{/+30°} = \frac{25.83}{\sqrt{3}}\underline{/-73.78° + 30°} = 14.91\underline{/-43.78°}\quad A$$

$$I_{BC} = 14.91\underline{/-163.78°}\quad A$$

$$I_{CA} = 14.91\underline{/+76.22°}\quad A \tag{2.6.19}$$

The voltages at the load terminals are

line to line load voltages

$$E_{AB} = Z_\Delta I_{AB} = (30\underline{/40°})(14.91\underline{/-43.78°}) = 447.3\underline{/-3.78°}$$

$$E_{BC} = 447.3\underline{/-123.78°}$$

$$E_{CA} = 447.3\underline{/116.22°}\quad \text{volts} \tag{2.6.20}$$

Figure 2.16

Circuit diagram for Example 2.5

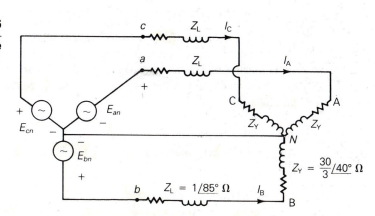

Equivalent Line-to-Neutral Diagrams

When working with balanced three-phase circuits, only one phase need be analyzed. Δ loads can be converted to Y loads, and all source and load neutrals can be connected with a zero-ohm neutral wire without changing the solution. Then one phase of the circuit can be solved. The voltages and currents in the other two phases are equal in magnitude to and $\pm 120°$ out of phase with those of the solved phase. Figure 2.17 shows an equivalent line-to-neutral diagram for one phase of the circuit in Example 2.5.

     When discussing three-phase systems in this text, voltages shall be rms line-to-line voltages unless otherwise indicated. This is standard industry practice.

Figure 2.17

Equivalent line-to-neutral diagram for the circuit of Example 2.5

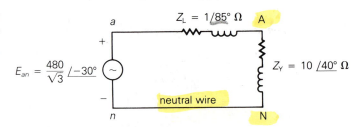

POWER IN BALANCED THREE-PHASE CIRCUITS

In this section we discuss instantaneous power and complex power for balanced three-phase generators and motors and for balanced-Y and Δ-impedance loads.

Instantaneous Power: Balanced Three-Phase Generators

Figure 2.18 shows a Y-connected generator represented by three voltage sources with their neutrals connected at bus n and by three identical generator impedances Z_g. Assume that the generator is operating under balanced steady-state conditions with the instantaneous generator terminal voltage given by

$$v_{an}(t) = \sqrt{2}\,V_{LN}\cos(\omega t + \delta) \quad \text{volts} \tag{2.7.1}$$

and with the instantaneous current leaving the positive terminal of phase a given by

$$i_a(t) = \sqrt{2}\,I_L\cos(\omega t + \beta) \quad \text{A} \tag{2.7.2}$$

where V_{LN} is the rms line-to-neutral voltage and I_L is the rms line current.

Figure 2.18

Y-connected generator

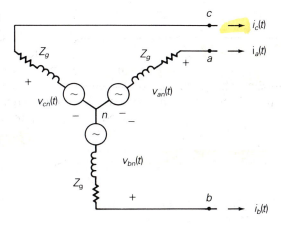

The instantaneous power $p_a(t)$ delivered by phase *a* of the generator is

$$p_a(t) = v_{an}(t)i_a(t)$$

$$= 2V_{LN}I_L \cos(\omega t + \delta)\cos(\omega t + \beta)$$

$$= V_{LN}I_L \cos(\delta - \beta) + V_{LN}I_L \cos(2\omega t + \delta + \beta) \quad W \qquad (2.7.3)$$

Assuming balanced operating conditions, the voltages and currents of phases *b* and *c* have the same magnitudes as those of phase *a* and are $\pm 120°$ out of phase with phase *a*. Therefore the instantaneous power delivered by phase *b* is

$$p_b(t) = 2V_{LN}I_L \cos(\omega t + \delta - 120°)\cos(\omega t + \beta - 120°)$$

$$= V_{LN}I_L \cos(\delta - \beta) + V_{LN}I_L \cos(2\omega t + \delta + \beta - 240°) \quad W \qquad (2.7.4)$$

and by phase *c*,

$$p_c(t) = 2V_{LN}I_L \cos(\omega t + \delta + 120°)\cos(\omega t + \beta + 120°)$$

$$= V_{LN}I_L \cos(\delta - \beta) + V_{LN}I_L \cos(2\omega t + \delta + \beta + 240°) \quad W \qquad (2.7.5)$$

The total instantaneous power $p_{3\phi}(t)$ delivered by the three-phase generator is the sum of the instantaneous powers delivered by each phase. Using (2.7.3)–(2.7.5):

$$p_{3\phi}(t) = p_a(t) + p_b(t) + p_c(t)$$

$$= 3V_{LN}I_L \cos(\delta - \beta) + V_{LN}I_L[\cos(2\omega t + \delta + \beta) +$$

$$\cos(2\omega t + \delta + \beta - 240°) + \cos(2\omega t + \delta + \beta + 240°)] \quad W$$

$$(2.7.6)$$

The three cosine terms within the brackets of (2.7.6) can be represented by a *balanced* set of three phasors. Therefore, the sum of these three terms is zero for any value of δ, for any value of β, and for all values of *t*. Equation (2.7.6) then reduces to

$$p_{3\phi}(t) = P_{3\phi} = 3V_{LN}I_L \cos(\delta - \beta) \quad W \qquad (2.7.7)$$

Equation (2.7.7) can be written in terms of the line-to-line voltage V_{LL} instead of the line-to-neutral voltage V_{LN}. Under balanced operating conditions,

line to line

$$V_{LN} = V_{LL}/\sqrt{3} \text{ and }$$

$$P_{3\phi} = \sqrt{3} V_{LL} I_L \cos(\delta - \beta) \quad W \tag{2.7.8}$$

Inspection of (2.7.8) leads to the following conclusion:

> The total instantaneous power delivered by a three-phase generator under balanced operating conditions is not a function of time, but a constant, $p_{3\phi}(t) = P_{3\phi}$.

Constant

Instantaneous Power: Balanced Three-Phase Motors and Impedance Loads

The total instantaneous power absorbed by a three-phase motor under balanced steady-state conditions is also a constant. Figure 2.18 can be used to represent a three-phase motor by reversing the line currents to enter rather than leave the positive terminals. Then (2.7.1)–(2.7.8), valid for power *delivered* by a generator, are also valid for power *absorbed* by a motor. These equations are also valid for the instantaneous power absorbed by a balanced three-phase impedance load.

"S"

Complex Power: Balanced Three-Phase Generators

The phasor representations of the voltage and current in (2.7.1) and (2.7.2) are

$$V_{an} = V_{LN}\underline{/\delta} \quad \text{volts} \tag{2.7.9}$$

$$I_a = I_L\underline{/\beta} \quad A \tag{2.7.10}$$

where I_a leaves positive terminal "a" of the generator. The complex power S_a delivered by phase a of the generator is

$$S_a = V_{an} I_a^* = V_{LN} I_L \underline{/(\delta - \beta)}$$
$$= V_{LN} I_L \cos(\delta - \beta) + j V_{LN} I_L \sin(\delta - \beta) \tag{2.7.11}$$

Under balanced operating conditions, the complex powers delivered by phases b and c are identical to S_a, and the total complex power $S_{3\phi}$ delivered by the generator is

$$S_{3\phi} = S_a + S_b + S_c = 3S_a$$
$$= 3V_{LN} I_L \underline{/(\delta - \beta)}$$
$$= 3V_{LN} I_L \cos(\delta - \beta) + j 3 V_{LN} I_L \sin(\delta - \beta) \tag{2.7.12}$$

In terms of the total real and reactive powers,

$$S_{3\phi} = P_{3\phi} + j Q_{3\phi} \tag{2.7.13}$$

where

$$P_{3\phi} = \text{Re}(S_{3\phi}) = 3V_{LN}I_L \cos(\delta - \beta)$$
$$= \sqrt{3}\,V_{LL}I_L \cos(\delta - \beta) \quad \text{W} \tag{2.7.14}$$

and

$$Q_{3\phi} = \text{Im}(S_{3\phi}) = 3V_{LN}I_L \sin(\delta - \beta)$$
$$= \sqrt{3}\,V_{LL}I_L \sin(\delta - \beta) \quad \text{var} \tag{2.7.15}$$

Also, the total apparent power is

$$S_{3\phi} = |S_{3\phi}| = 3V_{LN}I_L = \sqrt{3}\,V_{LL}I_L \quad \text{VA} \tag{2.7.16}$$

3 phases total

Complex Power: Balanced Three-Phase Motors

The preceding expressions for complex, real, reactive, and apparent power *delivered* by a three-phase generator are also valid for the complex, real, reactive, and apparent power *absorbed* by a three-phase motor.

Complex Power: Balanced-Y and -Δ Impedance Loads

Equations (2.7.13)–(2.7.16) are also valid for balanced-Y and -Δ impedance loads. For a balanced-Y load, the line-to-neutral voltage across the phase *a* load impedance and the current entering the positive terminal of that load impedance can be represented by (2.7.9) and (2.7.10). Then (2.7.11)–(2.7.16) are valid for the power absorbed by the balanced-Y load.

For a balanced-Δ load, the line-to-line voltage across the phase *a–b* load impedance and the current into the positive terminal of that load impedance can be represented by

$$V_{ab} = V_{LL}\underline{/\delta} \quad \text{volts} \tag{2.7.17}$$

$$I_{ab} = I_\Delta\underline{/\beta} \quad \text{A} \tag{2.7.18}$$

where V_{LL} is the rms line-to-line voltage and I_Δ is the rms Δ-load current. The complex power S_{ab} absorbed by the phase *a–b* load impedance is then

one phase

$$S_{ab} = V_{ab}I_{ab}^* = V_{LL}I_\Delta\underline{/(\delta - \beta)} \tag{2.7.19}$$

The total complex power absorbed by the Δ load is

all 3 phases

$$S_{3\phi} = S_{ab} + S_{bc} + S_{ca} = 3S_{ab}$$
$$= 3V_{LL}I_\Delta\underline{/(\delta - \beta)}$$
$$= 3V_{LL}I_\Delta\cos(\delta - \beta) + j\,3V_{LL}I_\Delta\sin(\delta - \beta) \tag{2.7.20}$$

Rewriting (2.7.19) in terms of the total real and reactive power,

$$S_{3\phi} = P_{3\phi} + jQ_{3\phi} \tag{2.7.21}$$

$$P_{3\phi} = \text{Re}(S_{3\phi}) = 3V_{LL}I_\Delta \cos(\delta - \beta)$$

$$= \sqrt{3}V_{LL}I_L \cos(\delta - \beta) \quad \text{W} \tag{2.7.22}$$

$$Q_{3\phi} = \text{Im}(S_{3\phi}) = 3V_{LL}I_\Delta \sin(\delta - \beta)$$

$$= \sqrt{3}V_{LL}I_L \sin(\delta - \beta) \quad \text{var} \tag{2.7.23}$$

where the Δ-load current I_Δ is expressed in terms of the line current $I_L = \sqrt{3}I_\Delta$ in (2.7.22) and (2.7.23). Also, the total apparent power is

$$S_{3\phi} = |S_{3\phi}| = 3V_{LL}I_\Delta = \sqrt{3}V_{LL}I_L \quad \text{VA} \tag{2.7.24}$$

Equations (2.7.21)–(2.7.24) developed for the balanced-Δ load are identical to (2.7.13)–(2.7.16). balanced Y

SECTION 2.8

ADVANTAGES OF BALANCED THREE-PHASE VERSUS SINGLE-PHASE SYSTEMS

Figure 2.19 shows three separate single-phase systems. Each single-phase system consists of the following identical components: (1) a generator represented by a voltage source and a generator impedance Z_g; (2) a forward and return conductor represented by two series line impedances Z_L; (3) a load represented by an impedance Z_Y. The three single-phase systems, although completely separated, are drawn in a Y configuration in the figure to illustrate two advantages of three-phase systems.

Each separate single-phase system requires that *both* the forward and return conductors have a current capacity (or *ampacity*) equal to or greater

Figure 2.19

Three single-phase systems

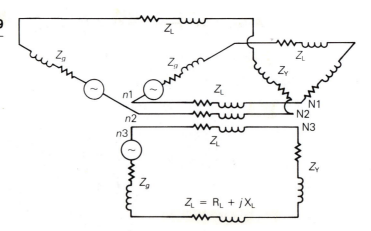

$$Z_L = R_L + jX_L$$

than the load current. However, if the source and load neutrals in Figure 2.19 are connected to form a three-phase system, and if the source voltages are balanced with equal magnitudes and with 120° displacement between phases, then the neutral current will be zero [see (2.6.10)] and the three neutral conductors can be removed. Thus, the balanced three-phase system, while delivering the same power to the three load impedances Z_Y, requires only half the number of conductors needed for the three separate single-phase systems. Also, the total I^2R line losses in the three-phase system are only half those of the three separate single-phase systems, and the line-voltage drop between the source and load in the three-phase system is half that of each single-phase system. Therefore, one advantage of balanced three-phase systems over separate single-phase systems is reduced capital and operating costs of transmission and distribution, as well as better voltage regulation.

Some three-phase systems such as Δ-connected systems and three-wire Y-connected systems do not have any neutral conductor. However, the majority of three-phase systems are four-wire Y-connected systems, where a grounded neutral conductor is used. Neutral conductors are used to reduce transient overvoltages, which can be caused by lightning strikes and by line-switching operations, and to carry unbalanced currents, which can occur during unsymmetrical short-circuit conditions. Neutral conductors for transmission lines are typically smaller in size and ampacity than the phase conductors because the neutral current is nearly zero under normal operating conditions. Thus, the cost of a neutral conductor is substantially less than that of a phase conductor. The capital and operating costs of three-phase transmission and distribution systems with or without neutral conductors are substantially less than those of separate single-phase systems.

A second advantage of three-phase systems is that the total instantaneous electric power delivered by a three-phase generator under balanced steady-state conditions is (nearly) constant, as shown in Section 2.7. A three-phase generator (constructed with its field winding on one shaft and with its three-phase windings equally displaced by 120° on the stator core) will also have a nearly constant mechanical input power under balanced steady-state conditions, since the mechanical input power equals the electrical output power plus the small generator losses. Furthermore, the mechanical shaft torque, which equals mechanical input power divided by mechanical radian frequency ($T_{mech} = P_{mech}/\omega_m$) is nearly constant.

On the other hand, the equation for the instantaneous electric power delivered by a single-phase generator under balanced steady-state conditions is the same as the instantaneous power delivered by one phase of a three-phase generator, given by $p_a(t)$ in (2.7.3). As shown in that equation, $p_a(t)$ has two components: a constant and a double-frequency sinusoid. Both the mechanical input power and the mechanical shaft torque of the single-phase generator will have corresponding double-frequency components that create shaft vibration and noise, which could cause shaft failure in large machines. Accordingly, most electric generators and motors rated 5 kVA and higher are constructed as three-phase machines in order to produce nearly constant torque and thereby minimize shaft vibration and noise.

PROBLEMS

Section 2.1

2.1 Given the complex numbers $A_1 = 5\underline{/60^\circ}$ and $A_2 = -3 - j4$, (a) convert A_1 to rectangular form; (b) convert A_2 to polar and exponential form; (c) calculate $A_3 = (A_1 + A_2)$, giving your answer in polar form; (d) calculate $A_4 = A_1 A_2$, giving your answer in rectangular form; (e) calculate $A_5 = A_1/(A_2^*)$, giving your answer in exponential form.

2.2 Convert the following instantaneous currents to phasors, using $\cos(\omega t)$ as the reference. Give your answers in both rectangular and polar form.
(a) $i(t) = 400\cos(\omega t - 30^\circ)$; (b) $i(t) = 5\sin(\omega t + 15^\circ)$;
(c) $i(t) = 4\cos(\omega t - 30^\circ) + 5\sin(\omega t + 15^\circ)$.

2.3 The instantaneous voltage across a circuit element is $v(t) = 678.8\sin(\omega t - 15^\circ)$ volts, and the instantaneous current entering the positive terminal of the circuit element is $i(t) = 200\cos(\omega t - 5^\circ)$ A. For both the current and voltage, determine: (a) the maximum value; (b) the rms value; (c) the phasor expression, using $\cos(\omega t)$ as the reference.

2.4 For the single-phase circuit shown in Figure 2.20, $I = 10\underline{/0^\circ}$ A. (a) Compute the phasors I_1, I_2, and V. (b) Draw a phasor diagram showing I, I_1, I_2, and V.

Figure 2.20

Circuit for Problem 2.4

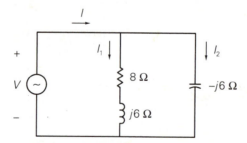

2.5 A 60-Hz, single-phase source with $V = 277\underline{/30^\circ}$ volts is applied to a circuit element. (a) Determine the instantaneous source voltage. Also determine the phasor and instantaneous currents entering the positive terminal if the circuit element is: (b) a 10-Ω resistor, (c) a 5-mH inductor, (d) a capacitor with 25-Ω reactance.

Section 2.2

2.6 For the circuit element of Problem 2.3, calculate: (a) the instantaneous power absorbed; (b) the real power (state whether it is delivered or absorbed); (c) the reactive power (state whether delivered or absorbed); (d) the power factor (state whether lagging or leading).

[*Note*: By convention the power factor $\cos(\delta - \beta)$ is positive. If $|\delta - \beta|$ is greater than 90°, then the reference direction for current may be reversed, resulting in a positive value of $\cos(\delta - \beta)$].

2.7 Referring to Problem 2.5, determine the instantaneous power, real power, and reactive power absorbed by: (a) the 10-Ω resistor, (b) the 5-mH inductor, (c) the capacitor with 25-Ω reactance. Also determine the source power factor and state whether lagging or leading.

2.8 The voltage $v(t) = 678.8\cos(\omega t + 45^\circ)$ volts is applied to a load consisting of a 10-Ω

resistor in parallel with a capacitive reactance $X_C = 25\,\Omega$. Calculate: (a) the instantaneous power absorbed by the resistor; (b) the instantaneous power absorbed by the capacitor; (c) the real power absorbed by the resistor; (d) the reactive power delivered by the capacitor; (e) the load power factor.

2.9 Repeat Problem 2.8 if the resistor and capacitor are connected in series.

2.10 A single-phase source is applied to a two-terminal, passive circuit with equivalent impedance $Z = 2.0\underline{/-45^\circ}\,\Omega$ measured from the terminals. The source current is $i(t) = 4\cos(\omega t)\,\text{kA}$. Determine the: (a) instantaneous power, (b) real power, and (c) reactive power delivered by the source. (d) Also determine the source power factor.

Section 2.3

2.11 Consider a single-phase load with an applied voltage $v(t) = 150\cos(\omega t + 10^\circ)$ volts and load current $i(t) = 5\cos(\omega t - 50^\circ)$ A. (a) Determine the power triangle. (b) Find the power factor and specify whether it is lagging or leading. (c) Calculate the reactive power supplied by capacitors in parallel with the load that correct the power factor to 0.9 lagging.

2.12 A circuit consists of two impedances, $Z_1 = 20\underline{/30^\circ}\,\Omega$ and $Z_2 = 14.14\underline{/-45^\circ}\,\Omega$, in parallel, supplied by a source voltage $V = 100\underline{/60^\circ}$ volts. Determine the power triangle for each of the impedances and for the source.

2.13 An industrial plant consisting primarily of induction motor loads absorbs $1000\,\text{kW}$ at 0.7 power factor lagging. (a) Compute the required kVA rating of a shunt capacitor to improve the power factor to 0.9 lagging. (b) If a synchronous motor rated $1000\,\text{hp}$ with 90% efficiency operating at rated load and at unity power factor is added to the plant instead of the capacitor, calculate the resulting power factor. Assume constant voltage. ($1\,\text{hp} = 0.746\,\text{kW}$)

2.14 The real power delivered by a source to two impedances, $Z_1 = 3 + j5\,\Omega$ and $Z_2 = 10\,\Omega$, connected in parallel, is $1500\,\text{W}$. Determine (a) the real power absorbed by each of the impedances and (b) the source current.

2.15 A single-phase source has a terminal voltage $V = 120\underline{/0^\circ}$ volts and a current $I = 25\underline{/30^\circ}$ A, which leaves the positive terminal of the source. Determine the real and reactive power, and state whether the source is delivering or absorbing each.

2.16 A source supplies power to the following three loads connected in parallel: (1) a lighting load drawing $10\,\text{kW}$, (2) an induction motor drawing $10\,\text{kVA}$ at 0.90 power factor lagging, and (3) a synchronous motor operating at $10\,\text{hp}$, 85% efficiency and 0.95 power factor leading ($1\,\text{hp} = 0.746\,\text{kW}$). Determine the real, reactive, and apparent power delivered by the source. Also, draw the source power triangle.

Section 2.4

2.17 For the circuit shown in Figure 2.21, convert the voltage sources to equivalent current sources and write nodal equations in matrix format using bus 0 as the reference bus. Do not solve the equations.

2.18 For the circuit shown in Figure 2.21, write a computer program that uses the sources, impedances, and bus connections as input data to: (a) compute the 2×2 bus admittance matrix Y; (b) convert the voltage sources to current sources and compute the vector of source currents into buses 1 and 2.

2.19 Determine the 4×4 bus admittance matrix and write nodal equations in matrix format for the circuit shown in Figure 2.22. Do not solve the equations.

Figure 2.21

Circuit diagram for Problems 2.17, 2.18, and 2.20

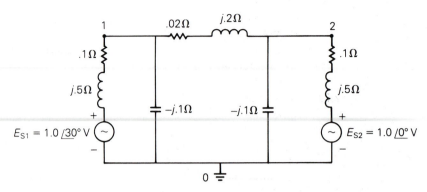

Figure 2.22

Circuit for Problem 2.19

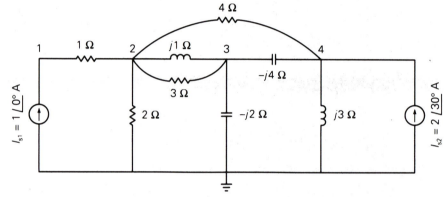

Section 2.5

2.20 For the circuit shown in Figure 2.21, use the CMI and CMM subroutines,* as well as the results of Problem 2.18, to compute the bus voltages V_{10} and V_{20}.

2.21 Use the CMI subroutine* to invert the 3×3 matrix A given by (3.1.8) to verify the resulting inverse matrix A^{-1} given by (3.1.10). Note that in (3.1.8), $a = 1/\underline{120°}$.

2.22 For the circuit shown in Figure 2.22, use the CMI and CMM subroutines,* as well as the results of Problem 2.19, to compute the bus voltages V_{10}, V_{20}, V_{30}, and V_{40}.

Sections 2.6 and 2.7

2.23 A balanced three-phase 480-V source supplies a balanced three-phase load. If the line current I_A is measured to be 10 A, and is in phase with the line-to-line voltage V_{BC}, find the per-phase load impedance if the load is (a) Y-connected, (b) Δ-connected.

2.24 A three-phase 25-kVA, 480-V, 60-Hz alternator, operating under balanced steady-state conditions, supplies a line current of 20 A per phase at a 0.8 lagging power factor and at rated voltage. Determine the power triangle for this operating condition.

2.25 A balanced Δ-connected impedance load with $(12 + j9)\,\Omega$ per phase is supplied by a balanced three-phase 60-Hz, 208-V source. (a) Calculate the line current, the total real and reactive power absorbed by the load, the load power factor, and the apparent load power. (b) Sketch a phasor diagram showing the line currents, the line-to-line source voltages, and the Δ-load currents. Assume positive sequence and use V_{ab} as the reference.

---

*Alternatively, use a general-purpose math program, for example Matlab, available for manipulating matrices.

2.26 Two balanced Y-connected loads, one drawing 10 kW at 0.8 p.f. lagging and the other 15 kW at 0.9 p.f. leading, are connected in parallel and supplied by a balanced three-phase Y-connected, 480-V source. (a) Determine the source current. (b) If the load neutrals are connected to the source neutral by a zero-ohm neutral wire through an ammeter, what will the ammeter read?

2.27 Three identical impedances $Z_\Delta = 20\underline{/60°}\,\Omega$ are connected in Δ to a balanced three-phase 208-V source by three identical line conductors with impedance $Z_L = (0.8 + j0.6)\,\Omega$ per line. (a) Calculate the line-to-line voltage at the load terminals. (b) Repeat part (a) when a Δ-connected capacitor bank with reactance $(-j20)\,\Omega$ per phase is connected in parallel with the load.

2.28 Two three-phase generators supply a three-phase load through separate three-phase lines. The load absorbs 30 kW at 0.8 p.f. lagging. The line impedance is $(1.4 + j1.6)\,\Omega$ per phase between generator G1 and the load, and $(0.8 + j1)\,\Omega$ per phase between generator G2 and the load. If generator G1 supplies 15 kW at 0.8 p.f. lagging, with a terminal voltage of 460 V line-to-line, determine: (a) the voltage at the load terminals; (b) the voltage at the terminals of generator G2; and (c) the real and reactive power supplied by generator G2. Assume balanced operation.

2.29 Two balanced Y-connected loads in parallel, one drawing 15 kW at 0.6 power factor lagging and the other drawing 10 kVA at 0.8 power factor leading, are supplied by a balanced, three-phase, 480-volt source. (a) Draw the power triangle for each load and for the combined load. (b) Determine the power factor of the combined load and state whether lagging or leading. (c) Determine the magnitude of the line current from the source. (d) Δ-connected capacitors are now installed in parallel with the combined load. What value of capacitive reactance is needed in each leg of the Δ to make the source power factor unity? Give your answer in Ω. (e) Compute the magnitude of the current in each capacitor and the line current from the source.

2.30 Figure 2.23 gives the general Δ–Y transformation. (a) Show that the general transformation reduces to that given in Figure 2.15 for a balanced three-phase load. (b) Determine the impedances of the equivalent Y for the following Δ impedances: $Z_{AB} = j10$, $Z_{BC} = j20$, and $Z_{CA} = -j25\,\Omega$.

Figure 2.23

General Δ–Y transformation

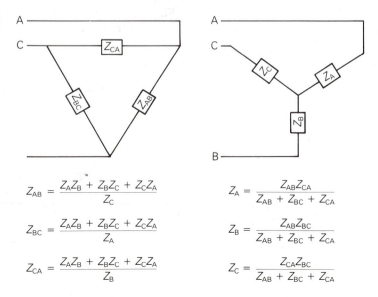

$$Z_{AB} = \frac{Z_A Z_B + Z_B Z_C + Z_C Z_A}{Z_C}$$

$$Z_{BC} = \frac{Z_A Z_B + Z_B Z_C + Z_C Z_A}{Z_A}$$

$$Z_{CA} = \frac{Z_A Z_B + Z_B Z_C + Z_C Z_A}{Z_B}$$

$$Z_A = \frac{Z_{AB} Z_{CA}}{Z_{AB} + Z_{BC} + Z_{CA}}$$

$$Z_B = \frac{Z_{AB} Z_{BC}}{Z_{AB} + Z_{BC} + Z_{CA}}$$

$$Z_C = \frac{Z_{CA} Z_{BC}}{Z_{AB} + Z_{BC} + Z_{CA}}$$

References

1.   W. H. Hayt, Jr. and J. E. Kemmerly, *Engineering Circuit Analysis*, 3d ed. (New York: McGraw-Hill, 1978).

2.   W. A. Blackwell and L. L. Grigsby, *Introductory Network Theory* (Boston: PWS, 1985).

3.   A. E. Fitzgerald, D. E. Higginbotham, and A. Grabel, *Basic Electrical Engineering* (New York: McGraw-Hill, 1981).

4.   W. D. Stevenson, Jr., *Elements of Power System Analysis*, 4th ed. (New York: McGraw-Hill, 1982).

CHAPTER 3

SYMMETRICAL COMPONENTS

Generator stator showing completed windings for a 757-MVA, 3600-RPM, 60-Hz synchronous generator
(Courtesy of General Electric)

The method of symmetrical components, first developed by C. L. Fortescue in 1918, is a powerful technique for analyzing unbalanced three-phase systems. Fortescue defined a linear transformation from phase components to a new set of components called *symmetrical components*. The advantage of this transformation is that for balanced three-phase networks the equivalent circuits obtained for the symmetrical components, called *sequence networks*, are separated into three uncoupled networks. Furthermore, for unbalanced three-phase systems, the three sequence networks are connected only at points of unbalance. As a result, sequence networks for many cases of unbalanced three-phase systems are relatively easy to analyze.

The symmetrical component method is basically a modeling technique that permits systematic analysis and design of three-phase systems. Decoupling a detailed three-phase network into three simpler sequence networks reveals complicated phenomena in more simplistic terms. Sequence network results can then be superposed to obtain three-phase network results. As an example, the application of symmetrical components to unsymmetrical short-circuit studies (see Chapter 9) is indispensable.

The objective of this chapter is to introduce the concept of symmetrical components in order to lay a foundation and provide a framework for later chapters covering both equipment models as well as power-system analysis and design methods. In Section 3.1, symmetrical components are defined. In Sections 3.2, 3.3, and 3.4 sequence networks of loads, series impedances, and rotating machines are presented. Complex power in sequence networks is discussed in Section 3.5, and symmetrical components computer programs are presented in Section 3.6. Although Fortescue's original work is valid for polyphase systems with *n* phases, only three-phase systems will be considered here.

SECTION 3.1

DEFINITION OF SYMMETRICAL COMPONENTS

Assume that a set of three-phase voltages designated V_a, V_b, and V_c is given. In accordance with Fortescue, these phase voltages are resolved into the following three sets of sequence components:

1. *Zero-sequence* components, consisting of three phasors with equal magnitudes and with zero phase displacement, as shown in Figure 3.1(a)

2. *Positive-sequence* components, consisting of three phasors with equal magnitudes, $\pm 120°$ phase displacement, and positive sequence, as in Figure 3.1(b)

3. *Negative-sequence* components, consisting of three phasors with equal magnitudes, $\pm 120°$ phase displacement, and negative sequence, as in Figure 3.1(c)

Figure 3.1

Resolving phase voltages into three sets of sequence components

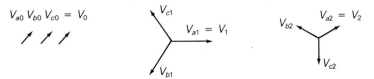

(a) Zero-sequence components   (b) Positive-sequence components   (c) Negative-sequence components

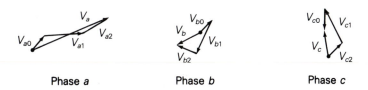

Phase *a*   Phase *b*   Phase *c*

In this text we will work only with the zero-, positive-, and negative-sequence components of phase *a*, which are V_{a0}, V_{a1}, and V_{a2}, respectively. For simplicity, we drop the subscript *a* and denote these sequence components as V_0, V_1, and V_2. They are defined by the following transformation:

$$\begin{bmatrix} V_a \\ V_b \\ V_c \end{bmatrix} = \begin{bmatrix} 1 & 1 & 1 \\ 1 & a^2 & a \\ 1 & a & a^2 \end{bmatrix} \begin{bmatrix} V_0 \\ V_1 \\ V_2 \end{bmatrix} \tag{3.1.1}$$

where

$$a = 1\underline{/120°} = \frac{-1}{2} + j\frac{\sqrt{3}}{2} \tag{3.1.2}$$

Table 3.1

Common identities involving $a = 1\underline{/120°}$

$a^4 = a = 1\underline{/120°}$
$a^2 = 1\underline{/240°}$
$a^3 = 1\underline{/0°}$
$1 + a + a^2 = 0$
$1 - a = \sqrt{3}\underline{/-30°}$
$1 - a^2 = \sqrt{3}\underline{/+30°}$
$a^2 - a = \sqrt{3}\underline{/270°}$
$ja = 1\underline{/210°}$
$1 + a = -a^2 = 1\underline{/60°}$
$1 + a^2 = -a = 1\underline{/-60°}$
$a + a^2 = -1 = 1\underline{/180°}$

Writing (3.1.1) as three separate equations:

$$V_a = V_0 + V_1 + V_2 \tag{3.1.3}$$

$$V_b = V_0 + a^2 V_1 + a V_2 \tag{3.1.4}$$

$$V_c = V_0 + a V_1 + a^2 V_2 \tag{3.1.5}$$

In (3.1.2) *a* is a complex number with unit magnitude and a 120° phase angle. When any phasor is multiplied by *a*, that phasor rotates by 120° (counterclockwise). Similarly, when any phasor is multiplied by $a^2 = (1\underline{/120°})(1\underline{/120°}) = 1\underline{/240°}$, the phasor rotates by 240°. Table 3.1 lists some common identities involving *a*.

The complex number *a* is similar to the well-known complex number $j = \sqrt{-1} = 1\underline{/90°}$. Thus the only difference between *j* and *a* is that the angle of *j* is 90°, and that of *a* is 120°.

Equation (3.1.1) can be rewritten more compactly using matrix notation. We define the following vectors V_p and V_s, and matrix A:

$$V_p = \begin{bmatrix} V_a \\ V_b \\ V_c \end{bmatrix} \tag{3.1.6}$$

$$V_s = \begin{bmatrix} V_0 \\ V_1 \\ V_2 \end{bmatrix} \tag{3.1.7}$$

$$A = \begin{bmatrix} 1 & 1 & 1 \\ 1 & a^2 & a \\ 1 & a & a^2 \end{bmatrix} \tag{3.1.8}$$

V_p is the column vector of phase voltages, V_s is the column vector of sequence voltages, and A is a 3×3 transformation matrix. Using these definitions, (3.1.1) becomes

$$V_p = AV_s \tag{3.1.9}$$

The inverse of the A matrix is

$$A^{-1} = \tfrac{1}{3} \begin{bmatrix} 1 & 1 & 1 \\ 1 & a & a^2 \\ 1 & a^2 & a \end{bmatrix} \tag{3.1.10}$$

Equation (3.1.10) can be verified by showing that the product AA^{-1} is the unit matrix. Also, premultiplying (3.1.9) by A^{-1} gives

$$V_s = A^{-1}V_p \tag{3.1.11}$$

Using (3.1.6), (3.1.7), and (3.1.10), then (3.1.11) becomes

zero seq
pos seq

$$\begin{bmatrix} V_0 \\ V_1 \\ V_2 \end{bmatrix} = \tfrac{1}{3} \begin{bmatrix} 1 & 1 & 1 \\ 1 & a & a^2 \\ 1 & a^2 & a \end{bmatrix} \begin{bmatrix} V_a \\ V_b \\ V_c \end{bmatrix} \tag{3.1.12}$$

Writing (3.1.12) as three separate equations,

$$V_0 = \tfrac{1}{3}(V_a + V_b + V_c) \tag{3.1.13}$$

$$V_1 = \tfrac{1}{3}(V_a + aV_b + a^2V_c) \tag{3.1.14}$$

$$V_2 = \tfrac{1}{3}(V_a + a^2V_b + aV_c) \tag{3.1.15}$$

no zero seq √
in bal ckt

Equation (3.1.13) shows that there is no zero-sequence voltage in a *balanced* three-phase system because the sum of three balanced phasors is zero. In an unbalanced three-phase system, line-to-neutral voltages may have a zero-sequence component. But line-to-line voltages never have a zero-sequence component, since by KVL their sum is always zero.

The symmetrical component transformation can also be applied to currents, as follows. Let

$$\mathbf{I}_p = A\mathbf{I}_s \qquad (3.1.16)$$

where $\mathbf{I}_p$ is a vector of phase currents,

$$\mathbf{I}_p = \begin{bmatrix} I_a \\ I_b \\ I_c \end{bmatrix} \qquad (3.1.17)$$

and $\mathbf{I}_s$ is a vector of sequence currents,

$$\mathbf{I}_s = \begin{bmatrix} I_0 \\ I_1 \\ I_2 \end{bmatrix} \qquad (3.1.18)$$

Also,

$$\mathbf{I}_s = A^{-1}\mathbf{I}_p \qquad (3.1.19)$$

Equations (3.1.16) and (3.1.19) can be written as separate equations as follows. The phase currents are

$$I_a = I_0 + I_1 + I_2 \qquad (3.1.20)$$

$$I_b = I_0 + a^2 I_1 + a I_2 \qquad (3.1.21)$$

$$I_c = I_0 + a I_1 + a^2 I_2 \qquad (3.1.22)$$

and the sequence currents are

$$I_0 = \tfrac{1}{3}(I_a + I_b + I_c) \qquad (3.1.23)$$

$$I_1 = \tfrac{1}{3}(I_a + a I_b + a^2 I_c) \qquad (3.1.24)$$

$$I_2 = \tfrac{1}{3}(I_a + a^2 I_b + a I_c) \qquad (3.1.25)$$

In a three-phase Y-connected system, the neutral current I_n is the sum of the line currents:

$$I_n = I_a + I_b + I_c \qquad (3.1.26)$$

Comparing (3.1.26) and (3.1.23),

$$I_n = 3I_0 \qquad (3.1.27)$$

The neutral current equals three times the zero-sequence current. In a

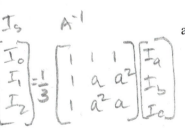

balanced Y-connected system, line currents have no zero-sequence component, since the neutral current is zero. Also, in any three-phase system with no neutral path, such as a Δ-connected system or a three-wire Y-connected system with an ungrounded neutral, line currents have no zero-sequence component.

The following three examples further illustrate symmetrical components.

EXAMPLE 3.1   **Sequence components: balanced line-to-neutral voltages**

Calculate the sequence components of the following balanced line-to-neutral voltages with *abc* sequence:

$$V_p = \begin{bmatrix} V_{an} \\ V_{bn} \\ V_{cn} \end{bmatrix} = \begin{bmatrix} 277\underline{/0°} \\ 277\underline{/-120°} \\ 277\underline{/+120°} \end{bmatrix} \quad \text{volts}$$

Solution   Using (3.1.13)–(3.1.15):

$$V_0 = \tfrac{1}{3}[277\underline{/0°} + 277\underline{/-120°} + 277\underline{/+120°}] = 0$$

$$V_1 = \tfrac{1}{3}[277\underline{/0°} + 277\underline{/(-120° + 120°)} + 277\underline{/(120° + 240°)}]$$

$$= 277\underline{/0°} \quad \text{volts} = V_{an}$$

$$V_2 = \tfrac{1}{3}[277\underline{/0°} + 277\underline{/(-120° + 240°)} + 277\underline{/(120° + 120°)}]$$

$$= \tfrac{1}{3}[277\underline{/0°} + 277\underline{/120°} + 277\underline{/240°}] = 0$$

This example illustrates the fact that balanced three-phase systems with *abc* sequence (or positive sequence) have no zero-sequence or negative-sequence components. For this example, the positive-sequence voltage V_1 equals V_{an}, and the zero-sequence and negative-sequence voltages are both zero. ∎

EXAMPLE 3.2   **Sequence components: balanced *acb* currents**

A Y-connected load has balanced currents with *acb* sequence given by

$$I_p = \begin{bmatrix} I_a \\ I_b \\ I_c \end{bmatrix} = \begin{bmatrix} 10\underline{/0°} \\ 10\underline{/+120°} \\ 10\underline{/-120°} \end{bmatrix} \quad \text{A}$$

Calculate the sequence currents.

Solution   Using (3.1.23)–(3.1.25):

$$I_0 = \tfrac{1}{3}[10\underline{/0°} + 10\underline{/120°} + 10\underline{/-120°}] = 0$$

$$I_1 = \tfrac{1}{3}[10\underline{/0°} + 10\underline{/(120° + 120°)} + 10\underline{/(-120° + 240°)}]$$

$$= \tfrac{1}{3}[10\underline{/0°} + 10\underline{/240°} + 10\underline{/120°}] = 0$$

$$I_2 = \tfrac{1}{3}[10\underline{/0°} + 10\underline{/(120° + 240°)} + 10\underline{/(-120° + 120°)}]$$

$$= 10\underline{/0°} \text{ A} = I_a$$

This example illustrates the fact that balanced three-phase systems with *acb* sequence (or negative sequence) have no zero-sequence or positive-sequence components. For this example the negative-sequence current I_2 equals I_a, and the zero-sequence and positive-sequence currents are both zero. ∎

EXAMPLE 3.3

Sequence components: unbalanced currents

A three-phase line feeding a balanced-Y load has one of its phases (phase *b*) open. The load neutral is grounded, and the unbalanced line currents are

pos seq

$$I_p = \begin{bmatrix} I_a \\ I_b \\ I_c \end{bmatrix} = \begin{bmatrix} 10\underline{/0°} \\ 0 \\ 10\underline{/120°} \end{bmatrix} \quad A$$

Calculate the sequence currents and the neutral current.

Solution   The circuit is shown in Figure 3.2. Using (3.1.23)–(3.1.25):

Zero seq →
$$I_0 = \tfrac{1}{3}[10\underline{/0°} + 0 + 10\underline{/120°}]$$
$$= 3.333\underline{/60°} \quad A$$

Pos seq →
$$I_1 = \tfrac{1}{3}[10\underline{/0°} + 0 + 10\underline{/(120° + 240°)}] = 6.667\underline{/0°} \quad A$$

neg seq
$$I_2 = \tfrac{1}{3}[10\underline{/0°} + 0 + 10\underline{/(120° + 120°)}]$$
$$= 3.333\underline{/-60°} \quad A$$

Using (3.1.26) the neutral current is

$$I_n = (10\underline{/0°} + 0 + 10\underline{/120°})$$
$$= 10\underline{/60°} \, A = 3I_0$$

This example illustrates the fact that *unbalanced* three-phase systems may have nonzero values for all sequence components. Also, the neutral current equals three times the zero-sequence current, as given by (3.1.27).

Figure 3.2

Circuit for Example 3.3

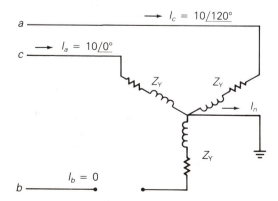

$I_c = 10\underline{/120°}$

a

$I_a = 10\underline{/0°}$

c

Z_Y Z_Y I_n

Z_Y

$I_b = 0$

b

SECTION 3.2

SEQUENCE NETWORKS OF IMPEDANCE LOADS

Figure 3.3 shows a balanced-Y impedance load. The impedance of each phase is designated Z_Y, and a neutral impedance Z_n is connected between the load neutral and ground. Note from Figure 3.3 that the line-to-ground voltage V_{ag} is

$$
\begin{aligned}
V_{ag} &= Z_Y I_a + Z_n I_n \\
&= Z_Y I_a + Z_n(I_a + I_b + I_c) \\
&= (Z_Y + Z_n)I_a + Z_n I_b + Z_n I_c
\end{aligned}
$$
(3.2.1)

Figure 3.3

Balanced-Y impedance load

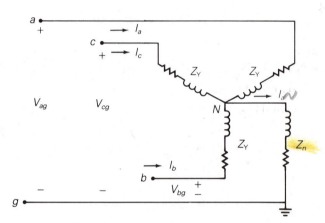

Similar equations can be written for V_{bg} and V_{cg}:

$$
V_{bg} = Z_n I_a + (Z_Y + Z_n)I_b + Z_n I_c
$$
(3.2.2)

$$
V_{cg} = Z_n I_a + Z_n I_b + (Z_Y + Z_n)I_c
$$
(3.2.3)

Equations (3.2.1)–(3.2.3) can be rewritten in matrix format:

$$
\begin{bmatrix} V_{ag} \\ V_{bg} \\ V_{cg} \end{bmatrix} = \begin{bmatrix} (Z_Y + Z_n) & Z_n & Z_n \\ Z_n & (Z_Y + Z_n) & Z_n \\ Z_n & Z_n & (Z_Y + Z_n) \end{bmatrix} \begin{bmatrix} I_a \\ I_b \\ I_c \end{bmatrix}
$$
(3.2.4)

Equation (3.2.4) is written more compactly as

$$
V_p = Z_p I_p
$$
(3.2.5)

where V_p is the vector of line-to-ground voltages (or phase voltages), I_p is the vector of line currents (or phase currents), and Z_p is the 3×3 phase impedance matrix shown in (3.2.4). Equations (3.1.9) and (3.1.16) can now be used in (3.2.5) to determine the relationship between the sequence voltages and currents, as follows:

$$
AV_s = Z_p A I_s
$$
(3.2.6)

Premultiplying both sides of (3.2.6) by A^{-1} gives

$$V_s = (A^{-1}Z_pA)I_s \tag{3.2.7}$$

or

$$V_s = Z_sI_s \tag{3.2.8}$$

where

$$Z_s = A^{-1}Z_pA \tag{3.2.9}$$

The impedance matrix Z_s defined by (3.2.9) is called the *sequence impedance matrix.* Using the definition of A, its inverse A^{-1}, and Z_p given by (3.1.8), (3.1.10), and (3.2.4), the sequence impedance matrix Z_s for the balanced-Y load is

$$Z_s = \tfrac{1}{3}\begin{bmatrix} 1 & 1 & 1 \\ 1 & a & a^2 \\ 1 & a^2 & a \end{bmatrix}\begin{bmatrix} (Z_Y + Z_n) & Z_n & Z_n \\ Z_n & (Z_Y + Z_n) & Z_n \\ Z_n & Z_n & (Z_Y + Z_n) \end{bmatrix}\begin{bmatrix} 1 & 1 & 1 \\ 1 & a^2 & a \\ 1 & a & a^2 \end{bmatrix} \tag{3.2.10}$$

Performing the indicated matrix multiplications in (3.2.10), and using the identity $(1 + a + a^2) = 0$,

$$Z_s = \tfrac{1}{3}\begin{bmatrix} 1 & 1 & 1 \\ 1 & a & a^2 \\ 1 & a^2 & a \end{bmatrix}\begin{bmatrix} (Z_Y + 3Z_n) & Z_Y & Z_Y \\ (Z_Y + 3Z_n) & a^2Z_Y & aZ_Y \\ (Z_Y + 3Z_n) & aZ_Y & a^2Z_Y \end{bmatrix}$$

$$Z_s = \begin{bmatrix} (Z_Y + 3Z_n) & 0 & 0 \\ 0 & Z_Y & 0 \\ 0 & 0 & Z_Y \end{bmatrix} \tag{3.2.11}$$

As shown in (3.2.11), the sequence impedance matrix Z_s for the balanced-Y load of Figure 3.3 is a diagonal matrix. Since Z_s is diagonal, (3.2.8) can be written as three *uncoupled* equations. Using (3.1.7), (3.1.18), and (3.2.11) in (3.2.8),

$$\begin{bmatrix} V_0 \\ V_1 \\ V_2 \end{bmatrix} = \begin{bmatrix} (Z_Y + 3Z_n) & 0 & 0 \\ 0 & Z_Y & 0 \\ 0 & 0 & Z_Y \end{bmatrix}\begin{bmatrix} I_0 \\ I_1 \\ I_2 \end{bmatrix} \tag{3.2.12}$$

Rewriting (3.2.12) as three separate equations,

$$V_0 = (Z_Y + 3Z_n)I_0 = Z_0I_0 \tag{3.2.13}$$

$$V_1 = Z_YI_1 = Z_1I_1 \tag{3.2.14}$$

$$V_2 = Z_YI_2 = Z_2I_2 \tag{3.2.15}$$

As shown in (3.2.13), the zero-sequence voltage V_0 depends only on the zero-sequence current I_0 and the impedance $(Z_Y + 3Z_n)$. This impedance is called the *zero-sequence impedance* and is designated Z_0. Also, the positive-sequence voltage V_1 depends only on the positive-sequence current I_1 and an impedance $Z_1 = Z_Y$ called the *positive-sequence impedance*. Similarly, V_2 depends only on I_2 and the *negative-sequence impedance* $Z_2 = Z_Y$.

Equations (3.2.13)–(3.2.15) can be represented by the three networks shown in Figure 3.4. These networks are called the *zero-sequence, positive-sequence,* and *negative-sequence networks*. As shown, each sequence network is separate, uncoupled from the other two. The separation of these sequence networks is a consequence of the fact that Z_s is a diagonal matrix for a balanced-Y load. This separation underlies the advantage of symmetrical components.

[handwritten: adv of symmetrical networks]

Figure 3.4

Sequence networks of a balanced-Y load

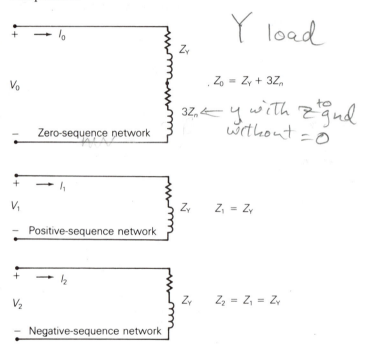

[handwritten annotations: Y load; $Z_0 = Z_Y + 3Z_n$; $3Z_n \leftarrow$ y with z to gnd without $= 0$]

Zero-sequence network

Positive-sequence network Z_Y $Z_1 = Z_Y$

Negative-sequence network Z_Y $Z_2 = Z_1 = Z_Y$

Note that the neutral impedance does not appear in the positive- and negative-sequence networks of Figure 3.4. This illustrates the fact that positive- and negative-sequence currents do not flow in neutral impedances. However, the neutral impedance is multiplied by 3 and placed in the zero-sequence network of the figure. The voltage $I_0(3Z_n)$ across the impedance $3Z_n$ is the voltage drop $(I_n Z_n)$ across the neutral impedance Z_n in Figure 3.3, since $I_n = 3I_0$.

[handwritten: $I_0 3Z_n = V_{Zn}$; $I_n = 3I_0$]

When the neutral of the Y load in Figure 3.3 has no return path, then the neutral impedance Z_n is infinite and the term $3Z_n$ in the zero-sequence network of Figure 3.4 becomes an open circuit. Under this condition of an open neutral, no zero-sequence current exists. However, when the neutral of the Y load is solidly grounded with a zero-ohm conductor, then the neutral impedance is zero and the term $3Z_n$ in the zero-sequence network becomes a short circuit. Under this condition of a solidly grounded neutral, zero-sequence current I_0 can exist when there is a zero-sequence voltage caused by unbalanced voltages applied to the load.

Figure 2.15 shows a balanced-Δ load and its equivalent balanced-Y load. Since the Δ load has no neutral connection, the equivalent Y load in Figure 2.15 has an open neutral. The sequence networks of the equivalent Y load corresponding to a balanced-Δ load are shown in Figure 3.5. As shown, the equivalent Y impedance $Z_Y = Z_\Delta/3$ appears in each of the sequence networks. Also, the zero-sequence network has an open circuit, since $Z_n = \infty$ corresponds to an open neutral. No zero-sequence current occurs in the equivalent Y load.

The sequence networks of Figure 3.5 represent the balanced-Δ load as viewed from its terminals, but they do not represent the internal load characteristics. The currents I_0, I_1, and I_2 in Figure 3.5 are the sequence components of the line currents feeding the Δ load, not the load currents within the Δ. The Δ load currents, which are related to the line currents by (2.6.14), are not shown in Figure 3.5.

Figure 3.5

Sequence networks for an equivalent Y representation of a balanced-Δ load

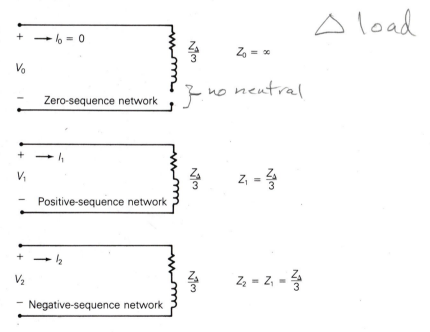

Δ load

$I_0 = 0$

V_0

$\frac{Z_\Delta}{3}$ $Z_0 = \infty$

Zero-sequence network

no neutral

I_1

V_1

$\frac{Z_\Delta}{3}$ $Z_1 = \frac{Z_\Delta}{3}$

Positive-sequence network

I_2

V_2

$\frac{Z_\Delta}{3}$ $Z_2 = Z_1 = \frac{Z_\Delta}{3}$

Negative-sequence network

EXAMPLE 3.4

Sequence networks: balanced-Y and -Δ loads

A balanced-Y load is in parallel with a balanced-Δ-connected capacitor bank. The Y load has an impedance $Z_Y = (3 + j4)\Omega$ per phase, and its neutral is grounded through an inductive reactance $X_n = 2\Omega$. The capacitor bank has a reactance $X_c = 30\Omega$ per phase. Draw the sequence networks for this load and calculate the load-sequence impedances.

Solution   The sequence networks are shown in Figure 3.6. As shown, the Y-load impedance in the zero-sequence network is in series with three times the neutral impedance. Also, the Δ-load branch in the zero-sequence network is open, since no zero-sequence current flows into the Δ load. In the positive- and negative-sequence circuits, the Δ-load impedance is divided by 3 and placed in parallel with the Y-load impedance. The equivalent sequence impedances are

$$Z_0 = Z_Y + 3Z_n = 3 + j4 + 3(j2) = 3 + j10 \quad \Omega$$

$$Z_1 = Z_Y // (Z_\Delta/3) = \frac{(3 + j4)(-j30/3)}{3 + j4 - j(30/3)}$$

$$= \frac{(5\underline{/53.13°})(10\underline{/-90°})}{6.708\underline{/-63.43°}} = 7.454\underline{/26.57°} \quad \Omega$$

$$Z_2 = Z_1 = 7.454\underline{/26.57°} \quad \Omega$$

Figure 3.6

Sequence networks for Example 3.4

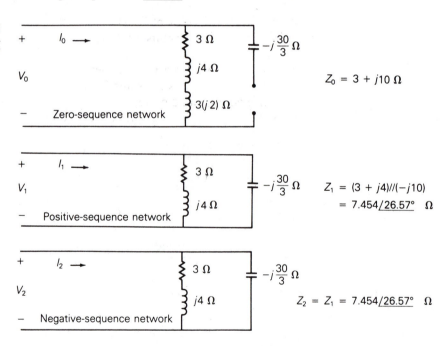

$Z_0 = 3 + j10 \ \Omega$

$Z_1 = (3 + j4)//(-j10)$
$= 7.454\underline{/26.57°} \ \Omega$

$Z_2 = Z_1 = 7.454\underline{/26.57°} \ \Omega$

Figure 3.7 shows a general three-phase linear impedance load. The load could represent a balanced load such as the balanced-Y or -Δ load, or an unbalanced impedance load. The general relationship between the line-to-ground voltages and line currents for this load can be written as

$$\begin{bmatrix} V_{ag} \\ V_{bg} \\ V_{cg} \end{bmatrix} = \begin{bmatrix} Z_{aa} & Z_{ab} & Z_{ac} \\ Z_{ab} & Z_{bb} & Z_{bc} \\ Z_{ac} & Z_{bc} & Z_{cc} \end{bmatrix} \begin{bmatrix} I_a \\ I_b \\ I_c \end{bmatrix} \tag{3.2.16}$$

Figure 3.7

General three-phase impedance load (linear, bilateral network, nonrotating equipment)

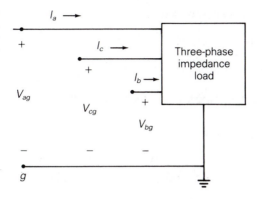

or

$$V_p = Z_p I_p \tag{3.2.17}$$

where V_p is the vector of line-to-neutral (or phase) voltages, I_p is the vector of line (or phase) currents, and Z_p is a 3×3 phase impedance matrix. It is assumed here that the load is nonrotating, and that Z_p is a symmetric matrix, which corresponds to a bilateral network.

Since (3.2.17) has the same form as (3.2.5), the relationship between the sequence voltages and currents for the general three-phase load of Figure 3.6 is the same as that of (3.2.8) and (3.2.9), which are rewritten here:

$$V_s = Z_s I_s \tag{3.2.18}$$

$$Z_s = A^{-1} Z_p A \tag{3.2.19}$$

The sequence impedance matrix Z_s given by (3.2.19) is a 3×3 matrix with nine sequence impedances, defined as follows:

$$Z_s = \begin{bmatrix} Z_0 & Z_{01} & Z_{02} \\ Z_{10} & Z_1 & Z_{12} \\ Z_{20} & Z_{21} & Z_2 \end{bmatrix} \tag{3.2.20}$$

The diagonal impedances Z_0, Z_1, and Z_2 in this matrix are the self-impedances of the zero-, positive-, and negative-sequence networks. The off-diagonal impedances are the mutual impedances between sequence networks. Using the definitions of A, A^{-1}, Z_p, and Z_s, (3.2.19) is

$$\begin{bmatrix} Z_0 & Z_{01} & Z_{02} \\ Z_{10} & Z_1 & Z_{12} \\ Z_{20} & Z_{21} & Z_2 \end{bmatrix} = \tfrac{1}{3} \begin{bmatrix} 1 & 1 & 1 \\ 1 & a & a^2 \\ 1 & a^2 & a \end{bmatrix} \begin{bmatrix} Z_{aa} & Z_{ab} & Z_{ac} \\ Z_{ab} & Z_{bb} & Z_{bc} \\ Z_{ac} & Z_{bc} & Z_{cc} \end{bmatrix} \begin{bmatrix} 1 & 1 & 1 \\ 1 & a^2 & a \\ 1 & a & a^2 \end{bmatrix} \tag{3.2.21}$$

Performing the indicated multiplications in (3.2.21), and using the identity $(1 + a + a^2) = 0$, the following separate equations can be obtained (see Problem 3.12):

Diagonal sequence impedances

$$Z_0 = \tfrac{1}{3}(Z_{aa} + Z_{bb} + Z_{cc} + 2Z_{ab} + 2Z_{ac} + 2Z_{bc}) \tag{3.2.22}$$

$$Z_1 = Z_2 = \tfrac{1}{3}(Z_{aa} + Z_{bb} + Z_{cc} - Z_{ab} - Z_{ac} - Z_{bc}) \tag{3.2.23}$$

Off-diagonal sequence impedances

$$Z_{01} = Z_{20} = \tfrac{1}{3}(Z_{aa} + a^2 Z_{bb} + a Z_{cc} - a Z_{ab} - a^2 Z_{ac} - Z_{bc}) \tag{3.2.24}$$

$$Z_{02} = Z_{10} = \tfrac{1}{3}(Z_{aa} + a Z_{bb} + a^2 Z_{cc} - a^2 Z_{ab} - a Z_{ac} - Z_{bc}) \tag{3.2.25}$$

$$Z_{12} = \tfrac{1}{3}(Z_{aa} + a^2 Z_{bb} + a Z_{cc} + 2a Z_{ab} + 2a^2 Z_{ac} + 2Z_{bc}) \tag{3.2.26}$$

$$Z_{21} = \tfrac{1}{3}(Z_{aa} + a Z_{bb} + a^2 Z_{cc} + 2a^2 Z_{ab} + 2a Z_{ac} + 2Z_{bc}) \tag{3.2.27}$$

A *symmetrical load* is defined as a load whose sequence impedance matrix is diagonal; that is, all the mutual impedances in (3.2.24)–(3.2.27) are zero. Equating these mutual impedances to zero and solving, the following conditions for a symmetrical load are determined. When both

and
$$\left.\begin{aligned} Z_{aa} &= Z_{bb} = Z_{cc} \\[6pt] Z_{ab} &= Z_{ac} = Z_{bc} \end{aligned}\right\} \begin{array}{l} \text{conditions for a} \\ \text{symmetrical load} \end{array}$$
$$\tag{3.2.28}$$
$$\tag{3.2.29}$$

then

$$Z_{01} = Z_{10} = Z_{02} = Z_{20} = Z_{12} = Z_{21} = 0 \tag{3.2.30}$$

$$Z_0 = Z_{aa} + 2Z_{ab} \tag{3.2.31}$$

$$Z_1 = Z_2 = Z_{aa} - Z_{ab} \tag{3.2.32}$$

The conditions for a symmetrical load are that the diagonal phase impedances be equal and that the off-diagonal phase impedances be equal. These conditions can be verified by using (3.2.28) and (3.2.29) with the identity $(1 + a + a^2) = 0$ in (3.2.24)–(3.2.27) to show that all the mutual sequence impedances are zero. Note that the positive- and negative-sequence impedances are equal for a symmetrical load, as shown by (3.2.32), and for a nonsymmetrical load, as shown by (3.2.23). This is always true for linear, symmetric impedances that represent nonrotating equipment such as transformers and transmission lines. However, the positive- and negative-sequence impedances of rotating equipment such as generators and motors are generally not equal. Note also that the zero-sequence impedance Z_0 is not equal to the positive- and negative-sequence impedances of a symmetrical load unless the mutual phase impedances $Z_{ab} = Z_{ac} = Z_{bc}$ are zero.

The sequence networks of a symmetrical impedance load are shown in Figure 3.8. Since the sequence impedance matrix $\mathbf{Z}_s$ is diagonal for a symmetrical load, the sequence networks are separate or uncoupled.

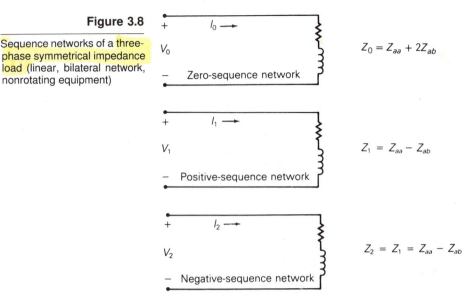

Figure 3.8

Sequence networks of a three-phase symmetrical impedance load (linear, bilateral network, nonrotating equipment)

$I_0 \rightarrow$

V_0

Zero-sequence network

$Z_0 = Z_{aa} + 2Z_{ab}$

$I_1 \rightarrow$

V_1

Positive-sequence network

$Z_1 = Z_{aa} - Z_{ab}$

$I_2 \rightarrow$

V_2

Negative-sequence network

$Z_2 = Z_1 = Z_{aa} - Z_{ab}$

SECTION 3.3

SEQUENCE NETWORKS OF SERIES IMPEDANCES

Figure 3.9 shows series impedances connected between two three-phase buses denoted abc and $a'b'c'$. Self-impedances of each phase are denoted Z_{aa}, Z_{bb}, and Z_{cc}. In general, the series network may also have mutual impedances between phases. The voltage drops across the series-phase impedances are given by

$$\begin{bmatrix} V_{an} - V_{a'n} \\ V_{bn} - V_{b'n} \\ V_{cn} - V_{c'n} \end{bmatrix} = \begin{bmatrix} V_{aa'} \\ V_{bb'} \\ V_{cc'} \end{bmatrix} = \begin{bmatrix} Z_{aa} & Z_{ab} & Z_{ac} \\ Z_{ab} & Z_{bb} & Z_{bc} \\ Z_{ac} & Z_{bc} & Z_{cc} \end{bmatrix} \begin{bmatrix} I_a \\ I_b \\ I_c \end{bmatrix} \tag{3.3.1}$$

Figure 3.9

Three-phase series impedances (linear, bilateral network, nonrotating equipment)

$I_a \rightarrow$ $+$ $V_{aa'}$ $-$

a — Z_{aa} — a'

$+$ V_{an} $+$

$I_c \rightarrow$ $+$ $V_{cc'}$ $-$

c — Z_{cc} — c' $V_{a'n}$

$+$ $+$

$I_b \rightarrow$ $+$ $V_{bb'}$ $-$

V_{cn} b — Z_{bb} — b' $V_{c'n}$

$+$ $+$

V_{bn} $V_{b'n}$

$-$ $-$ $-$ $-$ $-$ $-$

n —————————————— n

Both self-impedances and mutual impedances are included in (3.3.1). It is assumed that the impedance matrix is symmetric, which corresponds to a bilateral network. It is also assumed that these impedances represent non-rotating equipment. Typical examples are series impedances of transmission lines and of transformers. Equation (3.3.1) has the following form:

$$V_p - V_{p'} = Z_p I_p \tag{3.3.2}$$

where V_p is the vector of line-to-neutral voltages at bus abc, $V_{p'}$ is the vector of line-to-neutral voltages at bus $a'b'c'$, I_p is the vector of line currents, and Z_p is the 3×3 phase impedance matrix for the series network. Equation (3.3.2) is now transformed to the sequence domain in the same manner that the load-phase impedances were transformed in Section 3.2. Thus,

$$V_s - V_{s'} = Z_s I_s \tag{3.3.3}$$

where

$$Z_s = A^{-1} Z_p A \tag{3.3.4}$$

From the results of Section 3.2, this sequence impedance Z_s matrix is diagonal under the following conditions:

$$\left.
\begin{aligned}
Z_{aa} &= Z_{bb} = Z_{cc} \\
\text{and} \qquad & \\
Z_{ab} &= Z_{ac} = Z_{bc}
\end{aligned}
\right\}
\begin{aligned}
&\text{conditions for} \\
&\text{symmetrical} \\
&\text{series impedances}
\end{aligned} \tag{3.3.5}$$

When the phase impedance matrix Z_p of (3.3.1) has both equal self-impedances and equal mutual impedances, then (3.3.4) becomes

$$Z_s = \begin{bmatrix} Z_0 & 0 & 0 \\ 0 & Z_1 & 0 \\ 0 & 0 & Z_2 \end{bmatrix} \tag{3.3.6}$$

where

$$Z_0 = Z_{aa} + 2Z_{ab} \tag{3.3.7}$$

and

$$Z_1 = Z_2 = Z_{aa} - Z_{ab} \tag{3.3.8}$$

and (3.3.3) becomes three uncoupled equations, written as follows:

$$V_0 - V_{0'} = Z_0 I_0 \tag{3.3.9}$$
$$V_1 - V_{1'} = Z_1 I_1 \tag{3.3.10}$$
$$V_2 - V_{2'} = Z_2 I_2 \tag{3.3.11}$$

Equations (3.3.9)–(3.3.11) are represented by the three uncoupled sequence networks shown in Figure 3.10. From the figure it is apparent that for symmetrical series impedances, positive-sequence currents produce only positive-sequence voltage drops. Similarly, negative-sequence currents

Figure 3.10

Sequence networks of three-phase symmetrical series impedances (linear, bilateral network, nonrotating equipment)

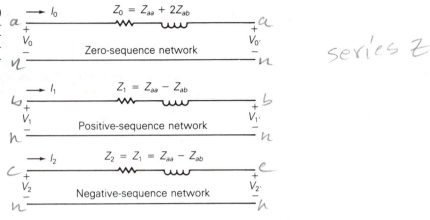

$Z_0 = Z_{aa} + 2Z_{ab}$

Zero-sequence network

$Z_1 = Z_{aa} - Z_{ab}$

Positive-sequence network

$Z_2 = Z_1 = Z_{aa} - Z_{ab}$

Negative-sequence network

series Z

produce only negative-sequence voltage drops, and zero-sequence currents produce only zero-sequence voltage drops. However, if the series impedances are not symmetrical, then Z_s is not diagonal, the sequence networks are coupled, and the voltage drop across any one sequence network depends on all three sequence currents.

SECTION 3.4

SEQUENCE NETWORKS OF ROTATING MACHINES

A Y-connected synchronous generator grounded through a neutral impedance Z_n is shown in Figure 3.11. The internal generator voltages are

Figure 3.11

Y-connected synchronous generator

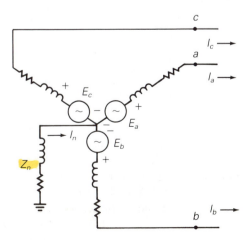

designated E_a, E_b, and E_c, and the generator line currents are designated I_a, I_b, and I_c.

The sequence networks of the generator are shown in Figure 3.12. Since a three-phase synchronous generator is designed to produce balanced internal phase voltages E_a, E_b, E_c with only a positive-sequence component, a source voltage E_{g1} is included only in the positive-sequence network. The sequence components of the line-to-ground voltages at the generator terminals are denoted V_0, V_1, and V_2 in Figure 3.12.

Figure 3.12

Sequence networks of a Y-connected synchronous generator

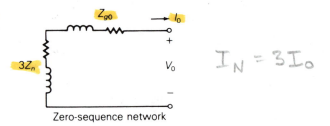

Zero-sequence network

$I_N = 3I_0$

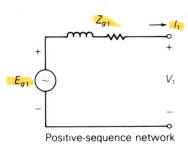

Positive-sequence network

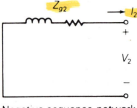

Negative-sequence network

The voltage drop in the generator neutral impedance is $Z_n I_n$, which can be written as $(3Z_n)I_0$, since, from (3.1.27), the neutral current is three times the zero-sequence current. Since this voltage drop is due only to zero-sequence current, an impedance $(3Z_n)$ is placed in the zero-sequence network of Figure 3.12 in series with the generator zero-sequence impedance Z_{g0}.

The sequence impedances of rotating machines are generally not equal. A detailed analysis of machine-sequence impedances is given in machine theory texts. Only a brief explanation is given here.

When a synchronous generator stator has balanced three-phase positive-sequence currents under steady-state conditions, the net mmf produced by these positive-sequence currents rotates at the synchronous rotor speed in the same direction as that of the rotor. Under this condition, a high value of magnetic flux penetrates the rotor, and the positive-sequence impedance Z_{g1} has a high value. Under steady-state conditions, the positive-sequence generator impedance is called the *synchronous impedance.*

When a synchronous generator stator has balanced three-phase negative-sequence currents, the net mmf produced by these currents rotates at synchronous speed in the direction opposite to that of the rotor. With respect to the rotor, the net mmf is not stationary but rotates at twice synchronous speed. Under this condition, currents are induced in the rotor windings that prevent the magnetic flux from penetrating the rotor. As such, the negative-sequence impedance Z_{g2} is less than the positive-sequence synchronous impedance.

When a synchronous generator has only zero-sequence currents, which are line (or phase) currents with equal magnitude and phase, then the net mmf produced by these currents is theoretically zero. The generator zero-sequence impedance Z_{g0} is the smallest sequence impedance and is due to leakage flux, end turns, and harmonic flux from windings that do not produce a perfectly sinusoidal mmf.

Typical values of machine-sequence impedances are listed in Table A.1 in the Appendix. The positive-sequence machine impedance is synchronous, transient, or subtransient. *Synchronous* impedances are used for steady-state conditions, such as in power-flow studies, which are described in Chapter 7. *Transient* impedances are used for stability studies, which are described in Chapter 12, and *subtransient* impedances are used for short-circuit studies, which are described in Chapters 8 and 9. Unlike the positive-sequence impedances, a machine has only one negative-sequence impedance and only one zero-sequence impedance.

The sequence networks for three-phase synchronous motors and for three-phase induction motors are shown in Figure 3.13. Synchronous motors have the same sequence networks as synchronous generators, except that the sequence currents for synchronous motors are referenced *into* rather than out of the sequence networks. Also, induction motors have the same sequence networks as synchronous motors, except that the positive-sequence voltage source E_{m1} is removed. Induction motors do not have a dc source of magnetic flux in their rotor circuits, and therefore E_{m1} is zero (or a short circuit).

The sequence networks shown in Figures 3.12 and 3.13 are simplified networks for rotating machines. The networks do not take into account such phenomena as machine saliency, saturation effects, and more complicated transient effects. These simplified networks, however, are in many cases accurate enough for power-system studies.

diff. btwn synch mtrs & induction mtrs

Figure 3.13

Sequence networks of three-phase motors

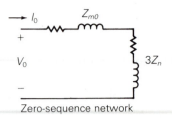

Zero-sequence network

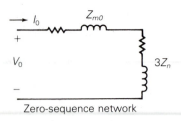

Zero-sequence network

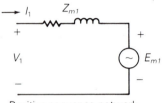

Positive-sequence network

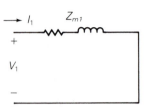

Positive-sequence network

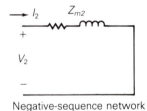

Negative-sequence network

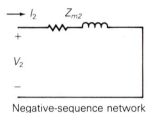

Negative-sequence network

(a) Synchronous motor

(b) Induction motor

EXAMPLE 3.5

Currents in sequence networks

Draw the sequence networks for the circuit of Example 2.5 and calculate the sequence components of the line current. Assume that the generator neutral is grounded through an impedance $Z_n = j\,10\,\Omega$, and that the generator sequence impedances are $Z_{g0} = j\,1\,\Omega$, $Z_{g1} = j\,15\,\Omega$, and $Z_{g2} = j\,3\,\Omega$.

Solution   The sequence networks are shown in Figure 3.14. They are obtained by interconnecting the sequence networks for a balanced-Δ load, for series-line impedances, and for a synchronous generator, which are given in Figures 3.5, 3.10, and 3.12.

It is clear from Figure 3.14 that $I_0 = I_2 = 0$ since there are no sources in the zero- and negative-sequence networks. Also, the positive-sequence generator terminal voltage V_1 equals the generator line-to-neutral terminal voltage. Therefore, from the positive-sequence network shown in the figure

Figure 3.14

Sequence networks for
Example 3.5

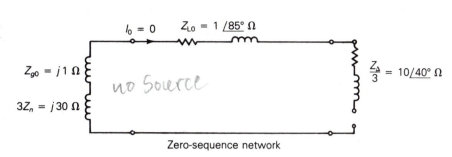

Zero-sequence network

Solidly gnd
$Z_n = 0$

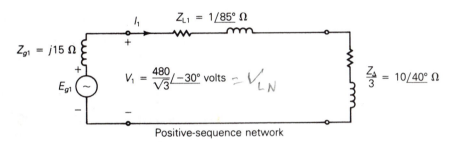

Positive-sequence network

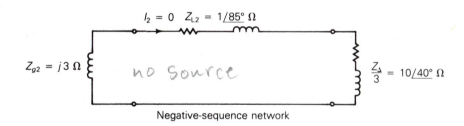

Negative-sequence network

and from the results of Example 2.5,

$$I_1 = \frac{V_1}{(Z_{L1} + \frac{1}{3}Z_\Delta)} = 25.83\underline{/-73.78°}\,\text{A} = I_a$$

Note that from (3.1.20), I_1 equals the line current I_a since $I_0 = I_2 = 0$. ∎

The following example illustrates the superiority of using symmetrical components for analyzing unbalanced systems.

EXAMPLE 3.6

Solving unbalanced three-phase networks using sequence components

A Y-connected voltage source with the following unbalanced voltage is applied to the balanced line and load of Example 2.5.

$$\begin{bmatrix} V_{ag} \\ V_{bg} \\ V_{cg} \end{bmatrix} = \begin{bmatrix} 277\underline{/0°} \\ 260\underline{/-120°} \\ 295\underline{/+115°} \end{bmatrix} \text{ volts}$$

The source neutral is solidly grounded. Using the method of symmetrical components, calculate the source currents I_a, I_b, and I_c.

Solution   Using (3.1.13)–(3.1.15), the sequence components of the source voltages are:

$$V_0 = \tfrac{1}{3}(277\underline{/0°} + 260\underline{/-120°} + 295\underline{/115°})$$

$$= 7.4425 + j14.065 = 15.912\underline{/62.11°} \quad \text{volts}$$

$$V_1 = \tfrac{1}{3}(227\underline{/0°} + 260\underline{/-120° + 120°} + 295\underline{/115° + 240°})$$

$$= \tfrac{1}{3}(277\underline{/0°} + 260\underline{/0°} + 295\underline{/-5°})$$

$$= 276.96 - j8.5703 = 277.1\underline{/-1.772°} \quad \text{volts}$$

$$V_2 = \tfrac{1}{3}(277\underline{/0°} + 260\underline{/-120° + 240°} + 295\underline{/115° + 120°})$$

$$= \tfrac{1}{3}(277\underline{/0°} + 260\underline{/120°} + 295\underline{/235°})$$

$$= -7.4017 - j5.4944 = 9.218\underline{/216.59°} \quad \text{volts}$$

These sequence voltages are applied to the sequence networks of the line and load, as shown in Figure 3.15. The sequence networks of this figure are

Figure 3.15

Sequence networks for
Example 3.6

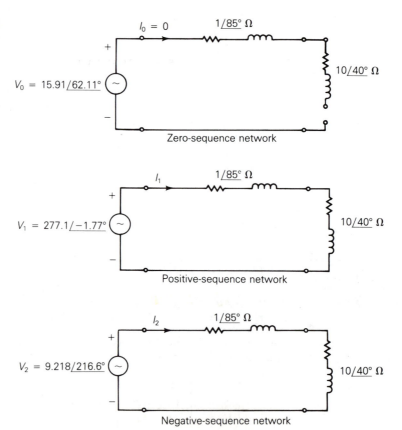

$I_0 = 0$ $1\underline{/85°}\ \Omega$

$V_0 = 15.91\underline{/62.11°}$ $10\underline{/40°}\ \Omega$

Zero-sequence network

I_1 $1\underline{/85°}\ \Omega$

$V_1 = 277.1\underline{/-1.77°}$ $10\underline{/40°}\ \Omega$

Positive-sequence network

I_2 $1\underline{/85°}\ \Omega$

$V_2 = 9.218\underline{/216.6°}$ $10\underline{/40°}\ \Omega$

Negative-sequence network

uncoupled, and the sequence components of the source currents are easily calculated as follows:

$$I_0 = 0$$

$$I_1 = \frac{V_1}{Z_{L1} + \dfrac{Z_\Delta}{3}} = \frac{277.1/-1.772°}{10.73/43.78°} = 25.82/-45.55° \quad A$$

$$I_2 = \frac{V_2}{Z_{L2} + \dfrac{Z_\Delta}{3}} = \frac{9.218/216.59°}{10.73/43.78°} = 0.8591/172.81° \quad A$$

P9 53

Using (3.1.20)–(3.1.22), the source currents are:

$$I_a = (0 + 25.82/-45.55° + 0.8591/172.81°)$$

$$= 17.23 - j\,18.32 = 25.15/-46.76° \quad A$$

$$I_b = (0 + 25.82/-45.55° + 240° + 0.8591/172.81° + 120°)$$

$$= (25.82/194.45° + 0.8591/292.81°)$$

$$= -24.67 - j7.235 = 25.71/196.34° \quad A$$

$$I_c = (0 + 25.82/-45.55° + 120° + 0.8591/172.81° + 240°)$$

$$= (25.82/74.45° + 0.8591/52.81°)$$

$$= 7.441 + j25.56 = 26.62/73.77° \quad A$$

The reader should calculate the line currents for this example without using symmetrical components, in order to verify this result and to compare the two solution methods (see Problem 3.20). Without symmetrical components, coupled KVL equations must be solved. With symmetrical components, the conversion from phase to sequence components decouples the networks as well as the resulting KVL equations, as shown above. ∎

SECTION 3.5

POWER IN SEQUENCE NETWORKS

The power delivered to a three-phase network can be determined from the power delivered to the sequence networks. Let S_p denote the total complex power delivered to the three-phase load of Figure 3.7, which can be calculated from

$$S_p = V_{ag}I_a^* + V_{bg}I_b^* + V_{cg}I_c^* \tag{3.5.1}$$

Equation (3.5.1) is also valid for the total complex power delivered by the three-phase generator of Figure 3.11, or for the complex power delivered to any three-phase bus. Rewriting (3.5.1) in matrix format,

$$S_p = [V_{ag}\, V_{bg}\, V_{cg}] \begin{bmatrix} I_a^* \\ I_b^* \\ I_c^* \end{bmatrix}$$

$$= V_p^{\mathrm{T}} I_p^* \tag{3.5.2}$$

where **T denotes transpose** and *** denotes complex conjugate.** Now, using (3.1.9) and (3.1.16),

$$S_p = (A V_s)^{\mathrm{T}} (A I_s)^*$$

$$= V_s^{\mathrm{T}} [A^{\mathrm{T}} A^*] I_s^* \tag{3.5.3}$$

Using the definition of A, which is (3.1.8), to calculate the term within the brackets of (3.5.3), and noting that a and a^2 are conjugates,

$$A^{\mathrm{T}} A^* = \begin{bmatrix} 1 & 1 & 1 \\ 1 & a^2 & a \\ 1 & a & a^2 \end{bmatrix}^{\mathrm{T}} \begin{bmatrix} 1 & 1 & 1 \\ 1 & a^2 & a \\ 1 & a & a^2 \end{bmatrix}^*$$

$$= \begin{bmatrix} 1 & 1 & 1 \\ 1 & a^2 & a \\ 1 & a & a^2 \end{bmatrix} \begin{bmatrix} 1 & 1 & 1 \\ 1 & a & a^2 \\ 1 & a^2 & a \end{bmatrix}$$

$$= \begin{bmatrix} 3 & 0 & 0 \\ 0 & 3 & 0 \\ 0 & 0 & 3 \end{bmatrix} = 3U \tag{3.5.4}$$

Equation (3.5.4) can now be used in (3.5.3) to obtain

$$S_p = 3 V_s^{\mathrm{T}} I_s^*$$

$$= 3[V_0 + V_1 + V_2] \begin{bmatrix} I_0^* \\ I_1^* \\ I_2^* \end{bmatrix} \tag{3.5.5}$$

$$S_p = 3(V_0 I_0^* + V_1 I_1^* + V_2 I_2^*)$$

$$= 3 S_s \tag{3.5.6}$$

Thus, the total complex power S_p delivered to a three-phase network equals *three* times the total complex power S_s delivered to the sequence networks.

The reason the factor of 3 occurs in (3.5.6) is that $A^{\mathrm{T}} A^* = 3U$, as shown by (3.5.4). It is possible to eliminate this factor of 3 by defining a new transformation matrix $A_1 = (1/\sqrt{3})A$ such that $A_1^{\mathrm{T}} A_1^* = U$, which means that A_1 is a *unitary* matrix. Using A_1 instead of A, the total complex power delivered to a three-phase networks would equal the total complex power delivered to the sequence networks. However, standard industry practice for symmetrical components is to use A, defined by (3.1.8).

| EXAMPLE 3.7 | **Power in sequence networks** |

Calculate S_p and S_s delivered by the three-phase source in Example 3.6. Verify that $S_p = 3S_s$.

Solution   Using (3.5.1),

All three phases
Apparent Power

$$S_p = (277\underline{/0°})(25.15\underline{/+46.76°}) + (260\underline{/-120°})(25.71\underline{/-196.34°})$$
$$+ (295\underline{/115°})(26.62\underline{/-73.77°})$$
$$= 6967\underline{/46.76°} + 6685\underline{/43.66°} + 7853\underline{/41.23°}$$
$$= 15,520 + j14,870 = 21,490\underline{/43.78°} \quad \text{VA}$$

In the sequence domain,

$$S_s = V_0 I_0^* + V_1 I_1^* + V_2 I_2^*$$
$$= 0 + (277.1\underline{/-1.77°})(25.82\underline{/45.55°}) + (9.218\underline{/216.59°})$$
$$(0.8591\underline{/-172.81°})$$
$$= 7155\underline{/43.78°} + 7.919\underline{/43.78°}$$
$$= 5172 + j4958 = 7163\underline{/43.78°} \quad \text{VA}$$

Also,

$$3S_s = 3(7163\underline{/43.78°}) = 21,490\underline{/43.78°} = S_p \qquad ∎$$

SECTION 3.6

PERSONAL COMPUTER PROGRAM: SYMMETRICAL COMPONENTS

The following symmetrical component subroutines are included in the personal computer software package that accompanies this text.

1. SEQVEC(V_p, V_s) computes the complex sequence vector $V_s = A^{-1}V_p$ for any three-phase complex vector V_p, where

$$V_p = \begin{bmatrix} V_a \\ V_b \\ V_c \end{bmatrix} \qquad V_s = \begin{bmatrix} V_0 \\ V_1 \\ V_2 \end{bmatrix}$$

   The complex vector V_p may be a three-phase voltage vector, a three-phase current vector, or any complex vector with three components.

2. PHAVEC(V_s, V_p) computes the complex phase vector $V_p = AV_s$ for any complex sequence vector V_s.

3. SEQIMP(Z_p, Z_s) computes the 3×3 complex sequence impedance matrix $Z_s = A^{-1}Z_pA$ for any 3×3 complex phase impedance matrix Z_p.

EXAMPLE 3.8   **Symmetrical components program**

A three-phase impedance load, as shown in Figure 3.7, is represented by the following phase impedance matrix:

$$Z_p = \begin{bmatrix} (10 + j30) & (5 + j20) & (5 + j20) \\ (5 + j20) & (10 + j30) & (5 + j20) \\ (5 + j20) & (5 + j20) & (10 + j30) \end{bmatrix} \ \Omega$$

A Y-connected source with the following unbalanced line-to-ground voltages is applied to the load:

$$V_p = \begin{bmatrix} 277\underline{/0°} \\ 260\underline{/-120°} \\ 295\underline{/+115°} \end{bmatrix} \ \text{volts}$$

The source neutral is solidly grounded. Use the personal computer software package to calculate the following:

a. Load-sequence impedance matrix Z_s
b. Load-sequence voltage vector V_s
c. Load-sequence current vector I_s
d. Load-phase current vector I_p
e. Total complex power delivered to the sequence networks, $S_s = V_s^T I_s^*$
f. Total complex power delivered to the three-phase load, $S_p = V_p^T I_p^*$

Solution   Table 3.2 shows the solution to Example 3.8, including the input data and output data listings. Note that the sequence impedance matrix is diagonal,

Table 3.2   Solution to Example 3.8

a.

$$Z_p = \begin{bmatrix} +1.000D+01 & j+3.000D+01 & +5.000D+00 & j+2.000D+01 & +5.000D+00 & j+2.000D+01 \\ +5.000D+00 & j+2.000D+01 & +1.000D+01 & j+3.000D+01 & +5.000D+00 & j+2.000D+01 \\ +5.000D+00 & j+2.000D+01 & +5.000D+00 & j+2.000D+01 & +1.000D+01 & j+3.000D+01 \end{bmatrix}$$

$$Z_s = \begin{bmatrix} +2.000D+01 & j+7.000D+01 & +1.000D-12 & j+0.000D+00 & -1.000D-12 & j+0.000D+00 \\ -2.000D-13 & j-7.000D-13 & +5.000D+00 & j+1.000D+01 & -3.000D-13 & j-3.000D-13 \\ -2.000D-13 & j-7.000D-13 & -1.600D-13 & j-1.000D-13 & +5.000D+00 & j+1.000D+01 \end{bmatrix}$$

b. $V_p = \begin{bmatrix} +2.770D+02 & j+0.000D+00 \\ -1.300D+02 & j-2.252D+02 \\ -1.247D+02 & j+2.674D+02 \end{bmatrix}$ $V_s = \begin{bmatrix} +7.443D+00 & j+1.406D+01 \\ +2.770D+02 & j-8.570D+00 \\ -7.402D+00 & j-5.494D+00 \end{bmatrix}$

c. $I_s = \begin{bmatrix} +2.138D-01 & j-4.526D-02 \\ +1.039D+01 & j-2.250D+01 \\ -7.356D-01 & j+3.724D-01 \end{bmatrix}$ d. $I_p = \begin{bmatrix} +9.868D+00 & j-2.217D+01 \\ -2.442D+01 & j+1.383D+00 \\ +1.519D+01 & j+2.065D+01 \end{bmatrix}$

e. $S_s = +3.075D+03 \quad j+6.154D+03$ f. $S_p = +9.224D+03 \quad j+1.846D+04$

since the conditions for a symmetrical load are satisfied; that is, $Z_{aa} = Z_{bb} = Z_{cc} = (10 + j30)\Omega$ and $Z_{ab} = Z_{ac} = Z_{bc} = (5 + j20)\Omega$. Note also that in addition to the subroutines listed in this section, the subroutines CMI, CMM, CMT, and CMC described in Section 2.5 are employed to compute I_s, S_s, and S_p. As shown in Table 3.2, the relationship $S_p = 3S_s$ is verified for this example.   ∎

PROBLEMS

Section 3.1

3.1 Using the operator $a = 1\underline{/120°}$, evaluate the following in polar form: (a) $(a + 1)/$ $(1 + a - a^2)$ (b) $(a^2 + a + j)/(ja - a^2)$ (c) $(1 - a)(1 + a^2)$ (d) $(a + a^2)(a^2 + 1)$.

3.2 Using $a = 1\underline{/120°}$, evaluate the following in rectangular form:

    a. a^{10}

    b. $(ja)^{10}$

    c. $(1 - a)^3$

    d. e^a.

    Hint for (d): $e^{(x + jy)} = e^x e^{jy} = e^x\underline{/y}$, where y is in radians.

3.3 Determine the symmetrical components of the following line currents: (a) $I_a = 5\underline{/90°}$, $I_b = 5\underline{/340°}$, $I_c = 5\underline{/200°}$ A (b) $I_a = 50$, $I_b = j50$, $I_c = 0$ A

3.4 Find the phase voltages V_{an}, V_{bn}, and V_{cn} whose sequence components are: $V_0 = 20\underline{/80°}$, $V_1 = 100\underline{/0°}$, $V_2 = 30\underline{/180°}$ V.

3.5 One line of a three-phase generator is open circuited, while the other two are short-circuited to ground. The line currents are $I_a = 0$, $I_b = 1000\underline{/90°}$, and $I_c = 1000\underline{/-30°}$ A. Find the symmetrical components of these currents. Also find the current into the ground.

3.6 Given the line-to-ground voltages $V_{ag} = 280\underline{/0°}$, $V_{bg} = 290\underline{/-130°}$, and $V_{cg} = 260\underline{/110°}$ volts, calculate (a) the sequence components of the line-to-ground voltages, denoted V_{Lg0}, V_{Lg1}, and V_{Lg2}; (b) line-to-line voltages V_{ab}, V_{bc}, and V_{ca}; and (c) sequence components of the line-to-line voltages V_{LL0}, V_{LL1}, and V_{LL2}. Also, verify the following general relation: $V_{LL0} = 0$, $V_{LL1} = \sqrt{3}V_{Lg1}\underline{/+30°}$, and $V_{LL2} = \sqrt{3}V_{Lg2}\underline{/-30°}$ volts.

Section 3.2

3.7 The currents in a Δ load are $I_{ab} = 10\underline{/0°}$, $I_{bc} = 20\underline{/-90°}$, and $I_{ca} = 15\underline{/90°}$ A. Calculate (a) the sequence components of the Δ-load currents, denoted I_{A0}, I_{A1}, I_{A2}; (b) the line currents I_a, I_b, and I_c, which feed the Δ load; and (c) sequence components of the line currents I_{L0}, I_{L1}, and I_{L2}. Also, verify the following general relation: $I_{L0} = 0$, $I_{L1} = \sqrt{3}I_{A1}\underline{/-30°}$, and $I_{L2} = \sqrt{3}I_{A2}\underline{/+30°}$ A.

3.8 The voltages given in Problem 3.6 are applied to a balaned-Y load consisting of $(12 + j16)$ ohms per phase. The load neutral is solidly grounded. Draw the sequence networks and calculate I_0, I_1, and I_2, the sequence components of the line currents. Then calculate the line currents I_a, I_b, and I_c.

3.9 Repeat Problem 3.8 with the load neutral open.

3.10 Repeat Problem 3.8 for a balaned-Δ load consisting of $(12 + j16)$ ohms per phase.

3.11 Repeat Problem 3.8 for the load shown in Example 3.4 (Figure 3.6).

3.12 Perform the indicated matrix multiplications in (3.2.21) and verify the sequence impedances given by (3.2.22)–(3.2.27).

3.13 The following unbalanced line-to-ground voltages are applied to the balanced-Y load shown in Figure 3.3: $V_{ag} = 100\underline{/0°}$, $V_{bg} = 75\underline{/180°}$, and $V_{cg} = 50\underline{/90°}$ volts. The Y load has $Z_Y = 3 + j4\,\Omega$ per phase with neutral impedance $Z_n = j1\,\Omega$. (a) Calculate the line currents I_a, I_b, and I_c without using symmetrical components. (b) Calculate the line currents I_a, I_b, and I_c using symmetrical components. Which method is easier?

3.14 The three-phase impedance load shown in Figure 3.7 has the following phase impedance matrix:

$$Z_p = \begin{bmatrix} (3+j5) & 0 & 0 \\ 0 & (3+j5) & 0 \\ 0 & 0 & (3+j5) \end{bmatrix} \Omega$$

Determine the sequence impedance matrix Z_s for this load. Is the load symmetrical?

ask Hanna → **3.15** The three-phase impedance load shown in Figure 3.7 has the following sequence impedance matrix:

$$Z_S = \begin{bmatrix} (4+j6) & 0 & 0 \\ 0 & 2 & 0 \\ 0 & 0 & 2 \end{bmatrix} \Omega$$

Determine the phase impedance matrix Z_p for this load. Is the load symmetrical?

Section 3.3

3.16 Repeat Problem 3.8 but include balanced three-phase line impedances of $(3 + j4)$ ohms per phase between the source and load.

Section 3.4

3.17 As shown in Figure 3.16, a balanced three-phase, positive-sequence source with $V_{AB} = 480\underline{/0°}$ volts is applied to an unbalanced Δ load. Note that one leg of the Δ is open. Determine: (a) the load currents I_{AB} and I_{BC}; (b) the line currents I_A, I_B, and I_C, which feed the Δ load; and (c) the zero-, positive-, and negative-sequence components of the line currents.

Figure 3.16

Problem 3.17

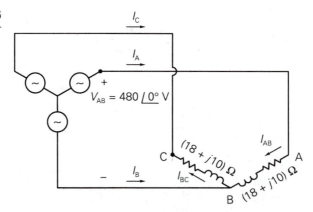

3.18 A balanced Y-connected generator with terminal voltage $V_{bc} = 208\underline{/90°}$ volts is connected to a balanced-Δ load whose impedance is $20\underline{/40°}$ ohms per phase. The line impedance between the source and load is $0.5\underline{/80°}$ ohm for each phase. The generator neutral is grounded through an impedance of $j5$ ohms. The generator sequence impedances are given by $Z_{g0} = j7$, $Z_{g1} = j15$, and $Z_{g2} = j10$ ohms. Draw the sequence networks for this system and determine the sequence components of the line currents.

3.19   In a three-phase system, a synchronous generator supplies power to a 480-volt synchronous motor through a line having an impedance of $0.5\underline{/80^\circ}$ ohm per phase. The motor draws 10 kW at 0.8 p.f. leading and at rated voltage. The neutrals of both the generator and motor are grounded through impedances of $j5$ ohms. The sequene impedances of both machines are $Z_0 = j5$, $Z_1 = j15$, and $Z_2 = j10$ ohms. Draw the sequence networks for this system and find the line-to-line voltage at the generator terminals. Assume balanced three-phase operation.

3.20   Calculate the source currents in Example 3.6 without using symmetrical components. Compare your solution method with that of Example 3.6. Which method is easier?

Section 3.5

3.21   For Problem 3.8, calculate the real and reactive power delivered to the three-phase load.

3.22   A three-phase impedance load consists of a balanced-Δ load in parallel with a balanced-Y load. The impedance of each leg of the Δ load is $Z_\Delta = 6 + j6\,\Omega$, and the impedance of each leg of the Y load is $Z_Y = 2 + j2\,\Omega$. The Y load is grounded through a neutral impedance $Z_n = j1\,\Omega$. Unbalanced line-to-ground source voltages V_{ag}, V_{bg}, and V_{cg} with sequence components $V_0 = 10\underline{/60^\circ}$, $V_1 = 100\underline{/0^\circ}$, and $V_2 = 15\underline{/200^\circ}$ volts are applied to the load. (a) Draw the zero-, positive-, and negative-sequence networks. (b) Determine the complex power delivered to each sequence network. (c) Determine the total complex power delivered to the three-phase load.

Section 3.6

3.23   Solve Problem 3.3 using the personal computer subroutine SEQVEC.

3.24   Solve Problem 3.4 using the personal computer subroutine PHAVEC.

3.25   Repeat Example 3.8, but include balanced three-phase line impedances of $(1 + j5)$ ohms per phase between the source and load.

References

1.   Westinghouse Electric Corporation, *Applied Protective Relaying* (Newark, NJ: Westinghouse, 1976).

2.   P. M. Anderson, *Analysis of Faulted Power Systems* (Ames, IA: Iowa State University Press, 1973).

3.   W. D. Stevenson, Jr., *Elements of Power System Analysis*, 4th ed. (New York: McGraw-Hill, 1982).

CHAPTER 4

POWER TRANSFORMERS

Core and coil assemblies of a three-phase 20.3 kVΔ/345 kVY step-up
transformer. This oil-immersed transformer is rated 325 MVA self-cooled
(OA)/542 MVA forced oil, forced air-cooled (FOA)/607 MVA forced oil,
forced air-cooled (FOA)
(Courtesy of General Electric)

The power transformer is a major power-system component that permits
economical power transmission with high efficiency and low series-voltage
drops. Since electric power is proportional to the product of voltage and
current, low current levels (and therefore low I^2R losses and low IZ voltage

drops) can be maintained for given power levels via high voltages. Power transformers transform ac voltage and current to optimum levels for generation, transmission, distribution, and utilization of electric power.

The development in 1885 by William Stanley of a commercially practical transformer was what made ac power systems more attractive than dc power systems. The ac system with a transformer overcame voltage problems encountered in dc systems as load levels and transmission distances increased. Today's modern power transformers have nearly 100% efficiency, with ratings up to and beyond 1300 MVA.

In this chapter, basic transformer theory is reviewed, and equivalent circuits for practical transformers operating under sinusoidal-steady-state conditions are developed. Models of single-phase two-winding, three-phase two-winding, and three-phase three-winding transformers, as well as autotransformers and regulating transformers are given. Also, the per-unit system, which simplifies power-system analysis by eliminating the ideal transformer winding in transformer equivalent circuits, is introduced in this chapter and used throughout the remainder of the text.

CASE STUDY   The following article illustrates what electric utility companies look for in power and distribution transformers. Transformer price, losses, noise, reliability, and manufacturer's support are discussed. Transformer evaluation programs of four utilities are described [8].

How Electric Utilities Buy Quality When They Buy Transformers

JOHN REASON

Because transformers are passive devices with few moving parts, it is difficult to evaluate the quality of one over another. But today, when the lifetime cost of transformer losses far exceeds the initial transformer purchase price and a significant percentage of transformer purchases is to replace units that have failed in service, utilities need a mechanism to weigh one manufacturer's offering against another's—often well before the transformer is actually built.

Power and distribution transformers present entirely different problems to the purchasing engineers charged with evaluating quality. Power transformers are generally custom-built and today they are often very different from any transformer the utility has bought before. Power transformers should be evaluated according to a wide range of quality factors, each of which has a different importance or weight, depending on the purchasing utility.

Reprinted with permission from Electrical World.

In contrast, distribution transformers are purchased in bulk and, provided detailed failure records are kept, the quality can be rather easily determined from computerized statistical programs.

Low losses mean high quality

One factor in the engineer's favor is that high-quality transformers are also low-loss transformers. In a sense, the cost of high quality is automatically paid for in the first few years of transformer life by reduced losses. To this benefit is added the fact that the lifetime of a transformer built today will actually be significantly longer than that of a transformer built only a few years ago.

Losses are divided into load and no-load losses and various formulas and/or computer programs are available to evaluate their lifetime impact. When individual utilities plug their cost factors into the formulas, the lifetime impacts they calculate vary widely. For example, the ratio of estimated costs of no-load to load losses can vary by a factor as much as 10 to one. The relative cost of load and no-load losses can also vary from year to year as regulatory pressures push utility management to emphasize different needs.

Noise is becoming an increasingly important factor in transformer selection. Again, this factor varies widely from utility to utility. The greatest need for a low-noise transformer is felt by utilities in highly developed areas where substations must be located close to residential neighborhoods.

Transformer noise is generated from three sources: (1) the magnetostrictive deformation of the core, (2) aerodynamic noise produced by cooling fans, and (3) the mechanical and flow noise from the oil-circulating pumps. The radiated core noise, consisting of a 120-Hz tone, is the most difficult to reduce and is also the noise that generates the most complaints from residents living near the transformer.

Fortunately, improved core-construction techniques and lower-loss core steel both tend to reduce transformer core noise. If further reduction in core noise is needed, it can only be achieved by increasing the cross-sectional area of the core to reduce the flux

density. This design change increases the construction cost of the transformer and decreases the core losses. However, a point of diminishing returns is reached at which the cost of increasing core size outweighs the savings in reduced losses (Fig. 1).

Installation costs are significant because a power transformer must generally be delivered partially disassembled and without oil in the tank. Today, the trend is for the manufacturer to assemble and fill the transformer on site, rather than leave it to the utility. This provides assurance that the transformer is correctly installed and minimizes the cost of lost parts, misunderstanding, etc.

Manufacturing facilities provide a key indication of the product quality (Fig. 2). Most utilities use plant visits as the first step in their evaluation process. Facility review should include the manufacturer's quality-assurance program, in-service and test reliability records, contract administration and order support, and technical strength.

Coating systems, especially for pad-mount transformers, are becoming increasingly important since the life of the transformer tank may be the limiting factor in transformer life. The problem of evaluating and comparing coating systems on pad-mount transformers from different manufacturers was

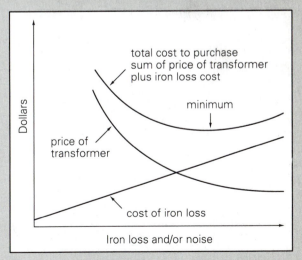

Figure 1 *Noise reduction requirements may require that iron losses are reduced below the optimum economic level*

Figure 2 *Manufacturing facilities are a good indicator of product quality. Here, electrical connections are made on core form power transformer*

Figure 3 *Coating systems are an important part of transformer quality. Here, tank is chemically cleaned prior to electrodeposition of coating*

eliminated with the introduction of ANSI Standard C57.12.28-1988. This is a functional standard that does not dictate to manufacturers how they should coat transformers, but prescribes a series of tests that the coating must withstand to meet the standard. A companion standard, C57.12.31 for poletop transformers, is now under development.

Tests prescribed by the standards include: Scratching to bare metal and exposing to salt spray for 1500 hours; cross-hatch scratching to check for adhesion, humidity exposure at 113°C, impact of 160 in.-lb with no paint chipping, oil and ultraviolet resistance, and 3000 cycles of abrasion resistance.

In response to this standard, most manufacturers have revamped or rebuilt their painting processes—

from surface preparation (Fig. 3) through application of primers, to finished coating systems. The most advanced painting processes now use electrodeposition methods—either as a dip process or with paint applied as a dry powder. These proesses not only ensure a uniform coating system to every part of the transformer tank but also, because they eliminate traditional solvent-based paints, more easily meet the Clean Air Act Amendments of 1990.

Hard evaluation factors are set down in the purchaser's technical specifications, which form the primary document to ensure that all suppliers' products meet a minimum standard. Technical specifications generally include an evaluation formula for no-load and load losses, price, noise level, and delivery date. Technical assistance during installation, warranty assistance, and the extent of warranty are additional hard evaluation factors.

Soft factors do not have a precise monetary value, but also may be important in comparing suppliers' bids. The [following] list suggests soft factors for buyers to include in a transformer-purchase decision. While they do not have a direct dollar value, it is valuable to assign a fixed dollar value or a percentage of bid value to these factors so that they can be used

how to write a spec.

in comparing suppliers' bids. A well-written specification places all potential suppliers on an equal footing.

Soft factors that should influence choice of supplier

- Wide choice of designs
- Computer-aided design procedures
- R&D directed at product improvement
- Participation in long-term R&D projects through industry groups
- Clean-room assembly facilities
- Availability of spare and replacement parts
- Wide range of field services
- Application assistance/coordination
- Ongoing communication with users

Tony Hartfield, ABB Power T&D Co, Power Transformer Div, St. Louis, Mo, says it is important to review technical specifications in detail with prospective suppliers before a request for bids is issued. "We attempt to resolve ambiguous terms such as 'substantial,' 'long-lasting,' or 'equal-to,' and replace them with functional requirements that clearly define what must be supplied.

"Many times, items are added to a specification to prevent recurrence of past problems. These can be counterproductive, particularly if the technology has advanced to a point where the source of the problem has been eliminated."

Good in-service records vital

Distribution transformers are purchased in large quantities under very competitive conditions where a unit-price change of a few cents can affect the choice of supplier. As a result, the most sophisticated programs used to guide purchasing policy are based on statistical records of units in service.

One example of a systematic failure-analysis program is that conducted by Wisconsin Public Service Corp (EW, September 1991, p 73). All transformers purchased by the utility since the mid 1980s and all

transformer failures are entered into a computerized record-keeping system. Failure rates and equivalent costs are calculated for each manufacturer on a four-year rolling window. According to Senior Standards Engineer Michael Radke, the system has substantially reduced failure rates, improved communications with transformer vendors, reduced costs, and reduced outages. The system has even helped some manufacturers to reduce failure rates.

Georgia Power Co's vendor evaluation program has been in place for about five years. This program looks at supplier and product separately, judging each according to pre-established criteria. The scores for each criterion are weighted and the overall score used to calculate a numerical multiplier, which is applied to initial bid price. David McClure, research manager, quality and support, explains that the program involves four departments: engineering, materials, quality assurance, and procurement. Each department is responsible for a portion of the evaluation and the results from each are entered into a computer program.

The evaluation involves objective and subjective factors. Compliance, for example, can be measured objectively, but customer service must be evaluated subjectively. Even so, reviewers follow a well-defined procedure to determine scores for each factor. This approach ensures that ratings are applied consistently to each vendor.

Public Service Co of Colorado (PSC) uses a numerical multiplier that is applied to the bid price. The multiplier incorporates several factors—including historical failure rate, delivery, and quality. Of these factors, historical failure rate is by far the most important, accounting for more than half of the multiplier penalty. For example, the average multiplier for pole-mounted transformers adds 6.3%, of which failure rate accounts for 4.9%; the average multiplier for single-phase pad-mount transformers adds 5.3%, of which failure rate accounts for 3.6%.

Failure rate is calculated using a computer program supplied by General Electric Co, Transformer Business Dept, Hickory, NC. It is based on failures of transformers purchased in the last 10 years. The cost of failure includes the cost of a replacement unit and the costs of changeout and downtime.

A delivery penalty is calculated by PSC, based on the difference in weeks between promised and actual delivery dates. Significantly, this penalty is calculated equally for early as for late delivery. Early delivery is considered disruptive. John Ainscough, senior engineer, automation analysis and research, reports that his department is planning to modify this factor to encourage both short lead-times and on-time delivery. Currently, the delivery factor does not incorporate the supplier's manufacturing cycle time.

PSC's quality factor is based on the percentage of an order that must be repaired or returned to the manufacturer; the accuracy with which products conform to the original specifications, including losses and impedance; and the number of days required to resolve complaints and warranty claims. Responsiveness to complaints is considered a soft evaluation factor and the number of days needed to resolve a complaint is a way of quantifying this factor. The utility is exploring ways to quantify other soft factors in the evaluation process.

According to Ainscough, the rating system in use at PSC seems to be effective for consistently selecting high-quality vendors and screening out those that offer low bids at the expense of product quality.

Another software program designed to help purchasers select the best available distribution transformer is a Lotus-compatible worksheet for evaluating distribution transformers offered by ABB Power T&D Co. The worksheet adjusts first-cost figures by a value factor that comprises criteria for reliability, quality, delivery/availability, and support. The lower the value factor, the lower is the effective first cost of the transformer. To the adjusted first cost is added the cost of losses, yielding a life-cycle cost for the transformer.

Suggested weightings, based on surveys of utilities, are provided for each criterion, but users can easily modify these criteria in light of their own experience and needs. According to ABB's Dorman Whitley, this ensures that the worksheet does not favor any one manufacturer. Users can also incorporate soft criteria (such as supplier's long-term commitment to the industry, or level of investment in R&D).

Losses influence reliability

As in the case of power transformers, the higher the quality of distribution transformers, the lower the losses. But the relationship between losses and reliability is even more pronounced: Low-loss transformers, because of the reduced internal heating, tend to be more reliable.

Utah Power & Light Co (UP&L) began monitoring transformer failures in 1973 and in 1985 established a rigorous auditing procedure to keep track of the failure rates of different manufacturers' products. Unsatisfactorily high failure rates were uncovered—including a rate of 33% for one manufacturer's 3-phase, pad-mounted unit. In this case, the high failure rate was eliminated by switching to amorphous-core, low-loss transformers.

Distribution Engineering Manager Dennis Horman reports that the highest failure rates occurred in seasonally loaded transformers serving irrigation pumps. These were shut off during the winter, when they were not in use, to conserve no-load losses. However, the action of shutting off the transformers apparently caused condensation which resulted in failures when the units were put back in service.

When these transformers were replaced with amorphous-core units, the utility found that they had such low losses that it was economical to keep them energized year-round. This avoided the problem of failures caused by condensation and eliminated the manpower cost of energizing and de-energizing them. Horman says that visits to three manufacturers of amorphous-core transformers and teardowns of delivered transformers convinced engineers at UP&L that all three were producing high-quality transformers. The utility now routinely purchases amorphous-core transformers from the three manufacturers and now accepts bids from other manufacturers without going through test procedures or plant visits.

SECTION 4.1

THE IDEAL TRANSFORMER

Figure 4.1 shows a basic single-phase two-winding transformer, where the two windings are wrapped around a magnetic core [1, 2, 3]. It is assumed here that the transformer is operating under sinusoidal-steady-state excitation. Shown in the figure are the phasor voltages E_1 and E_2 across the windings, and the phasor currents I_1 entering winding 1, which has N_1 turns, and I_2 leaving winding 2, which has N_2 turns. A phasor flux Φ_c set up in the core and a magnetic field intensity phasor H_c are also shown. The core has a cross-sectional area denoted A_c, a mean length of the magnetic circuit l_c, and a magnetic permeability μ_c, assumed constant.

Figure 4.1

Basic single-phase two-winding transformer

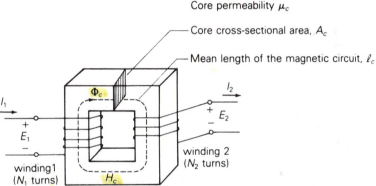

Core permeability μ_c

Core cross-sectional area, A_c

Mean length of the magnetic circuit, ℓ_c

For an ideal transformer, the following are assumed:

1. The windings have zero resistance; therefore, the I^2R losses in the windings are zero.
2. The core permeability μ_c is infinite, which corresponds to zero core reluctance.
3. There is no leakage flux; that is, the entire flux Φ_c is confined to the core and links both windings.
4. There are no core losses.

A schematic representation of a two-winding transformer is shown in Figure 4.2. Ampere's and Faraday's laws can be used along with the preceding assumptions to derive the ideal transformer relationships. Ampere's law states that the tangential component of the magnetic field intensity vector integrated along a closed path equals the net current enclosed by that path; that is,

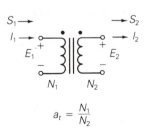

$$\oint H_{\tan} dl = I_{\text{enclosed}} \qquad (4.1.1)$$

Figure 4.2

Schematic representation of a single-phase two-winding transformer

If the core center line shown in Figure 4.1 is selected as the closed path, and if H_c is constant along the path as well as tangent to the path, then

(4.1.1) becomes

$$H_c l_c = N_1 I_1 - N_2 I_2 \tag{4.1.2}$$

Note that the current I_1 is enclosed N_1 times and I_2 is enclosed N_2 times, one time for each turn of the coils. Also, using the right-hand rule*, current I_1 contributes to clockwise flux but current I_2 contributes to counterclockwise flux. Thus, in (4.1.2) the net current enclosed is $N_1 I_1 - N_2 I_2$. For constant core permeability μ_c, the magnetic flux density B_c within the core, also constant, is

magnetic flux density

$$B_c = \mu_c H_c \quad \text{Wb/m}^2 \tag{4.1.3}$$

and the core flux Φ_c is

$$\Phi_c = B_c A_c \quad \text{Wb} \tag{4.1.4}$$

Using (4.1.3) and (4.1.4) in (4.1.2) yields

$$N_1 I_1 - N_2 I_2 = l_c B_c / \mu_c = \left(\frac{l_c}{\mu_c A_c}\right) \Phi_c \tag{4.1.5}$$

We define core reluctance R_c as

$$R_c = \frac{l_c}{\mu_c A_c} \tag{4.1.6}$$

Then (4.1.5) becomes

$$N_1 I_1 - N_2 I_2 = R_c \Phi_c \tag{4.1.7}$$

Equation (4.1.7) can be called "Ohm's law" for the magnetic circuit, wherein the net magnetomotive force mmf $= N_1 I_1 - N_2 I_2$ equals the product of the core reluctance R_c and the core flux Φ_c. Reluctance R_c, which impedes the establishment of flux in a magnetic circuit, is analogous to resistance in an electric circuit. For an ideal transformer, μ_c is assumed infinite, which, from (4.1.6), means that R_c is zero, and (4.1.7) becomes

$$N_1 I_1 = N_2 I_2 \tag{4.1.8}$$

In practice, power transformer windings and cores are contained within enclosures, and the winding directions are not visible. One way of conveying winding information is to place a dot at one end of each winding such that when current enters a winding at the dot, it produces an mmf acting in the *same* direction. This dot convention is shown in the schematic of Figure 4.2. The dots are conventionally called *polarity marks.*

Equation (4.1.8) is written for current I_1 *entering* its dotted terminal and current I_2 *leaving* its dotted terminal. As such, I_1 and I_2 are *in phase*, since $I_1 = (N_2/N_1)I_2$. If the direction chosen for I_2 were reversed, such that both currents entered their dotted terminals, then I_1 would be 180° *out of phase* with I_2.

---

*The right-hand rule for a coil is as follows: Wrap the fingers of your right hand around the coil in the direction of the current. Your right thumb then points in the direction of the flux.

Faraday's law states that the voltage $e(t)$ induced across an N-turn winding by a time-varying flux $\phi(t)$ linking the winding is

$$e(t) = N \frac{d\phi(t)}{dt} \qquad (4.1.9)$$

Assuming a sinusoidal-steady-state flux with constant frequency ω, and representing $e(t)$ and $\phi(t)$ by their phasors E and Φ, (4.1.9) becomes

$$E = N(j\omega)\Phi \qquad \textit{phasor notation} \qquad (4.1.10)$$

For an ideal transformer, the entire flux is assumed to be confined to the core, linking both windings. From Faraday's law, the induced voltages across the windings of Figure 4.1 are

$$E_1 = N_1(j\omega)\Phi_c \qquad (4.1.11)$$

$$E_2 = N_2(j\omega)\Phi_c \qquad (4.1.12)$$

Dividing (4.1.11) by (4.1.12) yields

$$\frac{E_1}{E_2} = \frac{N_1}{N_2} \qquad (4.1.13)$$

or

$$\frac{E_1}{N_1} = \frac{E_2}{N_2} \qquad (4.1.14)$$

The dots shown in Figure 4.2 indicate that the voltages E_1 and E_2, both of which have their $+$ polarities at the dotted terminals, are in phase. If the polarity chosen for one of the voltages in Figure 4.1 were reversed, then E_1 would be $180°$ out of phase with E_2.

The turns ratio a_t is defined as follows:

$$a_t = \frac{N_1}{N_2} \qquad (4.1.15)$$

Using a_t in (4.1.8) and (4.1.14), the basic relations for an ideal single-phase two-winding transformer are

$$E_1 = \left(\frac{N_1}{N_2}\right) E_2 = a_t E_2 \qquad (4.1.16)$$

$$I_1 = \left(\frac{N_2}{N_1}\right) I_2 = \frac{I_2}{a_t} \qquad (4.1.17)$$

Two additional relations concerning complex power and impedance can be derived from (4.1.16) and (4.1.17) as follows. The complex power entering winding 1 in Figure 4.2 is

$$S_1 = E_1 I_1^* \qquad (4.1.18)$$

Using (4.1.16) and (4.1.17),

$$S_1 = E_1 I_1^* = (a_t E_2)\left(\frac{I_2}{a_t}\right)^* = E_2 I_2^* = S_2 \tag{4.1.19}$$

note $S_1 = S_2$

As shown by (4.1.19), the complex power S_1 entering winding 1 equals the complex power S_2 leaving winding 2. That is, an ideal transformer has no real or reactive power loss.

If an impedance Z_2 is connected across winding 2 of the ideal transformer in Figure 4.2, then

$$Z_2 = \frac{E_2}{I_2} \quad \text{actual } Z_2 \tag{4.1.20}$$

This impedance, when measured from winding 1, is

$$Z_2' = \frac{E_1}{I_1} = \frac{a_t E_2}{I_2/a_t} = a_t^2 Z_2 = \left(\frac{N_1}{N_2}\right)^2 Z_2 \tag{4.1.21}$$

Z_2 seen from Z_1 is Z_2'

seen from = referred to

seen from

Thus, the impedance Z_2 connected to winding 2 is referred to winding 1 by multiplying Z_2 by a_t^2, the square of the turns ratio.

EXAMPLE 4.1 **Ideal, single-phase two-winding transformer**

A single-phase two-winding transformer is rated 20 kVA, 480/120 V, 60 Hz. A source connected to the 480-V winding supplies an impedance load connected to the 120-V winding. The load absorbs 15 kVA at 0.8 p.f. lagging when the load voltage is 118 V. Assume that the transformer is ideal and calculate the following:

S always has pf (lead/lag)

find

inductive load I is lagging V

a. The voltage across the 480-V winding
b. The load impedance
c. The load impedance referred to the 480-V winding
d. The real and reactive power supplied to the 480-V winding

Solution a. The circuit is shown in Figure 4.3, where winding 1 denotes the 480-V winding and winding 2 denotes the 120-V winding. Selecting the load voltage E_2 as the reference,

$$E_2 = 118\underline{/0°}\ \text{V}$$

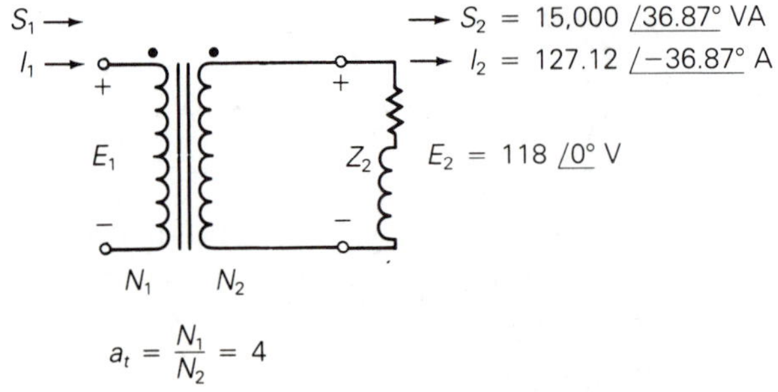

Figure 4.3

Circuit for Example 4.1

$S_1 \rightarrow$
$I_1 \rightarrow$
$\rightarrow S_2 = 15{,}000\ \underline{/36.87°}\ \text{VA}$
$\rightarrow I_2 = 127.12\ \underline{/-36.87°}\ \text{A}$
E_1
Z_2
$E_2 = 118\ \underline{/0°}\ \text{V}$
$N_1 \quad N_2$

$$a_t = \frac{N_1}{N_2} = 4$$

(handwritten note: voltage across primary dropped when load was applied)

The turns ratio is, from (4.1.13),

$$a_t = \frac{N_1}{N_2} = \frac{E_{1\text{rated}}}{E_{2\text{rated}}} = \frac{480}{120} = 4$$

and the voltage across winding 1 is

$$E_1 = a_t E_2 = 4(118\underline{/0^\circ}) = 472\underline{/0^\circ} \quad \text{V}$$

b. The complex power S_2 absorbed by the load is

$$S_2 = E_2 I_2^* = 118 I_2^* = 15,000/\cos^{-1}(0.8) = 15,000\underline{/36.87^\circ} \quad \text{VA}$$

Solving, the load current I_2 is

$$I_2 = 127.12\underline{/-36.87^\circ} \quad \text{A}$$

The load impedance Z_2 is

$$Z_2 = \frac{E_2}{I_2} = \frac{118\underline{/0^\circ}}{127.12\underline{/-36.87^\circ}} = 0.9283\underline{/36.87^\circ} \quad \Omega$$

c. From (4.1.21), the load impedance referred to the 480-V winding is

(handwritten: Z_2 seen from primary)

$$Z_2' = a_t^2 Z_2 = (4)^2(0.9283\underline{/36.87^\circ}) = 14.85\underline{/36.87^\circ} \quad \Omega$$

d. From (4.1.19)

$$S_1 = S_2 = 15,000\underline{/36.87^\circ} = 12,000 + j9000$$

Thus, the real and reactive powers supplied to the 480-V winding are

$$P_1 = \text{Re } S_1 = 12,000 \text{ W} = 12 \text{ kW}$$

$$Q_1 = \text{Im } S_1 = 9000 \text{ var} = 9 \text{ kvar} \qquad \blacksquare$$

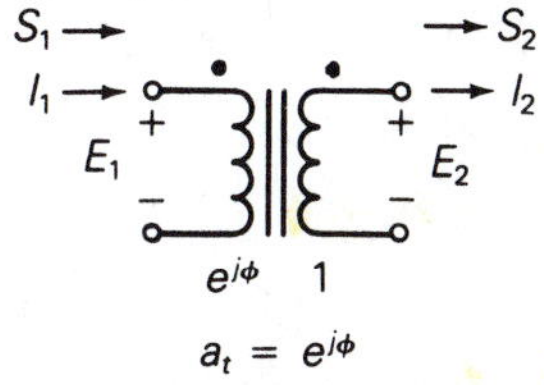

(handwritten: Power to $S_1 \rightarrow$ Power from $\rightarrow S_2$)

$e^{j\phi} \quad 1$

$a_t = e^{j\phi}$

$E_1 = a_t E_2 = e^{j\phi} E_2$

$I_1 = \dfrac{I_2}{a_t^*} = e^{j\phi} I_2$

$S_1 = S_2$

$Z_2' = Z_2$

Figure 4.4

Schematic representation of a conceptual single-phase, phase-shifting transformer

Figure 4.4 shows a schematic of a conceptual single-phase, phase-shifting transformer. This transformer is not an idealization of an actual transformer since it is physically impossible to obtain a complex turns ratio. It will be used later in this chapter as a mathematical model for representing phase shift of three-phase transformers. As shown in Figure 4.4, the complex turns ratio a_t is defined for the phase-shifting transformer as

$$a_t = \frac{e^{j\phi}}{1} = e^{j\phi} \tag{4.1.22}$$

where ϕ is the phase-shift angle. The transformer relations are then

$$E_1 = a_t E_2 = e^{j\phi} E_2 \qquad \text{(4.1.23)}$$

(handwritten: includes phase shift from primary to secondary)

$$I_1 = \frac{I_2}{a_t^*} = e^{j\phi} I_2 \qquad \text{(4.1.24)}$$

Note that the phase angle of E_1 leads the phase angle of E_2 by ϕ. Similarly, I_1 leads I_2 by the angle ϕ. But the magnitudes are unchanged. That is, $|E_1| = |E_2|$ and $|I_1| = |I_2|$.

From these two relations, the following two additional relations are derived:

$S =$ Complex Power

$$S_1 = E_1 I_1^* = \left(a_t E_2\right)\left(\frac{I_2}{a_t^*}\right)^* = E_2 I_2^* = S_2 \qquad (4.1.25)$$

$$Z_2' = \frac{E_1}{I_1} = \frac{a_t E_2}{\frac{1}{a_t^*}I_2} = |a_t|^2 Z_2 = Z_2 \qquad (4.1.26)$$

$S_1 = S_2$) if no pwr losses

Thus, per-unit impedance is unchanged when it is referred from one side of an ideal phase-shifting transformer to the other. Also, the ideal phase-shifting transformer has no real or reactive power losses since $S_1 = S_2$. Note that (4.1.23) and (4.1.24) for the phase-shifting transformer are the same as (4.1.16) and (4.1.17) for the ideal physical transformer except for the complex conjugate (*) in (4.1.24). The complex conjugate for the phase-shifting transformer is required to make $S_1 = S_2$ (complex power into winding 1 equals complex power out of winding 2), as shown in (4.1.25).

SECTION 4.2

EQUIVALENT CIRCUITS FOR PRACTICAL TRANSFORMERS

Figure 4.5 shows an equivalent circuit for a practical single-phase two-winding transformer, which differs from the ideal transformer as follows:

1. The windings have resistance.
2. The core permeability μ_c is finite.
3. The magnetic flux is not entirely confined to the core.
4. There are real and reactive power losses in the core.

The resistance R_1 is included in series with winding 1 of the figure to account for I^2R losses in this winding. A reactance X_1, called the leakage reactance of winding 1, is also included in series with winding 1 to account for the leakage flux of winding 1. This leakage flux is the component of the flux that links winding 1 but does not link winding 2; it causes a voltage drop $I_1(jX_1)$, which is proportional to I_1 and leads I_1 by 90°. There is also a

Figure 4.5

Equivalent circuit of a practical single-phase two-winding transformer

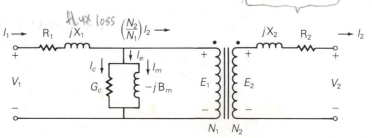

reactive power loss $I_1^2 X_1$ associated with this leakage reactance. Similarly, there is a resistance R_2 and a leakage reactance X_2 in series with winding 2.

Equation (4.1.7) shows that for finite core permeability μ_c, the total mmf is not zero. Dividing (4.1.7) by N_1 and using (4.1.11), we get

$$I_1 - \left(\frac{N_2}{N_1}\right) I_2 = \frac{R_c}{N_1} \Phi_c = \frac{R_c}{N_1} \left(\frac{E_1}{j\omega N_1}\right) = -j \left(\frac{R_c}{\omega N_1^2}\right) E_1 \qquad (4.2.1)$$

Defining the term on the right-hand side of (4.2.1) to be I_m, called *magnetizing current*, it is evident that I_m lags E_1 by 90°, and can be represented by a shunt inductor with susceptance $B_m = \left(\dfrac{R_c}{\omega N_1^2}\right)$ mhos.* However, in reality there is an additional shunt branch, represented by a resistor with conductance G_c mhos, which carries a current I_c, called the *core loss* current. I_c is in phase with E_1. When the core loss current I_c is included, (4.2.1) becomes

$$I_1 - \left(\frac{N_2}{N_1}\right) I_2 = I_c + I_m = (G_c - j B_m) E_1 \qquad (4.2.2)$$

The equivalent circuit of Figure 4.5, which includes the shunt branch with admittance $(G_c - j B_m)$ mhos, satisfies the KCL equation (4.2.2). Note that when winding 2 is open ($I_2 = 0$) and when a sinusoidal voltage V_1 is applied to winding 1, then (4.2.2) indicates that the current I_1 will have two components: the core loss current I_c and the magnetizing current I_m. Associated with I_c is a real power loss $I_c^2/G_c = E_1^2 G_c$ W. This real power loss accounts for both hysteresis and eddy current losses within the core. Hysteresis loss occurs because a cyclic variation of flux within the core requires energy dissipated as heat. As such, hysteresis loss can be reduced by the use of special high grades of alloy steel as core material. Eddy current loss occurs because induced currents called eddy currents flow within the magnetic core perpendicular to the flux. As such, eddy current loss can be reduced by constructing the core with laminated sheets of alloy steel. Associated with I_m is a reactive power loss $I_m^2/B_m = E_1^2 B_m$ var. This reactive power is required to magnetize the core. The phasor sum $(I_c + I_m)$ is called the *exciting* current I_e.

Figure 4.6 shows three alternative equivalent circuits for a practical single-phase two-winding transformer. In Figure 4.6(a) the resistance R_2 and leakage reactance X_2 of winding 2 are referred to winding 1 via (4.1.21). In Figure 4.6(b) the shunt branch is omitted, which corresponds to neglecting the exciting current. Since the exciting current is typically about 5% of rated current, neglecting it in power-system studies is often valid unless transformer efficiency or exciting current phenomena are of particular concern. For large power transformers rated more than 500 kVA, the winding resistances, which are small compared to the leakage reactances, can often be neglected, as shown in Figure 4.6(c).

Thus, a practical transformer operating in sinusoidal steady state is

---

*The units of admittance, conductance, and susceptance, which in the SI system are siemens (with symbol S), are also called mhos (with symbol ℧) or ohms^{-1} (with symbol Ω^{-1}).

Figure 4.6

Equivalent circuits for a practical single-phase two-winding transformer

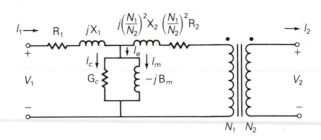

(a) R_2 and X_2 are referred to winding 1

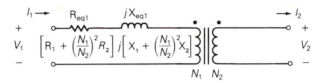

(b) Neglecting exciting current

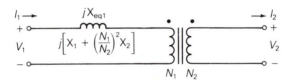

(c) Neglecting exciting current and I^2R winding loss

equivalent to an ideal transformer with external impedance and admittance branches, as shown in Figure 4.6. The external branches can be evaluated from short-circuit and open-circuit tests, as illustrated by the following example.

EXAMPLE 4.2   **Transformer short-circuit and open-circuit tests**

A single-phase two-winding transformer is rated 20 kVA, 480/120 volts, 60 Hz. During a short-circuit test, where rated current at rated frequency is applied to the 480-volt winding (denoted winding 1), with the 120-volt winding (winding 2) shorted, the following readings are obtained: $V_1 = 35$ volts, $P_1 = 300$ W. During an open-circuit test, where rated voltage is applied to winding 2, with winding 1 open, the following readings are obtained: $I_2 = 12$ A, $P_2 = 200$ W. o.c. test

> **a.** From the short-circuit test, determine the equivalent series impedance $Z_{eq1} = R_{eq1} + jX_{eq1}$ referred to winding 1. Neglect the shunt admittance.
>
> **b.** From the open-circuit test, determine the shunt admittance $Y_m = G_c - jB_m$ referred to winding 1. Neglect the series impedance.

Solution   **a.** The equivalent circuit for the short-circuit test is shown in Figure 4.7(a),

[handwritten margin notes:] S.C. test has secondary shorted and Primary rated current

Open ckt test has secondary open and rated voltage applied to secondary

Figure 4.7

Circuits for Example 4.2

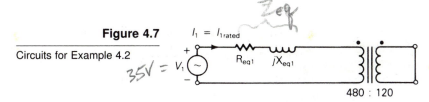

Zeq

$35V = V_1$

(a) Short-circuit test (neglecting shunt admittance)

open ckt test

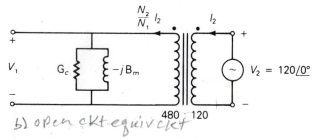

$V_2 = 120\underline{/0°}$

b) open ckt equivalent

where the shunt admittance branch is neglected. Rated current for winding 1 is

short ckt tested
used to get primary
equivalent parameters

$$I_{max} = I_{1rated} = \frac{S_{rated}}{V_{1rated}} = \frac{20 \times 10^3}{480} = 41.667 \text{ A}$$

note: rated values are max values

R_{eq1}, Z_{eq1}, and X_{eq1} are then determined as follows:

Real Power

pwr in winding 1

$$R_{eq1} = \frac{P_1}{I_{1rated}^2} = \frac{300}{(41.667)^2} = 0.1728 \text{ } \Omega$$

equivalents values

$$Z_{eq} = \frac{V_{primary}}{I_{rated}}$$
sec shorted

$$|Z_{eq1}| = \frac{V_1}{I_{1rated}} = \frac{35}{41.667} = 0.8400 \text{ } \Omega$$

s.c. tests $X_{eq1} = \sqrt{Z_{eq1}^2 - R_{eq1}^2} = 0.8220 \text{ } \Omega$

$$Z_{eq1} = R_{eq1} + jX_{eq1} = 0.1728 + j0.8220 = 0.8400\underline{/78.13°} \text{ } \Omega$$

b. The equivalent circuit for the open-circuit test is shown in Figure 4.7(b), where the series impedance is neglected. From (4.1.16),

$$V_1 = E_1 = a_t E_2 = \frac{N_1}{N_2} V_{2rated} = \frac{480}{120}(120) = 480 \text{ volts}$$

determined thru open ckt tests

G_c, Y_m, and B_m are then determined as follows:

$$G_c = \frac{P_2}{V_1^2} = \frac{200}{(480)^2} = 0.000868 \text{ S}$$

$P_2 = P_1$

V primary open
so, no I drops except
Ie = IB + Ic losses

o.c. tests $|Y_m| = \frac{\left(\frac{N_2}{N_1}\right)I_2}{V_1} = \frac{\left(\frac{120}{480}\right)(12)}{480} = 0.00625 \text{ S}$

$$B_m = \sqrt{Y_m^2 - G_c^2} = \sqrt{(0.00625)^2 - (0.000868)^2} = 0.00619 \text{ S}$$

$$Y_m = G_c - jB_m = 0.000868 - j0.00619 = 0.00625\underline{/-82.02°} \text{ S}$$

note {   Note that the equivalent series impedance is usually evaluated at rated current from a short-circuit test, and the shunt admittance is evaluated at rated voltage from an open-circuit test. For small variations in transformer operation near rated conditions, the impedance and admittance values are often assumed constant.   ■

The following are not represented by the equivalent circuit of Figure 4.5:

1. Saturation
2. Inrush current
3. Nonsinusoidal exciting current
4. Surge phenomena

They are briefly discussed in the following sections.

Saturation

In deriving the equivalent circuit of the ideal and practical transformers, we have assumed constant core permeability μ_c and the linear relationship $B_c = \mu_c H_c$ of (4.1.3). However, the relationship between B and H for ferromagnetic materials used for transformer cores is nonlinear and multivalued. Figure 4.8 shows a set of B–H curves for a grain-oriented electrical steel typically used in transformers. As shown, each curve is multivalued, which is caused by hysteresis. For many engineering applications, the B–H curves can be adequately described by the dashed line drawn through the curves in Figure 4.8. Note that as H increases, the core becomes saturated; that is, the curves flatten out as B increases above 1 Wb/m². If the magnitude of the

Figure 4.8

B–H curves for M-5 grain-oriented electrical steel 0.012 in. thick (Armco Inc.)

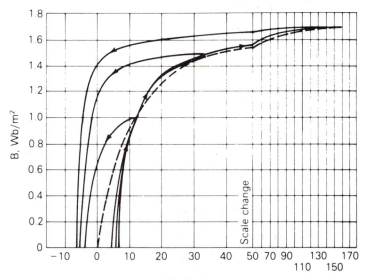

voltage applied to a transformer is too large, the core will saturate and a high magnetizing current will flow. In a well-designed transformer, the applied peak voltage causes the peak flux density in steady state to occur at the knee of the B–H curve, with a corresponding low value of magnetizing current.

field intensity
flux density

Inrush Current

When a transformer is first energized, a transient current much larger than rated transformer current can flow for several cycles. This current, called *inrush current,* is nonsinusoidal and has a large dc component. To understand the cause of inrush, assume that before energization, the transformer core is magnetized with a residual flux density $B(0) = 1.5 \, \text{Wb/m}^2$ (near the knee of the dotted curve in Figure 4.8). If the transformer is then energized when the source voltage is positive and increasing, Faraday's law, (4.1.9), will cause the flux density $B(t)$ to increase further, since

Faraday's Law

$$B(t) = \frac{\phi(t)}{A} = \frac{1}{NA} \int_0^t e(t) \, dt + B(0)$$

As $B(t)$ moves into the saturation region of the B–H curve, large values of $H(t)$ will occur, and, from Ampere's law, (4.1.1), corresponding large values of current $i(t)$ will flow for several cycles until it has dissipated. Since normal inrush currents can be as large as abnormal short-circuit currents in transformers, transformer protection schemes must be able to distinguish between these two types of currents.

note

Nonsinusoidal Exciting Current

When a sinusoidal voltage is applied to one winding of a transformer with the other winding open, the flux $\phi(t)$ and flux density $B(t)$ will, from Faraday's law, (4.1.9), be very nearly sinusoidal in steady state. But the magnetic field intensity $H(t)$ and the resulting exciting current will not be sinusoidal in steady state, due to the nonlinear B–H curve. If the exciting current is measured and analyzed by Fourier analysis techniques, one finds that it has a fundamental component and a set of odd harmonics. The principal harmonic is the third, whose rms value is typically about 40% of the total rms exciting current. However, the nonsinusoidal nature of exciting current is usually neglected unless harmonic effects are of direct concern, because the exciting current itself is only about 5% of rated current for power transformers.

note

exciting current

Surge Phenomena

When power transformers are subjected to transient overvoltages caused by lightning or switching surges, the capacitances of the transformer windings have important effects on transient response. Transformer winding capacitances and response to surges are discussed in Chapter 12.

SECTION 4.3

THE PER-UNIT SYSTEM

Power-system quantities such as voltage, current, power, and impedance are often expressed in per-unit or percent of specified base values. For example, if a base voltage of 20 kV is specified, then the voltage 18 kV is (18/20) = 0.9 per unit or 90%. Calculations can then be made with per-unit quantities rather than with the actual quantities.

One advantage of the per-unit system is that by properly specifying base quantities, the transformer equivalent circuit can be simplified. The ideal transformer winding can be eliminated, such that voltages, currents, and external impedances and admittances expressed in per-unit do not change when they are referred from one side of a transformer to the other. This can be a significant advantage even in a power system of moderate size, where hundreds of transformers may be encountered. The per-unit system allows us to avoid the possibility of making serious calculation errors when referring quantities from one side of a transformer to the other. Another advantage of the per-unit system is that the per-unit impedances of electrical equipment of similar type usually lie within a narrow numerical range when the equipment ratings are used as base values. Because of this, per-unit impedance data can be checked rapidly for gross errors by someone familiar with per-unit quantities. In addition, manufacturers usually specify the impedances of machines and transformers in per-unit or percent of nameplate rating.

Per-unit quantities are calculated as follows:

$$\text{per-unit quantity} = \frac{\text{actual quantity}}{\text{base value of quantity}} \tag{4.3.1}$$

where *actual quantity* is the value of the quantity in the actual units. The base value has the same units as the actual quantity, thus making the per-unit quantity dimensionless. Also, the base value is always a real number. Therefore, the angle of the per-unit quantity is the same as the angle of the actual quantity.

Two independent base values can be arbitrarily selected at one point in a power system. Usually the base voltage V_{baseLN} and base complex power $S_{base1\phi}$ are selected for either a single-phase circuit or for one phase of a three-phase circuit. Then, in order for electrical laws to be valid in the per-unit system, the following relations must be used for other base values:

$$P_{base1\phi} = Q_{base1\phi} = S_{base1\phi} \tag{4.3.2}$$

$$I_{base} = \frac{S_{base1\phi}}{V_{baseLN}} \tag{4.3.3}$$

$$Z_{base} = R_{base} = X_{base} = \frac{V_{baseLN}}{I_{base}} = \frac{V_{baseLN}^2}{S_{base1\phi}} \tag{4.3.4}$$

PU sys

admittance conductance susceptance

$$Y_{base} = G_{base} = B_{base} = \frac{1}{Z_{base}} \quad \text{no units} \quad \text{single phase} \tag{4.3.5}$$

line to real neutral

In (4.3.2)–(4.3.5) the subscripts LN and 1ϕ denote "line-to-neutral" and "per-phase," respectively, for three-phase circuits. These equations are also valid for single-phase circuits, where subscripts can be omitted.

By convention, we adopt the following two rules for base quantities:

1. The value of $S_{base1\phi}$ is the same for the entire power system of concern.

2. The ratio of the voltage bases on either side of a transformer is selected to be the same as the ratio of the transformer voltage ratings.

adv. of PU

With these two rules, a per-unit impedance remains unchanged when referred from one side of a transformer to the other.

EXAMPLE 4.3 **Per-unit impedance: single-phase transformer**

A single-phase two-winding transformer is rated 20 kVA, 480/120 volts, 60 Hz. The equivalent leakage impedance of the transformer referred to the 120-volt winding, denoted winding 2, is $Z_{eq2} = 0.0525\underline{/78.13°}\ \Omega$. Using the transformer ratings as base values, determine the per-unit leakage impedance referred to winding 2 and referred to winding 1.

Solution The values of S_{base}, $V_{base\,1}$, and $V_{base\,2}$ are, from the transformer ratings,

$$S_{base} = 20\,\text{kVA}, \qquad V_{base1} = 480\ \text{volts}, \qquad V_{base2} = 120\ \text{volts}$$

Using (4.3.4), the base impedance on the 120-volt side of the transformer is

$$Z_{base2} = \frac{V_{base2}^2}{S_{base}} = \frac{(120)^2}{20,000} = 0.72\ \Omega$$

Then, using (4.3.1), the per-unit leakage impedance referred to winding 2 is

$$Z_{eq2p.u.} = \frac{Z_{eq2}}{Z_{base2}} = \frac{0.0525\underline{/78.13°}}{0.72} = 0.0729\underline{/78.13°}\quad \text{per unit}$$

If Z_{eq2} is referred to winding 1,

$$Z_{eq1} = a_t^2 Z_{eq2} = \left(\frac{N_1}{N_2}\right)^2 Z_{eq2} = \left(\frac{480}{120}\right)^2 (0.0525\underline{/78.13°})$$

a_t^2 *Z_{eq2}*

$$= 0.84\underline{/78.13°}\ \Omega$$

$Z_{eq}1$

The base impedance on the 480-volt side of the transformer is

base Z, from ratings $$Z_{base1} = \frac{V_{base1}^2}{S_{base}} = \frac{(480)^2}{20,000} = 11.52\ \Omega$$

and the per-unit leakage reactance referred to winding 1 is

pu $Z_{eq}1$ $$Z_{eq1p.u.} = \frac{Z_{eq1}}{Z_{base1}} = \frac{0.84\underline{/78.13°}}{11.52} = 0.0729\underline{/78.13°}\ \text{per unit} = Z_{eq2p.u.}$$

Thus, the *per-unit* leakage impedance remains unchanged when referred from winding 2 to winding 1. This has been achieved by specifying

$$\frac{V_{base1}}{V_{base2}} = \frac{V_{rated1}}{V_{rated2}} = \left(\frac{480}{120}\right)$$

∎

Figure 4.9 shows three per-unit circuits of a single-phase two-winding transformer. The ideal transformer, shown in Figure 4.9(a), satisfies the per-unit relations $E_{1p.u.} = E_{2p.u.}$, and $I_{1p.u.} = I_{2p.u.}$, which can be derived as follows. First divide (4.1.16) by V_{base1}:

$$E_{1p.u.} = \frac{E_1}{V_{base1}} = \frac{N_1}{N_2} \times \frac{E_2}{V_{base1}} \tag{4.3.6}$$

Then, using $V_{base1}/V_{base2} = V_{rated1}/V_{rated2} = N_1/N_2$,

$$E_{1p.u.} = \frac{N_1}{N_2} \frac{E_2}{\left(\dfrac{N_1}{N_2}\right) V_{base2}} = \frac{E_2}{V_{base2}} = E_{2p.u.} \tag{4.3.7}$$

Similarly, divide (4.1.17) by I_{base1}:

$$I_{1p.u.} = \frac{I_1}{I_{base1}} = \frac{N_2}{N_1} \frac{I_2}{I_{base1}} \tag{4.3.8}$$

Figure 4.9

Per-unit equivalent circuits of a single-phase two-winding transformer

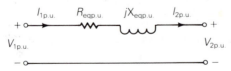

(a) Ideal transformer

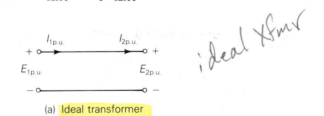

(b) Neglecting exciting current

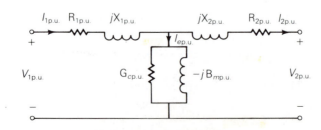

(c) Complete representation

Then, using $I_{base1} = S_{base}/V_{base1} = S_{base}/[(N_1/N_2)V_{base2}] = (N_2/N_1)I_{base2}$,

$$I_{1\text{p.u.}} = \frac{N_2}{N_1} \frac{I_2}{\left(\dfrac{N_2}{N_1}\right) I_{base2}} = \frac{I_2}{I_{base2}} = I_{2\text{p.u.}} \tag{4.3.9}$$

Thus, the ideal transformer winding in Figure 4.2 is eliminated from the per-unit circuit in Figure 4.9(a). The per-unit leakage impedance is included in Figure 4.9(b), and the per-unit shunt admittance branch is added in Figure 4.9(c) to obtain the complete representation.

When only one component, such as a transformer, is considered, the nameplate ratings of that component are usually selected as base values. When several components are involved, however, the system base values may be different from the nameplate ratings of any particular device. It is then necessary to convert the per-unit impedance of a device from its nameplate ratings to the system base values. To convert a per-unit impedance from "old" to "new" base values, use

$$Z_{\text{p.u.new}} = \frac{Z_{\text{actual}}}{Z_{\text{basenew}}} = \frac{Z_{\text{p.u.old}} Z_{\text{baseold}}}{Z_{\text{basenew}}} \tag{4.3.10}$$

or, from (4.3.4),

$$Z_{\text{p.u.new}} = Z_{\text{p.u.old}} \left(\frac{V_{\text{baseold}}}{V_{\text{basenew}}}\right)^2 \left(\frac{S_{\text{basenew}}}{S_{\text{baseold}}}\right) \tag{4.3.11}$$

EXAMPLE 4.4   **Per-unit circuit: three-zone single-phase network**

Three zones of a single-phase circuit are identified in Figure 4.10(a). The zones are connected by transformers T_1 and T_2, whose ratings are also shown. Using base values of 30 kVA and 240 volts in zone 1, draw the per-unit circuit and determine the per-unit impedances and the per-unit source voltage. Then calculate the load current both in per-unit and in amperes. Transformer winding resistances and shunt admittance branches are neglected.

Solution   First the base values in each zone are determined. S_{base} = 30 kVA is the same for the entire network. Also, V_{base1} = 240 volts, as specified for zone 1. When moving across a transformer, the voltage base is changed in proportion to the transformer voltage ratings. Thus,

$$V_{base2} = \left(\frac{480}{240}\right)(240) = 480 \quad \text{volts}$$

and

$$V_{base3} = \left(\frac{115}{460}\right)(480) = 120 \quad \text{volts}$$

The base impedances in zones 2 and 3 are

$$Z_{base2} = \frac{V_{base2}^2}{S_{base}} = \frac{480^2}{30,000} = 7.68 \quad \Omega$$

Figure 4.10

Circuits for Example 4.4

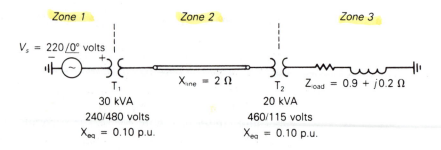

$V_s = 220\underline{/0°}$ volts

Zone 1   Zone 2   Zone 3

T_1
30 kVA
240/480 volts
$X_{eq} = 0.10$ p.u.

$X_{line} = 2\ \Omega$

T_2
20 kVA
460/115 volts
$X_{eq} = 0.10$ p.u.

$Z_{load} = 0.9 + j0.2\ \Omega$

(a) Single-phase circuit

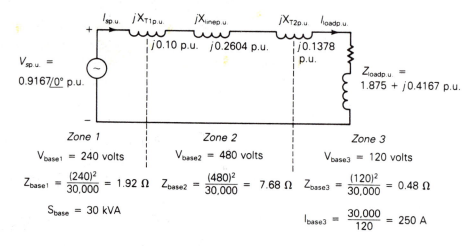

$I_{sp.u.}$ $jX_{T1p.u.}$ $jX_{linep.u.}$ $jX_{T2p.u.}$ $I_{loadp.u.}$

$j0.10$ p.u. $j0.2604$ p.u. $j0.1378$ p.u.

$V_{sp.u.} = 0.9167\underline{/0°}$ p.u.

$Z_{loadp.u.} = 1.875 + j0.4167$ p.u.

Zone 1   Zone 2   Zone 3

$V_{base1} = 240$ volts $V_{base2} = 480$ volts $V_{base3} = 120$ volts

$Z_{base1} = \dfrac{(240)^2}{30{,}000} = 1.92\ \Omega$ $Z_{base2} = \dfrac{(480)^2}{30{,}000} = 7.68\ \Omega$ $Z_{base3} = \dfrac{(120)^2}{30{,}000} = 0.48\ \Omega$

$S_{base} = 30$ kVA

$I_{base3} = \dfrac{30{,}000}{120} = 250$ A

(b) Per-unit circuit

and

$$Z_{base\,3} = \frac{V_{base\,3}^2}{S_{base}} = \frac{120^2}{30{,}000} = 0.48 \quad \Omega$$

and the base current in zone 3 is

$$I_{base\,3} = \frac{S_{base}}{V_{base\,3}} = \frac{30{,}000}{120} = 250 \quad A$$

Next, the per-unit circuit impedances are calculated using the system base values. Since $S_{base} = 30\,kVA$ is the same as the kVA rating of transformer T_1, and $V_{base\,1} = 240$ volts is the same as the voltage rating of the zone 1 side of transformer T_1, the per-unit leakage reactance of T_1 is the same as its nameplate value, $X_{T1p.u.} = 0.1$ per unit. But the per-unit leakage reactance of transformer T_2 must be converted from its nameplate rating to the system base. Using (4.3.11) and $V_{base\,2} = 480$ volts,

$$X_{T2p.u.} = (0.10)\left(\frac{460}{480}\right)^2\left(\frac{30{,}000}{20{,}000}\right) = 0.1378 \quad \text{per unit}$$

sys base

Alternatively, using $V_{base3} = 120$ volts,

Sys buse $\quad X_{T2p.u.} = \left(0.10\right)\left(\dfrac{115}{120}\right)^2\left(\dfrac{30,000}{20,000}\right) = 0.1378$ per unit

which gives the same result. The line, which is located in zone 2, has a **per-unit reactance**

$$X_{linep.u.} = \frac{X_{line}}{Z_{base\,2}} = \frac{2}{7.68} = 0.2604 \quad \text{per unit}$$

and the load, which is located in zone 3, has a per-unit impedance

$$Z_{loadp.u.} = \frac{Z_{load}}{Z_{base3}} = \frac{0.9 + j0.2}{0.48} = 1.875 + j0.4167 \quad \text{per unit}$$

The per-unit circuit is shown in Figure 4.10(b), where the base values for each zone, per-unit impedances, and the per-unit source voltage are shown. The per-unit load current is then easily calculated from Figure 4.10(b) as follows:

$$I_{loadp.u.} = I_{sp.u.} = \frac{V_{sp.u.}}{j(X_{T1p.u.} + X_{linep.u.} + X_{T2p.u.}) + Z_{loadp.u.}}$$

Source p.u.
or
Sys. p.u.
$$= \frac{0.9167\underline{/0°}}{j(0.10 + 0.2604 + 0.1378) + (1.875 + j0.4167)}$$

$$= \frac{0.9167\underline{/0°}}{1.875 + j0.9149} = \frac{0.9167\underline{/0°}}{2.086\underline{/26.01°}}$$

$$= 0.4395\underline{/-26.01°} \quad \text{per unit}$$

The actual load current is

$$I_{load} = (I_{loadp.u.})I_{base3} = (0.4395\underline{/-26.01°})(250) = 109.9\underline{/-26.01°} \quad \text{A}$$

Note that the per-unit equivalent circuit of Figure 4.10(b) is relatively easy to analyze, since ideal transformer windings have been eliminated by proper selection of base values.   ■

Balanced three-phase circuits can be solved in per-unit on a per-phase basis after converting Δ-load impedances to equivalent Y impedances. Base values can be selected either on a per-phase basis or on a three-phase basis. Equations (4.3.1)–(4.3.5) remain valid for three-phase circuits on a per-phase basis. Usually $S_{base3\phi}$ and V_{baseLL} are selected, where the subscripts 3ϕ and LL denote "three-phase" and "line-to-line," respectively. Then the following relations must be used for other base values:

$$S_{base1\phi} = \frac{S_{base3\phi}}{3} \tag{4.3.12}$$

$$V_{baseLN} = \frac{V_{baseLL}}{\sqrt{3}} \tag{4.3.13}$$

$$S_{\text{base}3\phi} = P_{\text{base}3\phi} = Q_{\text{base}3\phi} \qquad (4.3.14)$$

$$I_{\text{base}} = \frac{S_{\text{base}1\phi}}{V_{\text{baseLN}}} = \frac{S_{\text{base}3\phi}}{\sqrt{3}V_{\text{baseLL}}} \qquad (4.3.15)$$

$$Z_{\text{base}} = \frac{V_{\text{baseLN}}}{I_{\text{base}}} = \frac{V_{\text{baseLN}}^2}{S_{\text{base}1\phi}} = \frac{V_{\text{baseLL}}^2}{S_{\text{base}3\phi}} \qquad (4.3.16)$$

$$R_{\text{base}} = X_{\text{base}} = Z_{\text{base}} = \frac{1}{Y_{\text{base}}} \qquad (4.3.17)$$

EXAMPLE 4.5   **Per-unit and actual currents in balanced three-phase networks**

As in Example 2.5, a balanced-Y-connected voltage source with $E_{ab} = 480\underline{/0°}$ volts is applied to a balanced-Δ load with $Z_\Delta = 30\underline{/40°}\,\Omega$. The line impedance between the source and load is $Z_L = 1\underline{/85°}\,\Omega$ for each phase. Calculate the per-unit and actual current in phase a of the line using $S_{\text{base}3\phi} = 10\,\text{kVA}$ and $V_{\text{baseLL}} = 480$ volts.

Solution   First convert Z_Δ to an equivalent Z_Y; the equivalent line-to-neutral diagram is shown in Figure 2.17. The base impedance is, from (4.3.16),

P938

$$Z_{\text{base}} = \frac{V_{\text{baseLL}}^2}{S_{\text{base}3\phi}} = \frac{(480)^2}{10{,}000} = 23.04 \quad \Omega$$

page 38

The per-unit line and load impedances are

per unit line Z $\quad Z_{\text{Lp.u.}} = \dfrac{Z_L}{Z_{\text{base}}} = \dfrac{1\underline{/85°}}{23.04} = 0.04340\underline{/85°}$ per unit

and

30/3 (see page 38)

per unit load $\quad Z_{\text{Yp.u.}} = \dfrac{Z_Y}{Z_{\text{base}}} = \dfrac{10\underline{/40°}}{23.04} = 0.4340\underline{/40°}$ per unit

Also,

$$V_{\text{baseLN}} = \frac{V_{\text{baseLL}}}{\sqrt{3}} = \frac{480}{\sqrt{3}} = 277\underline{/3}\text{ volts}$$

and

$$E_{an\text{p.u.}} = \frac{E_{an}}{V_{\text{baseLN}}} = \frac{277\underline{/-30°}}{277} = 1.0\underline{/-30°} \quad \text{per unit}$$

The per-unit equivalent circuit is shown in Figure 4.11. The per-unit line current in phase a is then

per unit line I
per phase a

$$I_{a\text{p.u.}} = \frac{E_{an\text{p.u.}}}{Z_{\text{Lp.u.}} + Z_{\text{Yp.u.}}} = \frac{1.0\underline{/-30°}}{0.04340\underline{/85°} + 0.4340\underline{/40°}}$$

$$= \frac{1.0\underline{/-30°}}{(0.00378 + j0.04323) + (0.3325 + j0.2790)}$$

$$= \frac{1.0/-30°}{0.3362 + j0.3222} = \frac{1.0/-30°}{0.4657/43.78°}$$

$$I_{a\,p.u.} = 2.147/-73.78° \quad \text{per unit}$$

The base current is

$$I_{base} = \frac{S_{base\,3\phi}}{\sqrt{3}V_{base\,LL}} = \frac{10,000}{\sqrt{3}(480)} = 12.03 \quad A$$

and the actual phase a line current is

$$I_a = (2.147/-73.78°)(12.03) = 25.83/-73.78° \quad A$$

Figure 4.11

Circuit for Example 4.5

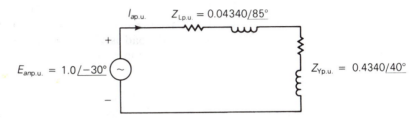

$E_{an\,p.u.} = 1.0/-30°$ $I_{a\,p.u.}$ $Z_{L\,p.u.} = 0.04340/85°$ $Z_{Y\,p.u.} = 0.4340/40°$

■

SECTION 4.4

THREE-PHASE TRANSFORMER CONNECTIONS AND PHASE SHIFT

Three identical single-phase two-winding transformers may be connected to form a three-phase bank. Four ways to connect the windings are Y–Y, Y–Δ, Δ–Y, and Δ–Δ. For example, Figure 4.12 shows a three-phase Y–Y bank. Figure 4.12(a) shows the core and coil arrangements. The American standard for marking three-phase transformers substitutes H1, H2, and H3 on the high-voltage terminals and X1, X2, and X3 on the low-voltage terminals in place of the polarity dots. Also, in this text, we will use uppercase letters *ABC* to identify phases on the high-voltage side of the transformer and lowercase letters *abc* to identify phases on the low-voltage side of the transformer. In Figure 4.12(a) the transformer high-voltage terminals H1, H2, and H3 are connected to phases *A*, *B*, and *C*, and the low-voltage terminals X1, X2, and X3 are connected to phases *a*, *b*, and *c*, respectively.

Figure 4.12(b) shows a schematic representation of the three-phase Y–Y transformer. Windings on the same core are drawn in parallel, and the phasor relationship for balanced positive-sequence operation is shown. For example, high-voltage winding H1–N is on the same magnetic core as low-voltage winding X1–n in Figure 4.12(b). Also, V_{AN} is in phase with V_{an}. Figure 4.12(c) shows a single-line diagram of a Y–Y transformer. A single-line diagram

Figure 4.12

Three-phase two-winding Y–Y transformer bank

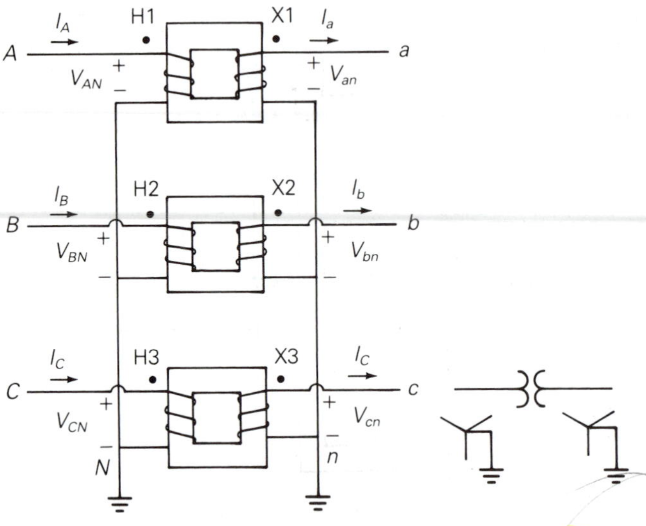

(a) Core and coil arrangements   (c) Single-line diagram

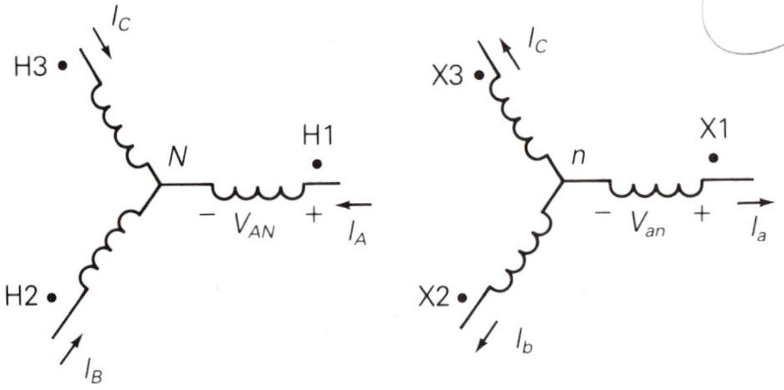

(b) Schematic representation showing phasor relationship for positive sequence operation

shows one phase of a three-phase network with the neutral wire omitted and with components represented by symbols rather than equivalent circuits.

The phases of a Y–Y or a Δ–Δ transformer can be labeled so there is no phase shift between corresponding quantities on the low- and high-voltage windings. But for Y–Δ and Δ–Y transformers, there is always a phase shift. Figure 4.13 shows a Y–Δ transformer. The labeling of the windings and the schematic representation are in accordance with the American standard, which is as follows:

> In either a Y–Δ or Δ–Y transformer, positive-sequence quantities on the high-voltage side shall lead their corresponding quantities on the low-voltage side by 30°.

As shown in Figure 4.13(b), V_{AN} leads V_{an} by 30°.

The positive-sequence phasor diagram shown in Figure 4.13(b) can be constructed via the following five steps, which are also indicated in Figure 4.13:

Figure 4.13

Three-phase two-winding Y–Δ
transformer bank

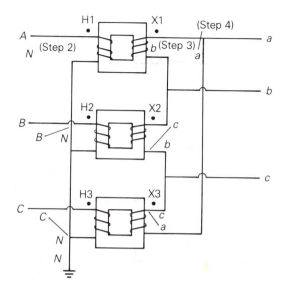

(a) Core and coil arrangement

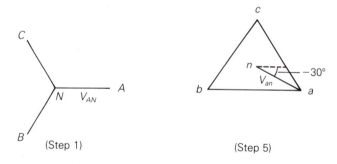

(b) Positive-sequence phasor diagram

Step 1 Assume that positive-sequence voltages are applied to the Y
winding. Draw the positive-sequence phasor diagram for these
voltages.

Step 2 Move phasor $A–N$ next to terminals $A–N$ in Figure 4.13(a).
Identify the ends of this line in the same manner as in the phasor
diagram. Similarly, move phasors $B–N$ and $C–N$ next to terminals $B–N$ and $C–N$ in Figure 4.13(a).

Step 3 For each single-phase transformer, the voltage across the low-
voltage winding must be in phase with the voltage across the
high-voltage winding, assuming an ideal transformer. Therefore, draw a line next to each low-voltage winding parallel to
the corresponding line already drawn next to the high-voltage
winding.

Step 4 Label the ends of the lines drawn in Step 3 by inspecting the polarity marks. For example, phase A is connected to dotted terminal H1, and A appears on the *right* side of line $A-N$. Therefore phase a, which is connected to dotted terminal X1, must be on the *right* side, and b on the left side of line $a-b$. Similarly, phase B is connected to dotted terminal H2, and B is *down* on line $B-N$. Therefore phase b, connected to dotted terminal X2, must be *down* on line $b-c$. Similarly, c is *up* on line $c-a$.

Step 5 Bring the three lines labeled in Step 4 together to complete the phasor diagram for the low-voltage Δ winding. Note that V_{AN} leads V_{an} by $30°$ in accordance with the American standard.

EXAMPLE 4.6 | **Phase shift in Δ–Y transformers**

Construct the negative-sequence diagram for the Y–Δ transformer shown in Figure 4.13. Determine the negative-sequence phase shift of this transformer.

Solution   The negative-sequence diagram, shown in Figure 4.14, is constructed from the following five steps, as outlined above:

Step 1 Draw the phasor diagram of negative-sequence voltages, which are assumed to be applied to the Y winding.

Step 2 Move the phasors $A-N$, $B-N$, and $C-N$ next to the high-voltage Y windings.

Step 3 For each single-phase transformer, draw a line next to the low-voltage winding that is parallel to the line drawn in Step 2 next to the high-voltage winding.

Step 4 Label the lines drawn in Step 3. For example, phase B, which is connected to dotted terminal H2, is shown *up* on line $B-N$; therefore phase b, which is connected to dotted terminal X2, must be *up* on line $b-c$.

Step 5 Bring the lines drawn in Step 4 together to form the negative-sequence phasor diagram for the low-voltage Δ winding.

As shown in Figure 4.14, the high-voltage phasors *lag* the low-voltage phasors by $30°$. Thus the negative-sequence phase shift is the reverse of the positive-sequence phase shift.   ∎

The Δ–Y transformer is commonly used as a generator step-up transformer, where the Δ winding is connected to the generator terminals and the Y winding is connected to a transmission line. One advantage of a high-voltage Y winding is that a neutral point N is provided for grounding on the high-voltage side. With a permanently grounded neutral, the insulation requirements for the high-voltage transformer windings are reduced. The high-voltage insulation can be graded or tapered from maximum insulation at terminals ABC to minimum insulation at grounded terminal N. One advantage of the Δ winding is that the undesirable third harmonic magnetiz-

Figure 4.14

Example 4.6–Construction of negative-sequence phasor diagram for Y–Δ transformer bank

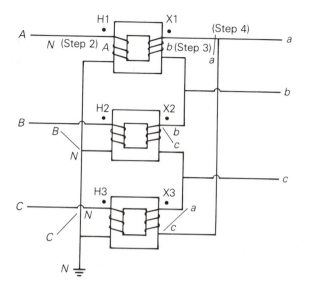

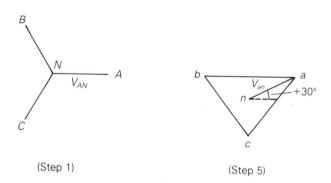

(Step 1)   (Step 5)

ing current, caused by the nonlinear core B–H characteristic, remains trapped inside the Δ winding. Third harmonic currents are (triple-frequency) zero-sequence currents, which cannot enter or leave a Δ connection, but can flow within the Δ. The Y–Y transformer is seldom used because of difficulties with third harmonic exciting current.

The Δ–Δ transformer has the advantage that one phase can be removed for repair or maintenance while the remaining phases continue to operate as a three-phase bank. This *open-*Δ connection permits balanced three-phase operation with the kVA rating reduced to 58% of the original bank (see Problem 4.18).

Instead of a bank of three single-phase transformers, all six windings may be placed on a common three-phase core to form a three-phase transformer, as shown in Figure 4.15. The three-phase core contains less iron than the three single-phase units; therefore it costs less, weighs less, requires less floor space, and has a slightly higher efficiency. However, a winding failure would require replacement of an entire three-phase transformer,

adv of 3φ over
3 single phases
tied together

disadv.

Figure 4.15

Transformer core
configurations

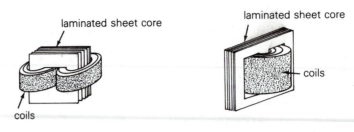

(a) Single-phase core type   (b) Single-phase shell type

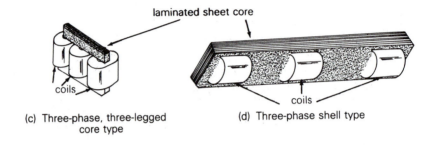

(c) Three-phase, three-legged   (d) Three-phase shell type
core type

compared to replacement of only one phase of a three-phase bank. In either case, the equivalent circuits developed here and subsequent analyses are the same.*

SECTION 4.5

PER-UNIT SEQUENCE MODELS OF THREE-PHASE TWO-WINDING TRANSFORMERS

Figure 4.16(a) is a schematic representation of an ideal Y–Y transformer grounded through neutral impedances Z_N and Z_n. Figure 4.16(b–d) show the per-unit sequence networks of this ideal transformer. Throughout the remainder of this text per-unit quantities will be used unless otherwise indicated. Also, the subscript "p.u.," used to indicate a per-unit quantity, will be omitted in most cases.

By convention, we adopt the following two rules for selecting base quantities:

1. A common S_{base} is selected for both the H and X terminals.
2. The ratio of the voltage bases V_{baseH}/V_{baseX} is selected to be equal to the ratio of the rated line-to-line voltages $V_{ratedHLL}/V_{ratedXLL}$.

---

*We note that the zero-sequence circuit of a three-phase shell-type transformer is not the same as the zero-sequence circuit of a three-phase core-type transformer [4].

Figure 4.16

Ideal Y–Y transformer

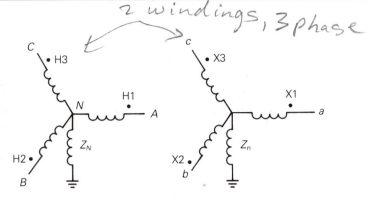

2 windings, 3 phase

(a) Schematic representation

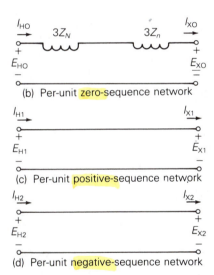

(b) Per-unit zero-sequence network

(c) Per-unit positive-sequence network

(d) Per-unit negative-sequence network

When balanced positive-sequence currents or balanced negative-sequence currents (as in Example 3.2) are applied to the transformer, the neutral currents are zero and there are no voltage drops across the neutral impedances. Therefore, the per-unit positive- and negative-sequence networks of the ideal Y–Y transformer, Figure 4.16(b) and (c), are the same as the per-unit single-phase ideal transformer, Figure 4.9(a).

Zero-sequence currents have equal magnitudes and equal phase angles. When per-unit sequence currents $I_{A0} = I_{B0} = I_{C0} = I_0$ are applied to the high-voltage windings of an ideal Y–Y transformer, the neutral current $I_N = 3I_0$ flows through the neutral impedance Z_N, with a voltage drop $(3Z_N)I_0$. Also, per-unit zero-sequence current I_0 flows in each low-voltage winding [from (4.3.9)], and therefore $3I_0$ flows through neutral impedance Z_n, with a voltage drop $(3I_0)Z_n$. The per-unit zero-sequence network, which includes the impedances $(3Z_N)$ and $(3Z_n)$, is shown in Figure 4.16(b).

Note that if either one of the neutrals of an ideal transformer is ungrounded, then no zero sequence can flow in either the high- or low-voltage windings. For example, if the high-voltage winding has an open neutral, then $I_N = 3I_0 = 0$, which in turn forces $I_0 = 0$ on the low-voltage side. This can be shown in the zero-sequence network of Figure 4.16(b) by making $Z_N = \infty$, which corresponds to an open circuit.

The per-unit sequence networks of a practical Y–Y transformer are shown in Figure 4.17(a). These networks are obtained by adding external

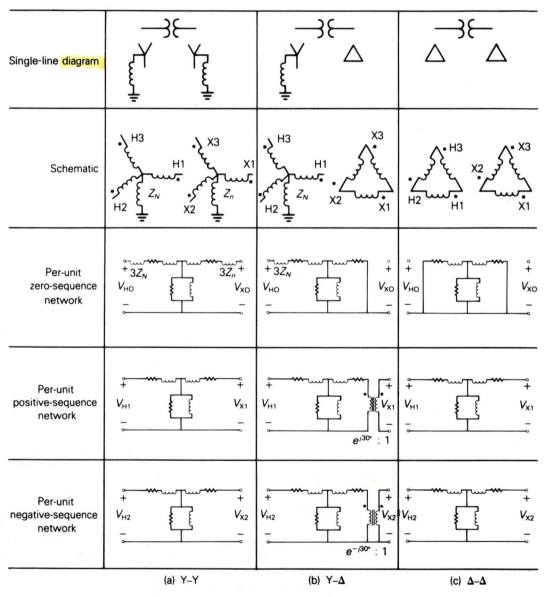

Figure 4.17 Per-unit sequence networks of practical Y–Y, Y–Δ, and Δ–Δ transformers

impedances to the sequence networks of the ideal transformer, as follows. The leakage impedances of the high-voltage windings are series impedances like the series impedances shown in Figure 3.9, with no coupling between phases ($Z_{ab} = 0$). If the phase a, b, and c windings have equal leakage impedances $Z_H = R_H + jX_H$, then the series impedances are *symmetrical* with sequence networks, as shown in Figure 3.10, where $Z_{H0} = Z_{H1} = Z_{H2} = Z_H$. Similarly, the leakage impedances of the low-voltage windings are symmetrical series impedances with $Z_{X0} = Z_{X1} = Z_{X2} = Z_X$. These series leakage impedances are shown in per-unit in the sequence networks of Figure 4.17(a).

The shunt branches of the practical Y–Y transformer, which represent exciting current, are equivalent to the Y load of Figure 3.3. Each phase in Figure 3.3 represents a core loss resistor in parallel with a magnetizing inductance. Assuming these are the same for each phase, then the Y load is *symmetrical*, and the sequence networks are shown in Figure 3.4. These shunt branches are also shown in Figure 4.17(a). Note that ($3Z_N$) and ($3Z_n$) have already been included in the zero-sequence network.

The per-unit positive- and negative-sequence transformer impedances of the practical Y–Y transformer in Figure 4.17(a) are identical, which is always true for nonrotating equipment. The per-unit zero-sequence network, however, depends on the neutral impedances Z_N and Z_n.

The per-unit sequence networks of the Y–Δ transformer, shown in Figure 4.17(b), have the following features:

1. The per-unit impedances do not depend on the winding connections. That is, the per-unit impedances of a transformer that is connected Y–Y, Y–Δ, Δ–Y, or Δ–Δ are the same. However, the base voltages do depend on the winding connections.

2. A phase shift is included in the per-unit positive- and negative-sequence networks. For the American standard, the positive-sequence voltages and currents on the high-voltage side of the Y–Δ transformer lead the corresponding quantities on the low-voltage side by 30°. For negative sequence, the high-voltage quantities lag by 30°.

3. Zero-sequence currents can flow in the Y winding if there is a neutral connection, and corresponding zero-sequence currents flow within the Δ winding. But no zero-sequence current enters or leaves the Δ winding.

The phase shifts in the positive- and negative-sequence networks of Figure 4.17(b) are represented by the phase-shifting transformer of Figure 4.4. Also, the zero-sequence network of Figure 4.17(b) provides a path on the Y side for zero-sequence current to flow, but no zero-sequence current can enter or leave the Δ side.

The per-unit sequence networks of the Δ–Δ transformer, shown in Figure 4.17(c), have the following features:

1. The positive- and negative-sequence networks, which are identical, are the same as those for the Y–Y transformer. It is assumed that the windings are labeled so there is no phase shift. Also, the per-unit impedances do not depend on the winding connections, but the base voltages do.

2. Zero-sequence currents *cannot* enter or leave either Δ winding, although they can circulate within the Δ windings.

EXAMPLE 4.7	**Voltage calculations: balanced Y–Y and Δ–Y transformers**

Three single-phase two-winding transformers, each rated 400 MVA, 13.8/199.2 kV, with leakage reactance $X_{eq} = 0.10$ per unit, are connected to form a three-phase bank. Winding resistances and exciting current are neglected. The high-voltage windings are connected in Y. A three-phase load operating under balanced positive-sequence conditions on the high-voltage side absorbs 1000 MVA at 0.90 p.f. lagging, with $V_{AN} = 199.2\underline{/0°}$ kV. Determine the voltage V_{an} at the low-voltage bus if the low-voltage windings are connected (a) in Y, (b) in Δ.

Solution Since balanced operation occurs for this example, zero- and negative-sequence currents and voltages are zero. The positive-sequence network is shown in Figure 4.18. Using the transformer bank ratings as base quantities, $S_{base3\phi} = 1200$ MVA, $V_{baseHLL} = 345$ kV, and $I_{baseH} = 1200/(345\sqrt{3}) = 2.008$ kA. The per-unit load voltage and load current are then

$$V_{H1} = V_{AN} = 1.0\underline{/0°} \quad \text{per unit}$$

$$I_{H1} = I_A = \frac{1000/(345\sqrt{3})}{2.008}\underline{/-\cos^{-1}0.9} = 0.8333\underline{/-25.84°} \quad \text{per unit}$$

Figure 4.18

Per-unit positive-sequence network for Example 4.7

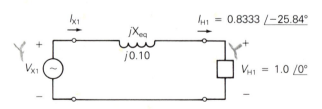

(a) Y-connected low-voltage windings

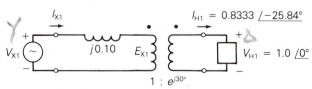

(b) Δ-connected low-voltage windings

a. For the Y–Y transformer, Figure 4.18(a),

$$I_{X1} = I_{H1} = 0.8333\underline{/-25.84°} \quad \text{per unit}$$

$$V_{X1} = V_{H1} + (jX_{eq})I_{X1}$$

$$= 1.0\underline{/0°} + (j0.10)(0.8333\underline{/-25.84°})$$

$$= 1.0 + 0.08333\underline{/64.16°} = 1.0363 + j0.0750 = 1.039\underline{/4.139°}$$

Since $V_{X0} = V_{X2} = 0$,

$$V_{an} = V_{X1} = 1.039\underline{/4.139°} \quad \text{per unit}$$

Further, since $V_{baseXLN} = 13.8\,\text{kV}$ for the low-voltage Y windings, $V_{an} = 1.039(13.8) = 14.34\,\text{kV}$, and

$$V_{an} = 14.34\underline{/4.139°} \quad \text{kV}$$

b. For the Δ–Y transformer, Figure 4.18(b),

$$E_{X1} = e^{-j30°}V_{H1} = 1.0\underline{/-30°} \quad \text{per unit}$$

$$I_{X1} = e^{-j30°}I_{H1} = 0.8333\underline{/-25.84° - 30°} = 0.8333\underline{/-55.84°}$$
$$\text{per unit}$$

$$V_{X1} = E_{X1} + (jX_{eq})I_{X1}$$

$$= 1.0\underline{/-30°} + (j0.10)(0.8333\underline{/-55.84°})$$

$$= 1.039\underline{/-25.861°} \quad \text{per unit}$$

Since $V_{X0} = V_{X2} = 0$,

$$V_{an} = V_{X1} = 1.039\underline{/-25.861°} \quad \text{per unit}$$

Further, since $V_{baseXLN} = 13.8/\sqrt{3} = 7.967\,\text{kV}$ for the low-voltage Δ windings, $V_{an} = (1.039)(7.967) = 8.278\,\text{kV}$, and

$$V_{an} = 8.278\underline{/-25.861°} \quad \text{kV} \qquad \blacksquare$$

EXAMPLE 4.8 | **Solving unbalanced three-phase networks with transformers using per-unit sequence components**

A 75-kVA, 480-volt Δ/208-volt Y transformer with a solidly grounded neutral is connected between the source and line of Example 3.6. The transformer leakage reactance is $X_{eq} = 0.10$ per unit; winding resistances and exciting current are neglected. Using the transformer ratings as base quantities, draw the per-unit sequence networks and calculate the phase a source current I_a.

Solution   The base quantities are $S_{base\,1\phi} = 75/3 = 25\,\text{kVA}$, $V_{baseHLN} = 480/\sqrt{3} = 277.1$ volts, $V_{baseXLN} = 208/\sqrt{3} = 120.1$ volts, and $Z_{baseX} = (120.1)^2/25,000 = 0.5770\,\Omega$. The sequence components of the actual source voltages are given in Figure 3.15. In per-unit, these voltages are

$$V_0 = \frac{15.91\underline{/62.11°}}{277.1} = 0.05742\underline{/62.11°} \quad \text{per unit}$$

$$V_1 = \frac{277.1\underline{/-1.772°}}{277.1} = 1.0\underline{/-1.772°} \quad \text{per unit}$$

$$V_2 = \frac{9.218\underline{/216.59°}}{277.1} = 0.03327\underline{/216.59°} \quad \text{per unit}$$

The per-unit line and load impedances, which are located on the low-voltage side of the transformer, are

$$Z_{L0} = Z_{L1} = Z_{L2} = \frac{1\underline{/85°}}{0.577} = 1.733\underline{/85°} \quad \text{per unit}$$

$$Z_{\text{load1}} = Z_{\text{load2}} = \frac{Z_\Delta}{3(0.577)} = \frac{10\underline{/40°}}{0.577} = 17.33\underline{/40°} \quad \text{per unit}$$

The per-unit sequence networks are shown in Figure 4.19. Note that the

Figure 4.19

Per-unit sequence networks
for Example 4.8

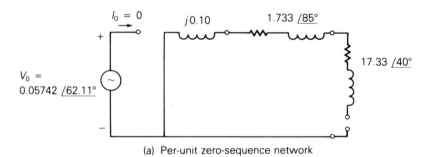

(a) Per-unit zero-sequence network

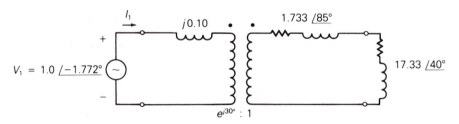

(b) Per-unit positive-sequence network

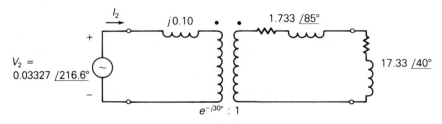

(c) Per-unit negative-sequence network

per-unit line and load impedances, when referred to the high-voltage side of the phase-shifting transformer, do not change (see (4.1.26)). Therefore, from Figure 4.19, the sequence components of the source currents are

$$I_0 = 0$$

$$I_1 = \frac{V_1}{jX_{eq} + Z_{L1} + Z_{load1}} = \frac{1.0/-1.772°}{j0.10 + 1.733/85° + 17.33/40°}$$

$$= \frac{1.0/-1.772°}{13.43 + j12.97} = \frac{1.0/-1.772°}{18.67/44.0°} = 0.05356/-45.77° \quad \text{per unit}$$

$$I_2 = \frac{V_2}{jX_{eq} + Z_{L2} + Z_{load2}} = \frac{0.03327/216.59°}{18.67/44.0°}$$

$$= 0.001782/172.59° \quad \text{per unit}$$

The phase a source current is then, using (3.1.20),

$$I_a = I_0 + I_1 + I_2$$

$$= 0 + 0.05356/-45.77° + 0.001782/172.59°$$

$$= 0.03511 - j0.03764 = 0.05216/-46.19° \quad \text{per unit}$$

$$\text{Using } I_{baseH} = \frac{75,000}{480\sqrt{3}} = 90.21 \text{ A},$$

$$I_a = (0.05216)(90.21)/-46.19° = 4.705/-46.19° \quad \text{A} \qquad \blacksquare$$

EXAMPLE 4.9 | **Per-unit voltage drop and per-unit fault current: balanced three-phase transformer**

A 200-MVA, 345-kVΔ/34.5-kV Y substation transformer has an 8% leakage reactance. The transformer acts as a connecting link between 345-kV transmission and 34.5-kV distribution. Transformer winding resistances and exciting current are neglected. The high-voltage bus connected to the transformer is assumed to be an ideal 345-kV positive-sequence source with negligible source impedance. Using the transformer ratings as base values, determine:

a. The per-unit magnitudes of transformer voltage drop and voltage at the low-voltage terminals when rated transformer current at 0.8 p.f. lagging enters the high-voltage terminals

b. The per-unit magnitude of the fault current when a three-phase-to-ground bolted short circuit occurs at the low-voltage terminals

Solution In both parts (a) and (b), only positive-sequence current will flow, since there are no imbalances; only the positive-sequence network need be considered. Also, as we are interested only in voltage and current magnitudes, the $\Delta-Y$ transformer phase shift can be omitted.

a. As shown in Figure 4.20(a),

$$V_{drop} = I_{rated}X_{eq} = (1.0)(0.08) = 0.08 \quad \text{per unit}$$

Figure 4.20

Circuits for Example 4.9

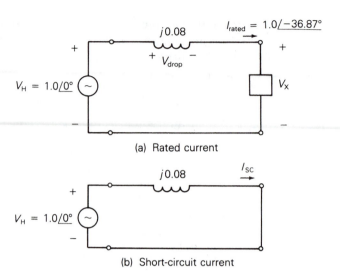

(a) Rated current

(b) Short-circuit current

and

$$V_X = V_H - (jX_{eq})I_{rated}$$

$$= 1.0\underline{/0^\circ} - (j0.08)(1.0\underline{/-36.87^\circ})$$

$$= 1.0 - (j0.08)(0.8 - j0.6) = 0.952 - j0.064$$

$$= 0.954\underline{/-3.85^\circ} \quad \text{per unit}$$

b. As shown in Figure 4.20(b),

$$I_{SC} = \frac{V_H}{X_{eq}} = \frac{1.0}{0.08} = 12.5 \quad \text{per unit}$$

Under rated current conditions [part (a)], the 0.08 per-unit voltage drop across the transformer leakage reactance causes the voltage at the low-voltage terminals to be 0.954 per unit. Also, under three-phase short-circuit conditions [part (b)], the fault current is 12.5 times the rated transformer current. This example illustrates a compromise in the design or specification of transformer leakage reactance. A low value is desired to minimize voltage drops, but a high value is desired to limit fault currents. Typical transformer leakage reactances are given in Table A.2 in the Appendix.   ∎

SECTION 4.6

THREE-WINDING TRANSFORMERS

Figure 4.21(a) shows a basic single-phase three-winding transformer. The ideal transformer relations for a two-winding transformer, (4.1.8) and (4.1.14), can easily be extended to obtain corresponding relations for an ideal

Figure 4.21

Single-phase three-winding
transformer

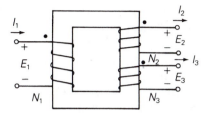

(a) Basic core and coil configuration

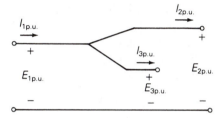

(b) Per-unit equivalent circuit—ideal transformer

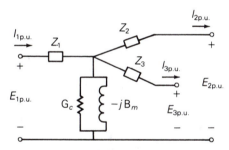

(c) Per-unit equivalent circuit—practical transformer

three-winding transformer. In actual units, these relations are:

$$N_1 I_1 = N_2 I_2 + N_3 I_3 \tag{4.6.1}$$

$$\frac{E_1}{N_1} = \frac{E_2}{N_2} = \frac{E_3}{N_3} \tag{4.6.2}$$

where I_1 enters the dotted terminal, I_2 and I_3 leave dotted terminals, and E_1, E_2, and E_3 have their + polarities at dotted terminals. In per-unit, (4.6.1) and (4.6.2) are:

$$I_{1\text{p.u.}} = I_{2\text{p.u.}} + I_{3\text{p.u.}} \tag{4.6.3}$$

$$E_{1\text{p.u.}} = E_{2\text{p.u.}} = E_{3\text{p.u.}} \tag{4.6.4}$$

where a common S_{base} is selected for all three windings, and voltage bases are selected in proportion to the rated voltages of the windings. These two per-unit relations are satisfied by the per-unit equivalent circuit shown in Figure

4.21(b). Also, external series impedance and shunt admittance branches are included in the practical three-winding transformer circuit shown in Figure 4.21(c). The shunt admittance branch, a core loss resistor in parallel with a magnetizing inductor, can be evaluated from an open-circuit test. Also, when one winding is left open, the three-winding transformer behaves as a two-winding transformer, and standard short-circuit tests can be used to evaluate per-unit leakage impedances, which are defined as follows:

Z_{12} = per-unit leakage impedance measured from winding 1, with winding 2 shorted and winding 3 open

Z_{13} = per-unit leakage impedance measured from winding 1, with winding 3 shorted and winding 2 open

Z_{23} = per-unit leakage impedance measured from winding 2, with winding 3 shorted and winding 1 open

From Figure 4.21(c), with winding 2 shorted and winding 3 open, the leakage impedance measured from winding 1 is, neglecting the shunt admittance branch,

$$Z_{12} = Z_1 + Z_2 \tag{4.6.5}$$

Similarly,

$$Z_{13} = Z_1 + Z_3 \tag{4.6.6}$$

and

$$Z_{23} = Z_2 + Z_3 \tag{4.6.7}$$

Solving (4.6.5)–(4.6.7),

$$Z_1 = \tfrac{1}{2}(Z_{12} + Z_{13} - Z_{23}) \tag{4.6.8}$$

$$Z_2 = \tfrac{1}{2}(Z_{12} + Z_{23} - Z_{13}) \tag{4.6.9}$$

$$Z_3 = \tfrac{1}{2}(Z_{13} + Z_{23} - Z_{12}) \tag{4.6.10}$$

Equations (4.6.8)–(4.6.10) can be used to evaluate the per-unit series impedances Z_1, Z_2, and Z_3 of the three-winding transformer equivalent circuit from the per-unit leakage impedances Z_{12}, Z_{13}, and Z_{23}, which, in turn, are determined from short-circuit tests.

Note that each of the windings on a three-winding transformer may have a *different* kVA rating. If the leakage impedances from short-circuit tests are expressed in per-unit based on winding ratings, they must first be converted to per-unit on a common S_{base} *before* they are used in (4.6.8)–(4.6.10).

| **EXAMPLE 4.10** | **Three-winding single-phase transformer: per-unit impedances** |

The ratings of a single-phase three-winding transformer are:

winding 1: 300 MVA, 13.8 kV

winding 2: 300 MVA, 199.2 kV

winding 3:   50 MVA, 19.92 kV

The leakage reactances, from short-circuit tests, are:

$X_{12} = 0.10$ per unit on a 300-MVA, 13.8-kV base

$X_{13} = 0.16$ per unit on a 50-MVA, 13.8-kV base

$X_{23} = 0.14$ per unit on a 50-MVA, 199.2-kV base

Winding resistances and exciting current are neglected. Calculate the impedances of the per-unit equivalent circuit using a base of 300 MVA and 13.8 kV for terminal 1.

Solution $S_{base} = 300$ MVA is the same for all three terminals. Also, the specified voltage base for terminal 1 is $V_{base1} = 13.8$ kV. The base voltages for terminals 2 and 3 are then $V_{base2} = 199.2$ kV and $V_{base3} = 19.92$ kV, which are the rated voltages of these windings. From the data given, $X_{12} = 0.10$ per unit was measured from terminal 1 using the same base values as those specified for the circuit. But $X_{13} = 0.16$ and $X_{23} = 0.14$ per unit on a 50-MVA base are first converted to the 300-MVA circuit base.

$$X_{13} = (0.16)\left(\frac{300}{50}\right) = 0.96 \quad \text{per unit}$$

$$X_{23} = (0.14)\left(\frac{300}{50}\right) = 0.84 \quad \text{per unit}$$

Then, from (4.6.8)–(4.6.10),

$X_1 = \tfrac{1}{2}(0.10 + 0.96 - 0.84) = \quad 0.11 \quad$ per unit

$X_2 = \tfrac{1}{2}(0.10 + 0.84 - 0.96) = -0.01 \quad$ per unit

$X_3 = \tfrac{1}{2}(0.84 + 0.96 - 0.10) = \quad 0.85 \quad$ per unit

The per-unit equivalent circuit of this three-winding transformer is shown in Figure 4.22. Note that X_2 is negative. This illustrates the fact that X_1, X_2, and X_3 are *not* leakage reactances, but instead are equivalent reactances derived from the leakage reactances. Leakage reactances are always positive.

Note also that the node where the three equivalent circuit reactances

Figure 4.22

Circuit for Example 4.10

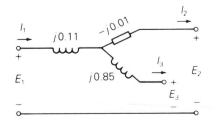

are connected does not correspond to any physical location within the transformer. Rather, it is simply part of the equivalent circuit representation. ∎

Three identical single-phase three-winding transformers can be connected to form a three-phase bank. Figure 4.23 shows the general per-unit sequence networks of a three-phase three-winding transformer. Instead of labeling the windings 1, 2, and 3, as was done for the single-phase transformer, the letters H, M, and X are used to denote the high-, medium-, and low-voltage windings, respectively. By convention, a common S_{base} is selected for the H, M, and X terminals, and voltage bases V_{baseH}, V_{baseM}, and V_{baseX} are selected in proportion to the rated line-to-line voltages of the transformer.

For the general zero-sequence network, Figure 4.23(a), the connection between terminals H and H′ depends on how the high-voltage windings are connected, as follows:

1. Solidly grounded Y—Short H to H′.
2. Grounded Y through Z_N—Connect $(3Z_N)$ from H to H′.
3. Ungrounded Y—Leave H–H′ open as shown.
4. Δ—Short H′ to the reference bus.

Terminals X–X′ and M–M′ are connected in a similar manner.

Figure 4.23

Per-unit sequence networks of a three-phase three-winding transformer

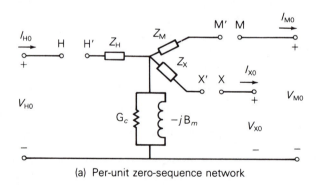

(a) Per-unit zero-sequence network

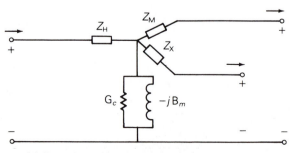

(b) Per-unit positive- or negative-sequence network (phase shift not shown)

The impedances of the per-unit negative-sequence network are the same as those of the per-unit positive-sequence network, which is always true for nonrotating equipment. Phase-shifting transformers, not shown in Figure 4.23(b), can be included to model phase shift between Δ and Y windings.

| EXAMPLE 4.11 | **Three-winding three-phase transformer: per-unit sequence networks** |

Three transformers, each identical to that described in Example 4.10, are connected as a three-phase bank in order to feed power from a 900-MVA, 13.8-kV generator to a 345-kV transmission line and to a 34.5-kV distribution line. The transformer windings are connected as follows:

    13.8-kV windings (X): Δ, to generator

    199.2-kV windings (H): solidly grounded Y, to 345-kV line

    19.92-kV windings (M): grounded Y through $Z_n = j0.10\,\Omega$, to 34.5-kV line

Figure 4.24

Per-unit sequence networks for Example 4.11

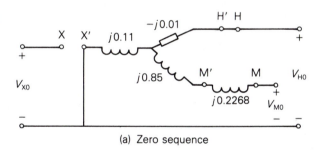

(a) Zero sequence

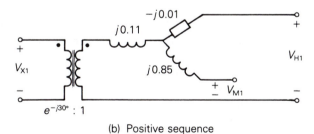

(b) Positive sequence

(c) Negative sequence

The positive-sequence voltages and currents of the high- and medium-voltage Y windings lead the corresponding quantities of the low-voltage Δ winding by 30°. Draw the per-unit sequence networks, using a three-phase base of 900 MVA and 13.8 kV for terminal X.

Solution   The per-unit sequence networks are shown in Figure 4.24. Since $V_{baseX} = 13.8$ kV is the rated line-to-line voltage of terminal X, $V_{baseM} = \sqrt{3}(19.92) = 34.5$ kV, which is the rated line-to-line voltage of terminal M. The base impedance of the medium-voltage terminal is then

$$Z_{baseM} = \frac{(34.5)^2}{900} = 1.3225 \quad \Omega$$

Therefore, the per-unit neutral impedance is

$$Z_n = \frac{j0.10}{1.3225} = j0.07561 \quad \text{per unit}$$

and $(3Z_n) = j0.2268$ is connected from terminal M to M′ in the per-unit zero-sequence network. Since the high-voltage windings have a solidly grounded neutral, H to H′ is shorted in the zero-sequence network. Also, phase-shifting transformers are included in the positive- and negative-sequence networks. ■

SECTION 4.7

AUTOTRANSFORMERS

A single-phase two-winding transformer is shown in Figure 4.25(a) with two separate windings, which is the usual two-winding transformer; the same transformer is shown in Figure 4.25(b) with the two windings connected in series, which is called an *autotransformer*. For the usual transformer [Figure 4.25(a)] the two windings are coupled magnetically via the mutual core flux. For the autotransformer [Figure 4.25(b)] the windings are both electrically and magnetically coupled. The autotransformer has smaller per-unit leakage impedances than the usual transformer; this results in both smaller series-

Figure 4.25

Ideal single-phase transformers

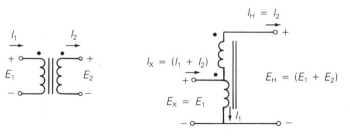

(a) Two-winding transformer   (b) Autotransformer

voltage drops (an advantage) and higher short-circuit currents (a disadvantage). The autotransformer also has lower per-unit losses (higher efficiency), lower exciting current, and lower cost if the turns ratio is not too large. The electrical connection of the windings, however, allows transient overvoltages to pass through the autotransformer more easily.

| **EXAMPLE 4.12** | **Autotransformer: single-phase** |

The single-phase two-winding 20-kVA, 480/120-volt transformer of Example 4.3 is connected as an autotransformer, as in Figure 4.25(b), where winding 1 is the 120-volt winding. For this autotransformer, determine (a) the voltage ratings E_X and E_H of the low- and high-voltage terminals, (b) the kVA rating, and (c) the per-unit leakage impedance.

Solution

a. Since the 120-volt winding is connected to the low-voltage terminal, $E_X = 120$ volts. When $E_X = E_1 = 120$ volts is applied to the low-voltage terminal, $E_2 = 480$ volts is induced across the 480-volt winding, neglecting the voltage drop across the leakage impedance. Therefore, $E_H = E_1 + E_2 = 120 + 480 = 600$ volts.

b. As a normal two-winding transformer rated 20 kVA, the rated current of the 480-volt winding is $I_2 = I_H = 20{,}000/480 = 41.667$ A. As an autotransformer, the 480-volt winding can carry the same current. Therefore, the kVA rating $S_H = E_H I_H = (600)(41.667) = 25$ kVA. Note also that when $I_H = I_2 = 41.667$ A, a current $I_1 = \dfrac{480}{120}(41.667) = 166.7$ A is induced in the 120-volt winding. Therefore, $I_X = I_1 + I_2 = 208.3$ A (neglecting exciting current) and $S_X = E_X I_X = (120)(208.3) = 25$ kVA, which is the same rating as calculated for the high-voltage terminal.

c. From Example 4.3, the leakage impedance is $0.0729\underline{/78.13°}$ per unit as a normal, two-winding transformer. As an autotransformer, the leakage impedance *in ohms* is the same as for the normal transformer, since the core and windings are the same for both (only the external winding connections are different). But the base impedances are different. For the high-voltage terminal, using (4.3.4),

$$Z_{\text{baseHold}} = \frac{(480)^2}{20{,}000} = 11.52 \ \Omega \quad \text{as a normal transformer}$$

$$Z_{\text{baseHnew}} = \frac{(600)^2}{25{,}000} = 14.4 \ \Omega \quad \text{as an autotransformer}$$

Therefore, using (4.3.10),

$$Z_{\text{p.u.new}} = (0.0729\underline{/78.13°}) \left(\frac{11.52}{14.4}\right) = 0.05832\underline{/78.13°} \quad \text{per unit}$$

For this example, the rating is 25 kVA, 120/600 volts as an autotransformer versus 20 kVA, 120/480 volts as a normal transformer. The autotransformer has both a larger kVA rating and a larger voltage ratio for the same cost.

Also, the per-unit leakage impedance of the autotransformer is smaller. However, the increased high-voltage rating as well as the electrical connection of the windings may require more insulation for both windings. ■

SECTION 4.8

TRANSFORMERS WITH OFF-NOMINAL TURNS RATIOS

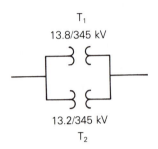

T$_1$

13.8/345 kV

13.2/345 kV

T$_2$

Figure 4.26

Two transformers connected in parallel

It has been shown that models of transformers that use per-unit quantities are simpler than those that use actual quantities. The ideal transformer winding is eliminated when the ratio of the selected voltage bases equals the ratio of the voltage ratings of the windings. In some cases, however, it is impossible to select voltage bases in this manner. For example, consider the two transformers connected in parallel in Figure 4.26. Transformer T$_1$ is rated 13.8/345 kV and T$_2$ is rated 13.2/345 kV. If we select V$_{\text{baseH}}$ = 345 kV, then transformer T$_1$ requires V$_{\text{baseX}}$ = 13.8 kV and T$_2$ requires V$_{\text{baseX}}$ = 13.2 kV. It is clearly impossible to select the appropriate voltage bases for both transformers.

To accommodate this situation, we will develop a per-unit model of a transformer whose voltage ratings are not in proportion to the selected base voltages. Such a transformer is said to have an "off-nominal turns ratio." Figure 4.27(a) shows a transformer with rated voltages V$_{1\text{rated}}$ and V$_{2\text{rated}}$, which satisfy

$$V_{1\text{rated}} = a_t V_{2\text{rated}} \tag{4.8.1}$$

where a_t is assumed, in general, to be either real or complex. Suppose the selected voltage bases satisfy

$$V_{\text{base}1} = b V_{\text{base}2} \tag{4.8.2}$$

Defining $c = a_t/b$, (4.8.1) can be rewritten as

$$V_{1\text{rated}} = b \left(\frac{a_t}{b}\right) V_{2\text{rated}} = bc\, V_{2\text{rated}} \tag{4.8.3}$$

Equation (4.8.3) can be represented by two transformers in series, as shown in Figure 4.27(b). The first transformer has the same ratio of rated winding voltages as the ratio of the selected base voltages, b. Therefore this transformer has a standard per-unit model, as shown in Figure 4.9 or 4.17. We will assume that the second transformer is ideal, all real and reactive losses being associated with the first transformer. The resulting per-unit model is shown in Figure 4.27(c), where, for simplicity, the shunt-exciting branch is neglected. Note that if $a_t = b$, then the ideal transformer winding shown in this figure can be eliminated, since its turns ratio $c = (a_t/b) = 1$.

The per-unit model shown in Figure 4.27(c) is perfectly valid, but it is not suitable for some of the computer programs presented in later chapters because these programs do not accommodate ideal transformer windings. An

Figure 4.27

Transformer with off-nominal
turns ratio

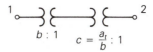

(a) Single-line diagram

(b) Represented as two
transformers in series

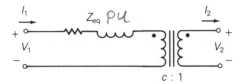

(c) Per-unit equivalent circuit
(Per-unit impedance is shown)

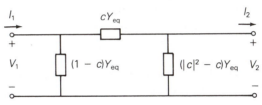

(d) π circuit representation for real c
$\left(\text{Per-unit admittances are shown; } Y_{eq} = \dfrac{1}{Z_{eq}}\right)$

alternative representation can be developed, however, by writing nodal
equations for this figure as follows:

$$
\begin{bmatrix} I_1 \\ -I_2 \end{bmatrix} = \begin{bmatrix} Y_{11} & Y_{12} \\ Y_{21} & Y_{22} \end{bmatrix} \begin{bmatrix} V_1 \\ V_2 \end{bmatrix}
\tag{4.8.4}
$$

where both I_1 and $-I_2$ are referenced *into* their nodes in accordance with the
nodal equation method (Section 2.4). Recalling two-port network theory, the
admittance parameters of (4.8.4) are, from Figure 4.27(c),

$$
Y_{11} = \left.\frac{I_1}{V_1}\right|_{V_2=0} = \frac{1}{Z_{eq}} = Y_{eq}
\tag{4.8.5}
$$

$$
Y_{22} = \left.\frac{-I_2}{V_2}\right|_{V_1=0} = \frac{1}{Z_{eq}/|c|^2} = |c|^2 Y_{eq}
\tag{4.8.6}
$$

$$
Y_{12} = \left.\frac{I_1}{V_2}\right|_{V_1=0} = \frac{-cV_2/Z_{eq}}{V_2} = -c Y_{eq}
\tag{4.8.7}
$$

$$
Y_{21} = \left.\frac{-I_2}{V_1}\right|_{V_2=0} = \frac{-c^* I_1}{V_1} = -c^* Y_{eq}
\tag{4.8.8}
$$

Equations (4.8.4)–(4.8.8) with real or complex c are convenient for
representing transformers with off-nominal turns ratios in the computer

programs presented later. Note that when c is complex, Y_{12} is not equal to Y_{21}, and the preceding admittance parameters cannot be synthesized with a passive RLC circuit. However, the π network shown in Figure 4.27(d), which has the same admittance parameters as (4.8.4)–(4.8.8), can be synthesized for real c. Note also that when $c = 1$, the shunt branches in this figure become open circuits (zero per unit mhos), and the series branch becomes Y_{eq} per unit mhos (or Z_{eq} per unit ohms).

EXAMPLE 4.13

Tap-changing three-phase transformer: per-unit positive-sequence network

A three-phase generator step-up transformer is rated 1000 MVA, 13.8 kV $\Delta/345$ kV Y with $Z_{eq} = j0.10$ per unit. The transformer high-voltage winding has $\pm 10\%$ taps. The system base quantities are

$$S_{base3\phi} = 500 \quad MVA$$

$$V_{baseXLL} = 13.8 \quad kV$$

$$V_{baseHLL} = 345 \quad kV$$

Determine the per-unit positive-sequence equivalent circuit for the following tap settings:

a. Rated tap

b. -10% tap (providing a 10% voltage decrease for the high-voltage winding)

Neglect transformer winding resistance, exciting current, and phase shift.

Solution   **a.** Using (4.8.1) and (4.8.2) with the low-voltage winding denoted winding 1,

$$a_t = \frac{13.8}{345} = 0.04 \qquad b = \frac{V_{baseXLL}}{V_{baseHLL}} = \frac{13.8}{345} = a_t \qquad c = 1$$

From (4.3.11)

$$Z_{p.u.new} = (j0.10)\left(\frac{500}{1000}\right) = j0.05 \quad \text{per unit}$$

The positive-sequence equivalent circuit, not including winding resistance, exciting current, and phase shift is:

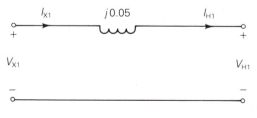

(Per-unit impedance is shown)

b. Using (4.8.1) and (4.8.2),

$$a_t = \frac{13.8}{345(0.9)} = 0.04444 \quad b = \frac{13.8}{345} = 0.04 \quad c = \frac{a_t}{b} = \frac{0.04444}{0.04} = 1.1111$$

From Figure 4.27(d),

$$cY_{eq} = 1.1111\left(\frac{1}{j0.05}\right) = -j22.22 \quad \text{per unit}$$

$$(1 - c)Y_{eq} = (-0.11111)(-j20) = +j2.222 \quad \text{per unit}$$

$$(|c|^2 - c)Y_{eq} = (1.2346 - 1.1)(-j20) = -j2.469 \quad \text{per unit}$$

The per-unit positive-sequence network is:

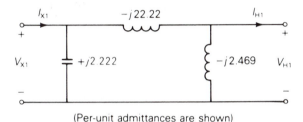

(Per-unit admittances are shown)   ∎

The three-phase regulating transformers shown in Figures 4.28 and 4.29 can be modeled as transformers with off-nominal turns ratios. For the voltage-magnitude-regulating transformer shown in Figure 4.28, adjustable voltages ΔV_{an}, ΔV_{bn}, and ΔV_{cn}, which have equal magnitudes ΔV and which are in phase with the phase voltages V_{an}, V_{bn}, and V_{cn}, are placed in the series

Figure 4.28

An example of a voltage-magnitude-regulating transformer

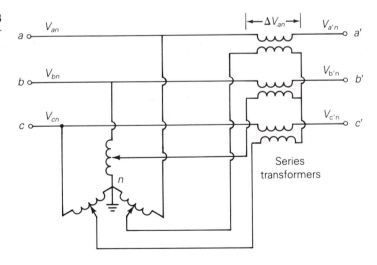

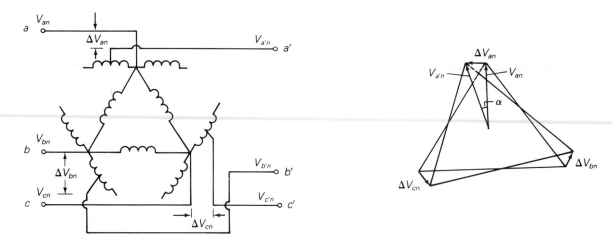

Figure 4.29   An example of a phase-angle-regulating transformer. Windings drawn in parallel are on the same core

link between buses a–a', b–b', and c–c'. Modeled as a transformer with an off-nominal turns ratio (see Figure 4.27), $c = (1 + \Delta V)$ for a voltage-magnitude increase toward bus abc, or $c = (1 + \Delta V)^{-1}$ for an increase toward bus $a'b'c'$.

   For the phase-angle-regulating transformer in Figure 4.29, the series voltages ΔV_{an}, ΔV_{bn}, and ΔV_{cn} are $\pm 90°$ out of phase with the phase voltages V_{an}, V_{bn}, and V_{cn}. The phasor diagram in Figure 4.29 indicates that each of the bus voltages $V_{a'n}$, $V_{b'n}$, and $V_{c'n}$, has a phase shift that is approximately proportional to the magnitude of the added series voltage. Modeled as a transformer with an off-nominal turns ratio (see Figure 4.27), $c \approx 1\underline{/\alpha}$ for a phase increase toward bus abc or $c \approx 1\underline{/-\alpha}$ for a phase increase toward bus $a'b'c'$.

| **EXAMPLE 4.14** | **Voltage-regulating and phase-shifting three-phase transformers** |

Two buses abc and $a'b'c'$ are connected by two parallel lines L1 and L2 with positive-sequence series reactances $X_{L1} = 0.25$ and $X_{L2} = 0.20$ per unit. A regulating transformer is placed in series with line L1 at bus $a'b'c'$. Determine the 2×2 positive-sequence bus admittance matrix when the regulating transformer (a) provides a 0.05 per-unit increase in voltage magnitude toward bus $a'b'c'$ and (b) advances the phase 3° toward bus $a'b'c'$. Assume that the regulating transformer is ideal. Also, the series resistance and shunt admittance of the lines are neglected.

Solution   The circuit is shown in Figure 4.30.

   a. For the voltage-magnitude-regulating transformer, $c = (1 + \Delta V)^{-1} = (1.05)^{-1} = 0.9524$ per unit. From (4.8.5)–(4.8.8), the admittance parameters of the regulating transformer in series with line L1 are

$$Y_{11L1} = \frac{1}{j0.25} = -j4.0$$

Figure 4.30

Positive-sequence circuit for Example 4.14

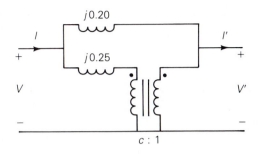

$$Y_{22L1} = (0.9524)^2(-j4.0) = -j3.628$$

$$Y_{12L1} = Y_{21L1} = (-0.9524)(-j4.0) = j3.810$$

For line L2 alone,

$$Y_{11L2} = Y_{22L2} = \frac{1}{j0.20} = -j5.0$$

$$Y_{12L2} = Y_{21L2} = -(-j5.0) = j5.0$$

Combining the above admittances in parallel,

$$Y_{11} = Y_{11L1} + Y_{11L2} = -j4.0 - j5.0 = -j9.0$$

$$Y_{22} = Y_{22L1} + Y_{22L2} = -j3.628 - j5.0 = -j8.628$$

$$Y_{12} = Y_{21} = Y_{12L1} + Y_{12L2} = j3.810 + j5.0 = j8.810 \quad \text{per unit}$$

b. For the phase-angle-regulating transformer, $c = 1\underline{/-\alpha} = 1\underline{/-3^\circ}$. Then, for this regulating transformer in series with line L1,

$$Y_{11L1} = \frac{1}{j0.25} = -j4.0$$

$$Y_{22L1} = |1.0\underline{/-3^\circ}|^2(-j4.0) = -j4.0$$

$$Y_{12L1} = -(1.0\underline{/-3^\circ})(-j4.0) = 4.0\underline{/87^\circ} = 0.2093 + j3.9945$$

$$Y_{21L1} = -(1.0\underline{/-3^\circ})^*(-j4.0) = 4.0\underline{/93^\circ} = -0.2093 + j3.9945$$

The admittance parameters for line L2 alone are given in part (a) above. Combining the admittances in parallel,

$$Y_{11} = Y_{22} = -j4.0 - j5.0 = -j9.0$$

$$Y_{12} = 0.2093 + j3.9945 + j5.0 = 0.2093 + j8.9945$$

$$Y_{21} = -0.2093 + j3.9945 + j5.0 = -0.2093 + j8.9945 \quad \text{per unit}$$

∎

Note that a voltage-magnitude-regulating transformer controls the *reactive* power flow in the series link in which it is installed, whereas a phase-angle-regulating transformer controls the *real* power flow (see Problem 4.31).

PROBLEMS

Section 4.1

4.1 Consider an ideal transformer with $N_1 = 3000$ and $N_2 = 500$ turns. Let winding 1 be connected to a source whose voltage is $e_1(t) = 100(1 - |t|)$ volts for $-1 \leqslant t \leqslant 1$ and $e_1(t) = 0$ for $|t| > 1$ second. A 3-farad capacitor is connected across winding 2. Sketch $e_1(t)$, $e_2(t)$, $i_1(t)$, and $i_2(t)$ versus time t.

4.2 A single-phase 100-kVA, 2400/240-volt, 60-Hz distribution transformer is used as a step-down transformer. The load, which is connected to the 240-volt secondary winding, absorbs 90 kVA at 0.8 power factor lagging and is at 230 volts. Assuming an ideal transformer, calculate the following: (a) primary voltage, (b) load impedance, (c) load impedance referred to the primary, and (d) the real and reactive power supplied to the primary winding.

4.3 Rework Problem 4.2 if the load connected to the 240-V secondary winding absorbs 110 kVA under short-term overload conditions at 0.85 power factor leading and at 230 volts.

4.4 For a conceptual single-phase, phase-shifting transformer, the primary voltage leads the secondary voltage by 30°. A load connected to the secondary winding absorbs 50 kVA at 0.9 power factor leading and at a voltage $E_2 = 277\underline{/0°}$ volts. Determine (a) the primary voltage, (b) primary and secondary currents, (c) load impedance referred to the primary winding, and (d) complex power supplied to the primary winding.

Section 4.2

4.5 The following data are obtained when open-circuit and short-circuit tests are performed on a single-phase, 50-kVA, 2400/240-volt, 60-Hz distribution transformer.

	VOLTAGE (volts)	CURRENT (amperes)	POWER (watts)
Measurements on low-voltage side with high-voltage winding open	240	5.97	213
Measurements on high-voltage side with low-voltage winding shorted	60.0	20.8	750

(a) Neglecting the series impedance, determine the exciting admittance referred to the high-voltage side. (b) Neglecting the exciting admittance, determine the equivalent series impedance referred to the high-voltage side. (c) Assuming equal series impedances for the primary and referred secondary, obtain an equivalent T-circuit referred to the high-voltage side.

4.6 A single-phase 100-kVA, 2400/240-volt, 60-Hz distribution transformer has a 2-ohm equivalent leakage reactance and a 5000-ohm magnetizing reactance referred to the high-voltage side. If rated voltage is applied to the high-voltage winding, calculate the open-circuit secondary voltage. Neglect I^2R and G_c^2V losses. Assume equal series leakage reactances for the primary and referred secondary.

4.7 A single-phase 50-kVA, 2400/240-volt, 60-Hz distribution transformer is used as a step-down transformer at the load end of a 2400-volt feeder whose series impedance is $(1.0 + j2.0)$ ohms. The equivalent series impedance of the transformer is $(1.0 + j2.5)$ ohms referred to the high-voltage (primary) side. The transformer is delivering rated load at 0.8 power factor lagging and at rated secondary voltage. Neglecting the transformer exciting current, determine (a) the voltage at the transformer primary terminals, (b) the voltage at the sending end of the feeder, and (c) the real and reactive power delivered to the sending end of the feeder.

4.8   Rework Problem 4.7 if the transformer is delivering rated load at rated secondary voltage and at: (a) unity power factor, (b) 0.8 power factor leading. Compare the results with those of Problem 4.7.

Section 4.3

4.9   Using the transformer ratings as base quantities, work Problem 4.6 in per-unit.

4.10   Using the transformer ratings as base quantities, work Problem 4.7 in per-unit.

4.11   Using base values of 20 kVA and 115 volts in zone 3, rework Example 4.4.

4.12   Rework Example 4.5 using $S_{base3\phi} = 100$ kVA and $V_{baseLL} = 600$ volts.

4.13   A balanced-Y-connected voltage source with $E_{ag} = 277\underline{/0°}$ volts is applied to a balanced-Y load in parallel with a balanced-Δ load, where $Z_Y = 30 + j10$ and $Z_\Delta = 45 - j25$ ohms. The Y load is solidly grounded. Using base values of $S_{base1\phi} = 10$ kVA and $V_{baseLN} = 277$ volts, calculate the source current I_a in per-unit and in amperes.

Section 4.4

4.14   Determine the positive- and negative-sequence phase shifts for the three-phase transformers shown in Figure 4.31.

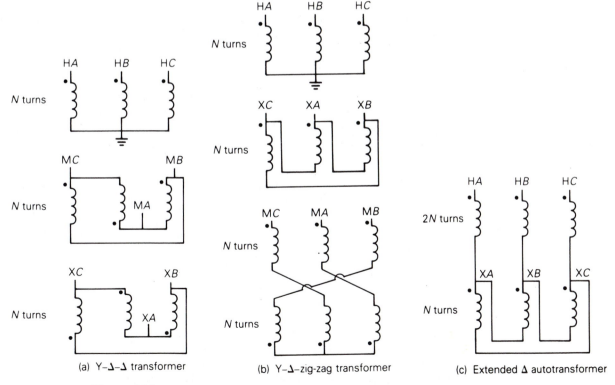

(a) Y–Δ–Δ transformer   (b) Y–Δ–zig-zag transformer   (c) Extended Δ autotransformer

Figure 4.31   Problems 4.14 and 4.28 (Coils drawn on the same vertical line are on the same core)

4.15 Consider the three single-phase two-winding transformers shown in Figure 4.32. The high-voltage windings are connected in Y. (a) For the low-voltage side, connect the windings in Δ, place the polarity marks, and label the terminals *a*, *b*, and *c* in accordance with the American standard. (b) Relabel the terminals *a'*, *b'*, and *c'* such that V_{AN} is 90° out of phase with $V_{a'n}$ for positive sequence.

Figure 4.32

Problem 4.15

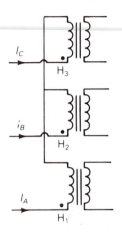

4.16 Three single-phase, two-winding transformers, each rated 450 MVA, 20 kV/288.7 kV, with leakage reactance $X_{eq} = 0.12$ per unit, are connected to form a three-phase bank. The high-voltage windings are connected in Y with a solidly grounded neutral. Draw the per-unit zero-, positive-, and negative-sequence networks if the low-voltage windings are connected: (a) in Δ with American standard phase shift, (b) in Y with an open neutral. Use the transformer ratings as base quantities. Winding resistances and exciting current are neglected.

4.17 Consider a bank of three single-phase two-winding transformers whose high-voltage terminals are connected to a three-phase, 13.8-kV feeder. The low-voltage terminals are connected to a three-phase substation load rated 2.1 MVA and 2.3 kV. Determine the required voltage, current, and MVA ratings of both windings of each transformer, when the high-voltage/low-voltage windings are connected (a) Y–Δ, (b) Δ–Y, (c) Y–Y, and (d) Δ–Δ.

4.18 Three single-phase two-winding transformers, each rated 25 MVA, 34.5/13.8 kV, are connected to form a three-phase Δ–Δ bank. Balanced positive-sequence voltages are applied to the high-voltage terminals, and a balanced, resistive Y load connected to the low-voltage terminals absorbs 75 MW at 13.8 kV. If one of the single-phase transformers is removed (resulting in an open-Δ connection) and the balanced load is simultaneously reduced to 43.3 MW (57.7% of the original value), determine (a) the load voltages V_{an}, V_{bn}, and V_{cn}; (b) load currents I_a, I_b, and I_c; and (c) the MVA supplied by each of the remaining two transformers. Are balanced voltages still applied to the load? Is the open-Δ transformer overloaded?

Section 4.5

4.19 The leakage reactance of a three-phase, 500-MVA, 345 Y/23 Δ-kV transformer is 0.09 per unit based on its own ratings. The Y winding has a solidly grounded neutral. Draw the sequence networks. Neglect the exciting admittance and assume American standard phase shift.

4.20 Choosing system bases to be 360/24 kV and 100 MVA, redraw the sequence networks for Problem 4.19.

4.21 Consider the single-line diagram of the power system shown in Figure 4.33. Equipment ratings are:

generator 1:	750 MVA, 18 kV, X″ = 0.2 per unit
generator 2:	750 MVA, 18 kV, X″ = 0.2
synchronous motor 3:	1500 MVA, 20 kV, X″ = 0.2
3-phase Δ–Y transformers T₁, T₂, T₃, T₄:	750 MVA, 500 kV Y/20 kV Δ, X = 0.1
3-phase Y–Y transformer T₅:	1500 MVA, 500 kV Y/20 kV Y, X = 0.1

Neglecting resistance, transformer phase shift, and magnetizing reactance, draw the positive-sequence reactance diagram. Use a base of 100 MVA and 500 kV for the 40-ohm line. Determine the per-unit reactances.

Figure 4.33

Problems 4.21, 4.22, 4.23

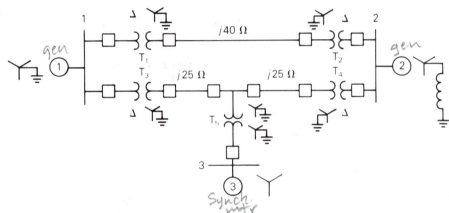

4.22 For the power system in Problem 4.21, the synchronous motor absorbs 1200 MW at 0.8 power factor leading with the bus 3 voltage at 18 kV. Determine the bus 1 and bus 2 voltages in kV. Assume that generators 1 and 2 deliver equal real powers and equal reactive powers. Also assume a balanced three-phase system with positive-sequence sources.

4.23 Draw the zero-sequence reactance diagram for the power system shown in Figure 4.33. The zero-sequence reactance of each generator and of the synchronous motor is 0.05 per unit based on equipment ratings. Generator 2 is grounded through a neutral reactor of 0.06 per unit on a 100-MVA, 18-kV base. The zero-sequence reactance of each transmission line is assumed to be three times its positive-sequence reactance. Use the same base as in Problem 4.21.

4.24 Three single-phase transformers, each rated 10 MVA, 66.4/12.5 kV, 60 Hz, with an equivalent series reactance of 0.1 per unit divided equally between primary and secondary, are connected in a three-phase bank. The high-voltage windings are Y connected and their terminals are directly connected to a 115-kV three-phase bus. The secondary terminals are all shorted together. Find the currents entering the high-voltage terminals and leaving the low-voltage terminals if the low-voltage windings are (a) Y connected, (b) Δ connected.

4.25   A 100-MVA, 13.2-kV three-phase generator, which has a positive-sequence reactance of 1.5 per unit on the generator base is connected to a 110-MVA, 13.2 Δ/115 Y-kV step-up transformer with a series impedance of $(0.005 + j0.1)$ per unit on its own base. (a) Calculate the per-unit generator reactance on the transformer base. (b) The load at the transformer terminals is 80 MW at unity power factor and at 115 kV. Choosing the transformer high-side voltage as the reference phasor, draw a phasor diagram for this condition. (c) For the condition of part (b), find the transformer low-side voltage and the generator internal voltage behind its reactance. Also compute the generator output power and power factor.

Section 4.6

4.26   A single-phase three-winding transformer has the following parameters: $Z_1 = Z_2 = Z_3 = 0 + j0.05$, $G_c = 0$, and $B_m = 0.2$ per unit. Three identical transformers, as described, are connected with their primaries in Y (solidly grounded neutral) and with their secondaries and tertiaries in Δ. Draw the per-unit sequence networks of this transformer bank.

4.27   The ratings of a three-phase three-winding transformer are:

Primary (1):   Y connected, 66 kV, 20 MVA
Secondary (2):   Y connected, 13.2 kV, 15 MVA
Tertiary (3):   Δ connected, 2.3 kV, 5 MVA

Neglecting winding resistances and exciting current, the per-unit leakage reactances are:

$X_{12} = 0.08$ on a 20-MVA, 66-kV base
$X_{13} = 0.10$ on a 20-MVA, 66-kV base
$X_{23} = 0.09$ on a 15-MVA, 13.2-kV base

(a) Determine the per-unit reactances X_1, X_2, X_3 of the equivalent circuit on a 20-MVA, 66-kV base at the primary terminals. (b) Purely resistive loads of 12 MW at 13.2 kV and 5 MW at 2.3 kV are connected to the secondary and tertiary sides of the transformer, respectively. Draw the per-unit impedance diagram, showing the per-unit impedances on a 20-MVA, 66-kV base at the primary terminals.

4.28   Draw the positive-, negative-, and zero-sequence circuits for the transformers shown in Figure 4.31. Include ideal phase-shifting transformers showing phase shifts determined in Problem 4.14. Assume that all windings have the same kVA rating and that the equivalent leakage reactance of any two windings with the third winding open is 0.10 per unit. Neglect the exciting admittance.

Section 4.7

4.29   A single-phase 15-kVA, 2400/240-volt, 60-Hz two-winding distribution transformer is connected as an autotransformer to step up the voltage from 2400 to 2640 volts. (a) Draw a schematic diagram of this arrangement, showing all voltages and currents when delivering full load at rated voltage. (b) Find the permissible kVA rating of the autotransformer if the winding currents and voltages are not to exceed the rated values as a two-winding transformer. How much of this kVA rating is transformed by magnetic induction? (c) The following data are obtained from tests carried out on the transformer when it is connected as a two-winding transformer:

Open-circuit test with the low-voltage terminals excited:
Applied voltage = 240 V, Input current = 0.68 A, Input power = 105 W.
Short-circuit test with the high-voltage terminals excited:
Applied voltage = 120 V, Input current = 6.25 A, Input power = 330 W.

Based on this data, compute the efficiency of the autotransformer corresponding to full load, rated voltage, and 0.8 power factor lagging. Comment on why the efficiency is higher as an autotransformer than as a two-winding transformer.

4.30   Three single-phase two-winding transformers, each rated 3 kVA, 220/110 volts, 60 Hz, with a 0.10 per-unit leakage reactance are connected as a three-phase extended Δ autotransformer bank, as shown in Figure 4.31(c). The low-voltage Δ winding has a 110 volt rating. (a) Draw the positive-sequence phasor diagram and show that the high-voltage winding has a 479.5 volt rating. (b) A three-phase load connected to the low-voltage terminals absorbs 6 kW at 110 volts and at 0.8 power factor lagging. Draw the per-unit impedance diagram and calculate the voltage and current at the high-voltage terminals. Assume positive-sequence operation.

Section 4.8

4.31   The two parallel lines in Example 4.14 supply a balanced load with a load current of $1.0 / -30°$ per unit. Determine the real and reactive power supplied to the load bus from each parallel line with (a) no regulating transformer, (b) the voltage-magnitude-regulating transformer in Example 4.14(a), and (c) the phase-angle-regulating transformer in Example 4.14(b). Assume that the voltage at bus abc is adjusted so that the voltage at bus $a'b'c'$ remains constant at $1.0 / 0°$ per unit. Also assume positive sequence. Comment on the effects of the regulating transformers.

4.32   Rework Example 4.13 for a +10% tap, providing a 10% increase for the high-voltage winding.

CASE STUDY QUESTIONS

A.   The case study describes transformer evaluation practices of four utilities: Wisconsin Public Service, Georgia Power, Public Service of Colorado, and Utah Power and Light. What are the differences in these practices among the four utilities?

B.   How do power transformer evaluation practices differ from distribution transformer practices?

References

1.   R. Feinberg, *Modern Power Transformer Practice* (New York: Wiley, 1979).

2.   A. C. Franklin and D. P. Franklin, *The J & P Transformer Book*, 11th ed. (London: Butterworths, 1983).

3.   W. D. Stevenson, Jr., *Elements of Power System Analysis*, 4th ed. (New York: McGraw-Hill, 1982).

4.   J. R. Neuenswander, *Modern Power Systems* (Scranton, PA: International Textbook Company, (1971).

5.   M. S. Sarma, *Electric Machines* (Dubuque, IA: Brown, 1985).

6. A. E. Fitzgerald, C. Kingsley, and S. Umans, *Electric Machinery*, 4th ed. (New York: McGraw-Hill, 1983).

7. O. I. Elgerd, *Electric Energy Systems: An Introduction* (New York: McGraw-Hill, 1982).

8. J. Reason, "How Electric Utilities Buy Quality When They Buy Transformers," *Electrical World*, *206*, 5 (May 1992), pp. 49–52.

CHAPTER 5

TRANSMISSION LINE PARAMETERS

765-kV transmission line with aluminum guyed-V towers
(Courtesy of American Electric Power)

In this chapter we discuss the four basic transmission-line parameters: series resistance, series inductance. shunt capacitance, and shunt conductance. We also investigate transmission-line electric and magnetic fields.

Series resistance accounts for ohmic (I^2R) line losses. Series impedance, including resistance and inductive reactance, gives rise to series-voltage drops along the line. Shunt capacitance gives rise to line-charging currents. Shunt conductance accounts for V^2G line losses due to leakage currents between conductors or between conductors and ground. Shunt conductance of overhead lines is usually neglected.

Although the ideas developed in this chapter can be applied to underground transmission and distribution, the primary focus here is on overhead lines. Underground transmission in the United States presently accounts for less than 1% of total transmission, and is found mostly in large cities or under waterways. There is, however, a large application for underground cable in distribution systems.

CASE STUDY   The following article covers many present-day factors considered in design and construction of overhead transmission [10]. Cost factors include cost of right-of-way, installation costs including materials and labor, and maintenance or life-cycle costs. Design factors include electrical clearances, lightning performance, grounding, and structural performance based on climatic loading (wind and ice) criteria. Other factors such as visual impact, land-use regulations and zoning laws, and environmental impacts including perceived EMF problems sometimes dictate the selection of right-of-way and transmission structure. Structure options include lattice steel towers, tubular steel towers, wood poles, and concrete poles. Methods to prevent corrosion of steel or decay of wood are discussed. The potential for future upgrade to higher voltages and larger conductor sizes is also considered.

Special Report— Transmission Structures

JOHN REASON

Almost every type of transmission structure ever invented is in use today, and most types are still being used to build new transmission lines. That's because so many factors impact the selection of transmission structures, it's impossible to say one type is better than another. And it's not necessarily advisable to select one type of tower for a given transmission line.

Reprinted with permission from Electrical World.

Parameters can vary so widely from one tower location to the next, that the transmission engineer often finalizes a design using a mix of structures.

Structure design is influenced by perceived problems caused by electromagnetic fields (EMF), by varying opinions about visual impact, by environmental considerations, and by land-use regulations and zoning laws that change as the transmission line passes from one jurisdiction to another. All these factors influence—and sometimes dictate—the transmission structure that's chosen for practically every location.

Manufacturers often claim that the type of structure a utility uses is largely determined by the prejudices of the senior engineers in the transmission department. In fact, established operating practices do strongly influence the types of structures that best meet the kaleidoscope of factors that determine what type of line will finally be built.

But there is still room for engineering, especially if ways can be found to factor the qualitative factors into the more precise calculations that show installation and life-cycle costs. Computer programs help because they can take all of the variables related to cost—such as installation, maintenance, right-of-way (ROW), discount rate, etc—and give the project manager a framework within which to consider more judgmental factors.

A set of computer programs offered by Valmont Industries Inc, Valley, NB, enable the engineer to compare the relative costs of proposed alternatives, as well as the optimum span length for the lowest total installed cost, taking into account such factors as: pole height, ROW width and clearance costs, and equipment and labor costs for the range of materials selected for structure shaft, crossarms, conductors, etc. Programs allow the user to input foundation cost separately for each structure, and find the total installed cost for each span length.

No options can be ruled out in transmission-line design, including the possibility of mixing old and new technologies or upgrading existing lines by replacing some of the structures with alternative types.

Self-supporting lattice, choice for remote lines

The traditional self-supporting, lattice transmission structure, because of its detailed engineering design, makes the most efficient use of material for any structure of comparable size and strength. This means that a lattice structure is light in weight and can be erected without the need for heavy equipment or access roads. However, lattice steel transmission structures are losing favor today because of visual and other limitations. Most people think that any lattice structure is ugly. It needs a wide ROW, and the engineering time needed to design it and the field time needed to erect it are very high.

Lattice structures are often used at highway and river crossings where extra long spans are needed for a line that otherwise consists of short spans between low-cost wood structures. But because of the high cost of designing lattice towers, these long-span structures are often built to standard designs that the utility has on file. As a result, they may be larger and stronger than necessary for the span they support—sacrificing some of the inherent benefits of the lattice design.

Custom-designed lattice towers still win out in long-distance, high-voltage lines crossing remote, rugged territory—particularly in the western US. There, they can be erected even in locations accessible only to four-wheel-drive vehicles. However, Ken Simpson, project manager, Sargent & Lundy, Chicago, IL, cautions that, "Density of development does not always correlate directly with the concern for aesthetics and ROW width. Endangered species, scenic vistas, and other factors in remote areas can sometimes create more concern over the appearance of the line and the amount of clearing required than for lines in some developed areas."

A typical example of the successful use of lattice towers is the 150-mi, 500-kV, single-circuit line now nearing completion between the Captain Jack substation in southern Oregon and the Olinda substation near Redding, CA. Nine structure types and 20 different combinations of conductor sizes, types, and subconductors per phase were considered.

The line crosses diverse topographical and climatic regions with ground elevations ranging from 500 ft near Olinda to over 4500 ft in Oregon, with some sections over 5000 ft above sea level. Much of the line crosses US Forest Service land; it crosses the Sacramento River once and the Pit River three times.

Because the line will be a significant part of the power transfer capacity into California, reliability was an important factor in developing design criteria. Sargent & Lundy engineers had to pay attention in particular to climatic loading criteria, lightning performance and grounding, and electrical clearances. Not only were initial and lifetime costs evaluated, but in some instances, issues such as environmental impact and reliability took precedence over cost.

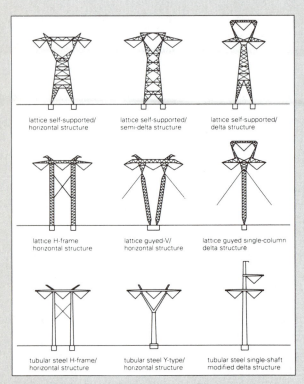

Figure 1 Nine structure types were considered by Sargent & Lundy for a 500-kV California/Oregon line built for the Transmission Agency of Northern California

The nine structure types considered are shown in Fig. 1. In order to evaluate all options, the Line Optimization Program originally developed by Power Technologies Inc, Schenectady, NY, and now part of the Electric Power Research Institute's (EPRI) TLWorkstation was used to find the initial cost and the present worth of losses over the life of the line.

The guyed delta structure was found to have the lowest initial cost and the lowest life-cycle cost on a present-worth basis. This results from the low weight of the structure and the narrow ROW width; ROW width was based on clearance under blowout conditions and on a maximum electric field strength of 0.61 kV/ft at the edge of the ROW. The narrower and higher delta configurations met these criteria with 45 to 55 ft less ROW width than the horizontal configurations.

However, in the final analysis, the self-supporting horizontal configuration lattice tower was chosen for the following reasons:

- The self-supporting structure was within about 5% of the present worth and initial cost of the guyed structure.

- The line crosses rugged terrain in which excessive guy lengths and other guying difficulties were anticipated. This means that self-supporting towers would be needed at some locations, which would increase design and detailing costs.

- Guying can be more disruptive to agricultural and forestry activities than self-supporting structures.

- The guyed delta structure is 20 to 30% taller than the self-supporting structure, which increases visibility.

Lattice easy to upgrade

The extent of field-assembly work weighs against lattice construction for new transmission lines, but can be an enormous benefit when line upgrades are planned. With the benefit of new calculational methods and sometimes new materials, structures that were designed for a specific voltage and conductor size can often be adapted to transmit significantly more power.

Earlier this year, helicopters were used to pluck the tops off 443 double-circuit 230-kV towers and fly them five miles away to an assembly yard, where they were converted from six-arm double-circuit configuration to three-arm, single-circuit configuration to support 500-kV conductor. Modified sections were then flown back and reattached to the tower bases (Figs. 2–4).

The transmission line is owned by the Western Area Power Administration. Helicopter service was supplied by Erickson Air-Crane, Central Point, OR. A special angle-guide system using a chain and two set screws was supplied by Erickson to facilitate the relocation of the tower tops on the bases.

During tower-top removal, 110 structures representing 26 miles of line were removed in one day.

Figure 2 *Top half of 230-kV, double-circuit structure is removed from base (seen at bottom) and flown to yard for modification*

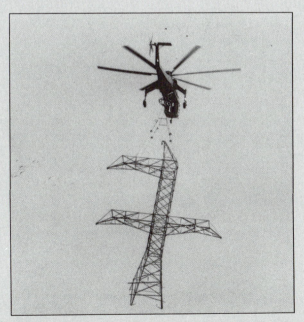

Figure 3 *Tower top, after modification to support 500-kV, single-circuit line, is flown back for reattachment to existing base*

Figure 4 *Guidance angles on existing structure's base ease the reconnection of top and bottom tower halves*

During the tower replacement, 70 structures were flown in one day.

Line upgrade to use porcelain and polymer insulators

Transmission towers with conventional suspension insulators can sometimes be operated at a higher voltage if the arms are removed and replaced with horizontal-V insulator asemblies. This procedure will be used when Baltimore Gas & Electric Co upgrades its Northern Ring line from the present 115 kV to 230 kV. Fig. 5 shows how the horizontal-V insulators (right) maintain greater minimum clearance between conductor and structure. They also raise the conductors by about five feet, providing more ground clearance. Conductors supported by suspension insulators are free to swing and have an unpredictable clearance from the grounded structure.

An interesting aspect of the study, which was performed by Black & Veatch, Kansas City, MO, is that the tension insulator in the vee will be polymer material, while the compression insulator may remain porcelain because of that material's greater compressive strength.

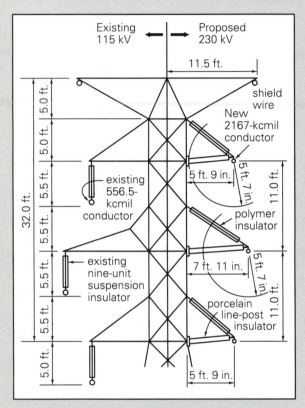

Existing 115 kV ← → Proposed 230 kV

11.5 ft.

shield wire

New 2167-kcmil conductor

existing 556.5-kcmil conductor

5 ft. 9 in.

5 ft. 7 in.

11.0 ft.

polymer insulator

existing nine-unit suspension insulator

7 ft. 11 in.

5 ft. 7 in.

11.0 ft.

porcelain line-post insulator

32.0 ft.

5.0 ft. | 5.0 ft. | 5.5 ft. | 5.5 ft. | 5.5 ft. | 5.5 ft. | 5.0 ft.

5 ft. 9 in.

Figure 5 *Horizontal-V insulators are used to replace metallic arms and upgrade voltage capacity of existing transmission structure*

Multiple environmental benefits push tubular steel

Tubular-steel transmission structures were first introduced in the early 1970s in urban and heavily populated areas where the appearance of lattice structures and the ROW width they demanded were just unacceptable to local residents. At about that time, members of the public and many groups impacted by transmission-line construction were discovering and exercising their power in the public permitting process.

The net result was that construction time was squeezed between the issuing of the permit and the need for the new transmission line. The reduced number of man-hours in the field was thus another impetus for utilities to choose the heavier, more costly tubular structures over more familiar lattice or wood towers.

Today, environmental and societal factors are continuing to promote the use of tubular steel and manufacturers are responding with design and construction developments that continue to push down both cost and weight, so that tubular-steel structures now account for 60 to 80% of new transmission towers. And while tubular-steel structures are among the most expensive in terms of installed cost, low maintenance gives them a competitive life-cycle cost.

Low design costs mean that they can be custom-designed to exactly fit every need, optimizing the use of material and minimizing weight. Bob Schultz, vice president, Meyer Industries Div, American Electric, Red Wing, MN, estimates that a tubular-steel structure can be designed in 10 hours vs about 1000 hours for an equivalent lattice structure.

The two structures in Fig. 6 are both part of the same 345-kV line built by Iowa-Illinois Gas & Electric Co. Where the ROW width is restricted, a single pole is used; where more space is available, less-expensive H-frames are installed.

ROW width needed for a tubular-steel structure is often less than for wood or lattice because a single shaft can replace an H-frame or a four-legged tower. Conductors are arranged vertically on arms or post insulators, closer to the centerline of the tower. Even if the structure is taller than the equivalent wood or lattice tower, it is often considered to be less obtrusive.

Strength can be custom-designed from virtually every tower location. If particularly slender towers are desired, thicker steel is used for fabrication. On the other hand, if a large-diameter shaft is acceptable, thinner steel can be used. Angle towers and deadends can often be engineered with a single shaft.

Foundations, as with most transmission structures, depend largely on soil conditions and significantly affect the cost of the structure (Fig. 7). Under ideal conditions, towers can be direct-embedded, with or without a crushed-stone backfill. With poor soil conditions or where the tower must support an angle

Figure 6 *Both tubular-steel structures above are used for the same 345-kV transmission line. Where ROW is restricted, single pole (above) is the choice*

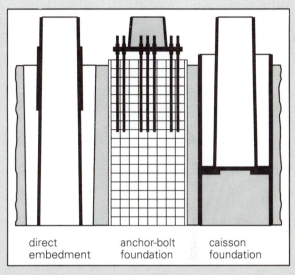

| direct embedment | anchor-bolt foundation | caisson foundation |

Figure 7 *Foundation type depends largely on soil conditions and significantly affects the cost of the tubular-steel structure*

or a deadend, an anchor-bolt foundation may be needed. This is a preclustered cage of anchor bolts encased in a reinforced concrete foundation. A flange at the base of the tower is bolted to the anchor bolts.

Another method of handling poor soil conditions is the caisson foundation. This is a steel tube that is driven into the ground to a depth determined by soil conditions. The transmission pole is then seated in the caisson, aligned, and the space filled with grout.

Three ways to prevent corrosion

Tubular-steel towers are protected from corrosion by one of three methods: weathering steel, galvanizing, or painting. Important factors in the design also help to ensure that the selected method of protection is not compromised.

Weathering steel, of which there are several grades and strengths, is an alloy of carbon steel with copper, chromium, nickel, and other elements. It is often known as Cor-Ten steel, a product of US Steel, Pittsburgh, PA. It differs from pure carbon steel because when it rusts, the oxide forms a protective layer over the surface that prevents further corrosion. Thus, under ideal conditions, a tower made of weathering steel needs no protective coating.

Weathering steel is the least expensive method of corrosion prevention. The rusty brown coloring is unobtrusive and, to the casual eye, is indistinguishable from wood. However, weathering steel is not maintenance-free and optimum performance depends largely on the design and assembly of the structure. This is because weathering steel only develops its protective oxide film if it is able to dry out completely after soaking with rain water. Thus if there are any joints or crevices in which water can accumulate or if parts of the tower are surrounded by vegetation that holds moisture on the surface, rusting will continue indefinitely and the benefits of the weathering steel will be lost.

Several utilities learned this fact the hard way a few years ago when they purchased lattice towers constructed of weathering steel (EW, October 1991, p 42). Moisture that collected in the bolted joints did not dry out between rainfalls and the buildup of rust packout continued until it threatened the structural integrity of the towers. Packout is now well understood by transmission-tower designers and should not occur on a properly designed tower. However, structures of bare weathering steel should be inspected, especially during the initial weathering period, to ensure that all joints and surfaces are performing satisfactorily.

Galvanizing involves dipping the entire tower in molten zinc, which forms an amalgam with the steel surface. It provides excellent corrosion resistance, though galvanized steel is more noticeable than weathering steel. Even if the zinc surface is scratched, galvanic action causes the zinc to migrate onto the exposed steel surface and heal the scratch.

Galvanizing adds between 5 and 15% to the cost of the steel tower (over the cost of weathering-steel). Actual amount depends largely on whether the tower manufacturer has a galvanizing process on site. Benefit of a galvanized surface is that it does not have to dry out between rainfalls to prevent corrosion. For this reason it is very suitable for transmission structures that are subject to salt spray or other contamination that might hold water on the surface.

Painting is used only on a very small proportion of tubular-steel transmission structures—usually when a specific color is requested by a developer. Because paint provides limited protection from corrosion, and repainting in the field is expensive, paint is usually applied over a galvanized surface or over weathering steel. Painting can be expected to add 6–7% to the cost of the transmission tower.

Foundations require special corrosion protection, especially on direct-imbedded poles. The area of most aggressive corrosion is at the ground line and virtually all tubular-steel towers are provided with a double thickness of steel, extending a foot or more above and below the ground line. In addition, the entire buried section of the pole is coated with an epoxy from the base to one foot above the ground line.

Making tubular-steel structures

The manufacture of tubular-steel structures is a highly computerized process that begins with flame cutting of the sheet steel (Fig. 8) followed by bending into a multi-faceted tube. Generally the tubes are made in two halves and then welded together with automatic, submerged-arc welding machines. Tube length is limited by the length of presses available and the permissible length for transportation—maximum is about 55 feet. However, as single tubular-steel struc-

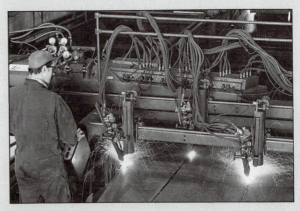

Figure 8 Steel plate is precut to needed shape by CAD/CAM-controlled oxy-acetylene or plasma-arc cutting machine

tures are almost invariably tapered, almost any length can be made up in the field by slip-jointing sections together without fasteners. If slip-jointing of complex structures is not possible, sections are pre-assembled with flanges for bolting together in the field.

All fittings, including brackets for the attachment of arms, base plate, climbing clips, etc, are welded to the tube in the shop according to a computer-generated drawing (Fig. 9). Normally, no cutting or welding in the field should be necessary.

A unique example of how the custom-built features of tubular steel can meet special requirements is shown in Fig. 10. This shows the top of a tangent 150-ft high structure for a 10-mi, 500-kV line built by Potomac Electric Power Co. Because of fierce community opposition and tough local zoning laws, this project has been tied up in litigation since 1976, when the line was first designed with simple H-frames on a 200-ft ROW.

The first of 42 structures, built by Meyer Industries, was finally erected last year. It is supported on a single tubular-steel tower capable of withstanding longitudinal loads of up to 42,000 lb caused by differential ice loading, plus a wind load of 31 lb/ft^2. The pentagonal design, considered to be more graceful than traditional structures, reduces EMF, radio interference, and transmission losses by holding the conductors in a delta configuration.

Figure 10 *Special designs, like this 150-ft delta structure, can be fabricated from tubular steel to accommodate visual and environmental demands*

Figure 9 *Seam weld is inspected and repaired by manual welding if necessary. All fittings are also manually welded in the shop before shipping*

Lightweight steel pushes wood

Transmission engineers who like the flexibility of wood and the ability to make final decisions about structure design in the field, now have the option of choosing lightweight steel poles. These poles are manufactured in traditional wood-pole sizes and can be drilled and assembled in the field, almost exactly like wood poles (EW, September 1991, p 116).

Pennsylvania Power & Light Co started using lightweight steel poles to replace wood for new 138-kV construction in the mid 1980s, when wood poles in classes 1 and H3 became difficult to obtain. The utility purchases steel poles up to 100 ft long in single pieces and stocks in 5-ft increments. Poles longer than 100 ft are assembled by flange-jointing a 70-ft top section onto bottom pieces of various sizes.

Handling is actually easier than wood, because lightweight steel poles weigh two-thirds those of wood. They are also slightly more flexible than wood, which may necessitate changes in stringing procedures.

Wood still popular for both cost and aesthetics

Few countries in the world make such extensive use of wood for construction as the US. The reason is that vast quantities of wood are available here. In the case of transmission structures, this bias is amplified by the fact that longer lengths of timber are available here than in most other countries.

However, as might be expected, suitable timber is becoming more difficult to obtain. But while environmental considerations are restricting the supply, property owners who must live with the transmission line often think that wooden structures are more acceptable than other materials.

Installed cost of wood structures is indisputably lower than for other material. Not only is the material cost low and the time to erect short, but most utilities have a wealth of experience with wood. The big question mark is maintenance—which varies widely from region to region—and hence life-cycle cost. Recently, a new unknown was introduced into the life-cycle-cost equation when it was suggested that some treated-wood poles may one day have to be disposed of as hazardous waste.

Some young engineers, who have limited experience with wood, are beginning to factor in all of the variables to show that a transmission line engineered with more costly material may have an overall lower life-cycle cost.

Will disposal become a problem?

Utility users of wood poles are anxiously watching the regulatory status of wood poles treated with pentachlorophenol (PCP), under the Resource Conservation and Recovery Act (RCRA), and with good reason. Classification of treated wood poles as hazardous

waste would increase disposal costs for the electric-utility industry by billions of dollars annually. But keep the facts in mind: (1) The use of wood poles is so extensive that any reduction in the use of wood (or even of PCP) would be of extraordinary benefit to the suppliers of competing products. Thus there are many organizations that benefit by increasing anxiety about the problem. (2) A study conducted by EPRI in accordance with the Environmental Protection Agency's Toxicity Characteristic Leaching Procedure concluded that PCP-treated poles have leachates well below the levels presently set by RCRA. Thus even if the regulatory levels are significantly lowered, it is not likely that used wood poles would become hazardous material.

Wood replaces wood in line upgrade

A 4.3-mi, 115-kV transmission line in Western Massachusetts Electric Co's (WMECO) service territory recently required uprating to accept the output from a cogeneration facility. The upgrade was accomplished by replacing the existing 336.4-kcmil ACSR conductors with 1272-kcmil ACSR conductors at the same voltage. Environmental disruption of the sensitive wetlands through which the line passes (Fig. 11) was minimized by replacing the existing single-circuit

Figure 11 *Wood replaces wood to upgrade this 115-kV transmission line through environmentally sensitive area in New England*

Figure 12 *Higher wood structures, supporting heavier conductors, are built alongside existing line, which is later removed*

wood H-frames with larger wood structures of similar design. Lattice towers on either side of a pond crossing were reused and the entire upgrade was maintained within the existing 200-ft ROW.

System studies were performed by Northeast Utilities Service Co, an affiliate of WMECO, and the upgrade was engineered by Chas T. Main Inc, Boston, MA. This work included arranging a complex series of outages intended to minimize the possibility of service disruption to local customers.

For most of the line's length, the new wood structures are similar to the structures on the old line, but are 15 ft taller to accommodate the heavier conductor. These structures are located 55 ft to the west of the old transmission line (Fig. 12), which was later removed.

Lattice steel towers at the Housatonic River/Woods Pond crossing were reused with a unique engineering approach. At the time of the upgrade, only three 336.4-kcmil ACSR conductors were installed on the double-circuit towers. To provide the additional capacity, a second circuit of the same weight conductors was installed on the unused side of the tower and bundled with the existing conductors.

Special wood transition structures were designed to accept six 336.4-kcmil conductors from the lattice towers and connect them to the new 1272-kcmil conductors on the upgraded line (Fig. 13).

Figure 13 *Transition structure converts from twin-bundled conductors (left) to heavier single conductors (right)*

Wood offers lowest life-cycle cost

Treated wood was chosen by engineers at Southwestern Public Service Co (SPS) for a 158-mi, 345-kV transmission line across rough, windswept ranchland and rolling farmland between Muleshoe, TX, and Artesia, NM (Fig. 14). The H-frame structure was designed to carry NESC heavy loading on 900-ft spans with 795-kcmil vertical bundled conductors. The wood poles are creosote- and Penta-treated Douglas fir, mostly Class 1, 90 ft long, with some longer poles (class H2). Laminated wood crossarms are used and about 90 structures are laminated wood.

David Stidham, principal engineer, transmission design, cites reasons for choosing wood H-frames over other structures:

- Life-cycle cost studies show that wood offers a 12% cost advantage over the nearest competing structure type.

- Long-term maintenance can be carried out by line

Figure 14 *Wood structure was chosen to carry 345-kV line across Texas ranchland after extensive evaluation of lifetime costs*

Figure 15 *Wood H-frames are easy to assemble and erect. Wood is favored by local ranchers and by the Bureau of Land Management*

workers climbing the poles. This reduces the need for heavy equipment, which on some parts of the line could damage archaeological sites or crop land.

- Farmers and ranchers in the area prefer the look of wood. Their preference is supported by local Bureau of Land Management officials.

- Wood structures are easy to erect (Fig. 15). SPS crews finished the line more than two months ahead of schedule and $2.5-million under the estimated cost.

Life-cycle cost analysis is based on concepts developed by Engineering Data Management Inc, Fort Collins, Colo, for the Western Wood Preservers Institute and further expanded by SPS. A 50-yr life is assumed, with initial groundline inspection and treatment scheduled after 20 years. Once the initial inspection is completed and any deteriorating poles replaced, an ongoing 10-yr cycle of inspection and replacement is planned. Replacement costs are calculated on an initial 3% replacement rate after the first 20 years, based on SPS data for treated wood poles in this service area.

SPS's Stidham is not worried about pole disposal. Replaced poles will be recycled either by the utility or

local landowners. SPS treatment specifications allow for both Penta and creosote. Penta is preferred by BLM personnel for aesthetic reasons.

To avoid the possibility of pole-supply problems, SPS notified potential suppliers of plans to construct the line in advance. All poles were delivered on time.

Glued lamination enhances wood's familiar benefits

Many utilities are taking a new look at glued laminated wood as old-growth lumber for transmission structures becomes more difficult to obtain. Clearly, laminated structural members can be produced to any required size from smaller, more-readily available pieces of lumber. Laminated wood can be cut and drilled in the field just like solid wood, but there are other benefits:

- Each lamination is nondestructively tested before layup to measure its modulus of elasticity. Only laminations that meet a predetermined quality level are used. Result is that the laminated pole is an engineered product with an established standard (ANSI 05.2) to define its strength and consistency.

- Appearance, especially for urban applications, is generally considered better than roughly shaped round poles. Various natural-wood color finishes are available.

- Purchase cost is lower than that of tubular steel, concrete, or solid cut lumber of the same species.

An important factor in the manufacturing process is that the laminations must be kiln dried before glueing. This sterilizes them and prevents warpage. Laminations are not treated prior to assembly of the poles. Instead, to ensure good penetration of preservative, the completed poles are incised from top to bottom and drilled 2 ft above and 3 ft below the ground line. Samples from the production line are taken periodically and tested to destruction to ensure the bond is stronger than the wood.

Guys avoided on angle poles

Recently, Central Iowa Power Cooperative (Cipco), needed to build a 69-kV tap through some scenic country using a county-road ROW. Most of the tangent poles in the two-mile line are regular 60–70-ft wood poles, but nine of the structures had to support angles from 3 to 18 deg. Guys were out of the question if the county ROW was to be used, so for these locations, an alternative, unguyed structure had to be chosen.

Marlon Vogt, supervisor of transmission engineering and construction, explains: "We have some experience with both concrete and tubular steel, but we wanted to check out laminated poles as another alternative. Apart from the problem of avoiding guys at the angle structures, we are concerned about the availability of large wood poles. Laminated poles are readily available. We priced laminated poles against concrete, Corten tubular steel, and galvanized steel. Laminated wood saved almost $800/pole."

The laminated poles were manufactured from Douglas fir by Bohemia Powerlam, Saginaw, OR, and shipped to Cipco, Cedar Rapids, IA, on a single truck. Laminated poles are generally supplied in standard sizes, but in this case, Cipco specified pole heights, angles, and stresses to which the poles were designed (Fig. 16).

Figure 16 *Laminated wood pole is specified simply as 70 ft high to support a 10-deg angle, without guys. Manufacturer assembles to suit*

Before choosing laminated poles, Cipco looked very carefully at field surveys of over 3000 laminated poles that have been in service for as long as 29 years. The conclusion was that any problems experienced with laminated poles in the past had been caused by one particular laminating process in which treatment had been attempted before glueing. Otherwise, all utilities contacted reported acceptable experience.

Concrete provides maximum resistance to environment

Concrete, as a material for transmission structures, becomes economical at pole sizes needed to accommodate lines between 69 and 230 kV (or over about 85 ft), for which wood is becoming difficult to obtain. Other significant benefits, especially in the semi-tropical southeastern US, is that concrete is immune to ground rot, of no interest to woodpeckers, and does not twist or warp. A concrete pole can be expected to last 50 years or more—two or three times the life of wood poles in Alabama.

The only problem is weight, which raises transportation cost. In general, the construction site must be within 400 miles of the concrete pole manufacturing plant—two separate manufacturing plants if management insists on a second source. Weight may also raise the cost of erection if the lifting equipment owned by the utility is not adequate and heavy cranes must be rented (EW, February 1991, p 32).

To use concrete transmission poles cost-effectively, they should be made to order for the specific job in hand and delivered by the manufacturer as close as possible to the erection site, as they are needed. This is quite unlike wood poles, which can be held in stock, transported to the site when needed, and drilled and assembled in-situ. Concrete poles can be drilled at the site in an emergency, but the preferred method is to cast holes at the time the poles are manufactured.

Poles are cast in a centrifuge, which is made up of 10-ft sections according to the pole length needed (Fig. 17). Reinforcing steel rods are first arranged in the centrifuge and prestressed. Plastic tubes for the hole locations are inserted (Fig. 18), along with inserts for attachment of climbing bolts etc. Concrete is then poured in and is spun until it sets. This produces a

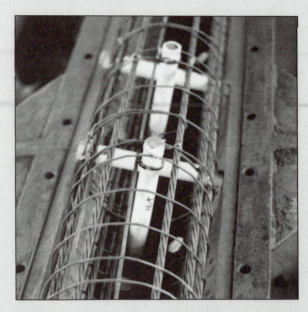

Figure 18 *Bolt holes are precast in concrete pole by placing plastic tubes in mold before pouring concrete*

hollow pole with a wall thickness of about 3 in., high strength, and rigidity.

Concrete-pole line set through swampland

Recently, Alabama Power Co needed a new 230-kV transmission line, for load growth in the Mobile (AL) area from its new Kushla switching station, five miles into North Mobile transmission substation. Bids were received from companies offering lattice towers, steel poles, and concrete poles. Concrete H-frame structures, ranging in height from 65 to 120 ft, were chosen (Fig. 19). The two-legged tangent structures (Fig. 20) and three-legged angle structures were supplied by Baldwin Utility Structures Inc, Bay Minette, AL.

According to Don Barker, supervisor of transmission lines, one of several reasons for choosing concrete was that much of the line passes through very swampy land, with a high potential for ground rot and corrosion.

In the swamp locations, the ground is so soft that mats had to be laid before the heavy lifting equipment could move the poles into place. Generally holes were

Figure 17 *Centrifuge for casting concrete poles is made up of 10-ft sections. Concrete sets while mold is rotating, producing a 3-in. wall thickness*

Figure 19 Concrete structures are chosen for 230-kV transmission line through swampland. Proximity of manufacturer helped keep cost down

Figure 20 Concrete-pole manufacturer delivers concrete poles as close as possible to the erection site as contractor needs them

augered with regular equipment, though in some cases, 48-in.-dia corrugated steel pipe had to be sunk to prevent the holes from caving in. Poles were set using standard setting depths, but in all swamp locations, the concrete structures were guyed for extra stability.

The poles were erected by a contractor who worked directly with Baldwin Utility Structures; structures were delivered as close as possible to each hole location as the work proceeded. In this case, distance from manufacturing plant to the work site was only about 30 miles, so transport did not add significantly to cost.

Guys cut structural cost, if you can use them

A guyed transmission tower or pole is generally cheaper to construct and erect than the equivalent unguyed structure. But if property owners object to a transmission structure, they'll object almost equally to each of the guy lines and anchors. On an easement along a street or rail line, it's often impossible to find room to bury the anchors. And even if space can be found to guy an angle tower or a dead-end, the guy line is one more item that needs periodic inspection. So for all these reasons, most electric utilities try to avoid the use of guys.

But in wide-open spaces, where the land is unused or is environmentally fragile, guyed structures are a natural choice. Not only are the structures much cheaper, but they can be erected in far less time than a self-supporting lattice, without the need for heavy equipment. Several western utilities now stock standard guyed towers for emergency replacement of damaged transmission towers (EW, May 1988, p 40).

When Tucson Electric Power Co needed to build 120 miles of 345-kV transmission line from Clifton, AZ, to the Vail Substation near Tucson, (Fig. 21) engineers hardly considered anything but single- and double-guyed lattice towers. In fact, the new line parallels 400 miles of identical structures that had given the utility trouble-free service for 20 years. Lattice towers were supplied by Lehigh Structural Steel Co, Allentown, Pa, and the new line went into service in June 1991.

Figure 21 Guyed structures have been used successfully by Tucson Electric Co for 20 years. Recently, second, identical line was added

Figure 22 Helicopter lifts assembled structure, complete with guy wires, into position

According to Senior Civil Engineer Ed Beck, the guyed structures save so much weight that helicopter installation is economical, even for locations that are accessible from the ground.

For this line, about 80% of the towers were lifted to the site in one piece by helicopter, complete with guy lines attached (Fig. 22). Anchors were set 10 to 30 ft into the ground, according to the load, by tracked or tired drilling rigs, and grouted in place. Adjustable dead-ends, supplied by A. B. Chance Co, Centralia, Mo, enabled guys to be set and maintained at the correct tension.

Self-supporting lattice towers are used at angles and also for a four-mile section where the line crosses farmland the utility was concerned that landowners would have problems with guys.

Beck points out that guyed towers are remarkably forgiving. "Once we lost a tower because a light plane flew into it. The towers on both sides stayed up and we had the line back in service in less than two days," he says.

SECTION 5.1

TRANSMISSION-LINE DESIGN CONSIDERATIONS

An overhead transmission line consists of conductors, insulators, support structures, and, in most cases, shield wires.

Conductors

Aluminum has replaced copper as the most common conductor metal for overhead transmission. Although a larger aluminum cross-sectional area is required to obtain the same loss as in a copper conductor, aluminum has a lower cost and lighter weight. Also, the supply of aluminum is abundant, while that of copper is limited.

One of the most common conductor types is aluminum conductor, steel-reinforced (ACSR), which consists of layers of aluminum strands surrounding a central core of steel strands (Figure 5.1). Stranded conductors are easier to

Figure 5.1

Typical ACSR conductor

54/7 Cardinal

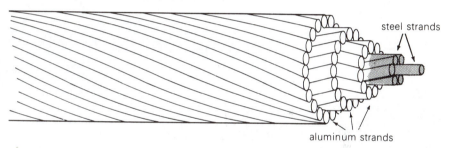

manufacture, since larger conductor sizes can be obtained by simply adding successive layers of strands. Stranded conductors are also easier to handle and more flexible than solid conductors, especially in larger sizes. The use of steel strands gives ACSR conductors a high strength-to-weight ratio. For purposes of heat dissipation, overhead transmission-line conductors are bare (no insulating cover).

Other conductor types include the all-aluminum conductor (AAC), all-aluminum-alloy conductor (AAAC), aluminum conductor alloy-reinforced (ACAR), and aluminum-clad steel conductor (Alumoweld). There is also a conductor known as "expanded ACSR," which has a filler such as fiber or paper between the aluminum and steel strands. The filler increases the conductor diameter, which reduces the electric field at the conductor surface, to control corona.

EHV lines often have more than one conductor per phase; these conductors are called a *bundle*. The 765-kV line in Figure 5.2 has four conductors per phase, and the 345-kV double-circuit line in Figure 5.3 has two conductors per phase. Bundle conductors have a lower electric strength at the conductor surfaces, thereby controlling corona. They also have a smaller series reactance.

Figure 5.2

A 765-kV transmission line with self-supporting lattice steel towers

(Courtesy of the American Electric Power Company)

Figure 5.3

A 345-kV double-circuit trans-mission line with self-support-ing lattice steel towers

(Courtesy of Boston Edison Company

Figure 5.4

Cut-away view of a standard insulator disc for suspension insulator strings

(Courtesy of Ohio Brass)

Insulators

Insulators for transmission lines above 69 kV are suspension-type insulators, which consist of a string of discs, typically porcelain. The standard disc (Figure 5.4) has a 10-in. (0.254-m) diameter, $5\frac{3}{4}$-in. (0.146-m) spacing between centers of adjacent discs, and a mechanical strength of 7500 kg. The 765-kV line in Figure 5.2 has two strings per phase in a V-shaped arrangement, which helps to restrain conductor swings. The 345-kV line in Figure 5.5 has one vertical string per phase. The number of insulator discs in a string increases with line voltage (Table 5.1). Other types of discs include larger units with higher mechanical strength and fog insulators for use in contaminated areas.

Support Structures

Transmission lines employ a variety of support structures. Figure 5.2 shows a self-supporting, lattice steel tower typically used for 500- and 765-kV lines. Double-circuit 345-kV lines usually have self-supporting steel towers with the phases arranged either in a triangular configuration to reduce tower height or in a vertical configuration to reduce tower width (Figure 5.3). Wood frame configurations are commonly used for voltages of 345 kV and below (Figure 5.5).

Shield Wires

Figure 5.5

Wood frame structure for a 345-kV line

(Courtesy of Boston Edison Company)

Shield wires located above the phase conductors protect the phase conductors against lightning. They are usually high- or extra-high-strength steel, Alumoweld, or ACSR with much smaller cross-section than the phase conductors. The number and location of the shield wires are selected so that almost all lightning strokes terminate on the shield wires rather than on the phase conductors. Figures 5.2, 5.3, and 5.5 have two shield wires. Shield wires are grounded to the tower. As such, when lightning strikes a shield wire, it flows harmlessly to ground, provided the tower impedance and tower footing resistance are small.

The decision to build new transmission is based on power-system

NOMINAL VOLTAGE	PHASE CONDUCTORS				
	NUMBER OF CONDUCTORS PER BUNDLE	ALUMINUM CROSS-SECTION AREA PER CONDUCTOR (ACSR)	BUNDLE SPACING	MINIMUM CLEARANCES	
				PHASE-TO-PHASE	PHASE-TO-GROUND
kV		kcmil	cm	m	m
69	1	–	–	–	–
138	1	300–700	–	4 to 5	–
230	1	400–1000	–	6 to 9	–
345	1	2000–2500	–	6 to 9	7.6 to 11
345	2	800–2200	45.7	6 to 9	7.6 to 11
500	2	2000–2500	45.7	9 to 11	9 to 14
500	3	900–1500	45.7	9 to 11	9 to 14
765	4	900–1300	45.7	13.7	12.2

NOMINAL VOLTAGE	SUSPENSION INSULATOR STRING		SHIELD WIRES		
	NUMBER OF STRINGS PER PHASE	NUMBER OF STANDARD INSULATOR DISCS PER SUSPENSION STRING	TYPE	NUMBER	DIAMETER
kV					cm
69	1	4 to 6	Steel	0, 1 or 2	–
138	1	8 to 11	Steel	0, 1 or 2	–
230	1	12 to 21	Steel or ACSR	1 or 2	1.1 to 1.5
345	1	18 to 21	Alumoweld	2	0.87 to 1.5
345	1 and 2	18 to 21	Alumoweld	2	0.87 to 1.5
500	2 and 4	24 to 27	Alumoweld	2	0.98 to 1.5
500	2 and 4	24 to 27	Alumoweld	2	0.98 to 1.5
765	2 and 4	30 to 35	Alumoweld	2	0.98

Table 5.1 Typical transmission-line characteristics [1, 2]

planning studies to meet future system requirements of load growth and new generation. The points of interconnection of each new line to the system, as well as the power and voltage ratings of each, are selected based on these studies. Thereafter, transmission-line design is based on optimization of electrical, mechanical, environmental, and economic factors.

Electrical Factors

Electrical design dictates the type, size, and number of bundle conductors per phase. Phase conductors are selected to have sufficient thermal capacity to meet continuous emergency overload and short-circuit current ratings. For EHV lines, the number of bundle conductors per phase is selected to control the voltage gradient at conductor surfaces, thereby reducing or eliminating corona.

Electrical design also dictates the number of insulator discs, vertical or V-shaped string arrangement, phase-to-phase clearance, and phase-to-tower clearance, all selected to provide adequate line insulation. Line insulation must withstand transient overvoltages due to lightning and switching surges, even when insulators are contaminated by fog, salt, or industrial pollution. Reduced clearances due to conductor swings during winds must also be accounted for.

The number, type, and location of shield wires are selected to intercept lightning strokes that would otherwise hit the phase conductors. Also, tower footing resistance can be reduced by using driven ground rods or a buried conductor (called *counterpoise*) running parallel to the line. Line height is selected to satisfy prescribed conductor-to-ground clearances and to control ground-level electric field and its potential shock hazard.

Conductor spacings, types, and sizes also determine the series impedance and shunt admittance. Series impedance affects line-voltage drops, I^2R losses, and stability limits (Chapters 6, 13). Shunt admittance, primarily capacitive, affects line-charging currents, which inject reactive power into the power system. Shunt reactors (inductors) are often installed on lightly loaded EHV lines to absorb part of this reactive power, thereby reducing overvoltages.

Mechanical Factors

Mechanical design focuses on the strength of the conductors, insulator strings, and support structures. Conductors must be strong enough to support a specified thickness of ice and a specified wind in addition to their own weight. Suspension insulator strings must be strong enough to support the phase conductors with ice and wind loadings from tower to tower (span length). Towers that satisfy minimum strength requirements, called suspension towers, are designed to support the phase conductors and shield wires with ice and wind loadings, and, in some cases, the unbalanced pull due to breakage of one or two conductors. Dead-end towers located every mile or so satisfy the maximum strength requirement of breakage of all conductors on one side of the tower. Angles in the line employ angle towers with intermediate strength. Conductor vibrations, which can cause conductor fatigue failure and damage to towers, are also of concern. Vibrations are controlled by adjustment of conductor tensions, use of vibration dampers, and—for bundle conductors—large bundle spacing and frequent use of bundle spacers.

Environmental Factors

Environmental factors include land usage and visual impact. When a line route is selected, the effect on local communities and population centers, land values, access to property, wildlife, and use of public parks and facilities must all be considered. Reduction in visual impact is obtained by aesthetic tower design and by blending the line with the countryside. Also, the biological effects of prolonged exposure to electric fields near transmission lines is of concern. Extensive research has been and continues to be done in this area.

Economic Factors

The optimum line design meets all the technical design criteria at lowest overall cost, which includes the total installed cost of the line as well as the cost of line losses over the operating life of the line. Many design factors affect cost. Utilities and consulting organizations use digital computer programs combined with specialized knowledge and physical experience to achieve optimum line design.

SECTION 5.2

RESISTANCE

The dc resistance of a conductor at a specified temperature T is

$$R_{dc,T} = \frac{\rho_T l}{A} \quad \Omega \tag{5.2.1}$$

where ρ_T = conductor resistivity at temperature T

l = conductor length

A = conductor cross-sectional area

Two sets of units commonly used for calculating resistance, SI and English units, are summarized in Table 5.2. In English units, conductor cross-sectional area is expressed in circular mils (cmil). One inch equals 1000 mils and 1 cmil equals $\pi/4$ sq mil. A circle with diameter D in., or (D in.) (1000 mil/in.) = 1000 D mil = d mil, has an area

$$A = \left(\frac{\pi}{4} D^2 \text{ in.}^2\right)\left(1000 \frac{\text{mil}}{\text{in.}}\right)^2 = \frac{\pi}{4}(1000 D)^2 = \frac{\pi}{4} d^2 \quad \text{sq mil}$$

or

$$A = \left(\frac{\pi}{4} d^2 \text{ sq mil}\right)\left(\frac{1 \text{ cmil}}{\pi/4 \text{ sq mil}}\right) = d^2 \quad \text{cmil} \tag{5.2.2}$$

Table 5.2

Comparison of SI and English units for calculating conductor resistance

QUANTITY	SYMBOL	SI UNITS	ENGLISH UNITS
Resistivity	ρ	Ωm	Ω-cmil/ft
Length	ℓ	m	ft
Cross-sectional area	A	m^2	cmil
dc resistance	$R_{dc} = \dfrac{\rho\ell}{A}$	Ω	Ω

Resistivity depends on the conductor metal. Annealed copper is the international standard for measuring resistivity ρ (or conductivity σ, where $\sigma = 1/\rho$). Resistivity of conductor metals is listed in Table 5.3. As shown, hard-drawn aluminum, which has 61% of the conductivity of the international standard, has a resistivity at 20°C of 17.00 Ω-cmil/ft or $2.83 \times 10^{-8}\,\Omega$m.

Table 5.3

% Conductivity, resistivity, and temperature constant of conductor metals

MATERIAL	% CONDUCTIVITY	$\rho_{20°C}$ RESISTIVITY AT 20°C		T TEMPERATURE CONSTANT
		Ωm × 10^{-8}	Ω-cmil/ft	°C
Copper:				
Annealed	100%	1.72	10.37	234.5
Hard-drawn	97.3%	1.77	10.66	241.5
Aluminum				
Hard-drawn	61%	2.83	17.00	228.1
Brass	20–27%	6.4–8.4	38–51	480
Iron	17.2%	10	60	180
Silver	108%	1.59	9.6	243
Sodium	40%	4.3	26	207
Steel	2 to 14%	12–88	72–530	180–980

Conductor resistance depends on the following factors:

1. Spiraling

2. Temperature

3. Frequency ("skin effect")

4. Current magnitude—magnetic conductors

These are described in the following paragraphs.

For stranded conductors, alternate layers of strands are spiraled in opposite directions to hold the strands together. Spiraling makes the strands 1 or 2% longer than the actual conductor length. As a result, the dc resistance of a stranded conductor is 1 or 2% larger than that calculated from (5.2.1) for a specified conductor length.

Resistivity of conductor metals varies linearly over normal operating temperatures according to

$$\rho_{T2} = \rho_{T1}\left(\frac{T_2 + T}{T_1 + T}\right) \tag{5.2.3}$$

where ρ_{T2} and ρ_{T1} are resistivities at temperatures T_2 and T_1°C, respectively. T is a temperature constant that depends on the conductor material, and is listed in Table 5.3.

The ac resistance or *effective* resistance of a conductor is

$$R_{ac} = \frac{P_{loss}}{|I|^2}\quad \Omega \tag{5.2.4}$$

where P_{loss} is the conductor real power loss in watts and I is the rms conductor current. For dc, the current distribution is uniform throughout the conductor cross section, and (5.2.1) is valid. However, for ac, the current distribution is nonuniform. As frequency increases, the current in a solid cylindrical conductor tends to crowd toward the conductor surface, with smaller current density at the conductor center. This phenomenon is called *skin effect*. A conductor with a large radius can even have an oscillatory current density versus the radial distance from the conductor center.

With increasing frequency, conductor loss increases, which, from (5.2.4), causes the ac resistance to increase. At power frequencies (60 Hz) the ac resistance is at most a few percent higher than the dc resistance. Conductor manufacturers normally provide dc, 50-Hz, and 60-Hz conductor resistance based on test data (see Appendix Tables A.3 and A.4).

For magnetic conductors, such as steel conductors used for shield wires, resistance depends on current magnitude. The internal flux linkages, and therefore the iron or magnetic losses, depend on the current magnitude. For ACSR conductors, the steel core has a relatively high resistivity compared to the aluminum strands, and therefore the effect of current magnitude on ACSR conductor resistance is small. Tables on magnetic conductors list resistance at two current levels (see Table A.4).

EXAMPLE 5.1   **Stranded conductor: dc and ac resistance**

Table A.3 lists a 4/0 copper conductor with 12 strands. Strand diameter is 0.1328 in. For this conductor:

   a. Verify the total copper cross-sectional area of 211,600 cmil.

   b. Verify the dc resistance at 50°C of 0.302 Ω/mi. Assume a 2% increase in resistance due to spiraling.

   c. From Table A.3, determine the percent increase in resistance at 60 Hz versus dc.

Solution   **a.** The strand diameter is $d = (0.1328 \text{ in.}) (1000 \text{ mil/in.}) = 132.8 \text{ mil}$, and, from (5.2.2), the strand area is d^2 cmil. Using four significant figures, the cross-sectional area of the 12-strand conductor is

$$A = 12d^2 = 12(132.8)^2 = 211,600 \quad \text{cmil}$$

which agrees with the value given in Table A.3.

b. Using (5.2.3) and hard-drawn copper data from Table 5.3,

$$\rho_{50°C} = 10.66 \left(\frac{50 + 241.5}{20 + 241.5} \right) = 11.88 \quad \Omega\text{-cmil/ft}$$

From (5.2.1), the dc resistance at 50°C for a conductor length of 1 mile (5280 ft) is

$$R_{dc,50°C} = \frac{(11.88)(5280 \times 1.02)}{211,600} = 0.302 \quad \Omega/\text{mi}$$

which agrees with the value listed in Table A.3.

c. From Table A.3,

$$\frac{R_{60Hz,50°C}}{R_{dc,50°C}} = \frac{0.303}{0.302} = 1.003 \qquad \frac{R_{60Hz,25°C}}{R_{dc,25°C}} = \frac{0.278}{0.276} = 1.007$$

Thus, the 60-Hz resistance of this conductor is about 0.3–0.7% higher than the dc resistance. The variation of these two ratios is due to the fact that resistance in Table A.3 is given to only three significant figures. ■

SECTION 5.3

CONDUCTANCE

Conductance accounts for real power loss between conductors or between conductors and ground. For overhead lines, this power loss is due to leakage currents at insulators and to corona. Insulator leakage current depends on the amount of dirt, salt, and other contaminants that have accumulated on insulators, as well as on meteorological factors, particularly the presence of moisture. Corona occurs when a high value of electric field strength at a conductor surface causes the air to become electrically ionized and to conduct. The real power loss due to corona, called *corona loss*, depends on meteorological conditions, particularly rain, and on conductor surface irregularities. Losses due to insulator leakage and corona are usually small compared to conductor I^2R loss. Conductance is usually neglected in power-system studies because it is a very small component of the shunt admittance.

SECTION 5.4

INDUCTANCE: SOLID CYLINDRICAL CONDUCTOR

The inductance of a magnetic circuit that has a constant permeability μ can be obtained by determining the following:

1. Magnetic field intensity H, from Ampere's law
2. Magnetic flux density B ($B = \mu H$)
3. Flux linkages λ
4. Inductance from flux linkages per ampere (L = λ/I)

As a step toward computing the inductances of more general conductors and conductor configurations, we first compute the internal, external, and total inductance of a solid cylindrical conductor. We also compute the flux linking one conductor in an array of current-carrying conductors.

Figure 5.6 shows a 1-meter section of a solid cylindrical conductor with

Figure 5.6

Internal magnetic field of a solid cylindrical conductor

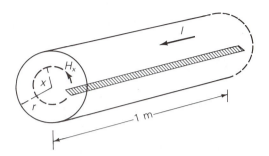

radius r, carrying current I. For simplicity, assume that the conductor (1) is sufficiently long that end effects are neglected, (2) is nonmagnetic ($\mu = \mu_0 = 4\pi \times 10^{-7}$ H/m), and (3) has a uniform current density (skin effect is neglected). From (4.1.1), Ampere's law states that

$$\oint H_{\text{tan}}\, dl = I_{\text{enclosed}} \tag{5.4.1}$$

To determine the magnetic field inside the conductor, select the dashed circle of radius $x < r$ shown in Figure 5.6 as the closed contour for Ampere's law. Due to symmetry, H_x is constant along the contour. Also, there is no radial component of H_x, so H_x is tangent to the contour. That is, the conductor has a concentric magnetic field. From (5.4.1), the integral of H_x around the selected contour is

$$H_x(2\pi x) = I_x \qquad \text{for } x < r \tag{5.4.2}$$

where I_x is the portion of the total current enclosed by the contour. Solving (5.4.2)

$$H_x = \frac{I_x}{2\pi x} \quad \text{A/m} \tag{5.4.3}$$

Now assume a uniform current distribution within the conductor, that is

$$I_x = \left(\frac{x}{r}\right)^2 I \qquad \text{for } x < r \tag{5.4.4}$$

Using (5.4.4) in (5.4.3)

$$H_x = \frac{xI}{2\pi r^2} \quad \text{A/m} \tag{5.4.5}$$

For a nonmagnetic conductor, the magnetic flux density B_x is

$$B_x = \mu_0 H_x = \frac{\mu_0 xI}{2\pi r^2} \quad \text{Wb/m}^2 \tag{5.4.6}$$

The differential flux $d\Phi$ per-unit length of conductor in the cross-hatched rectangle of width dx shown in Figure 5.6 is

$$d\Phi = B_x\, dx \qquad \text{Wb/m} \tag{5.4.7}$$

Computation of the differential flux linkage $d\lambda$ in the rectangle is tricky since only the fraction $(x/r)^2$ of the total current I is linked by the flux. That is,

$$d\lambda = \left(\frac{x}{r}\right)^2 d\Phi = \frac{\mu_0 I}{2\pi r^4} x^3 \, dx \quad \text{Wb-t/m} \tag{5.4.8}$$

Integrating (5.4.8) from $x = 0$ to $x = r$ determines the total flux linkages λ_{int} inside the conductor

$$\lambda_{\text{int}} = \int_0^r d\lambda = \frac{\mu_0 I}{2\pi r^4} \int_0^r x^3 \, dx = \frac{\mu_0 I}{8\pi} = \frac{1}{2} \times 10^{-7} I \quad \text{Wb-t/m} \tag{5.4.9}$$

The internal inductance L_{int} per-unit length of conductor due to this flux linkage is then

$$L_{\text{int}} = \frac{\lambda_{\text{int}}}{I} = \frac{\mu_0}{8\pi} = \frac{1}{2} \times 10^{-7} \quad \text{H/m} \tag{5.4.10}$$

Next, in order to determine the magnetic field outside the conductor, select the dashed circle of radius $x > r$ shown in Figure 5.7 as the closed contour for Ampere's law. Noting that this contour encloses the entire current I, integration of (5.4.1) yields

$$H_x(2\pi x) = I \tag{5.4.11}$$

which gives

$$H_x = \frac{I}{2\pi x} \quad \text{A/m} \quad x > r \tag{5.4.12}$$

Outside the conductor, $\mu = \mu_0$ and

$$B_x = \mu_0 H_x = (4\pi \times 10^{-7}) \frac{I}{2\pi x} = 2 \times 10^{-7} \frac{I}{x} \quad \text{Wb/m}^2 \tag{5.4.13}$$

$$d\Phi = B_x \, dx = 2 \times 10^{-7} \frac{I}{x} \, dx \quad \text{Wb/m} \tag{5.4.14}$$

Figure 5.7

External magnetic field of a solid cylindrical conductor

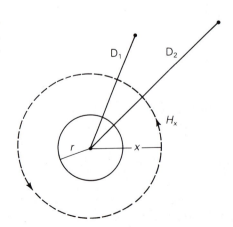

Since the entire current I is linked by the flux outside the conductor

$$d\lambda = d\Phi = 2 \times 10^{-7} \frac{I}{x} dx \quad \text{Wb-t/m} \tag{5.4.15}$$

Integrating (5.4.15) between two external points at distances D_1 and D_2 from the conductor center gives the external flux linkage λ_{12} between D_1 and D_2:

$$\lambda_{12} = \int_{D_1}^{D_2} d\lambda = 2 \times 10^{-7} I \int_{D_1}^{D_2} \frac{dx}{x}$$

$$= 2 \times 10^{-7} I \ln\left(\frac{D_2}{D_1}\right) \quad \text{Wb-t/m} \tag{5.4.16}$$

The external inductance L_{12} per-unit length due to the flux linkages between D_1 and D_2 is then

$$L_{12} = \frac{\lambda_{12}}{I} = 2 \times 10^{-7} \ln\left(\frac{D_2}{D_1}\right) \quad \text{H/m} \tag{5.4.17}$$

The total flux λ_P linking the conductor out to external point P at distance D is the sum of the internal flux linkage, (5.4.9), and the external flux linkage, (5.4.16) from $D_1 = r$ to $D_2 = D$. That is

$$\lambda_P = \frac{1}{2} \times 10^{-7} I + 2 \times 10^{-7} I \ln \frac{D}{r} \tag{5.4.18}$$

Using the identity $\frac{1}{2} = 2 \ln e^{1/4}$ in (5.4.18), a more convenient expression for λ_P is obtained:

$$\lambda_P = 2 \times 10^{-7} I \left(\ln e^{1/4} + \ln \frac{D}{r} \right)$$

$$= 2 \times 10^{-7} I \ln \frac{D}{e^{-1/4} r}$$

$$= 2 \times 10^{-7} I \ln \frac{D}{r'} \quad \text{Wb-t/m} \tag{5.4.19}$$

where

$$r' = e^{-1/4} r = 0.7788 r \tag{5.4.20}$$

Also, the total inductance L_P due to both internal and external flux linkages out to distance D is

$$L_P = \frac{\lambda_P}{I} = 2 \times 10^{-7} \ln\left(\frac{D}{r'}\right) \quad \text{H/m} \tag{5.4.21}$$

Finally, consider the array of M solid cylindrical conductors shown in Figure 5.8. Assume that each conductor m carries current I_m referenced out

Figure 5.8

Array of *M* solid cylindrical conductors

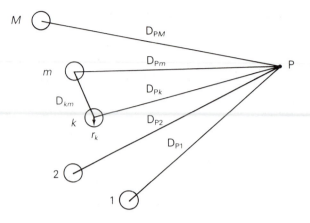

of the page. Also assume that the sum of the conductor currents is zero, that is,

$$I_1 + I_2 + \cdots + I_M = \sum_{m=1}^{M} I_m = 0 \qquad (5.4.22)$$

The flux linkage $\lambda_{k\text{P}k}$, which links conductor k out to point P due to current I_k, is, from (5.4.19),

$$\lambda_{k\text{P}k} = 2 \times 10^{-7} I_k \ln \frac{\text{D}_{\text{P}k}}{r'_k} \qquad (5.4.23)$$

Note that $\lambda_{k\text{P}k}$ includes both internal and external flux linkages due to I_k. The flux linkage $\lambda_{k\text{P}m}$, which links conductor k out to P due to I_m, is, from (5.4.16),

$$\lambda_{k\text{P}m} = 2 \times 10^{-7} I_m \ln \frac{\text{D}_{\text{P}m}}{\text{D}_{km}} \qquad (5.4.24)$$

In (5.4.24) we use D_{km} instead of $(\text{D}_{km} - r_k)$ or $(\text{D}_{km} + r_k)$, which is a valid approximation when D_{km} is much greater than r_k. It can also be shown that this is a good approximation even when D_{km} is small. Using superposition, the total flux linkage $\lambda_{k\text{P}}$, which links conductor k out to P due to all the currents, is

$$\lambda_{k\text{P}} = \lambda_{k\text{P}1} + \lambda_{k\text{P}2} + \cdots + \lambda_{k\text{P}M}$$

$$= 2 \times 10^{-7} \sum_{m=1}^{M} I_m \ln \frac{\text{D}_{\text{P}m}}{\text{D}_{km}} \qquad (5.4.25)$$

where we define $\text{D}_{kk} = r'_k = e^{-1/4} r_k$ when $m = k$ in the above summation. Equation (5.4.25) is separated into two summations:

$$\lambda_{k\text{P}} = 2 \times 10^{-7} \sum_{m=1}^{M} I_m \ln \frac{1}{\text{D}_{km}} + 2 \times 10^{-7} \sum_{m=1}^{M} I_m \ln \text{D}_{\text{P}m} \qquad (5.4.26)$$

Removing the last term from the second summation we get:

$$\lambda_{kP} = 2 \times 10^{-7} \left[\sum_{m=1}^{M} I_m \ln \frac{1}{D_{km}} + \sum_{m=1}^{M-1} I_m \ln D_{Pm} + I_M \ln D_{PM} \right] \quad (5.4.27)$$

From (5.4.22),

$$I_M = -(I_1 + I_2 + \cdots + I_{M-1}) = -\sum_{m=1}^{M-1} I_m \quad (5.4.28)$$

Using (5.4.28) in (5.4.27)

$$\lambda_{kP} = 2 \times 10^{-7} \left[\sum_{m=1}^{M} I_m \ln \frac{1}{D_{km}} + \sum_{m=1}^{M-1} I_m \ln D_{Pm} - \sum_{m=1}^{M-1} I_m \ln D_{PM} \right]$$

$$= 2 \times 10^{-7} \left[\sum_{m=1}^{M} I_m \ln \frac{1}{D_{km}} + \sum_{m=1}^{M-1} I_m \ln \frac{D_{Pm}}{D_{PM}} \right] \quad (5.4.29)$$

Now, let λ_k equal the total flux linking conductor k out to infinity. That is, $\lambda_k = \lim_{p \to \infty} \lambda_{kP}$. As $P \to \infty$, all the distances D_{Pm} become equal, the ratios D_{Pm}/D_{PM} become unity, and $\ln(D_{Pm}/D_{PM}) \to 0$. Therefore, the second summation in (5.4.29) becomes zero as $P \to \infty$, and

$$\lambda_k = 2 \times 10^{-7} \sum_{m=1}^{M} I_m \ln \frac{1}{D_{km}} \quad \text{Wb-t/m} \quad (5.4.30)$$

Equation (5.4.30) gives the total flux linking conductor k in an array of M conductors carrying currents $I_1, I_2, \ldots, I_M$, whose sum is zero. This equation is valid for either dc or ac currents. λ_k is a dc flux linkage when the currents are dc, and λ_k is a phasor flux linkage when the currents are phasor representations of sinusoids.

SECTION 5.5

INDUCTANCE: SINGLE-PHASE TWO-WIRE LINE AND THREE-PHASE THREE-WIRE LINE WITH EQUAL PHASE SPACING

The results of the previous section are used here to determine the inductances of two relatively simple transmission lines: a single-phase two-wire line and a three-phase three-wire line with equal phase spacing.

Figure 5.9(a) shows a single-phase two-wire line consisting of two solid cylindrical conductors x and y. Conductor x with radius r_x carries phasor current $I_x = I$ referenced out of the page. Conductor y with radius r_y carries return current $I_y = -I$. Since the sum of the two currents is zero, (5.4.30) is

Figure 5.9

Single-phase two-wire line

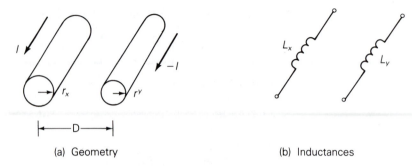

(a) Geometry   (b) Inductances

valid, from which the total flux linking conductor x is

$$\lambda_x = 2 \times 10^{-7} \left(I_x \ln \frac{1}{D_{xx}} + I_y \ln \frac{1}{D_{xy}} \right)$$

$$= 2 \times 10^{-7} \left(I \ln \frac{1}{r'_x} - I \ln \frac{1}{D} \right)$$

$$= 2 \times 10^{-7} I \ln \frac{D}{r'_x} \quad \text{Wb-t/m} \tag{5.5.1}$$

where $r'_x = e^{-1/4} r_x = 0.7788 r_x$.

The inductance of conductor x is then

$$L_x = \frac{\lambda_x}{I_x} = \frac{\lambda_x}{I} = 2 \times 10^{-7} \ln \frac{D}{r'_x} \quad \text{H/m per conductor} \tag{5.5.2}$$

Similarly, the total flux linking conductor y is

$$\lambda_y = 2 \times 10^{-7} \left(I_x \ln \frac{1}{D_{yx}} + I_y \ln \frac{1}{D_{yy}} \right)$$

$$= 2 \times 10^{-7} \left(I \ln \frac{1}{D} - I \ln \frac{1}{r'_y} \right)$$

$$= -2 \times 10^{-7} I \ln \frac{D}{r'_y} \tag{5.5.3}$$

and

$$L_y = \frac{\lambda_y}{I_y} = \frac{\lambda_y}{-I} = 2 \times 10^{-7} \ln \frac{D}{r'_y} \quad \text{H/m per conductor} \tag{5.5.4}$$

The total inductance of the single-phase circuit, also called *loop inductance*, is

$$L = L_x + L_y = 2 \times 10^{-7} \left(\ln \frac{D}{r'_x} + \ln \frac{D}{r'_y} \right)$$

$$= 2 \times 10^{-7} \ln \frac{D^2}{r'_x r'_y}$$

$$= 4 \times 10^{-7} \ln \frac{D}{\sqrt{r'_x r'_y}} \quad \text{H/m per circuit} \tag{5.5.5}$$

Also, if $r'_x = r'_y = r'$, the total circuit inductance is

$$L = 4 \times 10^{-7} \ln \frac{D}{r'} \quad \text{H/m per circuit} \tag{5.5.6}$$

The inductances of the single-phase two-wire line are shown in Figure 5.9(b).

Figure 5.10(a) shows a three-phase three-wire line consisting of three solid cylindrical conductors a, b, c, each with radius r, and with equal phase spacing D between any two conductors. To determine the positive-sequence inductance, assume positive-sequence currents I_a, I_b, I_c that satisfy $I_a + I_b + I_c = 0$. Then (5.4.30) is valid and the total flux linking the phase a conductor is

$$\lambda_a = 2 \times 10^{-7} \left(I_a \ln \frac{1}{r'} + I_b \ln \frac{1}{D} + I_c \ln \frac{1}{D} \right)$$

$$= 2 \times 10^{-7} \left[I_a \ln \frac{1}{r'} + (I_b + I_c) \ln \frac{1}{D} \right] \tag{5.5.7}$$

Using $(I_b + I_c) = -I_a$,

$$\lambda_a = 2 \times 10^{-7} \left(I_a \ln \frac{1}{r'} - I_a \ln \frac{1}{D} \right)$$

$$= 2 \times 10^{-7} I_a \ln \frac{D}{r'} \quad \text{Wb-t/m} \tag{5.5.8}$$

The inductance of phase a is then

$$L_a = \frac{\lambda_a}{I_a} = 2 \times 10^{-7} \ln \frac{D}{r'} \quad \text{H/m per phase} \tag{5.5.9}$$

Due to symmetry, the same result is obtained for $L_b = \lambda_b/I_b$ and for $L_c = \lambda_c/I_c$. However, only one phase need be considered for positive-sequence

Figure 5.10

Three-phase three-wire line with equal phase spacing

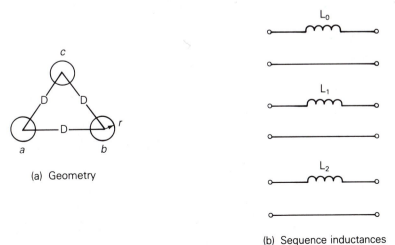

(a) Geometry

(b) Sequence inductances

operation of this line, since the flux linkages of each phase have equal magnitudes and 120° displacement. Thus, the inductance of (5.5.9) is the positive-sequence inductance L_1. It is also the negative-sequence inductance L_2, since the positive- and negative-sequence inductances of nonrotating equipment are equal. That is

$$L_1 = L_2 = 2 \times 10^{-7} \ln \frac{D}{r'} \quad \text{H/m} \tag{5.5.10}$$

Sequence inductances of the three-phase line with equal phase spacing are shown in Figure 5.10(b).

To determine the zero-sequence inductance L_0, the return paths including neutral conductors and earth return must be considered. In Section 5.7 we develop the equations for zero-sequence inductance as part of the series sequence impedances for more general three-phase lines.

SECTION 5.6

INDUCTANCE: COMPOSITE CONDUCTORS, UNEQUAL PHASE SPACING, BUNDLED CONDUCTORS

The results of Section 5.5 are extended here to include composite conductors, which consist of two or more solid cylindrical subconductors in parallel. A stranded conductor is one example of a composite conductor. For simplicity we assume that for each conductor, the subconductors are identical and share the conductor current equally.

Figure 5.11 shows a single-phase two-conductor line consisting of two composite conductors x and y. Conductor x has N identical subconductors, each with radius r_x and with current (I/N) referenced out of the page. Similarly, conductor y consists of M identical subconductors, each with

Figure 5.11

Single-phase two-conductor line with composite conductors

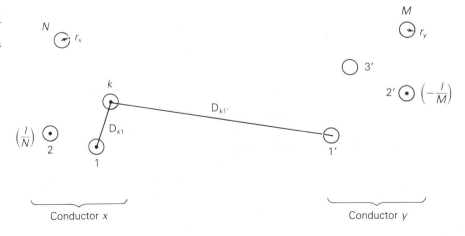

Conductor x   Conductor y

radius r_y and with return current $(-I/M)$. Since the sum of all the currents is zero, (5.4.30) is valid and the total flux Φ_k linking subconductor k of conductor x is

$$\Phi_k = 2 \times 10^{-7} \left[\frac{I}{N} \sum_{m=1}^{N} \ln \frac{1}{D_{km}} - \frac{I}{M} \sum_{m=1'}^{M} \ln \frac{1}{D_{km}} \right] \tag{5.6.1}$$

Since only the fraction $(1/N)$ of the total conductor current I is linked by this flux, the flux linkage λ_k of (the current in) subconductor k is

$$\lambda_k = \frac{\Phi_k}{N} = 2 \times 10^{-7} I \left[\frac{1}{N^2} \sum_{m=1}^{N} \ln \frac{1}{D_{km}} - \frac{1}{NM} \sum_{m=1'}^{M} \ln \frac{1}{D_{km}} \right] \tag{5.6.2}$$

The total flux linkage of conductor x is

$$\lambda_x = \sum_{k=1}^{N} \lambda_k = 2 \times 10^{-7} I \sum_{k=1}^{N} \left[\frac{1}{N^2} \sum_{m=1}^{N} \ln \frac{1}{D_{km}} - \frac{1}{NM} \sum_{m=1'}^{M} \ln \frac{1}{D_{km}} \right] \tag{5.6.3}$$

Using $\ln A^\alpha = \alpha \ln A$ and $\sum \ln A_k = \ln \Pi A_k$ (sum of lns = ln of products), (5.6.3) can be rewritten in the following form:

$$\lambda_x = 2 \times 10^{-7} I \ln \prod_{k=1}^{N} \frac{\left(\prod_{m=1'}^{M} D_{km} \right)^{1/NM}}{\left(\prod_{m=1}^{N} D_{km} \right)^{1/N^2}} \tag{5.6.4}$$

and the inductance of conductor x, $L_x = \dfrac{\lambda_x}{I}$, can be written as

$$L_x = 2 \times 10^{-7} \ln \frac{D_{xy}}{D_{xx}} \quad \text{H/m per conductor} \tag{5.6.5}$$

where

$$D_{xy} = \sqrt[MN]{\prod_{k=1}^{N} \prod_{m=1'}^{M} D_{km}} \tag{5.6.6}$$

$$D_{xx} = \sqrt[N^2]{\prod_{k=1}^{N} \prod_{m=1}^{N} D_{km}} \tag{5.6.7}$$

D_{xy}, given by (5.6.6), is the MNth root of the product of the MN distances from the subconductors of conductor x to the subconductors of conductor y. Associated with each subconductor k of conductor x are the M distances $D_{k1'}$, $D_{k2'}, \ldots, D_{kM}$ to the subconductors of conductor y. For N subconductors in conductor x, there are therefore MN of these distances. D_{xy} is called the *geometric mean distance* or GMD between conductors x and y.

Also, D_{xx}, given by (5.6.7), is the N^2 root of the product of the N^2 distances between the subconductors of conductor x. Associated with each subconductor k are the N distances $D_{k1}, D_{k2}, \ldots, D_{kk} = r', \ldots, D_{kN}$. For N

subconductors in conductor x, there are therefore N^2 of these distances. D_{xx} is called the *geometric mean radius* or GMR of conductor x.

Similarly, for conductor y,

$$L_y = 2 \times 10^{-7} \ln \frac{D_{xy}}{D_{yy}} \quad \text{H/m per conductor} \tag{5.6.8}$$

where

$$D_{yy} = \sqrt[M^2]{\prod_{k=1'}^{M} \prod_{m=1'}^{M} D_{km}} \tag{5.6.9}$$

D_{yy}, the GMR of conductor y, is the M^2 root of the product of the M^2 distances between the subconductors of conductor y. The total inductance L of the single-phase circuit is

$$L = L_x + L_y \quad \text{H/m per circuit} \tag{5.6.10}$$

EXAMPLE 5.2   **GMR, GMD, and inductance: single-phase two-conductor line**

Expand (5.6.6), (5.6.7), and (5.6.9) for $N = 3$ and $M = 2'$. Then evaluate L_x, L_y, and L in H/m for the single-phase two-conductor line shown in Figure 5.12.

Solution   For $N = 3$ and $M = 2'$, (5.6.6) becomes

$$D_{xy} = \sqrt[6]{\prod_{k=1}^{3} \prod_{m=1'}^{2'} D_{km}}$$

$$= \sqrt[6]{\prod_{k=1}^{3} D_{k1'} D_{k2'}}$$

$$= \sqrt[6]{(D_{11'} D_{12'})(D_{21'} D_{22'})(D_{31'} D_{32'})}$$

Similarly, (5.6.7) becomes

$$D_{xx} = \sqrt[9]{\prod_{k=1}^{3} \prod_{m=1}^{3} D_{km}}$$

$$= \sqrt[9]{\prod_{k=1}^{3} D_{k1} D_{k2} D_{k3}}$$

$$= \sqrt[9]{(D_{11} D_{12} D_{13})(D_{21} D_{22} D_{23})(D_{31} D_{32} D_{33})}$$

Figure 5.12

Single-phase two-conductor line for Example 5.2

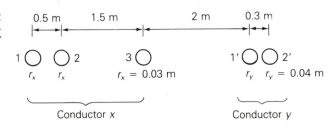

and (5.6.9) becomes

$$D_{yy} = \sqrt[4]{\prod_{k=1'}^{2'} \prod_{m=1'}^{2'} D_{km}}$$

$$= \sqrt[4]{\prod_{k=1'}^{2'} D_{k1'} D_{k2'}}$$

$$= \sqrt[4]{(D_{1'1'}D_{1'2'})(D_{2'1'}D_{2'2'})}$$

Evaluating D_{xy}, D_{xx}, and D_{yy} for the single-phase two-conductor line shown in Figure 5.12,

$$D_{11'} = 4\,\text{m} \qquad D_{12'} = 4.3\,\text{m} \quad D_{21'} = 3.5\,\text{m}$$

$$D_{22'} = 3.8\,\text{m} \qquad D_{31'} = 2\,\text{m} \qquad D_{32'} = 2.3\,\text{m}$$

$$D_{xy} = \sqrt[6]{(4)(4.3)(3.5)(3.8)(2)(2.3)} = 3.189\,\text{m}$$

$$D_{11} = D_{22} = D_{33} = r'_x = e^{-1/4}r_x = (0.7788)(0.03) = 0.02336\,\text{m}$$

$$D_{21} = D_{12} = 0.5\,\text{m}$$

$$D_{23} = D_{32} = 1.5\,\text{m}$$

$$D_{31} = D_{13} = 2.0\,\text{m}$$

$$D_{xx} = \sqrt[9]{(0.02336)^3(0.5)^2(1.5)^2(2.0)^2} = 0.3128\,\text{m}$$

$$D_{1'1'} = D_{2'2'} = r'_y = e^{-1/4}r_y = (0.7788)(0.04) = 0.03115\,\text{m}$$

$$D_{1'2'} = D_{2'1'} = 0.3\,\text{m}$$

$$D_{yy} = \sqrt[4]{(0.03115)^2(0.3)^2} = 0.09667\,\text{m}$$

Then, from (5.6.5), (5.6.8), and (5.6.10):

$$L_x = 2 \times 10^{-7} \ln\left(\frac{3.189}{0.3128}\right) = 4.644 \times 10^{-7} \quad \text{H/m per conductor}$$

$$L_y = 2 \times 10^{-7} \ln\left(\frac{3.189}{0.09667}\right) = 6.992 \times 10^{-7} \quad \text{H/m per conductor}$$

$$L = L_x + L_y = 1.164 \times 10^{-6} \quad \text{H/m per circuit} \qquad \blacksquare$$

It is seldom necessary to calculate GMR or GMD for standard lines. The GMR of standard conductors is provided by conductor manufacturers and can be found in various handbooks (see Appendix Tables A.3 and A.4). Also, if the distances between conductors are large compared to the distances between subconductors of each conductor, then the GMD between conductors is approximately equal to the distance between conductor centers.

EXAMPLE 5.3 | **Inductance and inductive reactance: single-phase line**

A single-phase line operating at 60 Hz consists of two 4/0 12-strand copper conductors with 5 ft spacing between conductor centers. The line length is 20 miles. Determine the total inductance in H and the total inductive reactance in Ω.

Solution   The GMD between conductor centers is $D_{xy} = 5$ ft. Also, from Table A.3, the GMR of a 4/0 12-strand copper conductor is $D_{xx} = D_{yy} = 0.01750$ ft. From (5.6.5) and (5.6.8),

$$L_x = L_y = 2 \times 10^{-7} \ln \left(\frac{5}{0.01750} \right) \frac{H}{m} \times 1609 \frac{m}{mi} \times 20 \text{ mi}$$

$$= 0.03639 \quad \text{H per conductor}$$

The total inductance is

$$L = L_x + L_y = 2 \times 0.03639 = 0.07279 \quad \text{H per circuit}$$

and the total inductive reactance is

$$X_L = 2\pi f L = (2\pi)(60)(0.07279) = 27.44 \quad \Omega \text{ per circuit} \qquad \blacksquare$$

To calculate $L_1 = L_2$ for three-phase lines with stranded conductors and equal phase spacing, r' is replaced by the conductor GMR in (5.5.10). If the spacings between phases are unequal, then positive-sequence flux linkages are not obtained from positive-sequence currents. Instead, unbalanced flux linkages occur, and the phase inductances are unequal. However, balance can be restored by exchanging the conductor positions along the line, a technique called *transposition*.

Figure 5.13 shows a completely transposed three-phase line. The line is transposed at two locations such that each phase occupies each position for one-third of the line length. Conductor positions are denoted 1, 2, 3 with distances D_{12}, D_{23}, D_{31} between positions. The conductors are identical, each with GMR denoted D_S. To calculate the positive-sequence inductance of this line, assume positive-sequence currents I_a, I_b, I_c, for which $I_a + I_b + I_c = 0$. Again, (5.4.30) is valid, and the total flux linking the phase a conductor while it is in position 1 is

$$\lambda_{a1} = 2 \times 10^{-7} \left[I_a \ln \frac{1}{D_S} + I_b \ln \frac{1}{D_{12}} + I_c \ln \frac{1}{D_{31}} \right] \quad \text{Wb-t/m} \qquad (5.6.11)$$

Figure 5.13

Completely transposed three-phase line

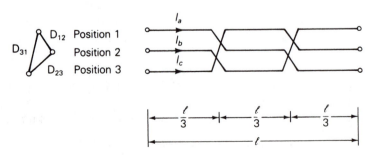

Similarly, the total flux linkage of this conductor while it is in positions 2 and 3 is

$$\lambda_{a2} = 2 \times 10^{-7} \left[I_a \ln \frac{1}{D_S} + I_b \ln \frac{1}{D_{23}} + I_c \ln \frac{1}{D_{12}} \right] \quad \text{Wb-t/m} \quad (5.6.12)$$

$$\lambda_{a3} = 2 \times 10^{-7} \left[I_a \ln \frac{1}{D_S} + I_b \ln \frac{1}{D_{31}} + I_c \ln \frac{1}{D_{23}} \right] \quad \text{Wb-t/m} \quad (5.6.13)$$

The average of the above flux linkages is

$$
\lambda_a = \frac{\lambda_{a1}\left(\frac{l}{3}\right) + \lambda_{a2}\left(\frac{l}{3}\right) + \lambda_{a3}\left(\frac{l}{3}\right)}{l} = \frac{\lambda_{a1} + \lambda_{a2} + \lambda_{a3}}{3}
$$

$$
= \frac{2 \times 10^{-7}}{3} \left[3I_a \ln \frac{1}{D_S} + I_b \ln \frac{1}{D_{12}D_{23}D_{31}} + I_c \ln \frac{1}{D_{12}D_{23}D_{31}} \right]
$$
$$(5.6.14)$$

Using $(I_b + I_c) = -I_a$ in (5.6.14),

$$
\lambda_a = \frac{2 \times 10^{-7}}{3} \left[3I_a \ln \frac{1}{D_S} - I_a \ln \frac{1}{D_{12}D_{23}D_{31}} \right]
$$

$$
= 2 \times 10^{-7} I_a \ln \frac{\sqrt[3]{D_{12}D_{23}D_{31}}}{D_S} \quad \text{Wb-t/m} \quad (5.6.15)
$$

and the average inductance of phase a is

$$
\text{L}_a = \frac{\lambda_a}{I_a} = 2 \times 10^{-7} \ln \frac{\sqrt[3]{D_{12}D_{23}D_{31}}}{D_S} \quad \text{H/m per phase} \quad (5.6.16)
$$

The same result is obtained for $\text{L}_b = \lambda_b/I_b$ and for $\text{L}_c = \lambda_c/I_c$. This inductance is the positive- (or negative-) sequence inductance of a completely transposed three-phase line. Defining

$$
\text{D}_{eq} = \sqrt[3]{D_{12}D_{23}D_{31}} \quad (5.6.17)
$$

we have

$$
\text{L}_1 = \text{L}_2 = 2 \times 10^{-7} \ln \frac{\text{D}_{eq}}{\text{D}_S} \quad \text{H/m} \quad (5.6.18)
$$

D_{eq}, the cube root of the product of the three-phase spacings, is the geometric mean distance between phases. Also, D_S is the conductor GMR for stranded conductors, or r' for solid cylindrical conductors.

| EXAMPLE 5.4 | **Positive-sequence inductance and inductive reactance: three-phase line** |

A completely transposed 60-Hz three-phase line has flat horizontal phase spacing with 10 m between adjacent conductors. The conductors are 1,590,000 cmil ACSR with 54/3 stranding. Line length is 200 km. Determine the positive-sequence inductance in H and the positive-sequence inductive reactance in Ω.

Solution   From Table A.4 the GMR of a 1,590,000 cmil 54/3 ACSR conductor is

$$D_S = 0.0520 \text{ ft } \frac{1 \text{ m}}{3.28 \text{ ft}} = 0.0159 \text{ m}$$

Also, from (5.6.17) and (5.6.18),

$$D_{eq} = \sqrt[3]{(10)(10)(20)} = 12.6 \text{ m}$$

$$L_1 = 2 \times 10^{-7} \ln \left(\frac{12.6}{0.0159} \right) \frac{\text{H}}{\text{m}} \times \frac{1000 \text{ m}}{\text{km}} \times 200 \text{ km}$$

$$= 0.267 \quad \text{H}$$

The positive-sequence inductive reactance is

$$X_1 = 2\pi f L_1 = 2\pi(60)(0.267) = 101 \quad \Omega \qquad \blacksquare$$

It is common practice for EHV lines to use more than one conductor per phase, a practice called *bundling*. Bundling reduces the electric field strength at the conductor surfaces, which in turn reduces or eliminates corona and its results: undesirable power loss, communications interference and audible noise. Bundling also reduces the series reactance of the line by increasing the GMR of the bundle.

Figure 5.14 shows common EHV bundles consisting of two, three, or four conductors. The three-conductor bundle has its conductors on the vertices of an equilateral triangle, and the four-conductor bundle has its conductors on the corners of a square. To calculate $L_1 = L_2$, D_S in (5.6.18) is replaced by the GMR of the bundle. Since the bundle constitutes a composite conductor, calculation of bundle GMR is, in general, given by (5.6.7). If the conductors are stranded and the bundle spacing d is large compared to the conductor outside radius, each stranded conductor is first replaced by an equivalent solid cylindrical conductor with GMR $= D_S$. Then the bundle is replaced by one equivalent conductor with GMR $= D_{SL}$, given by (5.6.7) with $n = 2$, 3, or 4 as follows:

Two-conductor bundle:

$$D_{SL} = \sqrt[4]{(D_S \times d)^2} = \sqrt{D_S d} \tag{5.6.19}$$

Three-conductor bundle:

$$D_{SL} = \sqrt[9]{(D_S \times d \times d)^3} = \sqrt[3]{D_S d^2} \tag{5.6.20}$$

Figure 5.14

Bundle conductor configurations

Four-conductor bundle:

$$D_{SL} = \sqrt[16]{(D_s \times d \times d \times d\sqrt{2})^4} = 1.091\sqrt[4]{D_s d^3} \tag{5.6.21}$$

The positive- (or negative-) sequence inductance is then

$$L_1 = L_2 = 2 \times 10^{-7} \ln \frac{D_{eq}}{D_{SL}} \quad \text{H/m} \tag{5.6.22}$$

If the phase spacings are large compared to the bundle spacing, then sufficient accuracy for D_{eq} is obtained by using the distances between bundle centers.

| **EXAMPLE 5.5** | **Positive-sequence inductive reactance:**
three-phase line with bundled conductors |

Each of the 1,590,000 cmil conductors in Example 5.4 is replaced by two 795,000 cmil ACSR 26/2 conductors, as shown in Figure 5.15. Bundle spacing is 0.40 m. Flat horizontal spacing is retained, with 10 m between adjacent bundle centers. Calculate the positive-sequence inductive reactance of the line and compare it with that of Example 5.4.

Solution   From Table A.4, the GMR of a 795,000 cmil 26/2 ACSR conductor is

$$D_s = 0.0375 \, \text{ft} \times \frac{1 \, \text{m}}{3.28 \, \text{ft}} = 0.0114 \, \text{m}$$

From (5.6.19), the two-conductor bundle GMR is

$$D_{SL} = \sqrt{(0.0114)(0.40)} = 0.0676 \quad \text{m}$$

Since $D_{eq} = 12.6 \, \text{m}$ is the same as in Example 5.4,

$$L_1 = 2 \times 10^{-7} \ln \left(\frac{12.6}{0.0676} \right) (1000)(200) = 0.209 \quad \text{H}$$

$$X_1 = 2\pi f L_1 = (2\pi)(60)(0.209) = 78.8 \quad \Omega$$

The reactance of the bundled line, 78.8 Ω, is 22% less than that of Example 5.4, even though the two-conductor bundle has the same amount of conductor material (that is, the same cmil per phase). One advantage of reduced series line reactance is smaller line-voltage drops. Also, the loadability of medium and long EHV lines is increased (see Chapter 6).

Figure 5.15

Three-phase bundled conductor line for Example 5.5

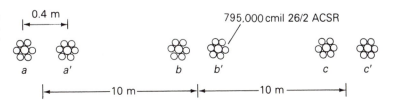

SERIES IMPEDANCES: THREE-PHASE LINE WITH NEUTRAL CONDUCTORS AND EARTH RETURN

In this section we develop equations suitable for computer calculation of the series impedances, including resistances and inductive reactances, for the three-phase line shown in Figure 5.16. This line has three phase conductors a, b, and c, where bundled conductors, if any, have already been replaced by equivalent conductors, as described in Section 5.6. The line also has N neutral conductors denoted $n1, n2, \ldots, nN$.* All the neutral conductors are connected in parallel and are grounded to the earth at regular intervals along the line.

Figure 5.16

Three-phase transmission line with earth replaced by earth return conductors

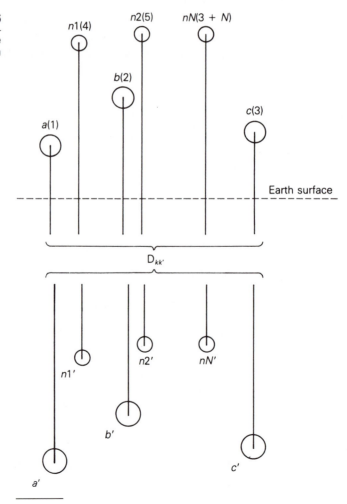

---

*Instead of *shield wire* we use the term *neutral conductor*, which applies to distribution as well as transmission lines.

Any isolated neutral conductors that carry no current are omitted. The phase conductors are insulated from each other and from earth.

If the phase currents are not balanced, there may be a return current in the grounded neutral wires and in the earth. The earth return current will spread out under the line, seeking the lowest impedance return path. A classic paper by Carson [4], later modified by others [5, 6], shows that the earth can be replaced by a set of "earth return" conductors located directly under the overhead conductors, as shown in Figure 5.16. Each earth return conductor carries the negative of its overhead conductor current, has a GMR denoted $D_{k'k'}$, distance $D_{kk'}$ from its overhead conductor, and resistance $R_{k'}$ given by:

$$D_{k'k'} = D_{kk} \quad \text{m} \tag{5.7.1}$$

$$D_{kk'} = 658.5\sqrt{\rho/f} \quad \text{m} \tag{5.7.2}$$

$$R_{k'} = 9.869 \times 10^{-7} f \quad \Omega/\text{m} \tag{5.7.3}$$

where ρ is the earth resistivity in ohm-meters and f is frequency in hertz. Table 5.4 lists earth resistivities and 60-Hz equivalent conductor distances for various types of earth. It is common practice to select $\rho = 100\,\Omega\text{m}$ when actual data is unavailable. Note that all the earth return conductors have the same distance $D_{kk'}$ from their overhead conductors and the same resistance $R_{k'}$. For simplicity we renumber the overhead conductors from 1 to $(3 + N)$, beginning with the phase conductors, then overhead neutral conductors, as shown in Figure 5.16.

Table 5.4

Earth resistivities and 60-Hz equivalent conductor distances

TYPE OF EARTH	RESISTIVITY (Ωm)	$D_{kk'}$ (m)
Sea water	0.01–1.0	8.50–85.0
Swampy ground	10–100	269–850
Average damp earth	100	850
Dry earth	1000	2690
Pure slate	10^7	269,000
Sandstone	10^9	2,690,000

Operating as a transmission line, the sum of the currents in all the conductors is zero. That is,

$$\sum_{k=1}^{(6+2N)} I_k = 0 \tag{5.7.4}$$

Equation (5.4.30) is therefore valid, and the flux linking overhead conductor k is

$$\lambda_k = 2 \times 10^{-7} \sum_{m=1}^{(3+N)} I_m \ln \frac{D_{km'}}{D_{km}} \quad \text{Wb-t/m} \tag{5.7.5}$$

In matrix format, (5.7.5) becomes

$$\lambda = \mathbf{L}I \tag{5.7.6}$$

where

λ is a $(3 + N)$ vector

I is a $(3 + N)$ vector

$\mathbf{L}$ is a $(3 + N) \times (3 + N)$ matrix whose elements are:

$$L_{km} = 2 \times 10^{-7} \ln \frac{D_{km'}}{D_{km}} \tag{5.7.7}$$

When $k = m$, D_{kk} in (5.7.7) is the GMR of (bundled) conductor k. When $k \neq m$, D_{km} is the distance between conductors k and m.

A circuit representation of a 1-meter section of the line is shown in Figure 5.17(a). Using this circuit, the vector of voltage drops across the

Figure 5.17

Circuit representation of series-phase impedances

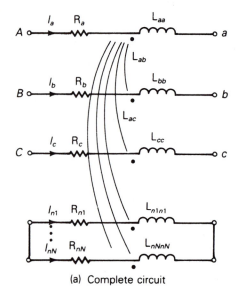

(a) Complete circuit

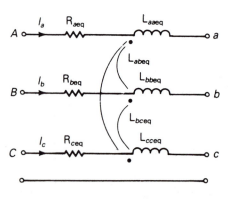

(b) Simplified circuit

conductors is:

$$
\begin{bmatrix}
E_{Aa} \\
E_{Bb} \\
E_{Cc} \\
0 \\
0 \\
\vdots \\
0
\end{bmatrix} = (\mathbf{R} + j\omega\mathbf{L})
\begin{bmatrix}
I_a \\
I_b \\
I_c \\
I_{n1} \\
\vdots \\
I_{nN}
\end{bmatrix}
\tag{5.7.8}
$$

where $\mathbf{L}$ is given by (5.7.7) and $\mathbf{R}$ is a $(3 + N) \times (3 + N)$ matrix of conductor resistances.

$$
\mathbf{R} =
\begin{bmatrix}
(\mathbf{R}_a + \mathbf{R}_{k'})\mathbf{R}_{k'}\cdots & & & \mathbf{R}_{k'} \\
\mathbf{R}_{k'}(\mathbf{R}_b + \mathbf{R}_{k'})\mathbf{R}_{k'}\cdots & & & \vdots \\
& (\mathbf{R}_c + \mathbf{R}_{k'})\mathbf{R}_{k'}\cdots & & \\
& (\mathbf{R}_{n1} + \mathbf{R}_{k'})\mathbf{R}_{k'}\cdots & & \\
& & \ddots & \\
\mathbf{R}_{k'n} & & & (\mathbf{R}_{nN} + \mathbf{R}_{k'})
\end{bmatrix}
\;\Omega/\text{m}
\tag{5.7.9}
$$

The resistance matrix of (5.7.9) includes the resistance $\mathbf{R}_k$ of each overhead conductor and a mutual resistance $\mathbf{R}_k$ due to the image conductors. $\mathbf{R}_k$ of each overhead conductor is obtained from conductor tables such as Appendix Table A.3 or A.4, for a specified frequency, temperature, and current. $\mathbf{R}_{k'}$ of all the image conductors is the same, as given by (5.7.3).

Our objective now is to reduce the $(3 + N)$ equations in (5.7.8) to three equations, thereby obtaining the simplified circuit representations shown in Figure 5.17(b). We partition (5.7.8) as follows.

$$
\begin{bmatrix}
E_{Aa} \\
E_{Bb} \\
E_{Cc} \\
0 \\
\cdots \\
0
\end{bmatrix}
\begin{bmatrix}
\overbrace{\begin{matrix} Z_{11} & Z_{12} & Z_{13} \\ Z_{21} & Z_{22} & Z_{23} \\ Z_{31} & Z_{32} & Z_{33} \end{matrix}}^{Z_A} & \Bigg| & \overbrace{\begin{matrix} Z_{14} & \cdots & Z_{1\,(3+N)} \\ Z_{24} & \cdots & Z_{2\,(3+N)} \\ Z_{34} & \cdots & Z_{3\,(3+N)} \end{matrix}}^{Z_B} \\
\hline
\underbrace{\begin{matrix} Z_{41} & Z_{42} & Z_{43} \\ Z_{(3+N)\,1} & Z_{(3+N)\,2} & Z_{(3+N)\,3} \end{matrix}}_{Z_C} & \Bigg| & \underbrace{\begin{matrix} Z_{44} & \cdots & Z_{4\,(3+N)} \\ Z_{(3+N)\,4} & \cdots & Z_{(3+N)(3+N)} \end{matrix}}_{Z_D}
\end{bmatrix}
\begin{bmatrix}
I_a \\
I_b \\
I_c \\
I_{n1} \\
I_{nN} \\
\vdots
\end{bmatrix}
\tag{5.7.10}
$$

The diagonal elements of this matrix are

$$
Z_{kk} = \mathbf{R}_k + \mathbf{R}_{k'} + j\omega 2 \times 10 \ln \frac{\mathbf{D}_{kk'}}{\mathbf{D}_{kk}} \quad \Omega/\text{m}
\tag{5.7.11}
$$

And the off-diagonal elements, for $k \neq m$, are

$$Z_{km} = R_{k'} + j\omega 2 \times 10 \ln \frac{D_{km'}}{D_{km}} \quad \Omega/m \tag{5.7.12}$$

Next, (5.7.10) is partitioned as shown above to obtain

$$\left[\frac{E_P}{0}\right] = \left[\begin{array}{c|c} Z_A & Z_B \\ \hline Z_C & Z_D \end{array}\right]\left[\frac{I_P}{I_n}\right] \tag{5.7.13}$$

where

$$E_P = \begin{bmatrix} E_{Aa} \\ E_{Bb} \\ E_{Cc} \end{bmatrix}; \quad I_P = \begin{bmatrix} I_a \\ I_b \\ I_c \end{bmatrix}; \quad I_n = \begin{bmatrix} I_{n1} \\ \vdots \\ I_{nN} \end{bmatrix}$$

E_P is the three-dimensional vector of voltage drops across the phase conductors (including the neutral voltage drop). I_P is the three-dimensional vector of phase currents and I_n is the N vector of neutral currents. Also, the $(3 + N) \times (3 + N)$ matrix in (5.7.10) is partitioned to obtain the following matrices:

Z_A with dimension 3×3

Z_B with dimension $3 \times N$

Z_C with dimension $N \times 3$

Z_D with dimension $N \times N$

Equation (5.7.13) is rewritten as two separate matrix equations:

$$E_P = Z_A I_P + Z_B I_n \tag{5.7.14}$$

$$0 = Z_C I_P + Z_D I_n \tag{5.7.15}$$

Solving (5.7.15) for I_n,

$$I_n = -Z_D^{-1} Z_C I_P \tag{5.7.16}$$

Using (5.7.16) in (5.7.14):

$$E_P = [Z_A - Z_B Z_D^{-1} Z_C] I_P \tag{5.7.17}$$

or

$$E_P = Z_P I_P \tag{5.7.18}$$

where

$$Z_P = Z_A - Z_B Z_D^{-1} Z_C \tag{5.7.19}$$

Equation (5.7.17), the desired result, relates the phase-conductor voltage drops (including neutral voltage drop) to the phase currents. Z_P given by

(5.7.19) is the 3×3 series-phase impedance matrix, whose elements are denoted

$$\mathbf{Z}_P = \begin{bmatrix} Z_{aaeq} & Z_{abeq} & Z_{aceq} \\ Z_{abeq} & Z_{bbeq} & Z_{bceq} \\ Z_{aceq} & Z_{bceq} & Z_{cceq} \end{bmatrix} \quad \Omega/m \tag{5.7.20}$$

In general, $\mathbf{Z}_P$ does not satisfy the conditions for a symmetrical impedance matrix, which require that the diagonal elements be equal and that the off-diagonal elements be equal. However, if the line is completely transposed, the diagonal and off-diagonal elements are averaged to obtain

$$\hat{\mathbf{Z}}_P = \begin{bmatrix} \hat{Z}_{aaeq} & \hat{Z}_{abeq} & \hat{Z}_{abeq} \\ \hat{Z}_{abeq} & \hat{Z}_{aaeq} & \hat{Z}_{abeq} \\ \hat{Z}_{abeq} & \hat{Z}_{abeq} & \hat{Z}_{aaeq} \end{bmatrix} \quad \Omega/m \tag{5.7.21}$$

where

$$\hat{Z}_{aaeq} = \tfrac{1}{3}(Z_{aaeq} + Z_{bbeq} + Z_{cceq}) \tag{5.7.22}$$

$$\hat{Z}_{abeq} = \tfrac{1}{3}(Z_{abeq} + Z_{aceq} + Z_{bceq}) \tag{5.7.23}$$

$\hat{\mathbf{Z}}_P$ is a symmetrical impedance matrix.

Equation (5.7.18) gives the series-voltage drops across the circuit in Figure 5.17(b), which is identical to the series impedance circuit shown in Figure 3.9. Therefore, (5.7.18) can be transformed to the sequence domain using the ideas of Section 3.3. From (3.3.3) and (3.3.4)

$$\begin{bmatrix} E_{00'} \\ E_{11'} \\ E_{22'} \end{bmatrix} = \mathbf{Z}_S \begin{bmatrix} I_0 \\ I_1 \\ I_2 \end{bmatrix} \tag{5.7.24}$$

where

$$\mathbf{Z}_S = A^{-1}\mathbf{Z}_P A \tag{5.7.25}$$

and A is the 3×3 transformation matrix given by (3.1.8). $E_{00'}$, $E_{11'}$, and $E_{22'}$ are the zero-, positive-, and negative-sequence series-voltage drops, and I_0, I_1, I_2 are the sequence currents. $\mathbf{Z}_S$ is the 3×3 series sequence impedance matrix whose elements are

$$\mathbf{Z}_S = \begin{bmatrix} Z_0 & Z_{01} & Z_{02} \\ Z_{10} & Z_1 & Z_{12} \\ Z_{20} & Z_{21} & Z_2 \end{bmatrix} \quad \Omega/m \tag{5.7.26}$$

In general $\mathbf{Z}_S$ is not diagonal. However, if the line is completely transposed,

$$\hat{\mathbf{Z}}_S = A^{-1}\hat{\mathbf{Z}}_P A = \begin{bmatrix} \hat{Z}_0 & 0 & 0 \\ 0 & \hat{Z}_1 & 0 \\ 0 & 0 & \hat{Z}_2 \end{bmatrix} \tag{5.7.27}$$

where, from (3.3.7) and (3.3.8),

$$\hat{Z}_0 = \hat{Z}_{aaeq} + 2\hat{Z}_{abeq} \tag{5.7.28}$$

$$\hat{Z}_1 = \hat{Z}_2 = \hat{Z}_{aaeq} - \hat{Z}_{abeq} \tag{5.7.29}$$

A circuit representation of the series sequence impedances of a completely transposed three-phase line is shown in Figure 5.18.

A personal computer program that computes the series phase impedance matrix and series sequence impedance matrix is described in Section 5.14.

Figure 5.18

Circuit representation of the series sequence impedances of a completely transposed three-phase line

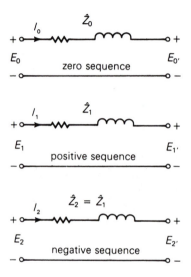

SECTION 5.8

ELECTRIC FIELD AND VOLTAGE: SOLID CYLINDRICAL CONDUCTOR

The capacitance between conductors in a medium with constant permittivity ε can be obtained by determining the following

1. Electric field strength E, from Gauss's law

2. Voltage between conductors

3. Capacitance from charge per unit volt ($C = q/V$)

As a step toward computing capacitances of general conductor configurations, we first compute the electric field of a uniformly charged, solid cylindrical conductor and the voltage between two points outside the conductor. We also compute the voltage between two conductors in an array of charged conductors.

Gauss's law states that the total electric flux leaving a closed surface equals the total charge within the volume enclosed by the surface. That is, the normal component of electric flux density integrated over a closed surface equals the charge enclosed:

$$\oiint D_\perp \, ds = \oiint \varepsilon E_\perp \, ds = Q_{\text{enclosed}} \qquad (5.8.1)$$

where $D_\perp$ denotes the normal component of electric flux density, $E_\perp$ denotes the normal component of electric field strength, and ds denotes the differential surface area. From Gauss's law, electric charge is a source of electric fields. Electric field lines originate from positive charges and terminate at negative charges.

Figure 5.19 shows a solid cylindrical conductor with radius r and with charge q coulombs per meter (assumed positive in the figure), uniformly distributed on the conductor surface. For simplicity, assume that the conductor is (1) sufficiently long that end effects are negligible, and (2) a perfect conductor (that is, zero resistivity, $\rho = 0$).

Inside the perfect conductor, Ohm's law gives $E_{\text{int}} = \rho J = 0$. That is, the internal electric field E_{int} is zero. To determine the electric field outside the conductor, select the cylinder with radius $x > r$ and with 1-meter length, shown in Figure 5.19, as the closed surface for Gauss's law. Due to the uniform charge distribution, the electric field strength E_x is constant on the cylinder. Also, there is no tangential component of E_x, so the electric field is radial to the conductor. Then, integration of (5.8.1) yields

$$\varepsilon E_x (2\pi x)(1) = q(1)$$

$$E_x = \frac{q}{2\pi\varepsilon x} \quad \text{V/m} \qquad (5.8.2)$$

where, for a conductor in free space, $\varepsilon = \varepsilon_0 = 8.854 \times 10^{-12} \, \text{F/m}$.

A plot of the electric field lines is also shown in Figure 5.19. The

Figure 5.19

Perfectly conducting solid cylindrical conductor with uniform charge distribution

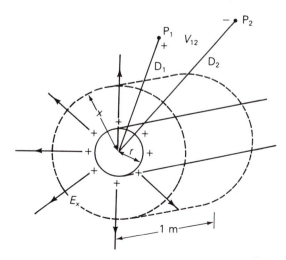

direction of the field lines, denoted by the arrows, is from the positive charges where the field originates, to the negative charges, which in this case are at infinity. If the charge on the conductor surface were negative, then the direction of the field lines would be reversed.

Concentric cylinders surrounding the conductor are constant potential surfaces. The potential difference between two concentric cylinders at distances D_1 and D_2 from the conductor center is

$$V_{12} = \int_{D_1}^{D_2} E_x \, dx \tag{5.8.3}$$

Using (5.8.2) in (5.8.1),

$$V_{12} = \int_{D_1}^{D_2} \frac{q}{2\pi\varepsilon x} \, dx = \frac{q}{2\pi\varepsilon} \ln \frac{D_2}{D_1} \quad \text{volts} \tag{5.8.4}$$

Equation (5.8.4) gives the voltage V_{12} between two points, P_1 and P_2, at distances D_1 and D_2 from the conductor center, as shown in Figure 5.19. Also, in accordance with our notation, V_{12} is the voltage at P_1 with respect to P_2. If q is positive and D_2 is greater than D_1, as shown in the figure, then V_{12} is positive; that is, P_1 is at a higher potential than P_2. Equation (5.8.4) is also valid for either dc or ac. For ac, V_{12} is a phasor voltage and q is a phasor representation of a sinusoidal charge.

Now apply (5.8.4) to the array of M solid cylindrical conductors shown in Figure 5.20. Assume that each conductor m has an ac charge q_m C/m uniformly distributed along the conductor. The voltage V_{kim} between condutors k and i due to the charge q_m acting alone is

$$V_{kim} = \frac{q_m}{2\pi\varepsilon} \ln \frac{D_{im}}{D_{km}} \quad \text{volts} \tag{5.8.5}$$

where $D_{mm} = r_m$ when $k = m$ or $i = m$. In (5.8.5) we have neglected the distortion of the electric field in the vicinity of the other conductors, caused by the fact that the other conductors themselves are constant potential surfaces. V_{kim} can be thought of as the voltage between cylinders with radii

Figure 5.20

Array of M solid cylindrical conductors

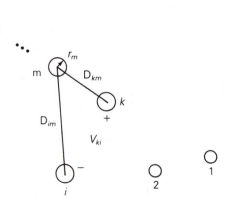

D_{km} and D_{im} concentric to conductor m at points on the cylinders remote from conductors, where there is no distortion.

Using superposition, the voltage V_{ki} between conductors k and i due to all the charges is

$$V_{ki} = \frac{1}{2\pi\varepsilon} \sum_{m=1}^{M} q_m \ln \frac{D_{im}}{D_{km}} \quad \text{volts} \tag{5.8.6}$$

SECTION 5.9

CAPACITANCE: SINGLE-PHASE TWO-WIRE LINE AND THREE-PHASE THREE-WIRE LINE WITH EQUAL PHASE SPACING

The results of the previous section are used here to determine the capacitances of the two relatively simple transmission lines considered in Section 5.5, a single-phase two-wire line and a three-phase three-wire line with equal phase spacing.

First we consider the single-phase two-wire line shown in Figure 5.9. Assume that the conductors are energized by a voltage source such that conductor x has a uniform charge q C/m and, assuming conservation of charge, conductor y has an equal quantity of negative charge $-q$. Using (5.8.6) with $k = x$, $i = y$, and $m = x$, y,

$$V_{xy} = \frac{1}{2\pi\varepsilon} \left[q \ln \frac{D_{yx}}{D_{xx}} - q \ln \frac{D_{yy}}{D_{xy}} \right]$$

$$= \frac{q}{2\pi\varepsilon} \ln \frac{D_{yx}D_{xy}}{D_{xx}D_{yy}} \tag{5.9.1}$$

Using $D_{xy} = D_{yx} = D$, $D_{xx} = r_x$, and $D_{yy} = r_y$, (5.9.1) becomes

$$V_{xy} = \frac{q}{\pi\varepsilon} \ln \frac{D}{\sqrt{r_x r_y}} \quad \text{volts} \tag{5.9.2}$$

For a 1-meter line length, the capacitance between conductors is

$$C_{xy} = \frac{q}{V_{xy}} = \frac{\pi\varepsilon}{\ln\left(\dfrac{D}{\sqrt{r_x r_y}}\right)} \quad \text{F/m line-to-line} \tag{5.9.3}$$

and if $r_x = r_y = r$,

$$C_{xy} = \frac{\pi\varepsilon}{\ln(D/r)} \quad \text{F/m line-to-line} \tag{5.9.4}$$

If the two-wire line is supplied by a transformer with a grounded center tap, then the voltage between each conductor and ground is one-half that given by (5.9.2). That is,

$$V_{xn} = V_{yn} = \frac{V_{xy}}{2} \tag{5.9.5}$$

and the capacitance from either line to the grounded neutral is

$$C_n = C_{xn} = C_{yn} = \frac{q}{V_{xn}} = 2C_{xy}$$

$$= \frac{2\pi\varepsilon}{\ln(D/r)} \quad \text{F/m line-to-neutral} \tag{5.9.6}$$

Circuit representations of the line-to-line and line-to-neutral capacitances are shown in Figure 5.21. Note that if the neutral is open in Figure 5.21(b), the two line-to-neutral capacitances combine in series to give the line-to-line capacitance.

Next consider the three-phase line with equal phase spacing shown in Figure 5.10. We shall neglect the effect of earth and neutral conductors here. To determine the positive-sequence capacitance, assume positive-sequence charges q_a, q_b, q_c such that $q_a + q_b + q_c = 0$. Using (5.8.6) with $k = a$, $i = b$, and $m = a, b, c$, the voltage V_{ab} between conductors a and b is

$$V_{ab} = \frac{1}{2\pi\varepsilon} \left[q_a \ln \frac{D_{ba}}{D_{aa}} + q_b \ln \frac{D_{bb}}{D_{ab}} + q_c \ln \frac{D_{bc}}{D_{ac}} \right] \tag{5.9.7}$$

Using $D_{aa} = D_{bb} = r$, and $D_{ab} = D_{ba} = D_{ca} = D_{cb} = D$, (5.9.7) becomes

$$V_{ab} = \frac{1}{2\pi\varepsilon} \left[q_a \ln \frac{D}{r} + q_b \ln \frac{r}{D} + q_c \ln \frac{D}{D} \right]$$

$$= \frac{1}{2\pi\varepsilon} \left[q_a \ln \frac{D}{r} + q_b \ln \frac{r}{D} \right] \quad \text{volts} \tag{5.9.8}$$

Note that the third term in (5.9.8) is zero because conductors a and b are equidistant from conductor c. Thus, conductors a and b lie on a constant potential cylinder for the electric field due to q_c.

Similarly, using (5.8.6) with $k = a$, $i = c$, and $m = a, b, c$, the voltage V_{ac} is

$$V_{ac} = \frac{1}{2\pi\varepsilon} \left[q_a \ln \frac{D_{ca}}{D_{aa}} + q_b \ln \frac{D_{cb}}{D_{ab}} + q_c \ln \frac{D_{cc}}{D_{ac}} \right]$$

$$= \frac{1}{2\pi\varepsilon} \left[q_a \ln \frac{D}{r} + q_b \ln \frac{D}{D} + q_c \ln \frac{r}{D} \right]$$

$$= \frac{1}{2\pi\varepsilon} \left[q_a \ln \frac{D}{r} + q_c \ln \frac{r}{D} \right] \quad \text{volts} \tag{5.9.9}$$

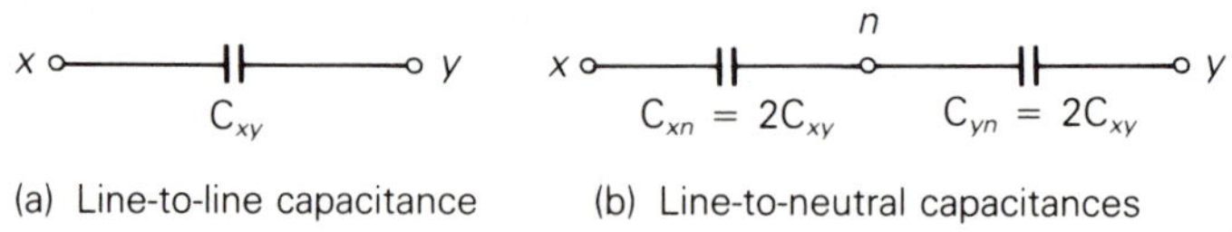

Figure 5.21

Circuit representation of capacitances for a single-phase two-wire line

(a) Line-to-line capacitance

(b) Line-to-neutral capacitances

Recall that for positive-sequence voltages,

$$V_{ab} = \sqrt{3} V_{an} \underline{/+30^\circ} = \sqrt{3} V_{an} \left[\frac{\sqrt{3}}{2} + j\frac{1}{2} \right]$$ (5.9.10)

$$V_{ac} = -V_{ca} = \sqrt{3} V_{an} \underline{/-30^\circ} = \sqrt{3} V_{an} \left[\frac{\sqrt{3}}{2} - j\frac{1}{2} \right]$$ (5.9.11)

Adding (5.9.10) and (5.9.11) yields

$$V_{ab} + V_{ac} = 3V_{an}$$ (5.9.12)

Using (5.9.8) and (5.9.9) in (5.9.12),

$$V_{an} = \frac{1}{3} \left(\frac{1}{2\pi\varepsilon} \right) \left[2q_a \ln \frac{D}{r} + (q_b + q_c) \ln \frac{r}{D} \right]$$ (5.9.13)

and with $q_b + q_c = -q_a$,

$$V_{an} = \frac{1}{2\pi\varepsilon} q_a \ln \frac{D}{r} \quad \text{volts}$$ (5.9.14)

The capacitance-to-neutral per line length is

$$C_{an} = \frac{q_a}{V_{an}} = \frac{2\pi\varepsilon}{\ln(D/r)} \quad \text{F/m line-to-neutral}$$ (5.9.15)

Due to symmetry the same result is obtained for $C_{bn} = q_b/V_{bn}$ and $C_{cn} = q_c/V_{cn}$. For positive sequence, however, only one phase need be considered. Thus, (5.9.15) gives the positive-sequence capacitance, which also equals the negative-sequence capacitance:

$$C_1 = C_2 = \frac{2\pi\varepsilon}{\ln(D/r)} \quad \text{F/m}$$ (5.9.16)

Circuit representations of the sequence capacitances are shown in Figure 5.22.

To determine the zero-sequence capacitance C_0, we must consider the neutral conductors and earth. The neutral conductors and earth also affect the positive- (and negative-) sequence capacitance, since they alter the electric field. We develop equations for the sequence capacitance matrix of more general three-phase lines in Section 5.11.

Figure 5.22

Circuit representation of the sequence capacitances of a three-phase line with equal phase spacing

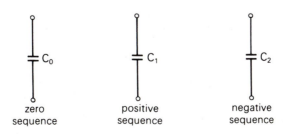

SECTION 5.10

CAPACITANCE: STRANDED CONDUCTORS, UNEQUAL PHASE SPACING, BUNDLED CONDUCTORS

Equations (5.9.6) and (5.9.16) are based on the assumption that the conductors are solid cylindrical conductors with zero resistivity. The electric field inside these conductors is zero, and the external electric field is perpendicular to the conductor surfaces. Practical conductors with resistivities similar to those listed in Table 5.3 have a small internal electric field. As a result, the external electric field is slightly altered near the conductor surfaces. Also, the electric field near the surface of a stranded conductor is not the same as that of a solid cylindrical conductor. However, it is normal practice when calculating line capacitance to replace a stranded conductor by a perfectly conduct-ing solid cylindrical conductor whose radius equals the outside radius of the stranded conductor. The resulting error in capacitance is small since only the electric field near the conductor surfaces is affected.

Also, (5.8.2) is based on the assumption that there is uniform charge distribution. But conductor charge distribution is nonuniform in the presence of other charged conductors. Therefore (5.9.6) and (5.9.16), which are derived from (5.8.2), are not exact. However, the nonuniformity of conductor charge distribution can be shown to have a negligible effect on line capacitance.

For three-phase lines with unequal phase spacing, positive-sequence voltages are not obtained with positive-sequence charges. Instead, unbalanced line-to-neutral voltages occur, and the phase-to-neutral capacitances are unequal. Balance can be restored by transposing the line such that each phase occupies each position for one-third of the line length. If equations similar to (5.9.7) for V_{ab} as well as for V_{ac} are written for each position in the transposition cycle, and are then averaged and used in (5.9.12)–(5.9.14), the resulting positive- and negative-sequence capacitance becomes

$$C_1 = C_2 = \frac{2\pi\varepsilon}{\ln(D_{eq}/r)} \quad \text{F/m} \tag{5.10.1}$$

where

$$D_{eq} = \sqrt[3]{D_{ab}D_{bc}D_{ac}} \tag{5.10.2}$$

Figure 5.23 shows a bundled conductor line with two conductors per bundle. To determine the positive-sequence capacitance of this line, assume

Figure 5.23

Three-phase line with two conductors per bundle

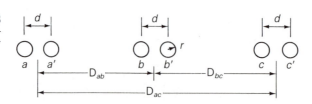

positive-sequence charges q_a, q_b, q_c for each phase such that $q_a + q_b + q_c = 0$. Assume that the conductors in each bundle, which are in parallel, share the charges equally. Thus conductors a and a' each have the charge $q_a/2$. Also assume that the phase spacings are much larger than the bundle spacings so that D_{ab} may be used instead of $(D_{ab} - d)$ or $(D_{ab} + d)$. Then, using (5.8.6) with $k = a$, $i = b$, $m = a, a', b, b', c, c'$,

$$V_{ab} = \frac{1}{2\pi\varepsilon}\left[\frac{q_a}{2}\ln\frac{D_{ba}}{D_{aa}} + \frac{q_a}{2}\ln\frac{D_{ba'}}{D_{aa'}} + \frac{q_b}{2}\ln\frac{D_{bb}}{D_{ab}}\right.$$

$$\left. + \frac{q_b}{2}\ln\frac{D_{bb'}}{D_{ab'}} + \frac{q_c}{2}\ln\frac{D_{bc}}{D_{ac}} + \frac{q_c}{2}\ln\frac{D_{bc'}}{D_{ac'}}\right]$$

$$= \frac{1}{2\pi\varepsilon}\left[\frac{q_a}{2}\left(\ln\frac{D_{ab}}{r} + \ln\frac{D_{ab}}{d}\right) + \frac{q_b}{2}\left(\ln\frac{r}{D_{ab}} + \ln\frac{d}{D_{ab}}\right)\right. \quad (5.10.3)$$

$$\left. + \frac{q_c}{2}\left(\ln\frac{D_{bc}}{D_{ac}} + \ln\frac{D_{bc}}{D_{ac'}}\right)\right]$$

$$= \frac{1}{2\pi\varepsilon}\left[q_a\ln\frac{D_{ab}}{\sqrt{rd}} + q_b\ln\frac{\sqrt{rd}}{D_{ab}} + q_c\ln\frac{D_{bc}}{D_{ac}}\right] \quad (5.10.3)$$

Equation (5.10.3) is the same as (5.9.7), except that D_{aa} and D_{bb} in (5.9.7) are replaced by $\sqrt{rd}$ in this equation. Therefore, for a transposed line, derivation of the positive- and negative-sequence capacitance would yield

$$C_1 = C_2 = \frac{2\pi\varepsilon}{\ln(D_{eq}/D_{SC})} \quad \text{F/m} \quad (5.10.4)$$

where

$$D_{SC} = \sqrt{rd} \quad \text{for a two-conductor bundle} \quad (5.10.5)$$

Similarly,

$$D_{SC} = \sqrt[3]{rd^2} \quad \text{for a three-conductor bundle} \quad (5.10.6)$$

$$D_{SC} = 1.091\sqrt[4]{rd^3} \quad \text{for a four-conductor bundle} \quad (5.10.7)$$

Equation (5.10.4) for capacitance is analogous to (5.6.22) for inductance. In both cases D_{eq}, given by (5.6.17) or (5.10.2), is the geometric mean of the distances between phases. Also, (5.10.5)–(5.10.7) for D_{SC} are analogous to (5.6.19)–(5.6.21) for D_{SL}, except that the conductor outside radius r replaces the conductor GMR D_S.

The current supplied to the transmission-line capacitance is called *charging current*. For a single-phase circuit operating at line-to-line voltage $V_{xy} = V_{xy}\underline{/0°}$, the charging current is

$$I_{chg} = Y_{xy}V_{xy} = j\omega C_{xy}V_{xy} \quad \text{A} \quad (5.10.8)$$

As shown in Chapter 2, a capacitor delivers reactive power. From (2.3.5), the reactive power delivered by the line-to-line capacitance is

$$Q_C = \frac{V_{xy}^2}{X_c} = Y_{xy}V_{xy}^2 = \omega C_{xy}V_{xy}^2 \quad \text{var} \tag{5.10.9}$$

For a completely transposed three-phase line that has positive-sequence voltages with $V_{an} = V_{LN}\underline{/0°}$, the phase a charging current is

$$I_{chg} = YV_{an} = j\omega C_1 V_{LN} \quad \text{A} \tag{5.10.10}$$

and the reactive power delivered by phase a is

$$Q_{C1\phi} = YV_{an}^2 = \omega C_1 V_{LN}^2 \quad \text{var} \tag{5.10.11}$$

The total reactive power supplied by the three-phase line is

$$Q_{C3\phi} = 3Q_{C1\phi} = 3\omega C_1 V_{LN}^2 = \omega C_1 V_{LL}^2 \quad \text{var} \tag{5.10.12}$$

EXAMPLE 5.6 | **Capacitance, admittance, and reactive power supplied: single-phase line**

For the single-phase line in Example 5.3, determine the line-to-line capacitance in F and the line-to-line admittance in S. If the line voltage is 20 kV, determine the reactive power in kvar supplied by this capacitance.

Solution   From Table A.3, the outside radius of a 4/0 12-strand copper conductor is

$$r = \frac{0.552}{2} \text{ in.} \times \frac{1 \text{ ft}}{12 \text{ in.}} = 0.023 \text{ ft}$$

and from (5.9.4),

$$C_{xy} = \frac{\pi(8.854 \times 10^{-12})}{\ln\left(\dfrac{5}{0.023}\right)} = 5.169 \times 10^{-12} \quad \text{F/m}$$

or

$$C_{xy} = 5.169 \times 10^{-12} \frac{\text{F}}{\text{m}} \times 1609 \frac{\text{m}}{\text{mi}} \times 20 \text{ mi} = 1.66 \times 10^{-7} \text{ F}$$

and the shunt admittance is

$$Y_{xy} = j\omega C_{xy} = j(2\pi60)(1.66 \times 10^{-7})$$
$$= j6.27 \times 10^{-5} \quad \text{S line-to-line}$$

From (5.10.9),

$$Q_C = (6.27 \times 10^{-5})(20 \times 10^3)^2 = 25.1 \quad \text{kvar}$$

EXAMPLE 5.7	**Positive-sequence capacitance and shunt admittance; charging current and reactive power supplied: three-phase line**

For the three-phase line in Example 5.5, determine the positive-sequence capacitance in F and the shunt admittance in S. If the line voltage is 345 kV, determine the charging current in kA per phase and the total reactive power in Mvar supplied by the line capacitance. Assume positive-sequence voltages.

Solution   From Table A.4, the outside radius of a 795,000 cmil 26/2 ACSR conductor is

$$r = \frac{1.108}{2} \text{ in.} \times 0.0254 \frac{\text{m}}{\text{in.}} = 0.0141 \text{ m}$$

From (5.10.5), the equivalent radius of the two-conductor bundle is

$$D_{SC} = \sqrt{(0.0141)(0.40)} = 0.0750 \text{ m}$$

$D_{eq} = 12.6$ m is the same as in Example 5.5. Therefore, from (5.10.4),

$$C_1 = \frac{(2\pi)(8.854 \times 10^{-12})}{\ln\left(\frac{12.6}{0.0750}\right)} \frac{\text{F}}{\text{m}} \times 1000 \frac{\text{m}}{\text{km}} \times 200 \text{ km}$$

$$= 2.17 \times 10^{-6} \text{ F}$$

The positive-sequence shunt admittance is

$$Y_1 = j\omega C_1 = j(2\pi 60)(2.17 \times 10^{-6})$$
$$= j8.19 \times 10^{-4} \text{ S}$$

From (5.10.10),

$$I_{chg} = |I_{chg}| = (8.19 \times 10^{-4})\left(\frac{345}{\sqrt{3}}\right) = 0.163 \text{ kA/phase}$$

and from (5.10.12),

$$Q_{C3\phi} = (8.19 \times 10^{-4})(345)^2 = 97.5 \text{ Mvar} \qquad \blacksquare$$

SECTION 5.11

SHUNT ADMITTANCES: LINES WITH NEUTRAL CONDUCTORS AND EARTH RETURN

In this section we develop equations suitable for computer calculation of the shunt admittances for the three-phase line shown in Figure 5.16. We approximate the earth surface as a perfectly conducting horizontal plane, even though the earth under the line may have irregular terrain and resistivities as shown in Table 5.4.

The effect of the earth plane is accounted for by the *method of images*, described as follows. Consider a single conductor with uniform charge distribution and with height H above a perfectly conducting earth plane, as shown in Figure 5.24(a). When the conductor has a positive charge, an equal quantity of negative charge is induced on the earth. The electric field lines will originate from the positive charges on the conductor and terminate at the negative charges on the earth. Also, the electric field lines are perpendicular to the surfaces of the conductor and earth.

Now replace the earth by the image conductor shown in Figure 5.24(b), which has the same radius as the original conductor, lies directly below the original conductor with conductor separation $H_{11} = 2H$, and has an equal quantity of negative charge. The electric field above the dotted line representing the location of the removed earth plane in Figure 5.24(b) is identical to the electric field above the earth plane in Figure 5.24(a). Therefore, the voltage between any two points above the earth is the same in both figures.

Figure 5.24

Method of images

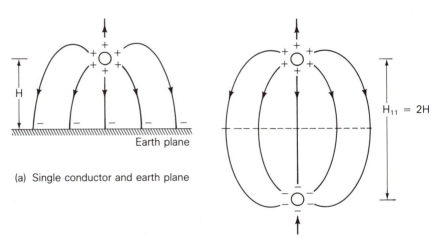

(a) Single conductor and earth plane

(b) Earth plane replaced by image conductor

EXAMPLE 5.8 | **Effect of earth on capacitance: single-phase line**

If the single-phase line in Example 5.6 has flat horizontal spacing with 18-ft average line height, determine the effect of the earth on capacitance. Assume a perfectly conducting earth plane.

Solution The earth plane is replaced by a separate image conductor for each overhead conductor, and the conductors are charged as shown in Figure 5.25. From

Figure 5.25

Single-phase line for Example 5.8

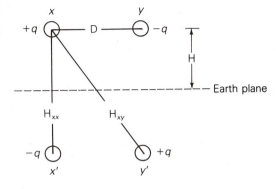

(5.8.6), the voltage between conductors x and y is

$$V_{xy} = \frac{q}{2\pi\varepsilon}\left[\ln\frac{D_{yx}}{D_{xx}} - \ln\frac{D_{yy}}{D_{xy}} - \ln\frac{H_{yx}}{H_{xx}} + \ln\frac{H_{yy}}{H_{xy}}\right]$$

$$= \frac{q}{2\pi\varepsilon}\left[\ln\frac{D_{yx}D_{xy}}{D_{xx}D_{yy}} - \ln\frac{H_{yx}H_{xy}}{H_{xx}H_{yy}}\right]$$

$$= \frac{q}{\pi\varepsilon}\left[\ln\frac{D}{r} - \ln\frac{H_{xy}}{H_{xx}}\right]$$

The line-to-line capacitance is

$$C_{xy} = \frac{q}{V_{xy}} = \frac{\pi\varepsilon}{\ln\dfrac{D}{r} - \ln\dfrac{H_{xy}}{H_{xx}}} \quad \text{F/m}$$

Using $D = 5\,\text{ft}$, $r = 0.023\,\text{ft}$, $H_{xx} = 2H = 36\,\text{ft}$, and $H_{xy} = \sqrt{(36)^2 + (5)^2} = 36.346\,\text{ft}$,

$$C_{xy} = \frac{\pi(8.854 \times 10^{-12})}{\ln\dfrac{5}{0.023} - \ln\dfrac{36.346}{36}} = 5.178 \times 10^{-12} \quad \text{F/m}$$

compared with 5.169×10^{-12} F/m in Example 5.6. The effect of the earth plane is to slightly increase the capacitance. Note that as the line height H increases, the ratio H_{xy}/H_{xx} approaches 1, $\ln(H_{xy}/H_{xx}) \to 0$, and the effect of the earth becomes negligible.   ■

For the three-phase line with N neutral conductors shown in Figure 5.26, the perfectly conducting earth plane is replaced by a separate image conductor for each overhead conductor. The overhead conductors a, b, c, $n1$, $n2, \ldots, nN$ carry charges q_a, q_b, q_c, $q_{n1}, \ldots, q_{nN}$, and the image conductors a', b', c', $n1', \ldots, nN'$ carry charges $-q_a$, $-q_b$, $-q_c$, $-q_{n1}, \ldots, -q_{nN}$. Applying (5.8.6) to determine the voltage $V_{kk'}$ between any conductor k and its image

Figure 5.26

Three-phase line with neutral conductors and with earth plane replaced by image conductors

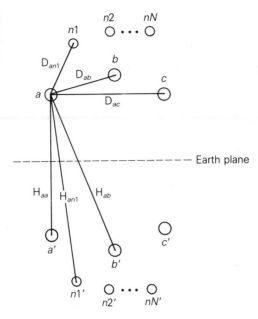

conductor k',

$$V_{kk'} = \frac{1}{2\pi\varepsilon}\left[\sum_{m=a}^{nN} q_m \ln \frac{H_{km}}{D_{km}} - \sum_{m=a}^{nN} q_m \ln \frac{D_{km}}{H_{km}}\right]$$

$$= \frac{2}{2\pi\varepsilon}\sum_{m=a}^{nN} q_m \ln \frac{H_{km}}{D_{km}} \tag{5.11.1}$$

where $D_{kk} = r_k$ and D_{km} is the distance between overhead conductors k and m. H_{km} is the distance between overhead conductor k and image conductor m. By symmetry, the voltage V_{kn} between conductor k and the earth is one-half of $V_{kk'}$.

$$V_{kn} = \frac{1}{2} V_{kk'} = \frac{1}{2\pi\varepsilon}\sum_{m=a}^{nN} q_m \ln \frac{H_{km}}{D_{km}} \tag{5.11.2}$$

where

$$k = a, b, c, n1, n2, \ldots, nN$$

$$m = a, b, c, n1, n2, \ldots, nN$$

Since all the neutral conductors are grounded to the earth,

$$V_{kn} = 0 \qquad \text{for } k = n1, n2, \ldots, nN \tag{5.11.3}$$

In matrix format, (5.11.2) and (5.11.3) are

$$
\begin{bmatrix} V_{an} \\ V_{bn} \\ V_{cn} \\ \hline 0 \\ \vdots \\ 0 \end{bmatrix} =
\begin{bmatrix}
\overbrace{\begin{matrix} P_{aa} & P_{ab} & P_{ac} \\ P_{ba} & P_{bb} & P_{bc} \\ P_{ca} & P_{cb} & P_{cc} \end{matrix}}^{\mathbf{P}_A} &
\overbrace{\begin{matrix} P_{an1} & \cdots & P_{anN} \\ P_{bn1} & \cdots & P_{bnN} \\ P_{cn1} & \cdots & P_{cnN} \end{matrix}}^{\mathbf{P}_B} \\
\hline
\underbrace{\begin{matrix} P_{n1a} & P_{n1b} & P_{n1c} \\ \vdots & & \\ P_{nNa} & P_{nNb} & P_{nNc} \end{matrix}}_{\mathbf{P}_C} &
\underbrace{\begin{matrix} P_{n1n1} & \cdots & P_{n1nN} \\ \vdots & & \\ P_{nNn1} & \cdots & P_{nNnN} \end{matrix}}_{\mathbf{P}_D}
\end{bmatrix}
\begin{bmatrix} q_a \\ q_b \\ q_c \\ \hline q_{n1} \\ \vdots \\ q_{nN} \end{bmatrix}
\tag{5.11.4}
$$

The elements of the $(3 + N) \times (3 + N)$ matrix **P** are

$$
P_{km} = \frac{1}{2\pi\varepsilon} \ln \frac{H_{km}}{D_{km}} \quad \text{m/F}
\tag{5.11.5}
$$

where

$$k = a, b, c, n1, \ldots, nN$$

$$m = a, b, c, n1, \ldots, nN$$

Equation (5.11.4) is now partitioned as shown above to obtain

$$
\begin{bmatrix} V_P \\ \hline 0 \end{bmatrix} = \begin{bmatrix} \mathbf{P}_A & \mathbf{P}_B \\ \hline \mathbf{P}_C & \mathbf{P}_D \end{bmatrix} \begin{bmatrix} q_P \\ \hline q_n \end{bmatrix}
\tag{5.11.6}
$$

V_P is the three-dimensional vector of phase-to-neutral voltages. q_P is the three-dimensional vector of phase-conductor charges and q_n is the N vector of neutral conductor charges. The $(3 + N) \times (3 + N)$ **P** matrix is partitioned as shown in (5.11.4) to obtain:

$\mathbf{P}_A$ with dimension 3×3

$\mathbf{P}_B$ with dimension $3 \times N$

$\mathbf{P}_C$ with dimension $N \times 3$

$\mathbf{P}_D$ with dimension $N \times N$

Equation (5.11.6) is rewritten as two separate equations:

$$
V_P = \mathbf{P}_A q_P + \mathbf{P}_B q_n
\tag{5.11.7}
$$

$$
0 = \mathbf{P}_C q_P + \mathbf{P}_D q_n
\tag{5.11.8}
$$

Then (5.11.8) is solved for q_n, which is used in (5.11.7) to obtain

$$
V_P = (\mathbf{P}_A - \mathbf{P}_B \mathbf{P}_D^{-1} \mathbf{P}_C) q_P
\tag{5.11.9}
$$

or

$$
q_P = \mathbf{C}_P V_P
\tag{5.11.10}
$$

where

$$
\mathbf{C}_P = (\mathbf{P}_A - \mathbf{P}_B \mathbf{P}_D^{-1} \mathbf{P}_C)^{-1} \quad \text{F/m}
\tag{5.11.11}
$$

Equation (5.11.10), the desired result, relates the phase-conductor charges to the phase-to-neutral voltages. $\mathbf{C_P}$ is the 3×3 matrix of phase capacitances whose elements are denoted

$$\mathbf{C_P} = \begin{bmatrix} C_{aa} & C_{ab} & C_{ac} \\ C_{ab} & C_{bb} & C_{bc} \\ C_{ac} & C_{bc} & C_{cc} \end{bmatrix} \quad \text{F/m} \tag{5.11.12}$$

It can be shown that $\mathbf{C_P}$ is a symmetric matrix whose diagonal terms C_{aa}, C_{bb}, C_{cc} are positive, and whose off-diagonal terms C_{ab}, C_{bc}, C_{ac} are negative. This indicates that when a positive line-to-neutral voltage is applied to one phase, a positive charge is induced on that phase and negative charges are induced on the other phases, which is physically correct.

In general, $\mathbf{C_P}$ does not satisfy the conditions for a symmetrical capacitance matrix. However, if the line is completely transposed, the diagonal and off-diagaonal elements of $\mathbf{C_P}$ are averaged to obtain

$$\hat{\mathbf{C}}_\mathbf{P} = \begin{bmatrix} \hat{C}_{aa} & \hat{C}_{ab} & \hat{C}_{ab} \\ \hat{C}_{ab} & \hat{C}_{aa} & \hat{C}_{ab} \\ \hat{C}_{ab} & \hat{C}_{ab} & \hat{C}_{aa} \end{bmatrix} \quad \text{F/m} \tag{5.11.13}$$

where

$$\hat{C}_{aa} = \tfrac{1}{3}(\mathbf{C}_{aa} + \mathbf{C}_{bb} + \mathbf{C}_{cc}) \quad \text{F/m} \tag{5.11.14}$$

$$\hat{C}_{ab} = \tfrac{1}{3}(\mathbf{C}_{ab} + \mathbf{C}_{bc} + \mathbf{C}_{ac}) \quad \text{F/m} \tag{5.11.15}$$

$\hat{\mathbf{C}}_\mathbf{P}$ is a symmetrical capacitance matrix.

The shunt phase admittance matrix is given by

$$\mathbf{Y_P} = j\omega\mathbf{C_P} = j(2\pi f)\mathbf{C_P} \quad \text{S/m} \tag{5.11.16}$$

or, for a completely transposed line,

$$\hat{\mathbf{Y}}_\mathbf{P} = j\omega\hat{\mathbf{C}}_\mathbf{P} = j(2\pi f)\hat{\mathbf{C}}_\mathbf{P} \quad \text{S/m} \tag{5.11.17}$$

Equation (5.11.16) can also be transformed to the sequence domain to obtain

$$\mathbf{Y_S} = \mathbf{A}^{-1}\mathbf{Y_P A} \tag{5.11.18}$$

where

$$\mathbf{Y_S} = \mathbf{G_S} + j(2\pi f)\mathbf{C_S} \tag{5.11.19}$$

$$\mathbf{C_S} = \begin{bmatrix} C_0 & C_{01} & C_{02} \\ C_{10} & C_1 & C_{12} \\ C_{20} & C_{21} & C_2 \end{bmatrix} \quad \text{F/m} \tag{5.11.20}$$

In general, $\mathbf{C_S}$ is not diagonal. However, for the completely transposed line,

$$\hat{\mathbf{Y}}_\mathbf{S} = \mathbf{A}^{-1}\hat{\mathbf{Y}}_\mathbf{P}\mathbf{A} = \begin{bmatrix} \hat{y}_0 & 0 & 0 \\ 0 & \hat{y}_1 & 0 \\ 0 & 0 & \hat{y}_2 \end{bmatrix} = j(2\pi f) \begin{bmatrix} \hat{C}_0 & 0 & 0 \\ 0 & \hat{C}_1 & 0 \\ 0 & 0 & \hat{C}_2 \end{bmatrix} \tag{5.11.21}$$

where

$$\hat{C}_0 = \hat{C}_{aa} + 2\hat{C}_{ab} \quad \text{F/m} \tag{5.11.22}$$

$$\hat{C}_1 = \hat{C}_2 = \hat{C}_{aa} - \hat{C}_{ab} \quad \text{F/m} \tag{5.11.23}$$

Since $\hat{C}_{ab}$ is negative, the zero-sequence capacitance $\hat{C}_0$ is usually much less than the positive- or negative-sequence capacitance.

Circuit representations of the phase and sequence capacitances of a completely transposed three-phase line are shown in Figure 5.27. A personal computer program that computes the shunt admittance matrices is described in Section 5.14.

Figure 5.27

Circuit representations of the capacitances of a completely transposed three-phase line

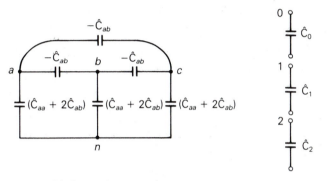

(a) Phase domain

(b) Sequence domain

SECTION 5.12

ELECTRIC FIELD STRENGTH AT CONDUCTOR SURFACES AND AT GROUND LEVEL

When the electric field strength at a conductor surface exceeds the breakdown strength of air, current discharges occur. This phenomenon, called corona, causes additional line losses (corona loss), communications interference, and audible noise. Although breakdown strength depends on many factors, a rough value is 30 kV/cm in a uniform electric field for dry air at atmospheric pressure. The presence of water droplets or rain can lower this value significantly. To control corona, transmission lines are usually designed to maintain calculated values of conductor surface electric field strength below $20\,\text{kV}_{\text{rms}}/\text{cm}$.

When line capacitances are determined and conductor voltages are known, the conductor charges can be calculated from (5.9.3) for a single-phase line or from (5.11.10) for a three-phase line. Then the electric field strength at the surface of one phase conductor, neglecting the electric fields due to charges on other phase conductors and neutral wires, is, from (5.8.2),

$$E_r = \frac{q}{2\pi\varepsilon r} \quad \text{V/m} \tag{5.12.1}$$

where r is the conductor outside radius.

For bundled conductors with N_b conductors per bundle and with charge q C/m per phase, the charge per conductor is q/N_b, and

$$E_{\text{rave}} = \frac{q/N_b}{2\pi\varepsilon r} \quad \text{V/m} \tag{5.12.2}$$

Equation (5.12.2) represents an average value for an individual conductor in a bundle. The maximum electric field strength at the surface of one conductor due to all charges in a bundle, obtained by the vector addition of electric fields (as shown in Figure 5.28), is as follows:

Two-conductor bundle ($N_b = 2$):

$$E_{r\text{max}} = \frac{q/2}{2\pi\varepsilon r} + \frac{q/2}{2\pi\varepsilon d} = \frac{q/2}{2\pi\varepsilon r}\left(1 + \frac{r}{d}\right)$$

$$= E_{\text{rave}}\left(1 + \frac{r}{d}\right) \tag{5.12.3}$$

Three-conductor bundle ($N_b = 3$):

$$E_{r\text{max}} = \frac{q/3}{2\pi\varepsilon}\left(\frac{1}{r} + \frac{2\cos 30°}{d}\right) = E_{\text{rave}}\left(1 + \frac{r\sqrt{3}}{d}\right) \tag{5.12.4}$$

Four-conductor bundle ($N_b = 4$):

$$E_{r\text{max}} = \frac{q/4}{2\pi\varepsilon}\left(\frac{1}{r} + \frac{1}{d\sqrt{2}} + \frac{2\cos 45°}{d}\right) = E_{\text{rave}}\left[1 + \frac{r}{d}(2.1213)\right] \tag{5.12.5}$$

Although the electric field strength at ground level is much less than at conductor surfaces where corona occurs, there are still capacitive coupling effects. Charges are induced on ungrounded equipment such as vehicles with rubber tires located near a line. If a person contacts the vehicle and ground, a discharge current will flow to ground. Transmission-line heights are designed to maintain discharge currents below prescribed levels for any equipment that may be on the right-of-way. Table 5.5 shows examples of maximum ground-level electric field strength.

Figure 5.28

Vector addition of electric fields at the surface of one conductor in a bundle

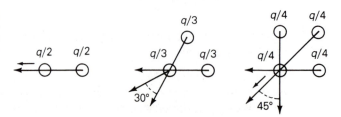

Table 5.5

Examples of maximum ground-level electric field strength versus transmission-line voltage [1]

LINE VOLTAGE kV$_{rms}$	MAXIMUM GROUND-LEVEL ELECTRIC FIELD STRENGTH kV$_{rms}$/m
23 (1ϕ)	0.01−0.025
23 (3ϕ)	0.01−0.05
115	0.1 −0.2
345	2.3 −5.0
345 (double circuit)	5.6
500	8.0
765	10.0

Figure 5.29

Ground-level electric field strength due to an overhead conductor and its image

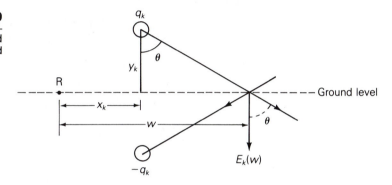

As shown in Figure 5.29, the ground-level electric field strength due to charged conductor k and its image conductor is perpendicular to the earth plane, with value

$$E_k(w) = \left(\frac{q_k}{2\pi\varepsilon}\right) \frac{2\cos\theta}{\sqrt{y_k^2 + (w - x_k)^2}}$$

$$= \left(\frac{q_k}{2\pi\varepsilon}\right) \frac{2y_k}{y_k^2 + (w - x_k)^2} \quad \text{V/m} \tag{5.12.6}$$

where (x_k, y_k) are the horizontal and vertical coordinates of conductor k with respect to reference point R, w is the horizontal coordinate of the ground-level point where the electric field strength is to be determined, and q_k is the charge on conductor k. The total ground-level electric field is the phasor sum of terms $E_k(w)$ for all overhead conductors. A lateral profile of ground-level electric field strength is obtained by varying w from the center of the line to the edge of the right-of-way.

EXAMPLE 5.9

Conductor surface and ground-level electric field strengths: single-phase line

For the single-phase line of Example 5.8, calculate the conductor surface electric field strength in kV$_{rms}$/cm. Also calculate the ground-level electric field in kV$_{rms}$/m directly under conductor x. The line voltage is 20 kV.

Solution   From Example 5.8, $C_{xy} = 5.178 \times 10^{-12}\,\text{F/m}$. Using (5.9.3) with $V_{xy} = 20\underline{/0°}\,\text{kV}$,

$$q_x = -q_y = (5.178 \times 10^{-12})(20 \times 10^3 \underline{/0°}) = 1.036 \times 10^{-7}\underline{/0°}\ \ \text{C/m}$$

From (5.12.1), the conductor surface electric field strength is, with $r = 0.023\,\text{ft} = 0.00701\,\text{m}$,

$$E_r = \frac{1.036 \times 10^{-7}}{(2\pi)(8.854 \times 10^{-12})(0.00701)}\frac{\text{V}}{\text{m}} \times \frac{\text{kV}}{1000\,\text{V}} \times \frac{\text{m}}{100\,\text{cm}}$$

$$= 2.66\ \ \text{kV}_{\text{rms}}/\text{cm}$$

Selecting the center of the line as the reference point R, the coordinates (x_x, y_x) for conductor x are $(-2.5\,\text{ft}, 18\,\text{ft})$ or $(-0.762\,\text{m}, 5.49\,\text{m})$ and $(+0.762\,\text{m}, 5.49\,\text{m})$ for conductor y. The ground-level electric field directly under conductor x, where $w = -0.762\,\text{m}$, is, from (5.12.6),

$$E(-0.762) = E_x(-0.762) + E_y(-0.762)$$

$$= \frac{1.036 \times 10^{-7}}{(2\pi)(8.85 \times 10^{-12})}\left[\frac{(2)(5.49)}{(5.49)^2} - \frac{(2)(5.49)}{(5.49)^2 + (0.762 + 0.762)^2}\right]$$

$$= 1.862 \times 10^3(0.364 - 0.338) = 48.5\underline{/0°}\ \text{V/m} = 0.0485\ \text{kV/m}$$

For this 20-kV line, the electric field strengths at the conductor surface and at ground level are low enough to be of relatively small concern. For EHV lines, electric field strengths and the possibility of corona and shock hazard are of more concern.   ∎

PARALLEL CIRCUIT THREE-PHASE LINES

If two parallel three-phase circuits are close together, either on the same tower as in Figure 5.3, or on the same right-of-way, there are mutual inductive and capacitive couplings between the two circuits. When calculating the equivalent series impedance and shunt admittance matrices, these couplings should not be neglected unless the spacing between the circuits is large.

Consider the double-circuit line shown in Figure 5.30. For simplicity, assume that the lines are not transposed. Since both are connected in parallel, they have the same series-voltage drop for each phase. Following the same

Figure 5.30

Single-line diagram of a double-circuit line

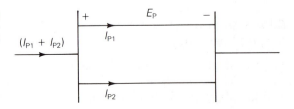

procedure as in Section 5.7, we can write $2(6 + N)$ equations similar to (5.7.6)–(5.7.9): six equations for the overhead phase conductors, N equations for the overhead neutral conductors, and $(6 + N)$ equations for the earth return conductors. After lumping the neutral voltage drop into the voltage drops across the phase conductors, and eliminating the neutral and earth return currents, we obtain

$$\begin{bmatrix} E_P \\ E_P \end{bmatrix} = Z_P \begin{bmatrix} I_{P1} \\ I_{P2} \end{bmatrix} \tag{5.13.1}$$

where E_P is the vector of phase-conductor voltage drops (including the neutral voltage drop), and I_{P1} and I_{P2} are the vectors of phase currents for lines 1 and 2. Z_P is a 6×6 impedance matrix. Solving (5.13.1)

$$\begin{bmatrix} I_{P1} \\ -- \\ I_{P2} \end{bmatrix} = Z_P^{-1} \begin{bmatrix} E_P \\ -- \\ E_P \end{bmatrix} = \left[\begin{array}{c|c} Y_A & Y_B \\ \hline Y_C & Y_D \end{array}\right] \begin{bmatrix} E_P \\ -- \\ E_P \end{bmatrix} = \begin{bmatrix} (Y_A + Y_B) \\ (Y_C + Y_D) \end{bmatrix} E_P \tag{5.13.2}$$

where Y_A, Y_B, Y_C, and Y_D are obtained by partitioning Z_P^{-1} into four 3×3 matrices. Adding I_{P1} and I_{P2},

$$(I_{P1} + I_{P2}) = (Y_A + Y_B + Y_C + Y_D)E_P \tag{5.13.3}$$

and solving for E_P,

$$E_P = Z_{Peq}(I_{P1} + I_{P2}) \tag{5.13.4}$$

where

$$Z_{Peq} = (Y_A + Y_B + Y_C + Y_D)^{-1} \tag{5.13.5}$$

Z_{Peq} is the equivalent 3×3 series phase impedance matrix of the double-circuit line. Note that in (5.13.5) the matrices Y_B and Y_C account for the inductive coupling between the two circuits.

An analogous procedure can be used to obtain the shunt admittance matrix. Following the ideas of Section 5.11, we can write $(6 + N)$ equations similar to (5.11.4). After eliminating the neutral wire charges, we obtain

$$\begin{bmatrix} q_{P1} \\ -- \\ q_{P2} \end{bmatrix} = C_P \begin{bmatrix} V_P \\ -- \\ V_P \end{bmatrix} = \left[\begin{array}{c|c} C_A & C_B \\ \hline C_C & C_D \end{array}\right] \begin{bmatrix} V_P \\ -- \\ V_P \end{bmatrix} = \begin{bmatrix} (C_A + C_B) \\ (C_C + C_D) \end{bmatrix} V_P \tag{5.13.6}$$

where V_P is the vector of phase-to-neutral voltages, and q_{P1} and q_{P2} are the vectors of phase-conductor charges for lines 1 and 2. C_P is a 6×6 capacitance matrix that is partitioned into four 3×3 matrices C_A, C_B, C_C, and C_D. Adding q_{P1} and q_{P2}

$$(q_{P1} + q_{P2}) = C_{Peq} V_P \tag{5.13.7}$$

where

$$C_{Peq} = (C_A + C_B + C_C + C_D) \tag{5.13.8}$$

Also,

$$Y_{Peq} = j\omega C_{Peq} \tag{5.13.9}$$

Y_{Peq} is the equivalent 3×3 shunt admittance matrix of the double-circuit line. The matrices C_B and C_C in (5.13.8) account for the capacitive coupling between the two circuits.

The corresponding sequence matrices of the double-circuit line are computed in the same manner as for the single-circuit line using (5.7.25) and (5.11.18). Also, these ideas can be extended in a straightforward fashion to more than two parallel circuits.

SECTION 5.14

PERSONAL COMPUTER PROGRAM: LINE CONSTANTS

The software package that accompanies this text includes a computer program entitled "LINE CONSTANTS" that calculates the series impedance and shunt admittance matrices of single- and double-circuit three-phase transmission lines. The program also calculates electric field strength at the surface of the phase conductors and a lateral profile of ground-level electric field strength.

Input data for the program consist of: (1) line voltage; (2) frequency; (3) earth resistivity; (4) right-of-way width; and, for each circuit, (5) outside radius, GMR, and resistance of each phase conductor and each neutral wire; (6) number of phase conductors in bundle and bundle spacing; and (7) horizontal and vertical position of each phase and of each neutral wire. Resistance and GMR data are given for a specified temperature, frequency, and current.

The output data consist of: the series phase impedance matrix Z_P, (5.7.19); the series sequence impedance matrix Z_S, (5.7.25); shunt phase admittance matrix Y_P, (5.11.16); shunt sequence admittance matrix Y_S, (5.11.18); conductor surface electric field strength (5.12.1)–(5.12.5); and the ground-level electric field strength from the center to the edge of the right-of-way, (5.12.6). The output also includes sequence impedances and admittances for completely transposed lines, (5.7.27) and (5.11.21). Output data for double-circuit lines are computed from the equations in Section 5.13.

EXAMPLE 5.10

LINE CONSTANTS program

Run the LINE CONSTANTS program for the 765-kV single-circuit line shown in Figure 5.31.

Solution   Program output data are given in Table 5.6.

Figure 5.31

Three-phase 765-kV line for Example 5.10

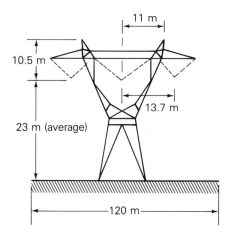

Neutrals:
2—Alumoweld 7 no. 8
Radius = 0.489 cm
GMR = 0.0636 cm
Resistance = 1.52 Ω/km

Phase conductors:
4—ACSR 954 kcmil, 54/7
Radius = 1.519 cm
GMR = 1.229 cm
Resistance = 0.0701 Ω/km
Bundle spacing = 45.7 cm

Earth resistivity = 100 Ωm
Frequency = 60 Hz
Voltage = 765 kV

Table 5.6 Output data for Example 5.10

Series phase impedance matrix Z_P Eq. 5.7.19 Units:Ohms/km

$$
\begin{bmatrix}
0.1181E + 00 + j5.532E - 01 & 0.1009E + 00 + j2.340E - 01 & 0.9813E - 01 + j1.842E - 01 \\
0.1009E + 00 + j2.339E - 01 & 0.1200E + 00 + j5.500E - 01 & 0.1009E + 00 + j2.339E - 01 \\
0.9813E - 01 + j1.842E - 01 & 0.1009E + 00 + j2.340E - 01 & 0.1181E + 00 + j5.532E - 01
\end{bmatrix}
$$

Series sequence impedance matrix Z_S Eq. (5.7.25 Units: Ohms/km

$$
\begin{bmatrix}
0.3187E + 00 + j9.869E - 01 & 0.1264E - 00 - j9.112E - 03 & -.1421E - 01 - j6.389E - 03 \\
-.1421E - 01 - j6.374E - 03 & 0.1875E - 01 + j3.347E - 01 & -.2903E - 01 + j1.814E - 02 \\
0.1262E - 01 - j9.117E - 03 & 0.3022E - 01 + j1.607E - 02 & 0.1875E - 01 + j3.347E - 01
\end{bmatrix}
$$

Shunt phase admittance matrix Y_P Eq. 5.11.16 Units: S/km

$$
\begin{bmatrix}
+ j4.311E - 06 & - j7.666E - 07 & -j2.167E - 07 \\
- j7.666E - 07 & + j4.439E - 06 & -j7.666E - 07 \\
- j2.167E - 07 & - j7.666E - 07 & +j4.311E - 06
\end{bmatrix}
$$

Shunt sequence admittance matrix Y Eq. 5.11. 20 Units: S/km

$$
\begin{bmatrix}
0.0000E + 00 & +j3.187E - 06 & -.1219E - 06 & +j7.036E - 08 & 0.1219E - 06 & +j7.036E - 08 \\
0.1219E - 06 & +j7.036E - 08 & -.3901E - 13 & +j4.937E - 06 & 0.3544E - 06 & -j2.046E - 07 \\
-.1219E - 06 & +j7.036E - 08 & -.3544E - 06 & -j2.046E - 07 & 0.3901E - 13 & +j4.937E - 06
\end{bmatrix}
$$

Conductor surface electric field strength Eqs. 5.12.1–5.12.5

$E_{r\max} = 19.3\,\text{kV}_{\text{rms}}/\text{cm}$

Lateral profile of ground-level electric field strength Eq. 5.12.6

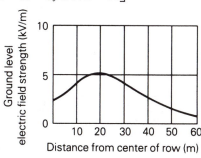

PROBLEMS

Section 5.2

5.1 The *Aluminum Electrical Conductor Handbook* lists a dc resistance of 0.01552 ohm per 1000 ft at 20°C and a 60-Hz resistance of 0.0951 ohm per mile at 50°C for the all-aluminum Marigold conductor, which has 61 strands and whose size is 1113 kcmil. Assuming an increase in resistance of 1.6% for spiraling, calculate and verify the dc resistance. Then calculate the dc resistance at 50°C, and determine the percentage increase due to skin effect.

5.2 One thousand circular mils or 1 kcmil is sometimes designated by the abbreviation MCM. Data for commercial bare aluminum electrical conductors lists a 60-Hz resistance of 0.0740 ohm per kilometer at 75°C for a 954-MCM AAC conductor.
(a) Determine the cross-sectional conducting area of this conductor in square meters.
(b) Find the 60-Hz resistance of this conductor in ohms per kilometer at 45°C.

5.3 A 60-Hz, 765-kV three-phase overhead transmission line has four ACSR 1113 kcmil 54/3 conductors per phase. Determine the 60-Hz resistance of this line in ohms per kilometer per phase at 50°C.

Sections 5.4 and 5.5

5.4 A 60-Hz single-phase, two-wire overhead line has solid cylindrical copper conductors with 1.5 cm diameter. The conductors are arranged in a horizontal configuration with 0.5 m spacing. Calculate in mH/km (a) the inductance of each conductor due to internal flux linkages only, (b) the inductance of each conductor due to both internal and external flux linkages, and (c) the total inductance of the line.

5.5 Rework Problem 5.4 if the diameter of each conductor is: (a) increased by 25% to 1.875 cm, (b) decreased by 25% to 1.125 cm, without changing the phase spacing. Compare the results with those of Problem 5.4.

5.6 A 60-Hz three-phase, three-wire overhead line has solid cylindrical conductors arranged in the form of an equilateral triangle with 4 ft conductor spacing. Conductor diameter is 0.5 in. Calculate the positive-sequence inductance in H/m and the positive-sequence inductive reactance in Ω/km.

5.7 Rework Problem 5.6 if the phase spacing is: (a) increased by 25% to 5 ft, (b) decreased by 25% to 3 ft. Compare the results with those of Problem 5.6.

Section 5.6

5.8 Find the GMR of a stranded conductor consisting of six outer strands surrounding and touching one central strand, all strands having the same radius r.

5.9 A bundle configuration for UHV lines (above 1000 kV) has identical conductors equally spaced around a circle, as shown in Figure 5.32. N_b is the number of conductors in the bundle, A is the circle radius, and D_S is the conductor GMR. Using the distance D_{1n} between conductors 1 and n given by $D_{1n} = 2A \sin[(n-1)\pi/N_b]$ for $n = 1, 2, \ldots, N_b$, and the following trigonometric identity:

$$[2\sin(\pi/N_b)][2\sin(2\pi/N_b)][2\sin(3\pi/N_b)]\cdots[2\sin\{(N_b - 1)\pi/N_b\}] = N_b$$

show that the bundle GMR, denoted D_{SL}, is

$$D_{SL} = [N_b D_S A^{(N_b - 1)}]^{(1/N_b)}$$

Also show that the above formula agrees with (5.6.19)–(5.6.21) for EHV lines with $N_b = 2, 3,$ and 4.

Figure 5.32

Bundle configuration for Problem 5.9

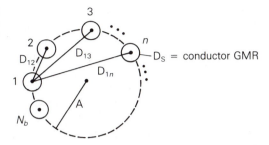

5.10 Determine the GMR of each of the unconventional stranded conductors shown in Figure 5.33. All strands have the same radius r.

Figure 5.33

Unconventional stranded conductors for Problem 5.10

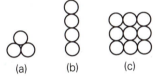

(a)   (b)   (c)

5.11 A 230-kV, 60-Hz, three-phase completely transposed overhead line has one ACSR 954-kcmil conductor per phase and flat horizontal phase spacing, with 7 m between adjacent conductors. Determine the positive-sequence inductance in H/m and the positive-sequence inductive reactance in Ω/km.

5.12 Rework Problem 5.11 if the phase spacing between adjacent conductors is: (a) increased by 10% to 7.7 m, (b) decreased by 10% to 6.3 m. Compare the results with those of Problem 5.11.

5.13 Calculate the positive-sequence inductive reactance in Ω/km of a bundled 500-kV, 60-Hz, three-phase completely transposed overhead line having three ACSR 1113-kcmil conductors per bundle, with 45.7 cm between conductors in the bundle. The horizontal phase spacings between bundle centers are 10, 10, and 20 m.

5.14 Rework Problem 5.13 if the bundled line has: (a) three ACSR, 1351-kcmil conductors per phase, (b) three ACSR, 900-kcmil conductors per phase, without changing the bundle spacing or the phase spacings between bundle centers. Compare the results with those of Problem 5.13.

Section 5.9

5.15 Calculate the capacitance-to-neutral in F/m and the admittance-to-neutral in S/km for the single-phase line in Problem 5.4. Neglect the effect of the earth plane.

5.16 Rework Problem 5.15 if the diameter of each conductor is: (a) increased by 25% to 1.875 cm, (b) decreased by 25% to 1.125 cm. Compare the results with those of Problem 5.15.

5.17 Calculate the positive-sequence shunt capacitance in F/m and the positive-sequence shunt admittance in S/km for the three-phase line in Problem 5.6. Neglect the effect of the earth plane.

5.18 Rework Problem 5.17 if the phase spacing is: (a) increased by 25% to 5 ft, (b) decreased by 25% to 3 ft. Compare the results with those of Problem 5.17.

Section 5.10

5.19   Calculate the positive-sequence shunt capacitance in F/m and the positive-sequence shunt admittance in S/km for the three-phase line in Problem 5.11. Also calculate the line-charging current in kA/phase if the line is 100 km in length and is operated at 230 kV. Neglect the effect of the earth plane.

5.20   Rework Problem 5.19 if the phase spacing between adjacent conductors is: (a) increased by 10% to 7.7 m, (b) decreased by 10% to 6.3 m. Compare the results with those of Problem 5.19.

5.21   Calculate the positive-sequence shunt capacitance in F/m and the positive-sequence shunt admittance in S/km for the line in Problem 5.13. Also calculate the toal reactive power in Mvar/km supplied by the line capacitance when it is operated at 500 kV. Neglect the effect of the earth plane.

5.22   Rework Problem 5.21 if the bundled line has: (a) three ACSR, 1351-kcmil conductors per phase, (b) three ACSR, 900-kcmil conductors per phase, without changing the bundle spacing or the phase spacings between bundle centers.

Section 5.11

5.23   For an average line height of 10 m, determine the effect of the earth on capacitance for the single-phase line in Problem 5.15. Assume a perfectly conducting earth plane.

Section 5.12

5.24   Calculate the conductor surface electric field strength in kV_{rms}/cm for the single-phase line in Problem 5.15 when the line is operating at 20 kV. Also calculate the ground-level electric field strength in kV_{rms}/m directly under one conductor. Assume a line height of 10 m.

5.25   Rework Problem 5.24 if the diameter of each conductor is: (a) increased by 25% to 1.875 cm, (b) decreased by 25% to 1.125 cm, without changing the phase spacings. Compare the results with those of Problem 5.24.

Section 5.14

5.26   Run the LINE CONSTANTS program for Example 5.10 and verify the computed results.

5.27   The phase conductors in Example 5.10 are replaced by 1113-kcmil 54/19 ACSR conductors, four per bundle. Use the LINE CONSTANTS program to compute the positive-sequence series impedance, positive-sequence shunt admittance, conductor surface electric field strength, and ground-level electric field strength profile. Compare the computed results with those of Problem 5.26.

5.28   Determine the effect of a 10% decrease as well as a 10% increase in bundle spacing on the positive-sequence series impedance and positive-sequence shunt admittance for the line in Example 5.10.

5.29   Determine the effect of a change in earth resistivity to 0.01, 1000, and $1.0 \times 10^9 \, \Omega$m on the zero-sequence series impedance for the line in Example 5.10.

5.30   Using the LINE CONSTANTS program, compute the series sequence impedance matrix for the three-phase line in Problem 5.11. Assume an average line height of 12 m, an earth resistivity of 100 Ωm, 60-Hz conductor resistances at 50°C, and no neutral wires. Compare the computed positive-sequence inductive reactance with the result calculated in Problem 5.11.

5.31 Rework Problem 5.30 with one neutral wire located 6 m directly above the center phase conductor. Compare the series sequence impedance matrix with that of Problem 5.30.

5.32 Using the LINE CONSTANTS program, compute the shunt sequence admittance matrix for the line in Problem 5.13. Assume an average line height of 20 m and no neutral wires. Compare the computed positive-sequence shunt admittance with the result calculated in Problem 5.21.

5.33 Rework Problem 5.32 with two neutral wires located 7 m above and 8 m to the left and right of the center bundle.

5.34 Using the LINE CONSTANTS program, compute the conductor surface electric field strength and the ground-level electric field strength profile for the line in Problem 5.33. Assume a 100 m right-of-way width.

5.35 Determine the effect of a 10% decrease as well as a 10% increase in phase spacing on the conductor surface electric field strength and on the ground-level electric field strength profile for the line in Problem 5.34.

5.36 Determine the effect of a 10% decrease as well as a 10% increase in the average line height on the conductor surface electric field strength as well as the ground-level electric field strength profile for the line in Problem 5.34.

5.37 Using the LINE CONSTANTS program, calculate the equivalent series sequence impedance matrix and the equivalent shunt sequence admittance matrix for the double-circuit, three-phase line shown in Figure 5.34 with phase arrangement I.

Figure 5.34

Double-circuit line for Problems 5.37 and 5.38

Phase conductors:
1—ACSR 2515 kcmil, 76/19
Radius = 2.388 cm
GMR   = 1.893 cm
Resistance = 0.0280 Ω/km

Neutrals:
2—Alumoweld 7 no. 8
Radius = 0.489 cm
GMR   = 0.0636 cm
Resistance = 1.52 Ω/km

Earth resistivity = 100 Ωm
Frequency = 60 Hz
Voltage = 345 kV

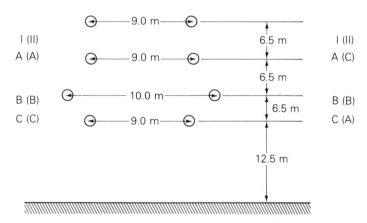

5.38 Rework Problem 5.37 for phase arrangement II shown in parentheses in Figure 5.34. Compare the computed results of the two phase arrangements.

CASE STUDY QUESTIONS

A. Are overhead transmission and distribution lines attractive, nondescript, or unsightly? What techniques are used to minimize visual impacts of transmission and distribution?

B. Which overhead transmission structures have the largest environmental impacts: steel, wood, or concrete? Why?

C. What are the maintenance requirements of overhead transmission with wood poles? Concrete poles? Steel towers?

References

1. General Electric Company, *Transmission Line Reference Book–345 kV and Above*, 2d ed. (Palo Alto, CA: Electric Power Research Institute, 1982).

2. Westinghouse Electric Corporation, *Electrical Transmission and Distribution Reference Book*, 4th ed. (East Pittsburgh, PA, 1964).

3. General Electric Company, *Electric Utility Systems and Practices*, 4th ed. (New York: Wiley, 1983).

4. John R. Carson, "Wave Propagation in Overhead Wires with Ground Return," *Bell System Tech. J.* 5 (1926): 539–554.

5. C. F. Wagner and R. D. Evans, *Symmetrical Components* (New York: McGraw-Hill, 1933).

6. Paul M. Anderson, *Analysis of Faulted Power Systems* (Ames, IA: Iowa State Press, 1973).

7. M. H. Hesse, "Electromagnetic and Electrostatic Transmission Line Parameters by Digital Computer," *Trans. IEEE* PAS-82 (1963): 282–291.

8. W. D. Stevenson, Jr., *Elements of Power System Analysis*, 4th ed. (New York: McGraw-Hill, 1982).

9. C. A. Gross, *Power System Analysis* (New York: Wiley, 1979).

10. John Reason, "Special Report—Transmission Structures," *Electrical World*, *206*, 3 (March 1992), pp. 31–49.

CHAPTER 6

TRANSMISSION LINES: STEADY-STATE OPERATION

345 kV substation with SF6 gas-insulated circuit breakers.
(Courtesy of New England Electric)

In this chapter we analyze the performance of single-phase and balanced three-phase transmission lines under normal steady-state operating conditions. Expressions for voltage and current at any point along a line are developed, where the distributed nature of the series impedance and shunt admittance is taken into account. A line is treated here as a two-port network for which the *ABCD* parameters and an equivalent π circuit are derived. Also, approximations are given for a medium-length line lumping the shunt admittance, for a short line neglecting the shunt admittance, and for a lossless line assuming zero series resistance and shunt conductance. The concepts of *surge impedance loading* and transmission-line *wavelength* are also presented.

An important issue discussed in this chapter is *voltage regulation*. Transmission-line voltages are generally high during light load periods and low during heavy load periods. Voltage regulation, defined in Section 6.1, refers to the change in line voltage as line loading varies from no-load to full load.

Another important issue discussed here is line loadability. Three major line-loading limits are: (1) the thermal limit; (2) the voltage-drop limit; and (3) the steady-state stability limit. Thermal and voltage-drop limits are discussed in Section 6.1. The theoretical steady-state stability limit, discussed in Section 6.4 for lossless lines and in Section 6.5 for lossy lines, refers to the ability of synchronous machines at the ends of a line to remain in synchronism. Practical line loadability is discussed in Section 6.6.

In Section 6.7 we discuss line compensation techniques for improving voltage regulation and for raising line loadings closer to the thermal limit. A personal computer program entitled "TRANSMISSION LINES—STEADY-STATE OPERATION," suitable for transmission-line analysis and design with respect to voltage regulation, line loadability, and transmission efficiency, is described in Section 6.8.

CASE STUDY Electric utilities have used fixed, series capacitor compensation for many years to increase the maximum power flow through long transmission lines. The following article describes a continuously adjustable series compensation installation that maintains network stability and optimizes losses

[6]. The series compensation consists of capacitor banks in parallel with a thyristor-switched inductor (reactor). Fast-acting thyristor switches provide direct control of power flow and improved damping of power oscillations. Transmission-line reactive compensation techniques including series compensation are discussed in Section 6.6.

Phase-controlled Reactor Bucks Series Capacitor

MARK R. WILHELM AND DUANE TORGERSON

Reprinted with permission from Electrical World.

In a few months, the Western Area Power Administration (Western) will commission the first transmission system with continuously variable series compensation. The project consists of two series capacitor banks, each rated at 165 Mvar with a single-phase impedance of 55 ohms. One bank is configured for conventional series compensation (see "Overview" section). The second bank is split; one segment is fixed and the other is adjustable compensation. This project is funded jointly by Western and Siemens Energy & Automation Inc, Alpharetta, GA.

Western is one of the five federal power-marketing administrations and transmits and markets power from federal water projects throughout the western US. The advanced series compensation (ASC) installation will be connected to Western's 230-kV network at the Kayenta substation in northeastern Arizona, which is between Shiprock, NM, and the Glen Canyon Dam in Arizona.

The compensation system at Kayenta will allow for rapid continuous control of the amount of compensation. Thus, power flow through the transmission line can be quickly, continuously, and directly controlled.

ASC works in the following way: If a reactor is connected in parallel with a conventional series compensation capacitor, the net series compensation seen by the transmission line is the impedance that results from the parallel combination of the reactor impedance and the capacitor impedance. By varying the impedance of the parallel reactor, the total impedance of the compensation can be varied.

In shunt compensation systems, the shunt-reactor current can be varied via a phase-controlled thyristor bridge connected in series with the shunt reactor. Changing the thyristor firing angle changes the point in the ac wave where the thyristor valve switches on. Therefore, it varies the amount of 60-Hz current flowing through the reactor. At 60 Hz, this thyristor-controlled reactor acts like a variable reactor.

A thyristor-controlled reactor also can be placed in parallel with series compensation capacitors and viewed—from a 60-Hz perspective—as a variable reactor; the thyristor-controlled reactor (TCR) draws varying amounts of 60-Hz reactive current depending on the thyristor firing angle. The net effect of the parallel combination of this TCR and series capacitor is a continuously and rapidly variable series compensation system. The system can be operated through a wide range of capacitive and inductive impedances (Fig. 1).

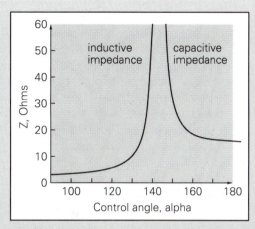

Figure 1 *Impedance for the ASC is plotted as a function of the firing angle*

A change in the total line impedance varies the magnitude of the line power vs angle curve. Assuming that the voltage angle between two electrical systems on either side of the transmission line remains fixed— a valid assumption for large or stiff electrical networks—the net effect of raising or lowering the power-transfer curve is to raise or lower the amount of power being transferred over the line. Thus, the ASC can be used to change or control the transmission-line power flow.

Overview of conventional series compensation

The maximum power that can be sent through a transmission line is determined by either the thermal capacity of the line or the line's series reactance. For relatively short lines, the thermal limit is reached before the line impedance becomes a factor. However, with longer lines, line impedance determines the upper limit. If a long transmission line is represented as its series inductive reactance (Xl), the amount of power through the line, is determined by the following equation:

$$P = (Vs*Vr/Xl)\sin(\hat{A})$$

where Vs is the sending-end voltage, Vr is the receiving end voltage, and $\hat{A}$ is the angular difference between the sending end and the receiving end voltages.

If Vs and Vr are held constant and Â is varied, a power vs angle curve is produced (Fig. 2). As Â is increased from zero, the power through the line increases until Â reaches 90 deg. Beyond this point, the transmitted power decreases with increasing Â. Therefore, power transmission reaches its maximum when Â is at 90 deg. The maximum power is (Vs*Vr/Xl).

Both Vs and Vr are fixed by transmission network design and operating practices. Thus, in this simplified transmission-line model, the only way maximum power transfer can be increased is to lower Xl. Capacitors connected in series with the transmission line result in capacitive impedance that, when added to the line inductive impedance, reduces the total line impedance and raises the power curve. This series capacitance is called series compensation. Fixed series compensation therefore can boost the power-transfer curve for a given line significantly. If this fixed compensation is switched in and out of the line in segments, several different power-transfer curves can be selected for a given line. However, with traditional series compensation, capacitance cannot be varied dynamically and can only be changed economically in relatively large, discrete steps.

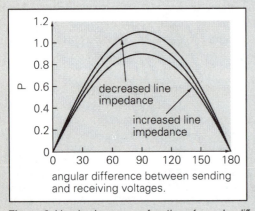

Figure 2 Line is shown as a function of angular difference

New features

By providing direct control of the power flow on a transmission line, an ASC system offers several advantages not present with conventional series compensation. These include:

- Direct control of power flow within the network. Flow can be changed quickly to accommodate changes in load patterns or outages. The capacity of an ASC line can be temporarily boosted to a much higher level than normal for emergency requirements.

- Network power swing damping capability. Damping is accomplished by modulating the power flow on the connecting line so the resulting dynamic exchange of energy between the two line ends acts to damp the system oscillation.

- Improved capacitor-bank protection. When a fault occurs on the transmission line, the resulting high currents produce an abnormally high voltage in the line's series capacitor bank. With conventional series compensation systems, the capacitors are protected from this overvoltage by metal oxide arresters connected in parallel with the bank and some means of bypassing the bank. With the ASC, parallel metal oxide arresters and bank bypassing also are used, but the thyristors can be fired to bypass the capacitor quickly if the energy accumulated by the arrester becomes excessive.

- Reduced dc offset. When conventional series capacitors are inserted or reinserted in the line, a dc offset, which dissipates slowly, exists in the capacitor voltage. With the ASC, this post-insertion offset is eliminated within a few ac cycles.

- Subsynchronous damping. The addition of any capacitance in series with a transmission line produces a tuned LC branch in the transmission network. To keep the 60-Hz line impedance inductive, the impedance of the series capacitance is always chosen to be less than the line inductive impedance. This means that a series compensated line is tuned to a frequency below 60 Hz.

Oscillations in a compensated line at its tuned frequency can excite mechanical oscillations in steam turbine/generators connected to the network. These mechanical oscillations can, in turn, further excite the line oscillation. Cases have occurred where subsynchronous oscillations have grown to where the mechanical system has been damaged. Concerns about this problem—called subsynchronous resonance—have inhibited significantly the application of series compensation.

One of the ASC's most promising advantages focuses on its behavior at subsynchronous frequencies and potential ability to damp subsynchronous oscillations. While the ASC acts like a variable reactor in parallel with a capacitor at 60 Hz, at subsynchronous frequencies it does not resemble the same parallel configuration because the TCR is a nonlinear, phase-controlled device and not a true variable reactor.

The Kayenta ASC has been simulated extensively along with the connected network. These simulations have shown—for most of its operating range—that the ASC resembles a resistor in series with an inductor at subsynchronous frequencies. For example, the Kayenta ASC looks to the 60 Hz system like a 30-ohm capacitive reactance when operated at a firing angle of 150 deg. However, at this same firing angle, the ASC responds to an injected 35-Hz current like a 6-ohm resistor in series with a 20-ohm inductance. A conventional series capacitor having a 30-ohm impedance at 60-Hz also looks like a capacitor at 35 Hz that has an impedance of 17.5 ohms. If the connected line had an inductive impedance that matched the conventional capacitor's impedance at 35 Hz, a subsynchronously tuned circuit would be formed. The ASC, however, is not a capacitor at this frequency and thus cannot form a tuned circuit by itself.

The ASC mitigates subsynchronous resonance in another way. Since the ASC can be used to control the line characteristic and, therefore, the power flow through the line, it can be used to actively damp subsynchronous oscillations. The damping is similar to the way it damps general power-system oscillations.

Required hardware

Series compensation capacitors usually are installed on a platform that, because it is at line voltage, must be supported on insulators. In a similar fashion, the Kayenta ASC is being built on an insulated platform (Fig. 3). Both the parallel reactor and the thyristor valve that will control the reactor also will reside on this platform.

Each of the three thyristor valves consists of 11 pairs of high-voltage thyristors. Each pair is connected in an antiparallel configuration to allow for bidirectional operation of the valves. The valves are designed with one redundant thyristor pair. If a thyristor fails, the ASC can remain in operation because the remaining 10 series-connected thyristor pairs are capable of withstanding fault-induced voltages. The valves are rated for a continuous current of 1000 amp and are able to withstand the full line short circuit current.

The valves are cooled with a deionized water/ethylene glycol mixture. Redundant pumps, heat exchangers, deionizers, and fluid-quality monitoring equipment are located at ground level. The cooling fluid is circulated to the valves through platform insulators. Thyristors within each valve are individually linked to the ASC control through a fiberoptic interface, which is used to send firing signals to the thyristors and monitor their status.

Figure 3 *The ASC will reside on a platform. Valves are made of 11 thyristor pairs*

Primary control of the ASC is performed in a multiprocessor-based computer. The ASC control can operate to set the system to a fixed impedance or can regulate to a constant line current. Controls are common for all three phases. Overload protection of the ASC valves and capacitors along with capacitor unbalance detection also are provided through the digital protection system. Provisions have been made to accommodate future control signals that may be used to provide system damping, subsynchronous resonance mitigation, or other advanced control functions.

Once the Kayenta ASC project begins operating, it will provide a valuable test bed for learning about the benefits of advanced compensation and dynamic compensation control.

Metal oxide arresters in parallel with the ASC capacitors provide primary capacitor overvoltage protection. The energy absorbed by the arresters is computed from arrester current information and monitored digitally. If arrester energy reaches excessive levels, the TCR is turned on completely to bypass the capacitors. Additionally, a close signal is sent to a breaker that bypasses the entire ASC.

SECTION 6.1

MEDIUM AND SHORT LINE APPROXIMATIONS

In this section we present short and medium-length transmission-line approximations as a means of introducing *ABCD* parameters. Some readers may prefer to start in Section 6.2, which presents the exact transmission-line equations.

It is convenient to represent a transmission line by the two-port network shown in Figure 6.1, where V_S and I_S are the sending-end voltage and current, and V_R and I_R are the receiving-end voltage and current.

Figure 6.1

Representation of two-port network

The relation between the sending-end and receiving-end quantities can be written as

$$V_S = AV_R + BI_R \quad \text{volts} \tag{6.1.1}$$

$$I_S = CV_R + DI_R \quad \text{A} \tag{6.1.2}$$

or, in matrix format,

$$\begin{bmatrix} V_S \\ I_S \end{bmatrix} = \begin{bmatrix} A & B \\ C & D \end{bmatrix} \begin{bmatrix} V_R \\ I_R \end{bmatrix} \tag{6.1.3}$$

where *A*, *B*, *C*, and *D* are parameters that depend on the transmission-line constants R, L, C, and G. The *ABCD* parameters are, in general, complex numbers. *A* and *D* are dimensionless. *B* has units of ohms, and *C* has units of siemens. Network theory texts [5] show that *ABCD* parameters apply to linear, passive, bilateral two-port networks, with the following general relation:

$$AD - BC = 1 \tag{6.1.4}$$

The circuit in Figure 6.2 represents a short transmission line, usually applied to overhead 60-Hz lines less than 80 km long. Only the series resistance and reactance are included. The shunt admittance is neglected. The circuit applies to either single-phase or completely transposed three-phase lines operating under balanced conditions. For a completely transposed three-phase line, *Z* is the positive-sequence series impedance, V_S and V_R are positive-sequence line-to-neutral voltages, and I_S and I_R are positive-sequence line currents. Although similar zero-sequence and negative-sequence circuits can be employed to represent unbalanced operating conditions, we consider only the positive-sequence network in this chapter.

Figure 6.2

Short transmission line

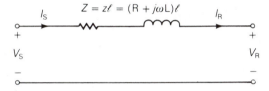

To avoid confusion between total series impedance and series impedance per unit length, we use the following notation:

$z = R + j\omega L$ Ω/m, series impedance per unit length

$y = G + j\omega C$ S/m, shunt admittance per unit length

$Z = zl$ Ω, total series impedance

$Y = yl$ S, total shunt admittance

$l = $ line length m

For three-phase lines, the subscript 1 indicating positive-sequence quantities is omitted in this chapter. Recall also that shunt conductance G is usually neglected for overhead transmission.

The *ABCD* parameters for the short line in Figure 6.2 are easily obtained by writing a KVL and KCL equation as

$$V_S = V_R + ZI_R \tag{6.1.5}$$

$$I_S = I_R \tag{6.1.6}$$

or, in matrix format,

$$\begin{bmatrix} V_S \\ I_S \end{bmatrix} = \begin{bmatrix} 1 & Z \\ 0 & 1 \end{bmatrix} \begin{bmatrix} V_R \\ I_R \end{bmatrix} \tag{6.1.7}$$

Comparing (6.1.7) and (6.1.3), the *ABCD* parameters for a short line are

$$A = D = 1 \quad \text{per unit} \tag{6.1.8}$$

$$B = Z \quad \Omega \tag{6.1.9}$$

$$C = 0 \quad \text{S} \tag{6.1.10}$$

For medium-length lines, typically ranging from 80 to 250 km at 60 Hz, it is common to lump the total shunt capacitance and locate half at each end of the line. Such a circuit, called a *nominal π circuit*, is shown in Figure 6.3.

Figure 6.3

Medium-length transmission line—nominal π circuit

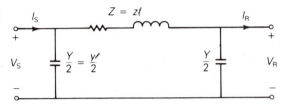

To obtain the *ABCD* parameters of the nominal π circuit, note first that the current in the series branch in Figure 6.3 equals $I_R + \dfrac{V_R Y}{2}$. Then, writing a KVL equation,

$$V_S = V_R + Z \left(I_R + \frac{V_R Y}{2} \right)$$

$$= \left(1 + \frac{YZ}{2} \right) V_R + Z I_R \tag{6.1.11}$$

Also, writing a KCL equation at the sending end,

$$I_S = I_R + \frac{V_R Y}{2} + \frac{V_S Y}{2} \tag{6.1.12}$$

Using (6.1.11) in (6.1.12),

$$I_S = I_R + \frac{V_R Y}{2} + \left[\left(1 + \frac{YZ}{2} \right) V_R + Z I_R \right] \frac{Y}{2} \tag{6.1.13}$$

$$= Y \left(1 + \frac{YZ}{4} \right) V_R + \left(1 + \frac{YZ}{2} \right) I_R$$

Writing (6.1.11) and (6.1.13) in matrix format,

$$
\begin{bmatrix} V_S \\ I_S \end{bmatrix} =
\begin{bmatrix} \left(1 + \dfrac{YZ}{2} \right) & Z \\ Y \left(1 + \dfrac{YZ}{4} \right) & \left(1 + \dfrac{YZ}{2} \right) \end{bmatrix}
\begin{bmatrix} V_R \\ I_R \end{bmatrix} \tag{6.1.14}
$$

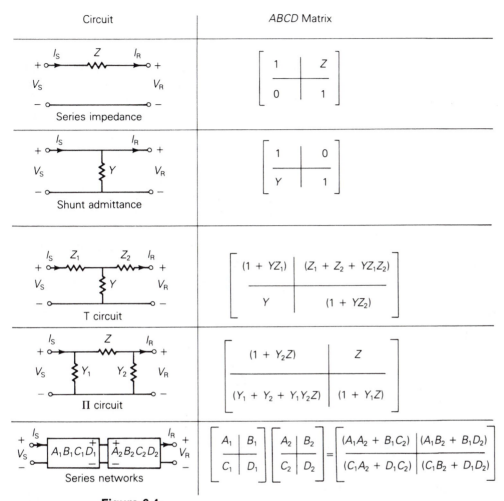

Circuit	*ABCD* Matrix

Series impedance: $\begin{bmatrix} 1 & Z \\ 0 & 1 \end{bmatrix}$

Shunt admittance: $\begin{bmatrix} 1 & 0 \\ Y & 1 \end{bmatrix}$

T circuit: $\begin{bmatrix} (1 + YZ_1) & (Z_1 + Z_2 + YZ_1Z_2) \\ Y & (1 + YZ_2) \end{bmatrix}$

Π circuit: $\begin{bmatrix} (1 + Y_2Z) & Z \\ (Y_1 + Y_2 + Y_1Y_2Z) & (1 + Y_1Z) \end{bmatrix}$

Series networks: $\begin{bmatrix} A_1 & B_1 \\ C_1 & D_1 \end{bmatrix} \begin{bmatrix} A_2 & B_2 \\ C_2 & D_2 \end{bmatrix} = \begin{bmatrix} (A_1A_2 + B_1C_2) & (A_1B_2 + B_1D_2) \\ (C_1A_2 + D_1C_2) & (C_1B_2 + D_1D_2) \end{bmatrix}$

Figure 6.4

ABCD parameters of common networks

Thus, comparing (6.1.14) and (6.1.3)

$$A = D = 1 + \frac{YZ}{2} \quad \text{per unit} \tag{6.1.15}$$

$$B = Z \quad \Omega \tag{6.1.16}$$

$$C = Y\left(1 + \frac{YZ}{4}\right) \quad \text{S} \tag{6.1.17}$$

Note that for both the short and medium-length lines, the relation $AD - BC = 1$ is verified. Note also that since the line is the same when viewed from either end, $A = D$.

Figure 6.4 gives the *ABCD* parameters for some common networks, including a series impedance network that approximates a short line and a π circuit that approximates a medium-length line. A medium-length line could also be approximated by the T circuit shown in Figure 6.4, lumping half of the series impedance at each end of the line. Also given are the *ABCD* parameters for networks in series, which are conveniently obtained by multiplying the *ABCD* matrices of the individual networks.

ABCD parameters can be used to describe the variation of line voltage with line loading. *Voltage regulation* is the change in voltage at the receiving end of the line when the load varies from no-load to a specified full load at a specified power factor, while the sending-end voltage is held constant. Expressed in percent of full-load voltage,

$$\text{percent VR} = \frac{|V_{RNL}| - |V_{RFL}|}{|V_{RFL}|} \times 100 \tag{6.1.18}$$

where percent VR is the percent voltage regulation, $|V_{RNL}|$ is the magnitude of the no-load receiving-end voltage, and $|V_{RFL}|$ is the magnitude of the full-load receiving-end voltage.

The effect of load power factor on voltage regulation is illustrated by the phasor diagrams in Figure 6.5 for short lines. The phasor diagrams are graphical representations of (6.1.5) for lagging and leading power factor loads.

Figure 6.5

Phasor diagrams for a short transmission line

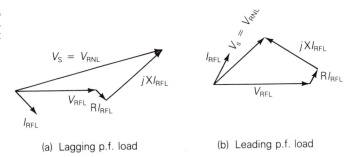

(a) Lagging p.f. load   (b) Leading p.f. load

Note that, from (6.1.5) at no-load, $I_{RNL} = 0$ and $V_S = V_{RNL}$ for a short line. As shown, the higher (worse) voltage regulation occurs for the lagging p.f. load, where V_{RNL} exceeds V_{RFL} by the larger amount. A smaller or even negative voltage regulation occurs for the leading p.f. load. In general, the no-load voltage is, from (6.1.1), with $I_{RNL} = 0$,

$$V_{RNL} = \frac{V_S}{A} \tag{6.1.19}$$

which can be used in (6.1.18) to determine voltage regulation.

In practice, transmission-line voltages decrease when heavily loaded and increase when lightly loaded. When voltages on EHV lines are maintained within $\pm 5\%$ of rated voltage, corresponding to about 10% voltage regulation, unusual operating problems are not encountered. Ten percent voltage regulation for lower voltage lines including transformer-voltage drops is also considered good operating practice.

In addition to voltage regulation, line loadability is an important issue.

Three major line-loading limits are: (1) the thermal limit; (2) the voltage-drop limit; and (3) the steady-state stability limit.

The maximum temperature of a conductor determines its thermal limit. Conductor temperature affects the conductor sag between towers and the loss of conductor tensile strength due to annealing. If the temperature is too high, prescribed conductor-to-ground clearances may not be met, or the elastic limit of the conductor may be exceeded such that it cannot shrink to its original length when cooled. Conductor temperature depends on the current magnitude and its time duration, as well as on ambient temperature, wind velocity, and conductor surface conditions. Appendix Tables A.3 and A.4 give approximate current-carrying capacities of copper and ACSR conductors. The loadability of short transmission lines (less than 80 km in length for 60-Hz overhead lines) is usually determined by the conductor thermal limit or by ratings of line terminal equipment such as circuit breakers.

For longer line lengths (up to 300 km), line loadability is often determined by the voltage-drop limit. Although more severe voltage drops may be tolerated in some cases, a heavily loaded line with $V_R/V_S \geqslant 0.95$ is usually considered safe operating practice. For line lengths over 300 km, steady-state stability becomes a limiting factor. Stability, discussed in Section 6.4, refers to the ability of synchronous machines on either end of a line to remain in synchronism.

| EXAMPLE 6.1 | ***ABCD* parameters and the nominal π circuit: medium-length line** |

A three-phase, 60-Hz, completely transposed 345-kV, 200-km line has two 795,000-cmil 26/2 ACSR conductors per bundle and the following positive-sequence line constants:

$$z = 0.032 + j0.35 \quad \Omega/\text{km}$$

$$y = j4.2 \times 10^{-6} \quad \text{S/km}$$

Full load at the receiving end of the line is 700 MW at 0.99 p.f. leading and at 95% of rated voltage. Assuming a medium-length line, determine the following:

   a. *ABCD* parameters of the nominal π circuit

   b. Sending-end voltage V_S, current I_S, and real power P_S

   c. Percent voltage regulation

   d. Thermal limit, based on the approximate current-carrying capacity listed in Table A.4

   e. Transmission-line efficiency at full load

Solution   **a.** The total series impedance and shunt admittance values are

$$Z = zl = (0.032 + j0.35)(200) = 6.4 + j70 = 70.29\underline{/84.78°} \quad \Omega$$

$$Y = yl = (j4.2 \times 10^{-6})(200) = 8.4 \times 10^{-4}\underline{/90°} \quad \text{S}$$

From (6.1.15)–(6.1.17),

$$A = D = 1 + (8.4 \times 10^{-4}\underline{/90°})(70.29\underline{/84.78°})(\tfrac{1}{2})$$

$$= 1 + 0.02952\underline{/174.78°}$$

$$= 0.9706 + j0.00269 = 0.9706\underline{/0.159°} \quad \text{per unit}$$

$$B = Z = 70.29\underline{/84.78°} \quad \Omega$$

$$C = (8.4 \times 10^{-4}\underline{/90°})(1 + 0.01476\underline{/174.78°})$$

$$= (8.4 \times 10^{-4}\underline{/90°})(0.9853 + j0.00134)$$

$$= 8.277 \times 10^{-4}\underline{/90.08°} \quad \text{S}$$

b. The receiving-end voltage and current quantities are

$$V_R = (0.95)(345) = 327.8 \quad \text{kV}_{LL}$$

$$V_R = \frac{327.8}{\sqrt{3}}\underline{/0°} = 189.2\underline{/0°} \quad \text{kV}_{LN}$$

$$I_R = \frac{700\underline{/\cos^{-1}0.99}}{(\sqrt{3})(0.95 \times 345)(0.99)} = 1.246\underline{/8.11°} \quad \text{kA}$$

From (6.1.1) and (6.1.2), the sending-end quantities are

$$V_S = (0.9706\underline{/0.159°})(189.2\underline{/0°}) + (70.29\underline{/84.78°})(1.246\underline{/8.11°})$$

$$= 183.6\underline{/0.159°} + 87.55\underline{/92.89°}$$

$$= 179.2 + j87.95 = 199.6\underline{/26.14°} \quad \text{kV}_{LN}$$

$$V_S = 199.6\sqrt{3} = 345.8\,\text{kV}_{LL} \approx 1.00 \quad \text{per unit}$$

$$I_S = (8.277 \times 10^{-4}\underline{/90.08°})(189.2\underline{/0°}) + (0.9706\underline{/0.159°})(1.246\underline{/8.11°})$$

$$= 0.1566\underline{/90.08°} + 1.209\underline{/8.27°}$$

$$= 1.196 + j0.331 = 1.241\underline{/15.5°} \quad \text{kA}$$

and the real power delivered to the sending end is

$$P_S = (\sqrt{3})(345.8)(1.241)\cos(26.14° - 15.5°)$$

$$= 730.5 \quad \text{MW}$$

c. From (6.1.19), the no-load receiving-end voltage is

$$V_{RNL} = \frac{V_S}{A} = \frac{345.8}{0.9706} = 356.3 \quad \text{kV}_{LL}$$

and, from (6.1.18),

$$\text{percent VR} = \frac{356.3 - 327.8}{327.8} \times 100 = 8.7\%$$

d. From Table A.4, the approximate current-carrying capacity of two 795,000-cmil 26/2 ACSR conductors is $2 \times 0.9 = 1.8\,\text{kA}$.

e. The full-load line losses are $P_S - P_R = 730.5 - 700 = 30.5\,\text{MW}$ and the full-load transmission efficiency is

$$\text{percent EFF} = \frac{P_R}{P_S} \times 100 = \frac{700}{730.5} \times 100 = 95.8\%$$

Since $V_S = 1.00$ per unit, the full-load receiving-end voltage of 0.95 per unit corresponds to $V_R/V_S = 0.95$, considered in practice to be the lowest operating voltage possible without encountering operating problems. Thus, for this 345-kV 200-km uncompensated line, voltage drop limits the full-load current to 1.246 kA at 0.99 p.f. leading, well below the thermal limit of 1.8 kA. ∎

SECTION 6.2

TRANSMISSION-LINE DIFFERENTIAL EQUATIONS

The line constants R, L, and C are derived in Chapter 5 as per-length values having units of Ω/m, H/m, and F/m. They are not lumped, but rather are uniformly distributed along the length of the line. In order to account for the distributed nature of transmission-line constants, consider the circuit shown in Figure 6.6, which represents a line section of length Δx. $V(x)$ and $I(x)$

Figure 6.6

Transmission-line section of length Δx

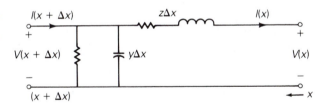

denote the voltage and current at position x, which is measured in meters from the right, or receiving end of the line. Similarly, $V(x + \Delta x)$ and $I(x + \Delta x)$ denote the voltage and current at position $(x + \Delta x)$. The circuit constants are

$$z = R + j\omega L \quad \Omega/\text{m} \tag{6.2.1}$$

$$y = G + j\omega C \quad \text{S/m} \tag{6.2.2}$$

where G is usually neglected for overhead 60-Hz lines. Writing a KVL equation for the circuit

$$V(x + \Delta x) = V(x) + (z\Delta x)I(x) \quad \text{volts} \tag{6.2.3}$$

Rearranging (6.2.3),

$$\frac{V(x + \Delta x) - V(x)}{\Delta x} = zI(x) \tag{6.2.4}$$

and taking the limit as Δx approaches zero,

$$\frac{dV(x)}{dx} = zI(x) \tag{6.2.5}$$

Similarly, writing a KCL equation for the circuit,

$$I(x + \Delta x) = I(x) + (y\Delta x)V(x + \Delta x) \quad \text{A} \tag{6.2.6}$$

Rearranging,

$$\frac{I(x + \Delta x) - I(x)}{\Delta x} = yV(x) \tag{6.2.7}$$

and taking the limit as Δx approaches zero,

$$\frac{dI(x)}{dx} = yV(x) \tag{6.2.8}$$

Equations (6.2.5) and (6.2.8) are two linear, first-order, homogeneous differential equations with two unknowns, $V(x)$ and $I(x)$. We can eliminate $I(x)$ by diffentiating (6.2.5) and using (6.2.8) as follows:

$$\frac{d^2V(x)}{dx^2} = z\frac{dI(x)}{dx} = zyV(x) \tag{6.2.9}$$

or

$$\frac{d^2V(x)}{dx^2} - zyV(x) = 0 \tag{6.2.10}$$

Equation (6.2.10) is a linear, second-order, homogeneous differential equation with one unknown, $V(x)$. By inspection, its solution is

$$V(x) = A_1 e^{\gamma x} + A_2 e^{-\gamma x} \quad \text{volts} \tag{6.2.11}$$

where A_1 and A_2 are integration constants and

$$\gamma = \sqrt{zy} \quad \text{m}^{-1} \tag{6.2.12}$$

γ, whose units are m^{-1}, is called the *propagation constant*. By inserting (6.2.11) and (6.2.12) into (6.2.10), the solution to the differential equation can be verified.

Next, using (6.2.11) in (6.2.5),

$$\frac{dV(x)}{dx} = \gamma A_1 e^{\gamma x} - \gamma A_2 e^{-\gamma x} = zI(x) \tag{6.2.13}$$

Solving for $I(x)$,

$$I(x) = \frac{A_1 e^{\gamma x} - A_2 e^{-\gamma x}}{z/\gamma} \tag{6.2.14}$$

Using (6.2.12), $z/\gamma = z/\sqrt{zy} = \sqrt{z/y}$, (6.2.14) becomes

$$I(x) = \frac{A_1 e^{\gamma x} - A_2 e^{-\gamma x}}{Z_c} \tag{6.2.15}$$

where

$$Z_c = \sqrt{\frac{z}{y}} \ \ \Omega \tag{6.2.16}$$

Z_c, whose units are Ω, is called the *characteristic impedance*.

Next, the integration constants A_1 and A_2 are evaluated from the boundary conditions. At $x = 0$, the receiving end of the line, the receiving-end voltage and current are

$$V_R = V(0) \tag{6.2.17}$$

$$I_R = I(0) \tag{6.2.18}$$

Also, at $x = 0$, (6.2.11) and (6.2.15) become

$$V_R = A_1 + A_2 \tag{6.2.19}$$

$$I_R = \frac{A_1 - A_2}{Z_c} \tag{6.2.20}$$

Solving for A_1 and A_2,

$$A_1 = \frac{V_R + Z_c I_R}{2} \tag{6.2.21}$$

$$A_2 = \frac{V_R - Z_c I_R}{2} \tag{6.2.22}$$

Substituting A_1 and A_2 into (6.2.11) and (6.2.15),

$$V(x) = \left(\frac{V_R + Z_c I_R}{2}\right) e^{\gamma x} + \left(\frac{V_R - Z_c I_R}{2}\right) e^{-\gamma x} \tag{6.2.23}$$

$$I(x) = \left(\frac{V_R + Z_c I_R}{2 Z_c}\right) e^{\gamma x} - \left(\frac{V_R - Z_c I_R}{2 Z_c}\right) e^{-\gamma x} \tag{6.2.24}$$

Rearranging (6.2.23) and (6.2.24),

$$V(x) = \left(\frac{e^{\gamma x} + e^{-\gamma x}}{2}\right) V_R + Z_c \left(\frac{e^{\gamma x} - e^{-\gamma x}}{2}\right) I_R \tag{6.2.25}$$

$$I(x) = \frac{1}{Z_c} \left(\frac{e^{\gamma x} - e^{-\gamma x}}{2}\right) V_R + \left(\frac{e^{\gamma x} + e^{-\gamma x}}{2}\right) I_R \tag{6.2.26}$$

Recognizing the hyperbolic functions cosh and sinh,

$$V(x) = \cosh(\gamma x) V_R + Z_c \sinh(\gamma x) I_R \tag{6.2.27}$$

$$I(x) = \frac{1}{Z_c} \sinh(\gamma x) V_R + \cosh(\gamma x) I_R \tag{6.2.28}$$

Equations (6.2.27) and (6.2.28) give the *ABCD* parameters of the distributed line. In matrix format,

$$
\begin{bmatrix} V(x) \\ I(x) \end{bmatrix} = \left[\begin{array}{c|c} A(x) & B(x) \\ \hline C(x) & D(x) \end{array} \right] \begin{bmatrix} V_R \\ I_R \end{bmatrix}
\tag{6.2.29}
$$

where

$$
A(x) = D(x) = \cosh(\gamma x) \quad \text{per unit}
\tag{6.2.30}
$$

$$
B(x) = Z_c \sinh(\gamma x) \quad \Omega
\tag{6.2.31}
$$

$$
C(x) = \frac{1}{Z_c} \sinh(\gamma x) \quad \text{S}
\tag{6.2.32}
$$

Equation (6.2.29) gives the current and voltage at any point x along the line in terms of the receiving-end voltage and current. At the sending end, where $x = l$, $V(l) = V_S$ and $I(l) = I_S$. That is,

$$
\begin{bmatrix} V_S \\ I_S \end{bmatrix} = \left[\begin{array}{c|c} A & B \\ \hline C & D \end{array} \right] \begin{bmatrix} V_R \\ I_R \end{bmatrix}
\tag{6.2.33}
$$

where

$$
A = D = \cosh(\gamma l) \quad \text{per unit}
\tag{6.2.34}
$$

$$
B = Z_c \sinh(\gamma l) \quad \Omega
\tag{6.2.35}
$$

$$
C = \frac{1}{Z_c} \sinh(\gamma l) \quad \text{S}
\tag{6.2.36}
$$

Equations (6.2.34)–(6.2.36) give the *ABCD* parameters of the distributed line. In these equations, the propagation constant γ is a complex quantity with real and imaginary parts denoted α and β. That is,

$$
\gamma = \alpha + j\beta \quad \text{m}^{-1}
\tag{6.2.37}
$$

The quantity γl is dimensionless. Also

$$
e^{\gamma l} = e^{(\alpha l + j\beta l)} = e^{\alpha l} e^{j\beta l} = e^{\alpha l} \underline{/\beta l}
\tag{6.2.38}
$$

Using (6.2.38) the hyperbolic functions cosh and sinh can be evaluated as follows:

$$
\cosh(\gamma l) = \frac{e^{\gamma l} + e^{-\gamma l}}{2} = \frac{1}{2}(e^{\alpha l} \underline{/\beta l} + e^{-\alpha l} \underline{/-\beta l})
\tag{6.2.39}
$$

and

$$
\sinh(\gamma l) = \frac{e^{\gamma l} - e^{-\gamma l}}{2} = \frac{1}{2}(e^{\alpha l} \underline{/\beta l} - e^{-\alpha l} \underline{/-\beta l})
\tag{6.2.40}
$$

Alternatively, the following identities can be used:

$$\cosh(\alpha l + j\beta l) = \cosh(\alpha l)\cos(\beta l) + j\sinh(\alpha l)\sin(\beta l) \qquad (6.2.41)$$

$$\sinh(\alpha l + j\beta l) = \sinh(\alpha l)\cos(\beta l) + j\cosh(\alpha l)\sin(\beta l) \qquad (6.2.42)$$

Note that in (6.2.39)–(6.2.42), the dimensionless quantity βl is in radians, not degrees.

The *ABCD* parameters given by (6.2.34)–(6.2.36) are exact parameters valid for any line length. For accurate calculations, these equations must be used for overhead 60-Hz lines longer than 250 km. The *ABCD* parameters derived in Section 6.1 are approximate parameters that are more conveniently used for hand calculations involving short and medium-length lines. Table 6.1 summarizes the *ABCD* parameters for short, medium, long, and lossless (see Section 6.4) lines.

Table 6.1

Summary: Transmission-line
ABCD parameters

PARAMETER	$A = D$	B	C
UNITS	Per Unit	Ω	S
Short line (less than 80 km)	1	Z	0
Medium line—nominal π circuit (80 to 250 km)	$1 + \dfrac{YZ}{2}$	Z	$Y\left(1 + \dfrac{YZ}{4}\right)$
Long line—equivalent π circuit (more than 250 km)	$\cosh(\gamma\ell)$ $= 1 + \dfrac{Y'Z'}{2}$	$Z_c\sinh(\gamma\ell)$ $= Z'$	$(1/Z_c)\sinh(\gamma\ell)$ $= Y'\left(1 + \dfrac{Y'Z'}{4}\right)$
Lossless line $(R = G = 0)$	$\cos(\beta\ell)$	$jZ_c\sin(\beta\ell)$	$\dfrac{j\sin(\beta\ell)}{Z_c}$

EXAMPLE 6.2   **Exact *ABCD* parameters: long line**

A three-phase 765-kV, 60-Hz, 300-km, completely transposed line has the following positive-sequence impedance and admittance:

$$z = 0.0165 + j0.3306 = 0.3310\underline{/87.14^\circ} \quad \Omega/\text{km}$$

$$y = j4.674 \times 10^{-6} \quad \text{S/km}$$

Assuming positive-sequence operation, calculate the exact *ABCD* parameters of the line. Compare the exact *B* parameter with that of the nominal π circuit.

Solution   From (6.2.12) and (6.2.16):

$$Z_c = \sqrt{\frac{0.3310\underline{/87.14^\circ}}{4.674 \times 10^{-6}\underline{/90^\circ}}} = \sqrt{7.082 \times 10^4\underline{/-2.86^\circ}}$$

$$= 266.1\underline{/-1.43^\circ} \quad \Omega$$

and

$$\gamma l = \sqrt{(0.3310\underline{/87.14°})(4.674 \times 10^{-6}\underline{/90°})} \times (300)$$

$$= \sqrt{1.547 \times 10^{-6}\underline{/177.14°}} \times (300)$$

$$= 0.3731\underline{/88.57°} = 0.00931 + j0.3730 \quad \text{per unit}$$

From (6.2.38),

$$e^{\gamma l} = e^{0.00931}e^{+j0.3730} = 1.0094\underline{/0.3730} \quad \text{radians}$$

$$= 0.9400 + j0.3678$$

and

$$e^{-\gamma l} = e^{-0.00931}e^{-j0.3730} = 0.9907\underline{/-0.3730} \quad \text{radians}$$

$$= 0.9226 - j0.3610$$

Then, from (6.2.39) and (6.2.40),

$$\cosh(\gamma l) = \frac{(0.9400 + j0.3678) + (0.9226 - j0.3610)}{2}$$

$$= 0.9313 + j0.0034 = 0.9313\underline{/0.209°}$$

$$\sinh(\gamma l) = \frac{(0.9400 + j0.3678) - (0.9226 - j0.3610)}{2}$$

$$= 0.0087 + j0.3644 = 0.3645\underline{/88.63°}$$

Finally, from (6.2.34)–(6.2.36),

$$A = D = \cosh(\gamma l) = 0.9313\underline{/0.209°} \quad \text{per unit}$$

$$B = (266.1\underline{/-1.43°})(0.3645\underline{/88.63°}) = 97.0\underline{/87.2°} \quad \Omega$$

$$C = \frac{0.3645\underline{/88.63°}}{266.1\underline{/-1.43°}} = 1.37 \times 10^{-3}\underline{/90.06°} \quad \text{S}$$

Using (6.1.16), the B parameter for the nominal π circuit is

$$B_{\text{nominal }\pi} = Z = (0.3310\underline{/87.14°})(300) = 99.3\underline{/87.14°} \quad \Omega$$

which is 2% larger than the exact value. ■

SECTION 6.3

EQUIVALENT π CIRCUIT

Many computer programs used in power-system analysis and design assume circuit representations of components such as transmission lines and transformers (see the power-flow program described in Chapter 7 as an example).

It is therefore convenient to represent the terminal characteristics of a transmission line by an equivalent circuit instead of its *ABCD* parameters.

Figure 6.7

Transmission-line equivalent π circuit

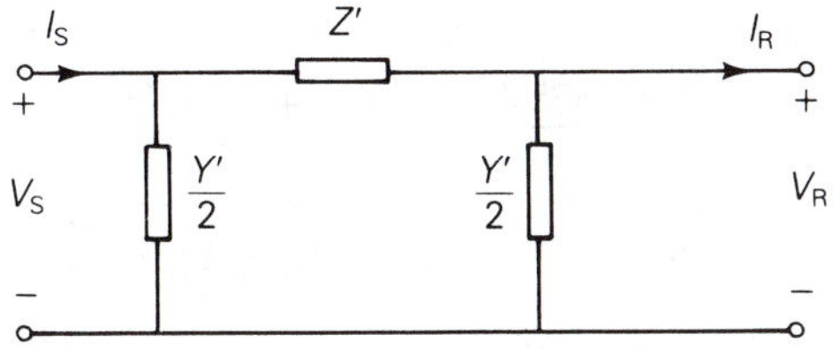

$$Z' = Z_c \sinh(\gamma\ell) = ZF_1 = Z\frac{\sinh(\gamma\ell)}{\gamma\ell}$$

$$\frac{Y'}{2} = \frac{\tanh(\gamma\ell/2)}{Z_c} = \frac{Y}{2}F_2 = \frac{Y}{2}\frac{\tanh(\gamma\ell/2)}{(\gamma\ell/2)}$$

The circuit shown in Figure 6.7 is called an *equivalent π circuit*. It is identical in structure to the nominal π circuit of Figure 6.3, except that Z' and Y' are used instead of Z and Y. Our objective is to determine Z' and Y' such that the equivalent π circuit has the same *ABCD* parameters as those of the distributed line, (6.2.34)–(6.2.36). The *ABCD* parameters of the equivalent π circuit, which has the same structure as the nominal π, are

$$A = D = 1 + \frac{Y'Z'}{2} \quad \text{per unit} \tag{6.3.1}$$

$$B = Z' \quad \Omega \tag{6.3.2}$$

$$C = Y'\left(1 + \frac{Y'Z'}{4}\right) \quad S \tag{6.3.3}$$

where we have replaced Z and Y in (6.1.15)–(6.1.17) with Z' and Y' in (6.3.1)–(6.3.3). Equating (6.3.2) to (6.2.35),

$$Z' = Z_c \sinh(\gamma l) = \sqrt{\frac{z}{y}}\sinh(\gamma l) \tag{6.3.4}$$

Rewriting (6.3.4) in terms of the nominal π circuit impedance $Z = zl$,

$$Z' = zl\left[\sqrt{\frac{z}{y}}\frac{\sinh(\gamma l)}{zl}\right] = zl\left[\frac{\sinh(\gamma l)}{\sqrt{zy}\,l}\right]$$

$$= ZF_1 \quad \Omega \tag{6.3.5}$$

where

$$F_1 = \frac{\sinh(\gamma l)}{\gamma l} \quad \text{per unit} \tag{6.3.6}$$

Similarly, equating (6.3.1) to (6.2.34),

$$1 + \frac{Y'Z'}{2} = \cosh(\gamma l)$$

$$\frac{Y'}{2} = \frac{\cosh(\gamma l) - 1}{Z'} \tag{6.3.7}$$

Using (6.3.4) and the identity $\tanh\left(\frac{\gamma l}{2}\right) = \frac{\cosh(\gamma l) - 1}{\sinh(\gamma l)}$, (6.3.7) becomes

$$\frac{Y'}{2} = \frac{\cosh(\gamma l) - 1}{Z_c \sinh(\gamma l)} = \frac{\tanh(\gamma l/2)}{Z_c} = \frac{\tanh(\gamma l/2)}{\sqrt{\dfrac{z}{y}}} \tag{6.3.8}$$

Rewriting (6.3.8) in terms of the nominal π circuit admittance $Y = yl$,

$$\frac{Y'}{2} = \frac{yl}{2}\left[\frac{\tanh(\gamma l/2)}{\sqrt{\dfrac{z}{y}}\dfrac{yl}{2}}\right] = \frac{yl}{2}\left[\frac{\tanh(\gamma l/2)}{\sqrt{zy}\,l/2}\right]$$

$$= \frac{Y}{2}\, F_2 \quad \text{S} \tag{6.3.9}$$

where

$$F_2 = \frac{\tanh(\gamma l/2)}{\gamma l/2} \quad \text{per unit} \tag{6.3.10}$$

Equations (6.3.6) and (6.3.10) give the correction factors F_1 and F_2 to convert Z and Y for the nominal π circuit to Z' and Y' for the equivalent π circuit.

EXAMPLE 6.3   **Equivalent π circuit: long line**

Compare the equivalent and nominal π circuits for the line in Example 6.2.

Solution   For the nominal π circuit,

$$Z = zl = (0.3310\underline{/87.14^\circ})(300) = 99.3\underline{/87.14^\circ} \quad \Omega$$

$$\frac{Y}{2} = \frac{yl}{2} = \left(\frac{j4.674 \times 10^{-6}}{2}\right)(300) = 7.011 \times 10^{-4}\underline{/90^\circ} \quad \text{S}$$

From (6.3.6) and (6.3.10), the correction factors are

$$F_1 = \frac{0.3645\underline{/88.63^\circ}}{0.3731\underline{/88.57^\circ}} = 0.9769\underline{/0.06^\circ} \quad \text{per unit}$$

$$F_2 = \frac{\tanh(\gamma l/2)}{\gamma l/2} = \frac{\cosh(\gamma l) - 1}{(\gamma l/2)\sinh(\gamma l)}$$

$$= \frac{0.9313 + j0.0034 - 1}{\left(\dfrac{0.3731}{2}\ \underline{/88.57°}\right)(0.3645\underline{/88.63°})}$$

$$= \frac{-0.0687 + j0.0034}{0.06800\underline{/177.20°}}$$

$$= \frac{0.06878\underline{/177.17°}}{0.06800\underline{/177.20°}} = 1.012\underline{/-0.03°}\quad\text{per unit}$$

Then, from (6.3.5) and (6.3.9), for the equivalent π circuit,

$$Z' = (99.3\underline{/87.14°})(0.9769\underline{/0.06°}) = 97.0\underline{/87.2°}\quad\Omega$$

$$\frac{Y'}{2} = (7.011 \times 10^{-4}\underline{/90°})(1.012\underline{/-0.03°}) = 7.095 \times 10^{-4}\underline{/89.97°}\quad\text{S}$$

$$= 3.7 \times 10^{-7} + j7.095 \times 10^{-4}\quad\text{S}$$

Comparing these nominal and equivalent π circuit values, Z' is about 2% smaller than Z, and $Y'/2$ is about 1% larger than $Y/2$. Although the circuit values are approximately the same for this line, the equivalent π circuit should be used for accurate calculations involving long lines. Note the small shunt conductance, $G' = 3.7 \times 10^{-7}$ S, introduced in the equivalent π circuit. G' is often neglected. ∎

SECTION 6.4

LOSSLESS LINES

In this section we discuss the following concepts for lossless lines: surge impedance, *ABCD* parameters, equivalent π circuit, wavelength, surge impedance loading, voltage profiles and steady-state stability limit.

When line losses are neglected, simpler expressions for the line parameters are obtained and the above concepts are more easily understood. Since transmission and distribution lines for power transfer generally are designed to have low losses, the equations and concepts developed here can be used for quick and reasonably accurate hand calculations leading to seat-of-the-pants analyses and to initial designs. More accurate calculations can then be made with the personal computer program given in Section 6.8 for follow-up analysis and design.

Surge Impedance

For a lossless line, $R = G = 0$, and

$$z = j\omega L \quad \Omega/m \tag{6.4.1}$$

$$y = j\omega C \quad S/m \tag{6.4.2}$$

From (6.2.12) and (6.2.16),

$$Z_c = \sqrt{\frac{z}{y}} = \sqrt{\frac{j\omega L}{j\omega C}} = \sqrt{\frac{L}{C}} \quad \Omega \tag{6.4.3}$$

and

$$\gamma = \sqrt{zy} = \sqrt{(j\omega L)(j\omega C)} = j\omega\sqrt{LC} = j\beta \quad \text{m}^{-1} \tag{6.4.4}$$

where

$$\beta = \omega\sqrt{LC} \quad \text{m}^{-1} \tag{6.4.5}$$

The characteristic impedance $Z_c = \sqrt{L/C}$, commonly called *surge* impedance for a lossless line, is pure real, that is, resistive. The propagation constant $\gamma = j\beta$ is pure imaginary.

ABCD Parameters

The *ABCD* parameters are, from (6.2.30)–(6.2.32),

$$A(x) = D(x) = \cosh(\gamma x) = \cosh(j\beta x)$$

$$= \frac{e^{j\beta x} + e^{-j\beta x}}{2} = \cos(\beta x) \quad \text{per unit} \tag{6.4.6}$$

$$\sinh(\gamma x) = \sinh(j\beta x) = \frac{e^{j\beta x} - e^{-j\beta x}}{2} = j\sin(\beta x) \quad \text{per unit} \tag{6.4.7}$$

$$B(x) = Z_c \sinh(\gamma x) = jZ_c \sin(\beta x) = j\sqrt{\frac{L}{C}} \sin(\beta x) \quad \Omega \tag{6.4.8}$$

$$C(x) = \frac{\sinh(\gamma x)}{Z_c} = \frac{j\sin(\beta x)}{\sqrt{\dfrac{L}{C}}} \quad \text{S} \tag{6.4.9}$$

$A(x)$ and $D(x)$ are pure real; $B(x)$ and $C(x)$ are pure imaginary.

A comparison of lossless versus lossy *ABCD* parameters is shown in Table 6.1.

Equivalent π Circuit

For the equivalent π circuit, using (6.3.4),

$$Z' = jZ_c \sin(\beta l) = jX' \quad \Omega \tag{6.4.10}$$

or, from (6.3.5) and (6.3.6),

$$Z' = (j\omega Ll)\left(\frac{\sin(\beta l)}{\beta l}\right) = jX' \quad \Omega \tag{6.4.11}$$

Also, from (6.3.9) and (6.3.10),

$$\frac{Y'}{2} = \frac{Y}{2}\frac{\tanh(j\beta l/2)}{j\beta l/2} = \frac{Y}{2}\frac{\sinh(j\beta l/2)}{(j\beta l/2)\cosh(j\beta l/2)}$$

$$= \left(\frac{j\omega Cl}{2}\right)\frac{j\sin(\beta l/2)}{(j\beta l/2)\cos(\beta l/2)} = \left(\frac{j\omega Cl}{2}\right)\frac{\tan(\beta l/2)}{\beta l/2}$$

$$= \left(\frac{j\omega C'l}{2}\right) \quad \text{S} \tag{6.4.12}$$

Z' and Y' are both pure imaginary. Also, for βl less than π radians, Z' is pure inductive and Y' is pure capacitive. Thus the equivalent π circuit for a lossless line, shown in Figure 6.8, is also lossless.

Figure 6.8

Equivalent π circuit for a lossless line ($\beta\ell$ less than π)

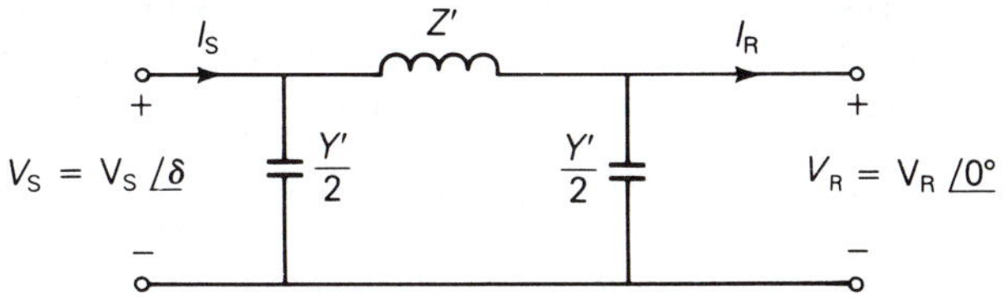

$$Z' = (j\omega L\ell)\left(\frac{\sin\beta\ell}{\beta\ell}\right) = jX' \ \Omega$$

$$\frac{Y'}{2} = \left(\frac{j\omega C\ell}{2}\right)\frac{\tan(\beta\ell/2)}{(\beta\ell/2)} = \frac{j\omega C'\ell}{2} \ \text{S}$$

Wavelength

A *wavelength* is the distance required to change the phase of the voltage or current by 2π radians or 360°. For a lossless line, using (6.2.29),

$$V(x) = A(x)V_R + B(x)I_R$$
$$= \cos(\beta x)V_R + jZ_c\sin(\beta x)I_R \tag{6.4.13}$$

and

$$I(x) = C(x)V_R + D(x)I_R$$
$$= \frac{j\sin(\beta x)}{Z_c}V_R + \cos(\beta x)I_R \tag{6.4.14}$$

From (6.4.13) and (6.4.14), $V(x)$ and $I(x)$ change phase by 2π radians when $x = 2\pi/\beta$. Denoting wavelength by λ, and using (6.4.5),

$$\lambda = \frac{2\pi}{\beta} = \frac{2\pi}{\omega\sqrt{LC}} = \frac{1}{f\sqrt{LC}} \quad \text{m} \tag{6.4.15}$$

or

$$f\lambda = \frac{1}{\sqrt{LC}} \tag{6.4.16}$$

It will be shown in Chapter 12 that the term $(1/\sqrt{LC})$ in (6.4.16) is the velocity of propagation of voltage and current waves along a lossless line. For overhead lines, $(1/\sqrt{LC}) \approx 3 \times 10^8$ m/s. And for $f = 60$ Hz, (6.4.14) gives

$$\lambda \approx \frac{3 \times 10^8}{60} = 5 \times 10^6 \text{ m} = 5000 \text{ km} = 3100 \text{ mi}$$

Typical power-line lengths are only a small fraction of the above 60-Hz wavelength.

Surge Impedance Loading

Surge impedance loading (SIL) is the power delivered by a lossless line to a load resistance equal to the surge impedance $Z_c = \sqrt{L/C}$. Figure 6.9 shows a lossless line terminated by a resistance equal to its surge impedance. This line represents either a single-phase line or one phase-to-neutral of a balanced three-phase line. At SIL, from (6.4.13),

$$V(x) = \cos(\beta x)V_R + jZ_c \sin(\beta x)I_R$$

$$= \cos(\beta x)V_R + jZ_c \sin(\beta x)\left(\frac{V_R}{Z_c}\right)$$

$$= (\cos \beta x + j \sin \beta x)V_R$$

$$= e^{j\beta x} V_R \quad \text{volts} \tag{6.4.17}$$

$$|V(x)| = |V_R| \quad \text{volts} \tag{6.4.18}$$

Thus, at SIL, the voltage profile is flat. That is, the voltage magnitude at any point x along a lossless line at SIL is constant.

Also from (6.4.14) at SIL,

$$I(x) = \frac{j \sin (\beta x)}{Z_c} V_R + (\cos \beta x) \frac{V_R}{Z_c}$$

$$= (\cos \beta x + j \sin \beta x) \frac{V_R}{Z_c}$$

$$= (e^{j\beta x}) \frac{V_R}{Z_c} \quad \text{A} \tag{6.4.19}$$

Using (6.4.17) and (6.4.19), the complex power flowing at any point x along the line is

$$S(x) = P(x) + jQ(x) = V(x)I^*(x)$$

$$= (e^{j\beta x} V_R)\left(\frac{e^{j\beta x} V_R}{Z_c}\right)^*$$

$$= \frac{|V_R|^2}{Z_c} \tag{6.4.20}$$

Figure 6.9

Lossless line terminated by its surge impedance

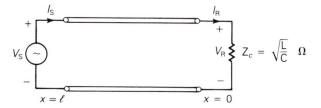

Thus the real power flow along a lossless line at SIL remains constant from the sending end to the receiving end. The reactive power flow is zero.

At rated line voltage, the real power delivered, or SIL, is, from (6.4.20),

$$\text{SIL} = \frac{V_{\text{rated}}^2}{Z_c} \qquad (6.4.21)$$

where rated voltage is used for a single-phase line and rated line-to-line voltage is used for the total real power delivered by a three-phase line. Table 6.2 lists surge impedance and SIL values for typical overhead 60-Hz three-phase lines.

Table 6.2

Surge impedance and SIL values for typical 60-Hz overhead lines [1, 2]

V_{rated}	$Z_c = \sqrt{L/C}$	$\text{SIL} = V_{\text{rated}}^2/Z_c$
kV	Ω	MW
69	366–400	12–13
138	366–405	47–52
230	365–395	134–145
345	280–366	325–425
500	233–294	850–1075
765	254–266	2200—2300

Voltage Profiles

In practice, power lines are not terminated by their surge impedance. Instead, loadings can vary from a small fraction of SIL during light load conditions up to multiples of SIL, depending on line length and line compensation, during heavy load conditions. If a line is not terminated by its surge impedance, then the voltage profile is not flat. Figure 6.10 shows voltage profiles of lines with a fixed sending-end voltage magnitude V_S for line lengths l up to a quarter wavelength. This figure shows four loading conditions: (1) no-load, (2) SIL, (3) short circuit, and (4) full load, which are described as follows:

1. At no-load, $I_{\text{RNL}} = 0$ and (6.4.13) yields

$$V_{\text{NL}}(x) = (\cos \beta x)V_{\text{RNL}} \qquad (6.4.22)$$

   The no-load voltage increases from $V_S = (\cos \beta l)V_{\text{RNL}}$ at the sending end to V_{RNL} at the receiving end (where $x = 0$).

2. From (6.4.18), the voltage profile at SIL is flat.

Figure 6.10

Voltage profiles of an uncompensated lossless line with fixed sending-end voltage for line lengths up to a quarter wavelength

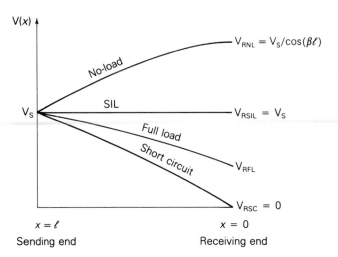

3. For a short circuit at the load, $V_{RSC} = 0$ and (6.4.13) yields

$$V_{SC}(x) = (Z_c \sin \beta x)I_{RSC} \qquad (6.4.23)$$

The voltage decreases from $V_S = (\sin \beta l)(Z_c I_{RSC})$ at the sending end to $V_{RSC} = 0$ at the receiving end.

4. The full-load voltage profile, which depends on the specification of full-load current, lies above the short-circuit voltage profile.

Figure 6.10 summarizes these results, showing a high receiving-end voltage at no-load and a low receiving-end voltage at full load. This voltage regulation problem becomes more severe as the line length increases. In Section 6.6, shunt compensation methods to reduce voltage fluctuations are discussed.

Steady-State Stability Limit

The equivalent π circuit of Figure 6.8 can be used to obtain an equation for the real power delivered by a lossless line. Assume that the voltage magnitudes V_S and V_R at the ends of the line are held constant. Also, let δ denote the voltage-phase angle at the sending end with respect to the receiving end. From KVL, the receiving-end current I_R is

$$I_R = \frac{V_S - V_R}{Z'} - \frac{Y'}{2} V_R$$

$$= \frac{V_S e^{j\delta} - V_R}{jX'} - \frac{j\omega C'l}{2} V_R \qquad (6.4.24)$$

and the complex power S_R delivered to the receiving end is

$$S_R = V_R I_R^* = V_R \left(\frac{V_S e^{j\delta} - V_R}{jX'}\right)^* + \frac{j\omega C'l}{2} V_R^2$$

$$= V_R \left(\frac{V_S e^{-j\delta} - V_R}{-jX'} \right) + \frac{j\omega Cl}{2} V_R^2$$

$$= \frac{jV_R V_S \cos\delta + V_R V_S \sin\delta - jV_R^2}{X'} + \frac{j\omega Cl}{2} V_R^2 \qquad (6.4.25)$$

The real power delivered is

$$P = P_S = P_R = \text{Re}(S_R) = \frac{V_R V_S}{X'} \sin\delta \quad \text{W} \qquad (6.4.26)$$

Note that since the line is lossless, $P_S = P_R$.

Figure 6.11

Real power delivered by a lossless line versus voltage angle across the line

Real power P

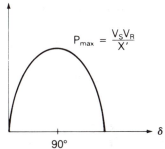

$$P_{max} = \frac{V_S V_R}{X'}$$

90°

Equation (6.4.26) is plotted in Figure 6.11. For fixed voltage magnitudes V_S and V_R, the phase angle δ increases from 0 to 90° as the real power delivered increases. The maximum power that the line can deliver, which occurs when $\delta = 90°$, is given by

$$P_{max} = \frac{V_S V_R}{X'} \quad \text{W} \qquad (6.4.27)$$

P_{max} represents the theoretical *steady-state stability* limit of a lossless line. If an attempt were made to exceed this steady-state stability limit, then synchronous machines at the sending end would lose synchronism with those at the receiving end. Stability is further discussed in Chapter 13.

It is convenient to express the steady-state stability limit in terms of SIL. Using (6.4.10) in (6.4.26),

$$P = \frac{V_S V_R \sin\delta}{Z_c \sin\beta l} = \left(\frac{V_S V_R}{Z_c} \right) \frac{\sin\delta}{\sin\left(\dfrac{2\pi l}{\lambda} \right)} \qquad (6.4.28)$$

Expressing V_S and V_R in per-unit of rated line voltage,

$$P = \left(\frac{V_S}{V_{rated}} \right) \left(\frac{V_R}{V_{rated}} \right) \left(\frac{V_{rated}^2}{Z_c} \right) \frac{\sin\delta}{\sin\left(\dfrac{2\pi l}{\lambda} \right)}$$

$$= V_{S.p.u.} V_{R.p.u.} (\text{SIL}) \frac{\sin\delta}{\sin\left(\dfrac{2\pi l}{\lambda} \right)} \quad \text{W} \qquad (6.4.29)$$

And for $\delta = 90°$, the theoretical steady-state stability limit is

$$P_{max} = \frac{V_{S.p.u.} V_{R.p.u.}(SIL)}{\sin\left(\dfrac{2\pi l}{\lambda}\right)} \quad W \tag{6.4.30}$$

Equations (6.4.27)–(6.4.30) reveal two important factors affecting the steady-state stability limit. First, from (6.4.27), it increases with the square of the line voltage. For example, a doubling of line voltage enables a fourfold increase in maximum power flow. Second, it decreases with line length. Equation (6.4.30) is plotted in Figure 6.12 for $V_{S.p.u.} = V_{R.p.u.} = 1$, $\lambda = 5000$ km, and line lengths up to 1100 km. As shown, the theoretical steady-state stability limit decreases from 4(SIL) for a 200-km line to about 2(SIL) for a 400-km line.

Figure 6.12

Transmission-line loadability curve for 60-Hz overhead lines—no series or shunt compensation

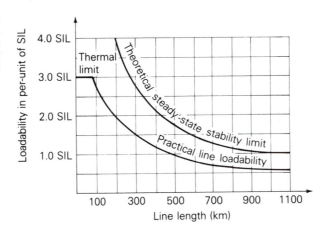

EXAMPLE 6.4   **Theoretical steady-state stability limit: long line**

Neglecting line losses, find the theoretical steady-state stability limit for the 300-km line in Example 6.2. Assume a 266.1-Ω surge impedance, a 5000-km wavelength, and $V_S = V_R = 765$ kV.

Solution   From (6.4.21),

$$SIL = \frac{(765)^2}{266.1} = 2199 \quad MW$$

From (6.4.30) with $l = 300$ km and $\lambda = 5000$ km,

$$P_{max} = \frac{(1)(1)(2199)}{\sin\left(\dfrac{2\pi \times 300}{5000}\right)} = (2.716)(2199) = 5974 \quad MW$$

Alternatively, from Figure 6.12, for a 300-km line, the theoretical steady-state stability limit is (2.72)SIL = (2.72)(2199) = 5980 MW, about the same as the above result.   ∎

SECTION 6.5

MAXIMUM POWER FLOW

Maximum power flow, discussed in Section 6.4 for lossless lines, is derived here in terms of the *ABCD* parameters for lossy lines. The following notation is used:

$$A = \cosh(\gamma l) = A\underline{/\theta_A}$$

$$B = Z' = Z'\underline{/\theta_Z}$$

$$V_S = V_S\underline{/\delta} \qquad V_R = V_R\underline{/0°}$$

Solving (6.2.33) for the receiving-end current,

$$I_R = \frac{V_S - AV_R}{B} = \frac{V_S e^{j\delta} - AV_R e^{j\theta_A}}{Z' e^{j\theta_Z}} \tag{6.5.1}$$

The complex power delivered to the receiving end is

$$S_R = P_R + jQ_R = V_R I_R^* = V_R \left[\frac{V_S e^{j(\delta - \theta_Z)} - AV_R e^{j(\theta_A - \theta_Z)}}{Z'}\right]^*$$

$$= \frac{V_R V_S}{Z'} e^{j(\theta_Z - \delta)} - \frac{AV_R^2}{Z'} e^{j(\theta_Z - \theta_A)} \tag{6.5.2}$$

The real and reactive power delivered to the receiving end are thus

$$P_R = \text{Re}(S_R) = \frac{V_R V_S}{Z'} \cos(\theta_Z - \delta) - \frac{AV_R^2}{Z'} \cos(\theta_Z - \theta_A) \tag{6.5.3}$$

$$Q_R = \text{Im}(S_R) = \frac{V_R V_S}{Z'} \sin(\theta_Z - \delta) - \frac{AV_R^2}{Z'} \sin(\theta_Z - \theta_A) \tag{6.5.4}$$

Note that for a lossless line, $\theta_A = 0°$, $B = Z' = jX'$, $Z' = X'$, $\theta_Z = 90°$, and (6.5.3) reduces to

$$P_R = \frac{V_R V_S}{X'} \cos(90 - \delta) - \frac{AV_R^2}{X'} \cos(90°)$$

$$= \frac{V_R V_S}{X'} \sin \delta \tag{6.5.5}$$

which is the same as (6.4.26).

The theoretical maximum real power delivered (or steady-state stability limit) occurs when $\delta = \theta_Z$ in (6.5.3):

$$P_{R\max} = \frac{V_R V_S}{Z'} - \frac{AV_R^2}{Z'} \cos(\theta_Z - \theta_A) \tag{6.5.6}$$

The second term in (6.5.6), and the fact that Z' is larger than X', reduce $P_{R\max}$ to a value somewhat less than that given by (6.4.27) for a lossless line.

EXAMPLE 6.5 | **Theoretical maximum power delivered: long line**

Determine the theoretical maximum power, in MW and in per-unit of SIL, that the line in Example 6.2 can deliver. Assume $V_S = V_R = 765$ kV.

Solution From Example 6.2,

$$A = 0.9313 \quad \text{per unit;} \quad \theta_A = 0.209°$$

$$B = Z' = 97.0 \quad \Omega; \qquad \theta_Z = 87.2°$$

$$Z_c = 266.1 \quad \Omega$$

From (6.5.6) with $V_S = V_R = 765$ kV,

$$P_{Rmax} = \frac{(765)^2}{97} - \frac{(0.9313)(765)^2}{97} \cos(87.2° - 0.209°)$$

$$= 6033 - 295 = 5738 \quad \text{MW}$$

From (6.4.20),

$$\text{SIL} = \frac{(765)^2}{266.1} = 2199 \quad \text{MW}$$

Thus

$$P_{Rmax} = \frac{5738}{2199} = 2.61 \quad \text{per unit}$$

This value is about 4% less than that found in Example 6.4, where losses were neglected. ∎

SECTION 6.6

LINE LOADABILITY

In practice, power lines are not operated to deliver their theoretical maximum power, which is based on rated terminal voltages and an angular displacement $\delta = 90°$ across the line. Figure 6.12 shows a practical line loadability curve plotted below the theoretical steady-state stability limit. This curve is based on the voltage-drop limit $V_R/V_S \geq 0.95$ and on a maximum angular displacement of 30 to 35° across the line (or about 45° across the line and equivalent system reactances), in order to maintain stability during transient disturbances [1, 3]. The curve is valid for typical overhead 60-Hz lines with no compensation. Note that for short lines less than 80 km long, loadability is limited by the thermal rating of the conductors or by terminal equipment ratings, not by voltage drop or stability considerations. In Section 6.7 we investigate series and shunt compensation techniques to increase the loadability of longer lines toward their thermal limit.

| EXAMPLE 6.6 | **Practical line loadability and percent voltage regulation: long line** |

The 300-km uncompensated line in Example 6.2 has four 1,272,000-cmil 54/3 ACSR conductors per bundle. The sending-end voltage is held constant at 1.0 per-unit of rated line voltage. Determine the following:

a. The practical line loadability (Assume an approximate receiving-end voltage $V_R = 0.95$ per unit and $\delta = 35°$ maximum angle across the line.)

b. The full-load current at 0.986 p.f. leading based on the above practical line loadability

c. The exact receiving-end voltage for the full-load current found in part (b)

d. Percent voltage regulation for the above full-load current

e. Thermal limit of the line, based on the approximate current-carrying capacity given in Table A.4

Solution a. From (6.5.3), with $V_S = 765$, $V_R = 0.95 \times 765\,\text{kV}$, and $\delta = 35°$, using the values of Z', θ_Z, A, and θ_A from Example 6.5,

$$P_R = \frac{(765)(0.95 \times 765)}{97.0} \cos(87.2° - 35°) -$$

$$\frac{(0.9313)(0.95 \times 765)^2}{97.0} \cos(87.2° - 0.209°)$$

$$= 3513 - 266 = 3247 \quad \text{MW}$$

$P_R = 3247\,\text{MW}$ is the practical line loadability, provided the thermal and voltage-drop limits are not exceeded. Alternatively, from Figure 6.12 for a 300-km line, the practical line loadability is $(1.49)\text{SIL} = (1.49)(2199) = 3277\,\text{MW}$, about the same as the above result.

b. For the above loading at 0.986 p.f. leading and at $0.95 \times 765\,\text{kV}$, the full-load receiving-end current is

$$I_{RFL} = \frac{P}{\sqrt{3}V_R(\text{p.f.})} = \frac{3247}{(\sqrt{3})(0.95 \times 765)(0.986)} = 2.616 \quad \text{kA}$$

c. From (6.1.1) with $I_{RFL} = 2.616\underline{/\cos^{-1}0.986} = 2.616\underline{/9.599°}\,\text{kA}$, using the A and B parameters from Example 6.2,

$$V_S = AV_{RFL} + BI_{RFL}$$

$$\frac{765}{\sqrt{3}}\underline{/\delta} = (0.9313\underline{/0.209°})(V_{RFL}\underline{/0°}) + (97.0\underline{/87.2°})(2.616\underline{/9.599°})$$

$$441.7\underline{/\delta} = (0.9313V_{RFL} - 30.04) + j(0.0034V_{RFL} + 251.97)$$

Taking the squared magnitude of the above equation,

$$(441.7)^2 = 0.8673 V_{RFL}^2 - 54.24 V_{RFL} + 64{,}391$$

Solving,

$$V_{RFL} = 420.7 \quad kV_{LN}$$

$$V_{RFL} = 420.7\sqrt{3} = 728.7 \quad kV_{LL} = 0.953 \quad \text{per unit}$$

d. From (6.1.19), the receiving-end no-load voltage is

$$V_{RNL} = \frac{V_S}{A} = \frac{765}{0.9313} = 821.4 \quad kV_{LL}$$

And from (6.1.18),

$$\text{percent VR} = \frac{821.4 - 728.7}{728.7} \times 100 = 12.72\%$$

e. From Table A.4, the approximate current-carrying capacity of four 1,272,000-cmil 54/3 ACSR conductors is $4 \times 1.2 = 4.8 \, kA$.

Since the voltages $V_S = 1.0$ and $V_{RFL} = 0.953$ per unit satisfy the voltage-drop limit $V_R/V_S \geqslant 0.95$, the factor that limits line loadability is steady-state stability for this 300-km uncompensated line. The full-load current of 2.616 kA corresponding to loadability is also well below the thermal limit of 4.8 kA. The 12.7% voltage regulation is too high because the no-load voltage is too high. Compensation techniques to reduce no-load voltages are discussed in Section 6.7. ∎

EXAMPLE 6.7

Selection of transmission line voltage and number of lines for power transfer

From a hydroelectric power plant 9000 MW are to be transmitted to a load center located 500 km from the plant. Based on practical line loadability criteria, determine the number of three-phase, 60-Hz lines required to transmit this power, with one line out of service, for the following cases: (a) 345-kV lines with $Z_c = 297 \, \Omega$; (b) 500-kV lines with $Z_c = 277 \, \Omega$; (c) 765-kV lines with $Z_c = 266 \, \Omega$. Assume $V_S = 1.0$ per unit, $V_R = 0.95$ per unit, and $\delta = 35°$. Also assume that the lines are uncompensated and widely separated such that there is negligible mutual coupling between them.

Solution   **a.** For 345-kV lines, (6.4.21) yields

$$SIL = \frac{(345)^2}{297} = 401 \quad MW$$

Neglecting losses, from (6.4.29), with $l = 500 \, km$ and $\delta = 35°$,

$$P = \frac{(1.0)(0.95)(401)\sin(35°)}{\sin\left(\dfrac{2\pi \times 500}{5000}\right)} = (401)(0.927) = 372 \quad MW/line$$

Alternatively, the practical line loadability curve in Figure 6.12 can be used to obtain $P = (0.93)$SIL for typical 500-km overhead 60-Hz uncompensated lines.

In order to transmit 9000 MW with one line out of service,

$$\#345\text{-kV lines} = \frac{9000\,\text{MW}}{372\,\text{MW/line}} + 1 = 24.2 + 1 \approx 26$$

b. For 500-kV lines,

$$\text{SIL} = \frac{(500)^2}{277} = 903 \quad \text{MW}$$

$$P = (903)(0.927) = 837 \quad \text{MW/line}$$

$$\#500\text{-kV lines} = \frac{9000}{837} + 1 = 10.8 + 1 \approx 12$$

c. For 765-kV lines,

$$\text{SIL} = \frac{(765)^2}{266} = 2200 \quad \text{MW}$$

$$P = (2200)(0.927) = 2039 \quad \text{MW/line}$$

$$\#765\text{-kV lines} = \frac{9000}{2039} + 1 = 4.4 + 1 \approx 6$$

Increasing the line voltage from 345 to 765 kV, a factor of 2.2, reduces the required number of lines from 26 to 6, a factor of 4.3. ∎

EXAMPLE 6.8

Effect of intermediate substations on number of lines required for power transfer

Can five instead of six 765-kV lines transmit the required power in Example 6.7 if there are two intermediate substations that divide each line into three 167-km line sections, and if only one line section is out of service?

Solution The lines are shown in Figure 6.13. For simplicity, we neglect line losses. The

Figure 6.13

Transmission-line configuration for Example 6.8

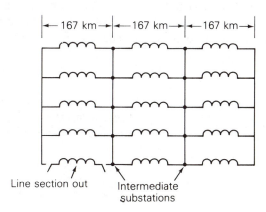

Line section out   Intermediate substations

equivalent π circuit of one 500-km, 765-kV line has a series reactance, from (6.4.10) and (6.4.15),

$$X' = (266) \sin \left(\frac{2\pi \times 500}{5000} \right) = 156.35 \quad \Omega$$

Combining series/parallel reactances in Figure 6.13, the equivalent reactance of five lines with one line section out of service is

$$X_{eq} = \frac{1}{5} \left(\frac{2}{3} X' \right) + \frac{1}{4} \left(\frac{X'}{3} \right) = 0.2167 X' = 33.88 \quad \Omega$$

Then, from (6.4.26) with $\delta = 35°$,

$$P = \frac{(765)(765 \times 0.95) \sin(35°)}{33.88} = 9412 \quad MW$$

Inclusion of line losses would reduce the above value by 3 or 4% to about 9100 MW. Therefore, the answer is yes. Five 765-kV, 500-km uncompensated lines with two intermediate substations and with one line section out of service will transmit 9000 MW. Intermediate substations are often economical if their costs do not outweigh the reduction in line costs. ∎

SECTION 6.7

REACTIVE COMPENSATION TECHNIQUES

Inductors and capacitors are used on medium-length and long transmission lines to increase line loadability and to maintain voltages near rated values. Shunt reactors (inductors) are commonly installed at selected points along EHV lines from each phase to neutral. The inductors absorb reactive power and reduce overvoltages during light load conditions. They also reduce transient overvoltages due to switching and lightning surges. However, shunt reactors can reduce line loadability if they are not removed under full-load conditions.

In addition to shunt reactors, shunt capacitors are sometimes used to deliver reactive power and increase transmission voltages during heavy load conditions. Another type of shunt compensation includes thyristor-switched reactors in parallel with capacitors. These devices, called *static var systems*, can absorb reactive power during light loads and deliver reactive power during heavy loads. Through automatic control of the thyristor switches, voltage fluctuations are minimized and line loadability is increased. Synchronous condensers (synchronous motors with no mechanical load) can also control their reactive power output, although more slowly than static var systems.

Series capacitors are sometimes used on long lines to increase line loadability. Capacitor banks are installed in series with each phase conductor at selected points along a line. Their effect is to reduce the net series impedance of the line in series with the capacitor banks, thereby reducing line-voltage drops and increasing the steady-state stability limit. A disadvantage of series capacitor banks is that automatic protection devices must be installed to bypass high currents during faults and to reinsert the capacitor banks after fault clearing. Also, the addition of series capacitors can excite low-frequency oscillations, a phenomenon called *subsynchronous resonance*, which may damage turbine-generator shafts. Studies have shown, however, that series capacitive compensation can increase the loadability of long lines at only a fraction of the cost of new transmission [1].

Figure 6.14 shows a schematic and an equivalent circuit for a compensated line section, where N_C is the amount of series capacitive compensation expressed in percent of the positive-sequence line impedance and N_L is the amount of shunt reactive compensation in percent of the positive-sequence line admittance. It is assumed in Figure 6.14 that half of the compensation is installed at each end of the line section. The following two examples illustrate the effect of compensation.

Figure 6.14

Compensated transmission-line section

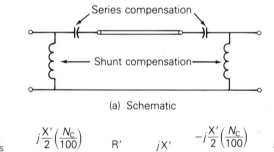

(a) Schematic

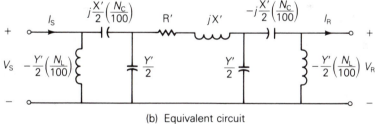

(b) Equivalent circuit

EXAMPLE 6.9

Shunt reactive compensation to improve transmission-line voltage regulation

Identical shunt reactors (inductors) are connected from each phase conductor to neutral at both ends of the 300-km line in Example 6.2 during light load conditions, providing 75% compensation. The reactors are removed during heavy load conditions. Full load is 1.90 kA at unity p.f. and at 730 kV. Assuming that the sending-end voltage is constant, determine the following:

    a. Percent voltage regulation of the uncompensated line

    b. The equivalent shunt admittance and series impedance of the compensated line

    c. Percent voltage regulation of the compensated line

Solution   **a.** From (6.1.1) with $I_{RFL} = 1.9\underline{/0°}\,kA$, using the A and B parameters from Example 6.2,

$$V_S = AV_{RFL} + BI_{RFL}$$

$$= (0.9313\underline{/0.209°})\left(\frac{730}{\sqrt{3}}\underline{/0°}\right) + (97.0\underline{/87.2°})(1.9\underline{/0°})$$

$$= 392.5\underline{/0.209°} + 184.3\underline{/87.2°}$$

$$= 401.5 + j185.5$$

$$= 442.3\underline{/24.8°}\quad kV_{LN}$$

$$V_S = 442.3\sqrt{3} = 766.0\quad kV_{LL}$$

The no-load receiving-end voltage is, from (6.1.19),

$$V_{RNL} = \frac{766.0}{0.9313} = 822.6\quad kV_{LL}$$

and the percent voltage regulation for the uncompensated line is, from (6.1.18),

$$\text{percent VR} = \frac{822.6 - 730}{730} \times 100 = 12.68\%$$

b. From Example 6.3, the shunt admittance of the equivalent π circuit without compensation is

$$Y' = 2(3.7 \times 10^{-7} + j7.094 \times 10^{-4})$$

$$= 7.4 \times 10^{-7} + j14.188 \times 10^{-4}\quad S$$

With 75% shunt compensation, the equivalent shunt admittance is

$$Y_{eq} = 7.4 \times 10^{-7} + j14.188 \times 10^{-4}(1 - \tfrac{75}{100})$$

$$= 3.547 \times 10^{-4}\underline{/89.88°}\quad S$$

Since there is no series compensation, the equivalent series impedance is the same as without compensation:

$$Z_{eq} = Z' = 97.0\underline{/87.2°}\quad \Omega$$

c. The equivalent A parameter for the compensated line is

$$A_{eq} = 1 + \frac{Y_{eq}Z_{eq}}{2}$$

$$= 1 + \frac{(3.547 \times 10^{-4}\underline{/89.88°})(97.0\underline{/87.2°})}{2}$$

$$= 1 + 0.0172\underline{/177.1°}$$

$$= 0.9828\underline{/0.05°}\quad \text{per unit}$$

Then, from (6.1.19),

$$V_{RNL} = \frac{766}{0.9828} = 779.4\quad kV_{LL}$$

Since the shunt reactors are removed during heavy load conditions, $V_{RFL} = 730\,kV$ is the same as without compensation. Therefore

$$\text{percent VR} = \frac{779.4 - 730}{730} \times 100 = 6.77\%$$

The use of shunt reactors at light loads improves the voltage regulation from 12.68% to 6.77% for this line. ∎

| EXAMPLE 6.10 | **Series capacitive compensation to increase transmission-line loadability** |

Identical series capacitors are installed in each phase at both ends of the line in Example 6.2, providing 30% compensation. Determine the theoretical maximum power that this compensated line can deliver and compare with that of the uncompensated line. Assume $V_S = V_R = 765\,kV$.

Solution   From Example 6.3, the equivalent series reactance without compensation is

$$X' = 97.0 \sin 87.2° = 96.88 \quad \Omega$$

Based on 30% series compensation, half at each end of the line, the impedance of each series capacitor is

$$Z_{cap} = -jX_{cap} = -j(\tfrac{1}{2})(0.30)(96.88) = -j14.53 \quad \Omega$$

From Figure 6.4, the *ABCD* matrix of this series impedance is

$$\begin{bmatrix} 1 & -j14.53 \\ 0 & 1 \end{bmatrix}$$

As also shown in Figure 6.4, the equivalent *ABCD* matrix of networks in series is obtained by multiplying the *ABCD* matrices of the individual networks. For this example there are three networks: the series capacitors at the sending end, the line, and the series capacitors at the receiving end. Therefore the equivalent *ABCD* matrix of the compensated line is, using the *ABCD* parameters, from Example 6.2,

$$\begin{bmatrix} 1 & -j14.53 \\ 0 & 1 \end{bmatrix} \begin{bmatrix} 0.9313\underline{/0.209°} & 97.0\underline{/87.2°} \\ 1.37 \times 10^{-3}\underline{/90.06°} & 0.9313\underline{/0.209°} \end{bmatrix} \begin{bmatrix} 1 & -j14.53 \\ 0 & 1 \end{bmatrix}$$

After performing these matrix multiplications, we obtain

$$\begin{bmatrix} A_{eq} & B_{eq} \\ C_{eq} & D_{eq} \end{bmatrix} = \begin{bmatrix} 0.9512\underline{/0.205°} & 69.70\underline{/86.02°} \\ 1.37 \times 10^{-3}\underline{/90.06°} & 0.9512\underline{/0.205°} \end{bmatrix}$$

Therefore

$$A_{eq} = 0.9512 \quad \text{per unit} \qquad \theta_{Aeq} = 0.205°$$

$$B_{eq} = Z'_{eq} = 69.70 \quad \Omega \qquad \theta_{Zeq} = 86.02°$$

From (6.5.6) with $V_S = V_R = 765 \, kV$,

$$P_{Rmax} = \frac{(765)^2}{69.70} - \frac{(0.9512)(765)^2}{69.70} \cos(86.02° - 0.205°)$$

$$= 8396 - 583 = 7813 \quad MW$$

which is 36.2% larger than the value of 5738 MW found in Example 6.5 without compensation. We note that the practical line loadability of this series compensated line is also about 35% larger than the value of 3247 MW found in Example 6.6 without compensation. ∎

SECTION 6.8

PERSONAL COMPUTER PROGRAM: TRANSMISSION LINES—STEADY-STATE OPERATION

The software package that accompanies this text includes the program entitled "TRANSMISSION LINES—STEADY-STATE OPERATION," which calculates: the *ABCD* parameters, equivalent π circuit values, sending-end quantities for a given receiving-end load, and voltage regulation and maximum power flow for single-phase or balanced three-phase lines, with or without compensation, operating under steady-state conditions.

The input data to the program consist of: (1) line constants R, ωL, G, and ωC; (2) rated line voltage, and line length; (3) receiving-end voltage, full-load MVA and power factor; (4) the number and location of intermediate substations; and (5) the percent shunt reactive (inductive) and percent series capacitive compensation installed at each line terminal and at each intermediate substation. The shunt reactors are removed at heavy loads.

The output data consist of (a) without compensation: (1) the characteristic impedance, (6.2.16); (2) propagation constant, (6.2.12); (3) wavelength, (6.4.15); (4) surge impedance loading, (6.4.21); (5) the equivalent π circuit impedance, (6.3.5) and (6.3.6), and admittance, (6.3.9) and (6.3.10); and (6) the *ABCD* parameters, (6.3.1)–(6.3.3); and (b) with compensation: (1) the equivalent *ABCD* parameters; (2) sending-end voltage, current, and power, (6.1.3); (3) percent voltage regulation, (6.1.18); and (4) the theoretical maximum real power delivered to the receiving end with rated terminal voltages, (6.5.6).

The following example compares the equivalent π circuit quantities R', X', and Y' with the nominal π circuit quantities R, X, and Y for some representative EHV lines.

| EXAMPLE 6.11 | **TRANSMISSION LINES—STEADY-STATE OPERATION program** |

The following constants represent typical 345-, 500-, and 765-kV, three-phase, 60-Hz overhead transmission lines:

LINE VOLTAGE	SERIES IMPEDANCE*	SHUNT ADMITTANCE*
kV	z Ω/km	y S/km
345	$0.047 + j0.37$	$j4.1 \times 10^{-6}$
500	$0.020 + j0.34$	$j4.5 \times 10^{-6}$
765	$0.012 + j0.33$	$j4.7 \times 10^{-6}$

*Positive sequence quantities are shown.
Resistances are given at 25°C.

Compute and plot the ratios (R'/R), (X'/X), and (Y'/Y) for the above three lines with line lengths varying from 0 to 1000 km. Neglect the conductances G' and G.

Solution The TRANSMISSION LINES — STEADY-STATE OPERATION program was run as a subprogram in a main program for the given three lines, with line lengths varying from 25 to 1000 km, in increments of 25 km. The outputs (R'/R), (X'/X), and (Y'/Y) are plotted in Figure 6.15. As shown in the figure,

Figure 6.15

Equivalent π to nominal π circuit ratios (R'/R), (X'/X), and (Y'/Y) for typical 345-, 500-, and 765-kV overhead 60-Hz lines

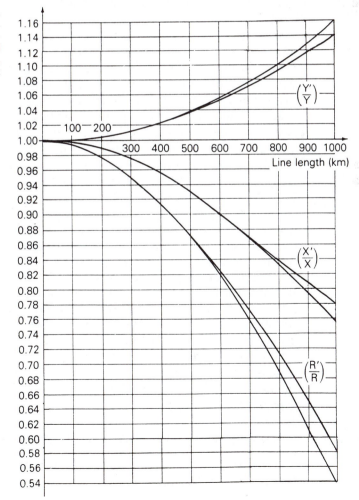

each curve falls within a reasonably narrow band for the three lines. These curves can be used for approximate hand calculations employing typical 60-Hz EHV lines (see Problem 6.25).

The curves also explain why the nominal π circuit gives reasonably accurate results for medium-length lines less than 250 km long. As the line length decreases, all the curves in Figure 6.15 approach unity; that is, the equivalent π circuit approaches the nominal π circuit.   ∎

PROBLEMS

Section 6.1

6.1   A 20-km 34.5-kV, 60-Hz three-phase line has a positive-sequence series impedance $z = 0.19 + j0.34\,\Omega/\text{km}$. The load at the receiving end absorbs 10 MVA at 33 kV. Assuming a short line, calculate: (a) the $ABCD$ parameters, (b) the sending-end voltage for a load power factor of 0.9 lagging, (c) the sending-end voltage for a load power factor of 0.9 leading.

6.2   A 150-km 230-kV, 60-Hz three-phase line has a positive-sequence series impedance $z = 0.08 + j0.48\,\Omega/\text{km}$ and a positive-sequence shunt admittance $y = j3.33 \times 10^{-6}$ S/km. At full load the line delivers 250 MW at 0.99 p.f. lagging and at 220 kV. Using the nominal π circuit, calculate: (a) the $ABCD$ parameters, (b) the sending-end voltage and current, and (c) the percent voltage regulation.

6.3   Rework Problem 6.2 in per-unit using 100-MVA (three-phase) and 230-kV (line-to-line) base values. Calculate: (a) the per-unit $ABCD$ parameters, (b) the per-unit sending-end voltage and current, and (c) the percent voltage regulation.

6.4   Derive the $ABCD$ parameters for the two networks in series, as shown in Figure 6.4.

6.5   Derive the $ABCD$ parameters for the T circuit shown in Figure 6.4.

6.6   The 100-km, 230-kV, 60-Hz, three-phase line in Problems 5.11 and 5.19 delivers 300 MVA at 218 kV to the receiving end at full load. Using the nominal π circuit, calculate the: (a) $ABCD$ parameters, sending-end voltage, and % voltage regulation when the receiving-end power factor is (b) 0.9 lagging; (c) unity; and (d) 0.9 leading. Assume a 50°C conductor temperature to determine the resistance of this line.

6.7   The 500-kV, 60-Hz, three-phase line in Problems 5.13 and 5.21 has a 200-km length and delivers 1600 MW at 475 kV and at 0.95 power factor leading to the receiving end at full load. Using the nominal π circuit, calculate the: (a) $ABCD$ parameters, (b) sending-end voltage and current, (c) sending-end power and power factor, (d) full-load line losses and efficiency, and (e) % voltage regulation. Assume a 50°C conductor temperature to determine the resistance of this line.

Section 6.2

6.8   Evaluate $\cosh(\gamma l)$ and $\tanh(\gamma l/2)$ for $\gamma l = 0.45\underline{/87°}$ per unit.

6.9   A 500-km 500-kV, 60-Hz three-phase uncompensated three-phase line has a positive-sequence series impedance $z = 0.03 + j0.35\,\Omega/\text{km}$ and a positive-sequence shunt admittance $y = j4.4 \times 10^{-6}$ S/km. Calculate: (a) Z_c, (b) (γl), and (c) the exact $ABCD$ parameters for this line.

6.10   At full load the line in Problem 6.9 delivers 1000 MW at unity power factor and at 475 kV. Calculate: (a) the sending-end voltage, (b) the sending-end current, (c) the sending-end power factor, (d) the full-load line losses, and (e) the percent voltage regulation.

6.11   The 500-kV, 60-Hz, three-phase line in Problems 5.13 and 5.21 has a 300-km length. Calculate: (a) Z_c, (b) (γl), and (c) the exact *ABCD* parameters for this line. Assume a 50°C conductor temperature.

6.12   At full load, the line in Problem 6.11 delivers 1500 MVA at 475 kV to the receiving-end load. Calculate the sending-end voltage and % voltage regulation when the receiving-end power factor is (a) 0.9 lagging, (b) unity, and (c) 0.9 leading.

Section 6.3

6.13   Determine the equivalent π circuit for the line in Problem 6.9 and compare it with the nominal π circuit.

6.14   Determine the equivalent π circuit for the line in Problem 6.11. Compare the equivalent π circuit with the nominal π circuit.

Section 6.4

6.15   A 300-km 500-kV, 60-Hz three-phase uncompensated line has a positive-sequence series reactance $x = 0.34\,\Omega/\text{km}$ and a positive-sequence shunt admittance $y = j4.5 \times 10^{-6}$ S/km. Neglecting losses, calculate: (a) Z_c, (b) (γl), (c) the *ABCD* parameters, (d) the wavelength λ of the line, in kilometers, and (e) the surge impedance loading in MW.

6.16   Determine the equivalent π circuit for the line in Problem 6.15.

6.17   Rated line voltage is applied to the sending end of the line in Problem 6.15. Calculate the receiving-end voltage when the receiving end is terminated by (a) an open circuit, (b) the surge impedance of the line, and (c) one-half of the surge impedance. (d) Also calculate the theoretical maximum real power that the line can deliver when rated voltage is applied to both ends of the line.

6.18   Rework Problems 6.6 and 6.11 neglecting the conductor resistance. Compare the results with and without losses.

6.19   From (5.6.22) and (5.10.4), the positive-sequence series inductance and shunt capacitance of a three-phase overhead line are

$$L_1 = 2 \times 10^{-7}\,\ln\,(D_{eq}/D_{SL}) = \frac{\mu_0}{2\pi}\,\ln(D_{eq}/D_{SL}) \quad \text{H/m}$$

$$C_1 = \frac{2\pi\varepsilon_0}{\ln(D_{eq}/D_{sc})} \quad \text{F/m}$$

where $\mu_0 = 4\pi \times 10^{-7}$ H/m   and $\varepsilon_0 = \left(\dfrac{1}{36\pi}\right) \times 10^{-9}$ F/m

Using these equations, determine formulas for surge impedance and velocity of propagation of an overhead lossless line. Then determine the surge impedance and velocity of propagation for the three-phase 765-kV line given in Example 5.10. Assume positive-sequence operation. Neglect line losses as well as the effects of the overhead neutral wires and the earth plane.

Section 6.5

6.20   The line in Problem 6.9 has three ACSR 1113-kcmil conductors per phase. Calculate the theoretical maximum real power that this line can deliver and compare with the thermal limit of the line. Assume $V_S = V_R = 1.0$ per unit and unity power factor at the receiving end.

6.21   Repeat Problems 6.9 and 6.20 if the line length is (a) 200 km, (b) 600 km.

6.22   For the 500-kV line given in Problem 6.11, (a) calculate the theoretical maximum real power that the line can deliver to the receiving end when rated voltage is applied to both ends. (b) Calculate the receiving-end reactive power and power factor at this theoretical loading.

Section 6.6

6.23   For the line in Problems 6.9 and 6.20, determine: (a) the practical line loadability in MW, assuming $V_S = 1.0$ per unit, $V_R \approx 0.95$ per unit, and $\delta_{max} = 35°$; (b) the full-load current at 0.99 p.f. leading, based on the above practical line loadability; (c) the exact receiving-end voltage for the full-load current in (b) above; and (d) the percent voltage regulation. For this line, is loadability determined by the thermal limit, the voltage-drop limit, or steady-state stability?

6.24   Repeat Problem 6.23 for the 500-kV line given in Problem 6.7.

6.25   Determine the practical line loadability in MW and in per-unit of SIL for the line in Problem 6.9 if the line length is (a) 200 km, (b) 600 km. Assume $V_S = 1.0$ per unit, $V_R = 0.95$ per unit, $\delta_{max} = 35°$, and 0.99 leading power factor at the receiving end.

6.26   It is desired to transmit 2000 MW from a power plant to a load center located 300 km from the plant. Determine the number of 60-Hz three-phase, uncompensated transmission lines required to transmit this power with one line out of service for the following cases: (a) 345-kV lines, $Z_c = 300 \, \Omega$, (b) 500-kV lines, $Z_c = 275 \, \Omega$, (c) 765-kV lines, $Z_c = 260 \, \Omega$. Assume that $V_S = 1.0$ per unit, $V_R = 0.95$ per unit, and $\delta_{max} = 35°$.

6.27   Repeat Problem 6.26 if it is desired to transmit: (a) 3000 MW to a load center located 300 km from the plant, (b) 2000 MW to a load center located 400 km from the plant.

Section 6.7

6.28   Recalculate the percent voltage regulation in Problem 6.10 when identical shunt reactors are installed at both ends of the line during light loads, providing 70% total shunt compensation. The reactors are removed at full load. Also calculate the impedance of each shunt reactor.

6.29   Rework Problem 6.12 when identical shunt reactors are installed at both ends of the line, providing 50% total shunt compensation. The reactors are removed at full load.

6.30   Identical series capacitors are installed at both ends of the line in Problem 6.9, providing 40% total series compensation. Determine the equivalent $ABCD$ parameters of this compensated line. Also calculate the impedance of each series capacitor.

6.31   Identical series capacitors are installed at both ends of the line in Problem 6.11, providing 30% total series compensation. (a) Determine the equivalent $ABCD$ parameters for this compensated line. (b) Determine the theoretical maximum real power that this series-compensated line can deliver when $V_S = V_R = 1.0$ per unit. Compare your result with that of Problem 6.22.

6.32   Determine the theoretical maximum real power that the series-compensated line in Problem 6.30 can deliver when $V_S = V_R = 1.0$ per unit. Compare your result with that of Problem 6.20.

6.33   What is the minimum amount of series capacitive compensation N_C in percent of the positive-sequence line reactance needed to reduce the number of 765-kV lines in Example 6.8 from five to four. Assume two intermediate substations with one line section out of service. Also, neglect line losses and assume that the series compensation is sufficiently distributed along the line so as to effectively reduce the series reactance of the equivalent π circuit to $X'(1 - N_C/100)$.

6.34 Determine the equivalent *ABCD* parameters for the line in Problem 6.9 if it has 70% shunt reactive (inductors) compensation and 40% series capacitive compensation. Half of this compensation is installed at each end of the line, as in Figure 6.14.

Section 6.8

6.35 Using the TRANSMISSION LINES—STEADY-STATE OPERATION program, verify the curves in Figure 6.15 for the 500-kV line in Example 6.11 with 100-, 300-, 500-, and 1000-km line lengths.

6.36 Using the curves in Figure 6.15, determine an approximate equivalent π circuit for the line in Problem 6.9 and compare it with the exact equivalent π circuit determined in Problem 6.13. Also determine approximate equivalent π circuits for 200- and 600-km line lengths.

6.37 Using the TRANSMISSION LINES—STEADY-STATE OPERATION program, rework Problems 6.9, 6.10, 6.13, and 6.20.

6.38 Using the TRANSMISSION LINES—STEADY-STATE OPERATION program, vary the receiving-end load at unity power factor and at $V_R = 0.95$ per unit for the line in Problem 6.9 until $V_R/V_S = 0.95$ or $\delta = 35°$, whichever occurs first. Determine whether voltage drop or steady-state stability limits the loadability of this line.

6.39 Repeat Problem 6.38 if the line has 35% total series capacitive compensation, half installed at each end of the line.

6.40 Using the TRANSMISSION LINES—STEADY-STATE OPERATION program, rework Problems 6.11, 6.12, 6.14, and 6.22. Assume unity receiving-end power factor.

6.41 Identical shunt reactors are installed at both ends of the line in Problems 6.11 and 6.12. Determine the % total shunt compensation required for 10% voltage regulation. The reactors are removed at full load. Assume unity receiving-end power factor.

6.42 A 650-km 765-kV, 60-Hz three-phase transmission line has a positive-sequence series impedance $z = 0.0185 + j0.3170 \, \Omega/\text{km}$ and a positive-sequence shunt admittance $y = j4.875 \times 10^{-6} \, \text{S/km}$. Full load at the receiving end is 2500 MVA at 0.99 p.f. leading and at 95% of rated voltage. Two intermediate substations divide the line into three equal line sections. Twenty-two percent total series capacitive compensation and 65% total shunt reactive (inductive) compensation is installed with equal compensation at the line ends and at the intermediate substations. The shunt reactors are removed during heavy loads. Run the TRANSMISSION LINES—STEADY-STATE OPERATION program to determine (a) without compensation: Z_c, γ, λ, the *ABCD* parameters, and the equivalent π circuit impedance and admittance; (b) with compensation: the equivalent *ABCD* parameters, sending-end voltage, sending-end current, sending-end power factor, percent voltage regulation, and the theoretical maximum real power delivered to the receiving end with rated terminal voltages.

CASE STUDY QUESTIONS

A. Dispatchers at power system control centers have limited control of power flows in today's transmission and distribution networks. Dispatch options include rescheduling of generation and intertie flows, adjusting generator excitation and transformer taps, removing lines or equipment from service, and bringing standby equipment on-line. However, Ohm's law seems to have the greatest influence on network flows. That is, power (more correctly, energy) flows from generators to customer loads through paths of lowest impedance. Does continuously variable series line compensation provide the opportunity for central dispatch (control) of flow on each compensated line?

B. What are the benefits of controlling power flows on individual lines? What are the risks?

References

1. General Electric Company, *Transmission Line Reference Book—345 kV and Above*, 2nd ed. (Palo Alto, CA. Electric Power Research Institute, 1982).

2. Westinghouse Electric Corporation, *Electrical Transmission and Distribution Reference Book*, 4th ed. (East Pittsburgh, PA, 1964).

3. R. D. Dunlop, R. Gutman, and P. P. Marchenko, "Analytical Development of Loadability Characteristics for EHV and UHV Lines," *IEEE Trans. PAS*, Vol. PAS-98, No. 2 (March/April 1979): pp. 606–607.

4. W. D. Stevenson, Jr., *Elements of Power System Analysis*, 4th ed. (New York: McGraw-Hill, 1982).

5. W. H. Hayt, Jr., and J. E. Kemmerly, *Engineering Circuit Analysis*, 3rd ed. (New York: McGraw-Hill, 1978).

6. Mark Wilhelm and Duane Torgerson, "Phase-controlled Reactor Bucks Series Capacitor," *Electrical World*, *206*, 6 (June 1992), pp. 60–70.

CHAPTER 7

POWER FLOWS

Supervisory control and data acquisition (SCADA) center
(Courtesy of Boston Edison)

Successful power-system operation under normal balanced three-phase steady-state conditions requires the following:

1. Generation supplies the demand (load) plus losses.

2. Bus voltage magnitudes remain close to rated values.

3. Generators operate within specified real and reactive power limits.

4. Transmission lines and transformers are not overloaded.

The power-flow computer program (commonly called *load flow*) is the basic tool for investigating these requirements. This program computes the voltage magnitude and angle at each bus in a power system under balanced three-phase steady-state conditions. Real and reactive power flows for all equipment interconnecting the buses, as well as equipment losses, are also computed. Both existing power systems and proposed changes including new generation and transmission to meet projected load growth are of interest.

Conventional nodal or loop analysis is not suitable for power-flow studies because the input data for loads are normally given in terms of power, not impedance. Also, generators are considered as power sources, not voltage or current sources. The power-flow problem is therefore formulated as a set of nonlinear algebraic equations suitable for computer solution.

In Sections 7.1–7.3 we review some basic methods, including direct and iterative techniques for solving algebraic equations. Then in Sections 7.4–7.6 the power-flow problem is formulated, computer input data are specified, and two solution methods, Gauss–Seidel and Newton–Raphson, are presented. Sparsity techniques for reducing computer storage and time requirements are introduced in Section 7.7, and the POWER FLOW personal computer program, which accompanies this text, is described in Section 7.8. A fast decoupled power-flow method is described in Section 7.9, and means for controlling power flows are discussed in Section 7.10.

Since balanced three-phase steady-state conditions are assumed, only positive-sequence networks are used in this chapter. Also, all power-flow equations and input/output data are given in per-unit.

CASE STUDY

Power-flow programs are used to study power systems under both normal operating conditions and disturbance conditions. One disturbance condition is the voltage collapse phenomenon, which has recently become a significant issue in the electric utility industry. This phenomenon involves severe, uncontrollable drops in system voltages following a large disturbance such as a sharp rise in system load or a major outage. One of the factors contributing to voltage collapse is the increasingly large reactive power absorbed by air-conditioning motors and other types of motors at lower voltages. The following report describes three voltage collapse incidents [9].

Survey of the Voltage Collapse Phenomenon

NORTH AMERICAN ELECTRIC RELIABILITY COUNCIL

Voltage collapse is becoming the subject of more frequent discussions by utility personnel. Both operations and planning engineers are taking much more interest in this phenomenon. It is a very complex problem and is not well understood within the industry. It is an emerging area, with differing views on the modeling techniques and analysis methods for determining the critical voltage.

There have been disturbances involving voltage collapse over the last twenty years with the majority of the disturbances occurring since 1982. The time frame for these types of disturbances varies from a few seconds to thirty minutes or longer. This long range of time is quite different from disturbances resulting from transient stability. The 1987 Tokyo and 1987 Memphis voltage collapse disturbances and the voltage stability incident without collapse in England in 1986 are reviewed in this report.

Steady-state analysis

Voltage stability (longer-term voltage stability) is usually studied as a steady-state problem using conventional or extended power flow programs. The load characteristics represented in the power flow are very important and include:

- Voltage sensitivity of loads.
- Representation of voltage regulating devices in subtransmission and distribution systems that are part of the "load"—LTC transformers, distribution voltage regulators, and switched capacitors.
- Equivalent impedance of subtransmission and distribution network.
- Constant energy nature of some loads such as thermostatically controlled loads.

Reprinted with permission from the North American Electric Reliability Council

For simplicity, voltage stability power flow simulations usually assume constant power loads modeled on the high side busses of bulk power delivery substations. For classical voltage stability, a point in time one to five minutes following a major disturbance is simulated. Tap changing action is assumed to have restored low-side voltage—so that voltage sensitive loads are restored to pre-disturbance levels.

Some utilities, such as Bonneville Power Administration, are developing very detailed load representations. They are modeling the voltage sensitivity of loads, along with tap changing and thermostat control effects.

Another critical aspect of voltage stability is field current limiting at generators. The usual model is a P-V bus with a Q_{max} limit. Q_{max} may be viewed as a function of real power loading and, less frequently, as a function of terminal voltage.

Incidents

The Power failure in Tokyo—July 23, 1987

The event occurred on a hot summer afternoon and resulted in the loss of most of the 500 kV system of the Tokyo Electric Power Company as well as 8,000 MW of load to almost three million customers for about three hours.

The event was initiated by a sharp increase in the demand which was estimated to reach about 38,000 MW with a 33°C ambient temperature. It was assumed that a 3.8% margin was adequate. By afternoon, however, the temperature had risen to 39°C (over 100°F) and the load as well so that the margin was wiped out. The 500 kV voltages began to sink (to 460 kV) and eventually the 500 kV lines were opened (on overcurrent) and seven 500 kV substations were without supply.

It is interesting that all this time there was no fault indication or abnormal operation which would have alerted the operators to the impending disaster. It was simply the relentless mathematical accuracy of the famous regulation curve and the system just moved on this curve the way the Titanic approached the iceberg.

The only indication that something out of the ordinary was happening was that the rate of rise of the load was 400 MW/minute, which was twice as much as ever recorded before.

The use of modern air conditioner motor controls results in very high reactive requirements under conditions of low voltage. It is believed that the fairly extensive use of these air conditioners could be an important factor in explaining the voltage collapse in Japan.

Cascading voltage collapse in West Tennessee—August 22, 1987

On a hot summer Saturday within the Tennessee Valley Authority (TVA) service area, a 115 kV switch belonging to Memphis Light, Gas, & Water Division (MLG&W) arced and flashed phase-to-phase while the MLG&W operator was attempting to isolate a damaged airblast breaker. Because the faulted bus lacked a bus differential protection scheme, the fault continued for more than a second and was eventually cleared by backup relays at remote locations. Due to the long fault duration, motor loads in Memphis and the surrounding TVA area began to stall and draw large amounts of reactive power even after the fault was cleared. A depressed voltage condition developed on both the 161 kV and 500 kV systems in southwestern Tennessee and continued for 10 to 15 seconds. During this time, reverse zone 3 relays at several remote 161 kV substations began to trip. This started a cascading effect that eventually tripped all source lines into TVA's South Jackson, Milan, and Covington substations. In addition, the low and unbalanced voltages in the area caused backup distance or ground relays to trip transformer banks at three other 161 kV substations. MLG&W lost 700 MW of load during the disturbance. A good portion of this was lost as thermal and overload protection on individual

pump and air conditioning loads tripped. TVA lost an additional 565 MW of load after thirteen 161 kV substations were deenergized.

(For additional information, refer to NERC's *1987 System Disturbances* report and the paper "Cascade Voltage Collapse in West Tennessee, August 22, 1987," Georgia Institute of Technology, 44th Annual Protective Relaying Conference, May 2–4, 1990.)

Near voltage collapse of the British 400 kV system—May 20, 1986

This event is a non-event because the approaching voltage collapse was averted by the inherent design of the system (for this purpose) and the alert action of the CEGB Operating Department.

The event began with the loss of six 400 kV lines from the north to London as a result of a thunderstorm. Actually, the British system is noted for these multiple outages (called bunching) and so is prepared to act. The lines in question were carrying almost 6,000 MW to various load centers and their loss resulted in a voltage drop to 350 kV and overload on the 275 kV system.

The 400 kV transformers to the various load centers (area boards) have taps with different time settings so that there is a kind of inherent load management at work that mitigates the effect of the sinking 400 kV voltage.

At the same time, the operators switched on 1,000 MW of gas turbines that helped to forestall the collapse of the 400 kV voltage. Subsequent fault analysis showed that when the gas turbines were switched on, the system was only one minute away from voltage collapse.

The tapchanger behavior seems to have been the controlling influence providing system operators with valuable additional time to avert the approaching collapse condition.

SECTION 7.1

DIRECT SOLUTIONS TO LINEAR ALGEBRAIC EQUATIONS: GAUSS ELIMINATION

Consider the following set of linear algebraic equations in matrix format:

$$\begin{bmatrix} A_{11} & A_{12} & \cdots & A_{1N} \\ A_{21} & A_{22} & \cdots & A_{2N} \\ \vdots & & \vdots & \\ A_{N1} & A_{N2} & \cdots & A_{NN} \end{bmatrix} \begin{bmatrix} x_1 \\ x_2 \\ \vdots \\ x_N \end{bmatrix} = \begin{bmatrix} y_1 \\ y_2 \\ \vdots \\ y_N \end{bmatrix} \tag{7.1.1}$$

or

$$\mathbf{Ax} = \mathbf{y} \tag{7.1.2}$$

where $\mathbf{x}$ and $\mathbf{y}$ are N vectors and $\mathbf{A}$ is an $N \times N$ square matrix. The components of $\mathbf{x}$, $\mathbf{y}$, and $\mathbf{A}$ may be real or complex. Given $\mathbf{A}$ and $\mathbf{y}$, it is desired to solve for $\mathbf{x}$. We assume the $\det(\mathbf{A})$ is nonzero, so a unique solution to (7.1.1) exists.

The solution $\mathbf{x}$ can easily be obtained when $\mathbf{A}$ is an upper triangular matrix with nonzero diagonal elements. Then (7.1.1) has the form

$$\begin{bmatrix} A_{11} & A_{12} \cdots & & A_{1N} \\ 0 & A_{22} \cdots & & A_{2N} \\ \vdots & & & \\ 0 & 0 \cdots & A_{N-1,N-1} & A_{N-1,N} \\ 0 & 0 \cdots 0 & & A_{NN} \end{bmatrix} \begin{bmatrix} x_1 \\ x_2 \\ \vdots \\ x_{N-1} \\ x_N \end{bmatrix} = \begin{bmatrix} y_1 \\ y_2 \\ \vdots \\ y_{N-1} \\ y_N \end{bmatrix} \tag{7.1.3}$$

Since the last equation in (7.1.3) involves only x_N,

$$x_N = \frac{y_N}{A_{NN}} \tag{7.1.4}$$

After x_N is computed, the next-to-last equation can be solved:

$$x_{N-1} = \frac{y_{N-1} - A_{N-1,N} x_N}{A_{N-1,N-1}} \tag{7.1.5}$$

In general, with $x_N, x_{N-1}, \dots, x_{k+1}$ already computed, the kth equation can be solved

$$x_k = \frac{y_k - \displaystyle\sum_{n=k+1}^{N} A_{kn} x_n}{A_{kk}} \qquad k = N, N-1, \dots, 1 \tag{7.1.6}$$

This procedure for solving (7.1.3) is called *back substitution*.

If $\mathbf{A}$ is not upper triangular, (7.1.1) can be transformed to an equivalent equation with an upper triangular matrix. The transformation, called *Gauss elimination*, is described by the following $(N-1)$ steps. During step 1, we use the first equation in (7.1.1) to eliminate x_1 from the remaining equations. That is, equation 1 is multiplied by A_{n1}/A_{11} and then subtracted from equation n, for $n = 2, 3, \ldots, N$. After completing step 1, we have

$$
\begin{bmatrix}
A_{11} & A_{12} & \cdots & A_{1N} \\
0 & \left(A_{22} - \dfrac{A_{21}}{A_{11}} A_{12}\right) & \cdots & \left(A_{2N} - \dfrac{A_{21}}{A_{11}} A_{1N}\right) \\
0 & \left(A_{32} - \dfrac{A_{31}}{A_{11}} A_{12}\right) & \cdots & \left(A_{3N} - \dfrac{A_{31}}{A_{11}} A_{1N}\right) \\
\vdots & \vdots & & \vdots \\
0 & \left(A_{N2} - \dfrac{A_{N1}}{A_{11}} A_{12}\right) & \cdots & \left(A_{NN} - \dfrac{A_{N1}}{A_{11}} A_{1N}\right)
\end{bmatrix}
\begin{bmatrix}
x_1 \\ x_2 \\ x_3 \\ \vdots \\ x_N
\end{bmatrix}
=
\begin{bmatrix}
y_1 \\ y_2 - \dfrac{A_{21}}{A_{11}} y_1 \\ y_3 - \dfrac{A_{31}}{A_{11}} y_1 \\ \vdots \\ y_N - \dfrac{A_{N1}}{A_{11}} y_1
\end{bmatrix}
$$

$$(7.1.7)$$

Equation (7.1.7) has the following form:

$$
\begin{bmatrix}
A_{11}^{(1)} & A_{12}^{(1)} & \cdots & A_{1N}^{(1)} \\
0 & A_{22}^{(1)} & \cdots & A_{2N}^{(1)} \\
0 & A_{32}^{(1)} & \cdots & A_{3N}^{(1)} \\
\vdots & \vdots & & \vdots \\
0 & A_{N2}^{(1)} & \cdots & A_{NN}^{(1)}
\end{bmatrix}
\begin{bmatrix}
x_1 \\ x_2 \\ x_3 \\ \vdots \\ x_N
\end{bmatrix}
=
\begin{bmatrix}
y_1^{(1)} \\ y_2^{(1)} \\ y_3^{(1)} \\ \vdots \\ y_N^{(1)}
\end{bmatrix}
$$

$$(7.1.8)$$

where the superscript (1) denotes step 1 of Gauss elimination.

During step 2 we use the second equation in (7.1.8) to eliminate x_2 from the remaining (third, fourth, fifth, and so on) equations. That is, equation 2 is multiplied by $A_{n2}^{(1)}/A_{22}^{(1)}$ and subtracted from equation n, for $n = 3, 4, \ldots, N$. After step 2, we have

$$
\begin{bmatrix}
A_{11}^{(2)} & A_{12}^{(2)} & A_{13}^{(2)} & \cdots & A_{1N}^{(2)} \\
0 & A_{22}^{(2)} & A_{23}^{(2)} & \cdots & A_{2N}^{(2)} \\
0 & 0 & A_{33}^{(2)} & \cdots & A_{3N}^{(2)} \\
0 & 0 & A_{43}^{(2)} & \cdots & A_{4N}^{(2)} \\
\vdots & \vdots & \vdots & & \vdots \\
0 & 0 & A_{N3}^{(2)} & \cdots & A_{NN}^{(2)}
\end{bmatrix}
\begin{bmatrix}
x_1 \\ x_2 \\ x_3 \\ x_4 \\ \vdots \\ x_N
\end{bmatrix}
=
\begin{bmatrix}
y_1^{(2)} \\ y_2^{(2)} \\ y_3^{(2)} \\ y_4^{(2)} \\ \vdots \\ y_N^{(2)}
\end{bmatrix}
$$

$$(7.1.9)$$

During step k, we start with $\mathbf{A}^{(k-1)}\mathbf{x} = \mathbf{y}^{(k-1)}$. The first k of these equations, already triangularized, are left unchanged. Also, equation k is multiplied by $A_{nk}^{(k-1)}/A_{kk}^{(k-1)}$ and then subtracted from equation n, for $n = k + 1, k + 2, \ldots, N$.

After $(N - 1)$ steps, we arrive at the equivalent equation $\mathbf{A}^{(N-1)}\mathbf{x} = \mathbf{y}^{(N-1)}$, where $\mathbf{A}^{(N-1)}$ is upper triangular.

EXAMPLE 7.1

Gauss elimination and back substitution: direct solution to linear algebraic equations

Solve

$$
\left[
\begin{array}{c|c}
10 & 5 \\
\hline
2 & 9
\end{array}
\right]
\left[
\begin{array}{c}
x_1 \\
x_2
\end{array}
\right]
=
\left[
\begin{array}{c}
6 \\
3
\end{array}
\right]
$$

using Gauss elimination and back substitution.

Solution   Since $N = 2$ for this example, there is $(N - 1) = 1$ Gauss elimination step. Multiplying the first equation by $A_{21}/A_{11} = 2/10$ and then subtracting from the second,

$$
\left[
\begin{array}{c|c}
10 & 5 \\
\hline
0 & 9 - \dfrac{2}{10}\,(5)
\end{array}
\right]
\left[
\begin{array}{c}
x_1 \\
x_2
\end{array}
\right]
=
\left[
\begin{array}{c}
6 \\
3 - \dfrac{2}{10}\,(6)
\end{array}
\right]
$$

or

$$
\left[
\begin{array}{c|c}
10 & 5 \\
\hline
0 & 8
\end{array}
\right]
\left[
\begin{array}{c}
x_1 \\
x_2
\end{array}
\right]
=
\left[
\begin{array}{c}
6 \\
1.8
\end{array}
\right]
$$

which has the form $\mathbf{A}^{(1)}\mathbf{x} = \mathbf{y}^{(1)}$, where $\mathbf{A}^{(1)}$ is upper triangular. Now, using back substitution, (7.1.6) gives, for $k = 2$:

$$
x_2 = \frac{y_2^{(1)}}{A_{22}^{(1)}} = \frac{1.8}{8} = 0.225
$$

and, for $k = 1$,

$$
x_1 = \frac{y_1^{(1)} - A_{12}^{(1)} x_2}{A_{11}^{(1)}} = \frac{6 - (5)(0.225)}{10} = 0.4875
$$

∎

EXAMPLE 7.2

Gauss elimination: triangularizing a matrix

Use Gauss elimination to triangularize

$$
\left[
\begin{array}{ccc}
2 & 3 & -1 \\
-4 & 6 & 8 \\
10 & 12 & 14
\end{array}
\right]
\left[
\begin{array}{c}
x_1 \\
x_2 \\
x_3
\end{array}
\right]
=
\left[
\begin{array}{c}
5 \\
7 \\
9
\end{array}
\right]
$$

Solution   There are $(N - 1) = 2$ Gauss elimination steps. During step 1, we subtract $A_{21}/A_{11} = -4/2 = -2$ times equation 1 from equation 2, and we subtract

$A_{31}/A_{11} = 10/2 = 5$ times equation 1 from equation 3, to give

$$\begin{bmatrix} 2 & 3 & -1 \\ 0 & 6-(-2)(3) & 8-(-2)(-1) \\ 0 & 12-(5)(3) & 14-(5)(-1) \end{bmatrix} \begin{bmatrix} x_1 \\ x_2 \\ x_3 \end{bmatrix} = \begin{bmatrix} 5 \\ 7-(-2)(5) \\ 9-(5)(5) \end{bmatrix}$$

or

$$\begin{bmatrix} 2 & 3 & -1 \\ 0 & 12 & 6 \\ 0 & -3 & 19 \end{bmatrix} \begin{bmatrix} x_1 \\ x_2 \\ x_3 \end{bmatrix} = \begin{bmatrix} 5 \\ 17 \\ -16 \end{bmatrix}$$

which is $A^{(1)}x = y^{(1)}$. During step 2, we subtract $A_{32}^{(1)}/A_{22}^{(1)} = -3/12 = -0.25$ times equation 2 from equation 3, to give

$$\begin{bmatrix} 2 & 3 & -1 \\ 0 & 12 & 6 \\ 0 & 0 & 19-(-.25)(6) \end{bmatrix} \begin{bmatrix} x_1 \\ x_2 \\ x_3 \end{bmatrix} = \begin{bmatrix} 5 \\ 17 \\ -16-(-.25)(17) \end{bmatrix}$$

or

$$\begin{bmatrix} 2 & 3 & -1 \\ 0 & 12 & 6 \\ 0 & 0 & 20.5 \end{bmatrix} \begin{bmatrix} x_1 \\ x_2 \\ x_3 \end{bmatrix} = \begin{bmatrix} 5 \\ 17 \\ -11.75 \end{bmatrix}$$

which is triangularized. The solution **x** can now be easily obtained via back substitution. ∎

Computer storage requirements for Gauss elimination and back substitution include N^2 memory locations for **A** and N locations for **y**. If there is no further need to retain **A** and **y**, then $A^{(k)}$ can be stored in the location of **A**, and $y^{(k)}$, as well as the solution **x**, can be stored in the location of **y**. Additional memory is also required for DO loops, arithmetic statements, and working space.

Computer time requirements can be evaluated by determining the number of arithmetic operations required for Gauss elimination and back substitution. One can show (see Problem 7.3) that Gauss elimination requires $(N^3 - N)/3$ multiplications, $(N)(N-1)/2$ divisions, and $(N^3 - N)/3$ subtractions. Also, back substitution (see Problem 7.4) requires $(N)(N-1)/2$ multiplications, N divisions, and $(N)(N-1)/2$ subtractions. Therefore, for very large N, the approximate computer time for solving (7.1.1) by Gauss elimination and back substitution is the time required to perform $N^3/3$ multiplications and $N^3/3$ subtractions.

For example, consider a digital computer with a 10^{-6} s multiplication or division time and 10^{-7} s addition or subtraction time. Solving $N = 1000$ equations would require approximately

$$\tfrac{1}{3}N^3(10^{-6}) + \tfrac{1}{3}N^3(10^{-7}) = \tfrac{1}{3}(1000)^3(10^{-6} + 10^{-7}) = 366.7 \quad \text{s}$$

plus some additional bookkeeping time for indexing and managing DO loops.

For matrices that have only a few nonzero elements, special techniques can be employed to significantly reduce computer storage and time requirements. These techniques are discussed in Section 7.7.

SECTION 7.2

ITERATIVE SOLUTIONS TO LINEAR ALGEBRAIC EQUATIONS: JACOBI AND GAUSS–SEIDEL

A general iterative solution to (7.1.1) proceeds as follows. First select an initial guess $x(0)$. Then use

$$\mathbf{x}(i + 1) = \mathbf{g}[\mathbf{x}(i)] \qquad i = 0, 1, 2, \ldots \tag{7.2.1}$$

where $\mathbf{x}(i)$ is the ith guess and $\mathbf{g}$ is an N vector of functions that specify the iteration method. The procedure continues until the following stopping condition is satisfied:

$$\left| \frac{x_k(i + 1) - x_k(i)}{x_k(i)} \right| < \varepsilon \qquad \text{for all } k = 1, 2, \ldots, N \tag{7.2.2}$$

where $x_k(i)$ is the kth component of $\mathbf{x}(i)$ and ε is a specified *tolerance level*.

The following questions are pertinent:

1. Will the iteration procedure converge to the unique solution?
2. What is convergence rate (how many iterations are required)?
3. When using a digital computer, what are the computer storage and time requirements?

These questions are addressed for two specific iteration methods: *Jacobi* and *Gauss–Seidel*.* The Jacobi method is obtained by considering the kth equation of (7.1.1), as follows:

$$y_k = A_{k1}x_1 + A_{k2}x_2 + \cdots + A_{kk}x_k + \cdots + A_{kN}x_N \tag{7.2.3}$$

---

*The Jacobi method is also called the Gauss method.

Solving for x_k,

$$x_k = \frac{1}{A_{kk}} [y_k - (A_{k1}x_1 + \cdots + A_{k,k-1}x_{k-1} + A_{k,k+1}x_{k+1} + \cdots + A_{kN}x_N)]$$

$$= \frac{1}{A_{kk}} \left[y_k - \sum_{n=1}^{k-1} A_{kn}x_n - \sum_{n=k+1}^{N} A_{kn}x_n \right] \qquad (7.2.4)$$

The Jacobi method uses the "old" values of $\mathbf{x}(i)$ at iteration i on the right side of (7.2.4) to generate the "new" value $x_k(i+1)$ on the left side of (7.2.4). That is,

$$x_k(i+1) = \frac{1}{A_{kk}} \left[y_k - \sum_{n=1}^{k-1} A_{kn}x_n(i) - \sum_{n=k+1}^{N} A_{kn}x_n(i) \right] \quad k = 1, 2, \ldots, N$$

$$(7.2.5)$$

The Jacobi method given by (7.2.5) can also be written in the following matrix format:

$$\mathbf{x}(i+1) = \mathbf{M}\mathbf{x}(i) + \mathbf{D}^{-1}\mathbf{y} \qquad (7.2.6)$$

where

$$\mathbf{M} = \mathbf{D}^{-1}(\mathbf{D} - \mathbf{A}) \qquad (7.2.7)$$

and

$$\mathbf{D} = \begin{bmatrix} A_{11} & 0 & 0 & \cdots & 0 \\ 0 & A_{22} & 0 & \cdots & 0 \\ 0 & \vdots & \vdots & & \vdots \\ \vdots & & & & 0 \\ 0 & 0 & 0 & \cdots & A_{NN} \end{bmatrix} \qquad (7.2.8)$$

For Jacobi, $\mathbf{D}$ consists of the diagonal elements of the $\mathbf{A}$ matrix.

EXAMPLE 7.3

Jacobi method: iterative solution to linear algebraic equations

Solve Example 7.1 using the Jacobi method. Start with $x_1(0) = x_2(0) = 0$ and continue until (7.2.2) is satisfied for $\varepsilon = 10^{-4}$.

Solution   From (7.2.5) with $N = 2$,

$$k = 1 \qquad x_1(i+1) = \frac{1}{A_{11}} [y_1 - A_{12}x_2(i)] = \tfrac{1}{10}[6 - 5x_2(i)]$$

$$k = 2 \qquad x_2(i+1) = \frac{1}{A_{22}} [y_2 - A_{21}x_1(i)] = \tfrac{1}{9}[3 - 2x_1(i)]$$

Alternatively, in matrix format, using (7.2.6)–(7.2.8)

$$
\mathbf{D}^{-1} = \left[\begin{array}{c|c} 10 & 0 \\ \hline 0 & 9 \end{array}\right]^{-1} = \left[\begin{array}{c|c} \frac{1}{10} & 0 \\ \hline 0 & \frac{1}{9} \end{array}\right]
$$

$$
\mathbf{M} = \left[\begin{array}{c|c} \frac{1}{10} & 0 \\ \hline 0 & \frac{1}{9} \end{array}\right] \left[\begin{array}{c|c} 0 & -5 \\ \hline -2 & 0 \end{array}\right] = \left[\begin{array}{c|c} 0 & -\frac{5}{10} \\ \hline -\frac{2}{9} & 0 \end{array}\right]
$$

$$
\left[\begin{array}{c} x_1(i+1) \\ x_2(i+1) \end{array}\right] = \left[\begin{array}{c|c} 0 & -\frac{5}{10} \\ \hline -\frac{2}{9} & 0 \end{array}\right]\left[\begin{array}{c} x_1(i) \\ x_2(i) \end{array}\right] + \left[\begin{array}{c|c} \frac{1}{10} & 0 \\ \hline 0 & \frac{1}{9} \end{array}\right]\left[\begin{array}{c} 6 \\ 3 \end{array}\right]
$$

The above two formulations are identical. Starting with $x_1(0) = x_2(0) = 0$, the iterative solution is given in the following table:

Jacobi

i	0	1	2	3	4	5	6	7	8	9	10
$x_1(i)$	0	0.60000	0.43334	0.50000	0.48148	0.48889	0.48683	0.48766	0.48743	0.48752	0.48749
$x_2(i)$	0	0.33333	0.20000	0.23704	0.22222	0.22634	0.22469	0.22515	0.22496	0.22502	0.22500

As shown, the Jacobi method converges to the unique solution obtained in Example 7.1. The convergence criterion is satisfied at the 10th iteration, since

$$
\left|\frac{x_1(10) - x_1(9)}{x_1(9)}\right| = \left|\frac{0.48749 - 0.48752}{0.48749}\right| = 6.2 \times 10^{-5} < \varepsilon
$$

and

$$
\left|\frac{x_2(10) - x_2(9)}{x_2(9)}\right| = \left|\frac{0.22500 - 0.22502}{0.22502}\right| = 8.9 \times 10^{-5} < \varepsilon \qquad \blacksquare
$$

The Gauss–Seidel method is given by

$$
x_k(i+1) = \frac{1}{A_{kk}}\left[y_k - \sum_{n=1}^{k-1} A_{kn}x_n(i+1) - \sum_{n=k+1}^{N} A_{kn}x_n(i)\right] \qquad (7.2.9)
$$

Comparing (7.2.9) with (7.2.5), note that Gauss–Seidel is similar to Jacobi except that during each iteration, the "new" values, $x_n(i+1)$, for $n < k$ are used on the right side of (7.2.9) to generate the "new" value $x_k(i+1)$ on the left side.

The Gauss–Seidel method of (7.2.9) can also be written in the matrix format of (7.2.6) and (7.2.7), where

$$
\mathbf{D} =
\begin{bmatrix}
A_{11} & 0 & 0 & \cdots & 0 \\
A_{21} & A_{22} & 0 & \cdots & 0 \\
\vdots & \vdots & & & \vdots \\
A_{N1} & A_{N2} & & \cdots A_{NN}
\end{bmatrix}
\tag{7.2.10}
$$

For Gauss–Seidel, $\mathbf{D}$ in (7.2.10) is the lower triangular portion of $\mathbf{A}$, whereas for Jacobi, $\mathbf{D}$ in (7.2.8) is the diagonal portion of $\mathbf{A}$.

EXAMPLE 7.4

Gauss–Seidel method: iterative solution to linear algebraic equations

Rework Example 7.3 using the Gauss–Seidel method.

Solution From (7.2.9),

$$
k = 1 \qquad x_1(i + 1) = \frac{1}{A_{11}} [y_1 - A_{12}x_2(i)] = \tfrac{1}{10}[6 - 5x_2(i)]
$$

$$
k = 2 \qquad x_2(i + 1) = \frac{1}{A_{22}} [y_2 - A_{21}x_1(i + 1)] = \tfrac{1}{9}[3 - 2x_1(i + 1)]
$$

Using this equation for $x_1(i + 1)$, $x_2(i + 1)$ can also be written as

$$
x_2(i + 1) = \tfrac{1}{9}\{3 - \tfrac{2}{10}[6 - 5x_2(i)]\}
$$

Alternatively, in matrix format, using (7.2.10), (7.2.6), and (7.2.7):

$$
\mathbf{D}^{-1} =
\left[
\begin{array}{c|c}
10 & 0 \\
\hline
2 & 9
\end{array}
\right]^{-1}
=
\left[
\begin{array}{c|c}
\tfrac{1}{10} & 0 \\
\hline
-\tfrac{2}{90} & \tfrac{1}{9}
\end{array}
\right]
$$

$$
\mathbf{M} =
\left[
\begin{array}{c|c}
\tfrac{1}{10} & 0 \\
\hline
-\tfrac{2}{90} & \tfrac{1}{9}
\end{array}
\right]
\left[
\begin{array}{c|c}
0 & -5 \\
\hline
0 & 0
\end{array}
\right]
=
\left[
\begin{array}{c|c}
0 & -\tfrac{1}{2} \\
\hline
0 & \tfrac{1}{9}
\end{array}
\right]
$$

$$
\begin{bmatrix}
x_1(i + 1) \\
x_2(i + 1)
\end{bmatrix}
=
\left[
\begin{array}{c|c}
0 & -\tfrac{1}{2} \\
\hline
0 & \tfrac{1}{9}
\end{array}
\right]
\begin{bmatrix}
x_1(i) \\
x_2(i)
\end{bmatrix}
+
\left[
\begin{array}{c|c}
\tfrac{1}{10} & 0 \\
\hline
-\tfrac{2}{90} & \tfrac{1}{9}
\end{array}
\right]
\begin{bmatrix}
6 \\
3
\end{bmatrix}
$$

These two formulations are identical. Starting with $x_1(0) = x_2(0) = 0$, the solution is given in the following table:

Gauss–Seidel

i	0	1	2	3	4	5	6
$x_1(i)$	0	0.60000	0.50000	0.48889	0.48765	0.48752	0.48750
$x_2(i)$	0	0.20000	0.22222	0.22469	0.22497	0.22500	0.22500

For this example, Gauss–Seidel converges in 6 iterations, compared to 10 iterations with Jacobi. ∎

The convergence rate is faster with Gauss–Seidel for some **A** matrices, but faster with Jacobi for other **A** matrices. In some cases, one method diverges while the other converges. In other cases both methods diverge, as illustrated by the next example.

EXAMPLE 7.5   **Divergence of Gauss–Seidel method**

Using the Gauss–Seidel method with $x_1(0) = x_2(0) = 0$, solve

$$\left[\begin{array}{c|c} 5 & 10 \\ \hline 9 & 2 \end{array}\right]\left[\begin{array}{c} x_1 \\ x_2 \end{array}\right] = \left[\begin{array}{c} 6 \\ 3 \end{array}\right]$$

Solution   Note that these equations are the same as those in Example 7.1, except that x_1 and x_2 are interchanged. Using (7.2.9),

$$k = 1 \quad x_1(i + 1) = \frac{1}{A_{11}}[y_1 - A_{12}x_2(i)] = \tfrac{1}{5}[6 - 10x_2(i)]$$

$$k = 2 \quad x_2(i + 1) = \frac{1}{A_{22}}[y_2 - A_{21}x_1(i + 1)] = \tfrac{1}{2}[3 - 9x_1(i + 1)]$$

Successive calculations of x_1 and x_2 are shown in the following table:

Gauss–Seidel

i	0	1	2	3	4	5
$x_1(i)$	0	1.2	9	79.2	711	6397
$x_2(i)$	0	−3.9	−39	−354.9	−3198	−28786

The unique solution by matrix inversion is

$$\left[\begin{array}{c} x_1 \\ x_2 \end{array}\right] = \left[\begin{array}{c|c} 5 & 10 \\ \hline 9 & 2 \end{array}\right]^{-1}\left[\begin{array}{c} 6 \\ 3 \end{array}\right] = \frac{-1}{80}\left[\begin{array}{c|c} 2 & -10 \\ \hline -9 & 5 \end{array}\right]\left[\begin{array}{c} 6 \\ 3 \end{array}\right] = \left[\begin{array}{c} 0.225 \\ 0.4875 \end{array}\right]$$

As shown, Gauss–Seidel does not converge to the unique solution; instead it diverges. We could show that Jacobi also diverges for this example. ∎

If any diagonal element A_{kk} equals zero, then Jacobi and Gauss–Seidel are undefined, because the right-hand sides of (7.2.5) and (7.2.9) are divided by A_{kk}. Also, if any one diagonal element has too small a magnitude, these methods will diverge. In Examples 7.3 and 7.4, Jacobi and Gauss–Seidel converge, since the diagonals (10 and 9) are both large; in Example 7.5, however, the diagonals (5 and 2) are small compared to the off-diagonals, and the methods diverge.

In general, convergence of Jacobi or Gauss–Seidel can be evaluated by recognizing that (7.2.6) represents a digital filter with input **y** and output **x**(i).

The z-transform of (7.2.6) may be employed to determine the filter transfer function and its poles. The output $\mathbf{x}(i)$ converges if and only if all the filter poles have magnitudes less than 1 (see Problems 7.10, 7.11, and 7.12).

Rate of convergence is also established by the filter poles. Fast convergence is obtained when the magnitudes of all the poles are small. In addition, experience with specific $\mathbf{A}$ matrices has shown that more iterations are required for Jacobi and Gauss–Seidel as the dimension N increases.

Computer storage requirements for Jacobi include N^2 memory locations for the $\mathbf{A}$ matrix and $3N$ locations for the vectors $\mathbf{y}$, $\mathbf{x}(i)$ and $\mathbf{x}(i + 1)$. Storage space is also required for DO loops, arithmetic statements, and working space to compute (7.2.5). Gauss–Seidel requires N fewer memory locations, since for (7.2.9) the new value $x_k(i + 1)$ can be stored in the location of the old value $x_k(i)$.

Computer time per iteration is relatively small for Jacobi and Gauss–Seidel. Inspection of (7.2.5) or (7.2.9) shows that N^2 multiplications/divisions and $N(N - 1)$ subtractions per iteration are required [one division, $(N - 1)$ multiplications, and $(N - 1)$ subtractions for each $k = 1, 2, \ldots, N$]. For example, consider a digital computer with a 10^{-6} s multiplication or division time and a 10^{-7} s addition or subtraction time. Solving $N = 1000$ equations by Jacobi or Gauss–Seidel would require $N^2 \times 10^{-6} + N(N - 1) \times 10^{-7} = (1000)^2 \times 10^{-6} + 1000(999) \times 10^{-7} = 1.1$ s per iteration plus some additional bookkeeping time for indexing and managing DO loops.

SECTION 7.3

ITERATIVE SOLUTIONS TO NONLINEAR ALGEBRAIC EQUATIONS: NEWTON–RAPHSON

A set of nonlinear algebraic equations in matrix format is given by

$$\mathbf{f}(\mathbf{x}) = \begin{bmatrix} f_1(\mathbf{x}) \\ f_2(\mathbf{x}) \\ \vdots \\ f_N(\mathbf{x}) \end{bmatrix} = \mathbf{y} \tag{7.3.1}$$

where $\mathbf{y}$ and $\mathbf{x}$ are N vectors and $\mathbf{f}(\mathbf{x})$ is an N vector of functions. Given $\mathbf{y}$ and $\mathbf{f}(\mathbf{x})$, we want to solve for $\mathbf{x}$. The iterative methods described in Section 7.2 can be extended to nonlinear equations as follows. Rewriting (7.3.1),

$$\mathbf{0} = \mathbf{y} - \mathbf{f}(\mathbf{x}) \tag{7.3.2}$$

Adding $\mathbf{Dx}$ to both sides of (7.3.2), where $\mathbf{D}$ is a square $N \times N$ invertible matrix,

$$\mathbf{Dx} = \mathbf{Dx} + \mathbf{y} - \mathbf{f}(\mathbf{x}) \tag{7.3.3}$$

Premultiplying by $\mathbf{D}^{-1}$,

$$\mathbf{x} = \mathbf{x} + \mathbf{D}^{-1}[\mathbf{y} - \mathbf{f}(\mathbf{x})] \tag{7.3.4}$$

The old values $\mathbf{x}(i)$ are used on the right side of (7.3.4) to generate the new values $\mathbf{x}(i + 1)$ on the left side. That is,

$$\mathbf{x}(i + 1) = \mathbf{x}(i) + \mathbf{D}^{-1}\{\mathbf{y} - \mathbf{f}[\mathbf{x}(i)]\} \tag{7.3.5}$$

For linear equations, $\mathbf{f}(\mathbf{x}) = \mathbf{A}\mathbf{x}$ and (7.3.5) reduces to

$$\mathbf{x}(i + 1) = \mathbf{x}(i) + \mathbf{D}^{-1}[\mathbf{y} - \mathbf{A}\mathbf{x}(i)] = \mathbf{D}^{-1}(\mathbf{D} - \mathbf{A})\mathbf{x}(i) + \mathbf{D}^{-1}\mathbf{y} \tag{7.3.6}$$

which is identical to the Jacobi and Gauss–Seidel methods of (7.2.6). For nonlinear equations, the matrix $\mathbf{D}$ in (7.3.5) must be specified.

One method for specifying $\mathbf{D}$, called *Newton–Raphson*, is based on the following Taylor series expansion of $\mathbf{f}(\mathbf{x})$ about an operating point $\mathbf{x}_0$.

$$\mathbf{y} = \mathbf{f}(\mathbf{x}_0) + \frac{d\mathbf{f}}{d\mathbf{x}}\bigg|_{\mathbf{x}=\mathbf{x}_0} (\mathbf{x} - \mathbf{x}_0) \quad \cdots \tag{7.3.7}$$

Neglecting the higher order terms in (7.3.7) and solving for $\mathbf{x}$,

$$\mathbf{x} = \mathbf{x}_0 + \left[\frac{d\mathbf{f}}{d\mathbf{x}}\bigg|_{\mathbf{x}=\mathbf{x}_0}\right]^{-1} [\mathbf{y} - \mathbf{f}(\mathbf{x}_0)] \tag{7.3.8}$$

The Newton–Raphson method replaces $\mathbf{x}_0$ by the old value $\mathbf{x}(i)$ and $\mathbf{x}$ by the new value $\mathbf{x}(i + 1)$ in (7.3.8), Thus,

$$\mathbf{x}(i + 1) = \mathbf{x}(i) + \mathbf{J}^{-1}(i)\{\mathbf{y} - \mathbf{f}[\mathbf{x}(i)]\} \tag{7.3.9}$$

where

$$\mathbf{J}(i) = \frac{d\mathbf{f}}{d\mathbf{x}}\bigg|_{\mathbf{x}=\mathbf{x}(i)} = \begin{bmatrix} \dfrac{\partial f_1}{\partial x_1} & \dfrac{\partial f_1}{\partial x_2} & \cdots & \dfrac{\partial f_1}{\partial x_N} \\[2ex] \dfrac{\partial f_2}{\partial x_1} & \dfrac{\partial f_2}{\partial x_2} & \cdots & \dfrac{\partial f_2}{\partial x_N} \\[2ex] \vdots & \vdots & & \vdots \\[2ex] \dfrac{\partial f_N}{\partial x_1} & \dfrac{\partial f_N}{\partial x_2} & \cdots & \dfrac{\partial f_N}{\partial x_N} \end{bmatrix}_{\mathbf{x}=\mathbf{x}(i)} \tag{7.3.10}$$

The $N \times N$ matrix $\mathbf{J}(i)$, whose elements are the partial derivatives shown in (7.3.10), is called the Jacobian matrix. The Newton–Raphson method is similar to extended Gauss–Seidel, except that $\mathbf{D}$ in (7.3.5) is replaced by $\mathbf{J}(i)$ in (7.3.9).

| EXAMPLE 7.6 |

Newton–Raphson method: solution to polynomial equations

Solve the scalar equation $f(x) = y$, where $y = 9$ and $f(x) = x^2$. Starting with $x(0) = 1$, use (a) Newton–Raphson and (b) extended Gauss–Seidel with $\mathbf{D} = 3$ until (7.2.2) is satisfied for $\varepsilon = 10^{-4}$. Compare the two methods.

Solution   **a.** Using (7.3.10) with $f(x) = x^2$,

$$J(i) = \frac{d}{dx}(x^2)\bigg|_{x=x(i)} = 2x\bigg|_{x=x(i)} = 2x(i)$$

Using J(i) in (7.3.9),

$$x(i+1) = x(i) + \frac{1}{2x(i)}[9 - x^2(i)]$$

Starting with $x(0) = 1$, successive calculations of the Newton–Raphson equation are shown in the following table:

Newton – Raphson

i	0	1	2	3	4	5
$x(i)$	1	5.00000	3.40000	3.02353	3.00009	3.00000

b. Using (7.3.5) with $D = 3$, the Gauss–Seidel method is

$$x(i+1) = x(i) + \tfrac{1}{3}[9 - x^2(i)]$$

The corresponding Gauss–Seidel calculations are as follows:

Gauss – Seidel (D = 3)

i	0	1	2	3	4	5	6
$x(i)$	1	3.66667	2.18519	3.59351	2.28908	3.54245	2.35945

As shown, Gauss–Seidel oscillates about the solution, slowly converging, whereas Newton–Raphson converges in five iterations to the solution $x = 3$. Note that if $x(0)$ is negative, Newton–Raphson converges to the negative solution $x = -3$. Also, it is assumed that the matrix inverse J^{-1} exists. Thus the initial value $x(0) = 0$ should be avoided for this example. ∎

EXAMPLE 7.7   **Newton – Raphson method: solution to nonlinear algebraic equations**

Solve

$$\begin{bmatrix} x_1 + x_2 \\ \\ x_1 x_2 \end{bmatrix} = \begin{bmatrix} 15 \\ \\ 50 \end{bmatrix} \qquad \mathbf{x}(0) = \begin{bmatrix} 4 \\ \\ 9 \end{bmatrix}$$

Use the Newton–Raphson method starting with the above $\mathbf{x}(0)$ and continue until (7.2.2) is satisfied with $\varepsilon = 10^{-4}$.

Solution   Using (7.3.10) with $f_1 = (x_1 + x_2)$ and $f_2 = x_1 x_2$,

$$J(i)^{-1} = \begin{bmatrix} \dfrac{\partial f_1}{\partial x_1} & \dfrac{\partial f_1}{\partial x_2} \\ \\ \dfrac{\partial f_2}{\partial x_1} & \dfrac{\partial f_2}{\partial x_2} \end{bmatrix}_{x=x(i)}^{-1} = \begin{bmatrix} 1 & 1 \\ \\ x_2(i) & x_1(i) \end{bmatrix}^{-1} = \dfrac{\begin{bmatrix} x_1(i) & -1 \\ \\ -x_2(i) & 1 \end{bmatrix}}{x_1(i) - x_2(i)}$$

Using $\mathbf{J}(i)^{-1}$ in (7.3.9),

$$
\begin{bmatrix} x_1(i+1) \\ x_2(i+1) \end{bmatrix} = \begin{bmatrix} x_1(i) \\ x_2(i) \end{bmatrix} + \frac{\begin{bmatrix} x_1(i) & -1 \\ -x_2(i) & 1 \end{bmatrix}}{x_1(i) - x_2(i)} \begin{bmatrix} 15 - x_1(i) - x_2(i) \\ 50 - x_1(i)x_2(i) \end{bmatrix}
$$

Writing the preceding as two separate equations,

$$
x_1(i+1) = x_1(i) + \frac{x_1(i)[15 - x_1(i) - x_2(i)] - [50 - x_1(i)x_2(i)]}{x_1(i) - x_2(i)}
$$

$$
x_2(i+1) = x_2(i) + \frac{-x_2(i)[15 - x_1(i) - x_2(i)] + [50 - x_1(i)x_2(i)]}{x_1(i) - x_2(i)}
$$

Successive calculations of these equations are shown in the following table:

Newton–Raphson

i	0	1	2	3	4
$x_1(i)$	4	5.20000	4.99130	4.99998	5.00000
$x_2(i)$	9	9.80000	10.00870	10.00002	10.00000

Newton–Raphson converges in four iterations for this example.   ∎

Equation (7.3.9) contains the matrix inverse $\mathbf{J}^{-1}$. Instead of computing $\mathbf{J}^{-1}$, (7.3.9) can be rewritten as follows:

$$
\mathbf{J}(i)\,\Delta\mathbf{x}(i) = \Delta\mathbf{y}(i) \tag{7.3.11}
$$

where

$$
\Delta\mathbf{x}(i) = \mathbf{x}(i+1) - \mathbf{x}(i) \tag{7.3.12}
$$

and

$$
\Delta\mathbf{y}(i) = \mathbf{y} - \mathbf{f}[\mathbf{x}(i)] \tag{7.3.13}
$$

Then, during each iteration, the following four steps are completed:

Step 1 Compute $\Delta\mathbf{y}(i)$ from (7.3.13).
Step 2 Compute $\mathbf{J}(i)$ from (7.3.10).
Step 3 Using Gauss elimination and back substitution, solve (7.3.11) for $\Delta\mathbf{x}(i)$.
Step 4 Compute $\mathbf{x}(i+1)$ from (7.3.12).

EXAMPLE 7.8 | **Newton–Raphson method in four steps**

Complete the above four steps for the first iteration of Example 7.7.

Solution

Step 1 $\Delta\mathbf{y}(0) = \mathbf{y} - \mathbf{f}[\mathbf{x}(0)] = \begin{bmatrix} 15 \\ 50 \end{bmatrix} - \begin{bmatrix} 4+9 \\ (4)(9) \end{bmatrix} = \begin{bmatrix} 2 \\ 14 \end{bmatrix}$

Step 2 $\mathbf{J}(0) = \begin{bmatrix} 1 & 1 \\ \hline x_2(0) & x_1(0) \end{bmatrix} = \begin{bmatrix} 1 & 1 \\ \hline 9 & 4 \end{bmatrix}$

Step 3 Using $\Delta\mathbf{y}(0)$ and $\mathbf{J}(0)$, (7.3.11) becomes

$$\begin{bmatrix} 1 & 1 \\ \hline 9 & 4 \end{bmatrix} \begin{bmatrix} \Delta x_1(0) \\ \Delta x_2(0) \end{bmatrix} = \begin{bmatrix} 2 \\ 14 \end{bmatrix}$$

Using Gauss elimination, subtract $J_{21}/J_{11} = 9/1 = 9$ times the first equation from the second equation, giving

$$\begin{bmatrix} 1 & 1 \\ \hline 0 & -5 \end{bmatrix} \begin{bmatrix} \Delta x_1(0) \\ \Delta x_2(0) \end{bmatrix} = \begin{bmatrix} 2 \\ -4 \end{bmatrix}$$

Solving by back substitution,

$$\Delta x_2(0) = \frac{-4}{-5} = 0.8$$

$$\Delta x_1(0) = 2 - 0.8 = 1.2$$

Step 4 $\mathbf{x}(1) = \mathbf{x}(0) + \Delta\mathbf{x}(0) = \begin{bmatrix} 4 \\ 9 \end{bmatrix} + \begin{bmatrix} 1.2 \\ 0.8 \end{bmatrix} = \begin{bmatrix} 5.2 \\ 9.8 \end{bmatrix}$

This is the same as computed in Example 7.7. ∎

Experience from power-flow studies has shown that Newton–Raphson converges in many cases where Jacobi and Gauss–Seidel diverge. Furthermore, the number of iterations required for convergence is independent of the dimension N for Newton–Raphson, but increases with N for Jacobi and Gauss–Seidel. Most Newton–Raphson power-flow problems converge in less than ten iterations [1].

The disadvantage of Newton–Raphson compared to Jacobi and Gauss–Seidel is that more computer storage and computer time per iteration are required. Newton–Raphson requires the storage space of Jacobi plus N^2 memory locations for $\mathbf{J}(i)$ and working space to solve (7.3.11). Also, as shown in Section 7.1, Gauss elimination and back substitution require approximately $\frac{1}{3}N^3$ multiplications and $\frac{1}{3}N^3$ subtractions to solve (7.3.11) for large N. For example, solving 1000 equations on a digital computer with a 10^{-6}-s multiplication or division time and a 10^{-7}-s addition or subtraction time would require 366.7 s per iteration plus some additional bookkeeping time for Newton–Raphson. In Section 7.7, we discuss sparsity techniques that significantly reduce the storage and time requirements of Newton–Raphson.

SECTION 7.4

THE POWER-FLOW PROBLEM

The power-flow problem is the computation of voltage magnitude and phase angle at each bus in a power system under balanced three-phase steady-state conditions. As a by-product of this calculation, real and reactive power flows in equipment such as transmission lines and transformers, as well as equipment losses, can be computed.

The starting point for a power-flow problem is a single-line diagram of the power system, from which the input data for computer solutions can be obtained. Input data consist of bus data, transmission line data, and transformer data.

Figure 7.1

Bus variables V_k, δ_k, P_k, and Q_k

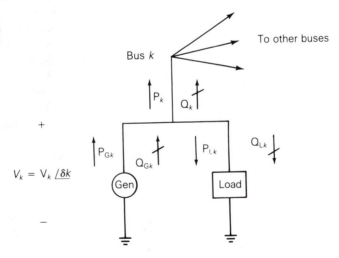

As shown in Figure 7.1, the following four variables are associated with each bus k: voltage magnitude V_k, phase angle δ_k, net real power P_k, and reactive power Q_k supplied to the bus. At each bus, two of these variables are specified as input data, and the other two are unknowns to be computed by the power-flow program. For convenience, the power delivered to bus k in Figure 7.1 is separated into generator and load terms. That is,

$$P_k = P_{Gk} - P_{Lk}$$

$$Q_k = Q_{Gk} - Q_{Lk}$$

(7.4.1)

Each bus k is categorized into one of the following three bus types:

1. **Swing bus**—There is only one swing bus, which for convenience is numbered bus 1 in this text. The swing bus is a reference bus for which $V_1/\underline{\delta_1}$, typically $1.0/\underline{0°}$ per unit, is input data. The power-flow program computes P_1 and Q_1.

2. **Load bus**—P_k and Q_k are input data. The power-flow program computes V_k and δ_k. Most buses in a typical power-flow program are load buses.

3. **Voltage controlled bus**—P_k and V_k are input data. The power-flow program computes Q_k and δ_k. Examples are buses to which generators, switched shunt capacitors, or static var systems are connected. Maximum and minimum var limits Q_{Gkmax} and Q_{Gkmin} that this equipment can supply are also input data. Another example is a bus to which a tap-changing transformer is connected; the power-flow program then computes the tap setting.

Note that when bus k is a load bus with no generation, $P_k = -P_{Lk}$ is negative; that is, the real power supplied to bus k in Figure 7.1 is negative. If the load is inductive, $Q_k = -Q_{Lk}$ is negative.

Transmission lines are represented by the equivalent π circuit, shown in Figure 6.7. Transformers are also represented by equivalent circuits, as shown in Figure 4.9 for a two-winding transformer, Figure 4.21 for a three-winding transformer, or Figure 4.27 for a tap-changing transformer.

Input data for each transmission line include the per-unit equivalent π circuit series impedance Z' and shunt admittance Y', the two buses to which the line is connected, and maximum MVA rating. Similarly, input data for each transformer include per-unit winding impedances Z, the per-unit exciting branch admittance Y, the buses to which the windings are connected, and maximum MVA ratings. Input data for tap-changing transformers also include maximum tap settings.

The bus admittance matrix Y_{bus} can be constructed from the line and transformer input data. From (2.4.3) and (2.4.4), the elements of Y_{bus} are:

Diagonal elements: Y_{kk} = sum of admittances connected to bus k

Off-diagonal elements: Y_{kn} = −(sum of admittances connected between buses k and n) $k \neq n$ (7.4.2)

| EXAMPLE 7.9 | **Power-flow input data and Y_{bus}** |

Figure 7.2 shows a single-line diagram of a five-bus power system. Input data are given in Tables 7.1, 7.2, and 7.3. As shown in Table 7.1, bus 1, to which a generator is connected, is the swing bus. Bus 3, to which a generator and a load are connected, is a voltage-controlled bus. Buses 2, 4, and 5 are load buses. Note that the loads at buses 2 and 3 are inductive since $Q_2 = -Q_{L2} = -0.7$ and $-Q_{L3} = -0.1$ are negative.

For each bus k, determine which of the variables V_k, δ_k, P_k, and Q_k are

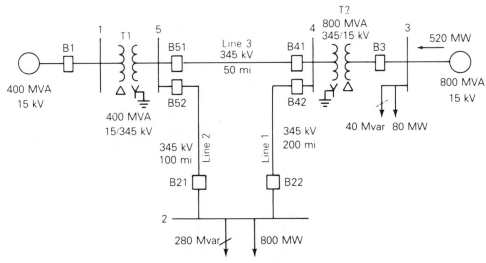

Figure 7.2   Single-line diagram for Example 7.9

BUS	TYPE	V	δ	P_G	Q_G	P_L	Q_L	Q_{Gmax}	Q_{Gmin}
		per unit	degrees	per unit	per unit	per unit	per unit	per unit	per unit
1	Swing	1.0	0	—	—	0	0	—	—
2	Load	—	—	0	0	2.0	0.7	—	—
3	Constant voltage	1.05	—	1.3	—	0.2	0.1	1.0	−0.7
4	Load	—	—	0	0	0	0	—	—
5	Load	—	—	0	0	0	0	—	—

Table 7.1

Bus input data for Example 7.9*

*S_{base} = 400 MVA, V_{base} = 15 kV at buses 1, 3, and 345 kV at buses 2, 4, 5

BUS-TO-BUS	R'	X'	G'	B'	MAXIMUM MVA
	per unit	per unit	per unit	per unit	per unit
2–4	0.036	0.4	0	0.43	3.0
2–5	0.018	0.2	0	0.22	3.0
4–5	0.009	0.1	0	0.11	3.0

Table 7.2

Line input data for Example 7.9

BUS-TO-BUS	R	X	G_c	B_m	MAXIMUM MVA	MAXIMUM TAP SETTING
	per unit	per unit	per unit	per unit	per unit	per unit
1–5	0.006	0.08	0	0	1.5	—
3–4	0.003	0.04	0	0	2.5	—

Table 7.3

Transformer input data for Example 7.9

input data and which are unknowns. Also, compute the elements of the second row of Y_{bus}.

Solution   The input data and unknowns are listed in Table 7.4. For bus 1, the swing

Table 7.4

Input data and unknowns for Example 7.9

BUS	INPUT DATA	UNKNOWNS
1	$V_1 = 1.0$, $\delta_1 = 0$	P_1, Q_1
2	$P_2 = P_{G2} - P_{L2} = -2$	V_2, δ_2
	$Q_2 = Q_{G2} - Q_{L2} = -0.7$	
3	$V_3 = 1.05$	Q_3, δ_3
	$P_3 = P_{G3} - P_{L3} = 1.1$	
4	$P_4 = 0$, $Q_4 = 0$	V_4, δ_4
5	$P_5 = 0$, $Q_5 = 0$	V_5, δ_5

bus, P_1 and Q_1 are unknowns. For bus 3, a voltage-controlled bus, Q_3 and δ_3 are unknowns. For buses 2, 4, and 5, load buses, V_2, V_4, V_5 and δ_2, δ_4, δ_5 are unknowns.

The elements of Y_{bus} are computed from (7.4.2). Since buses 1 and 3 are not directly connected to bus 2,

$$Y_{21} = Y_{23} = 0$$

Using (7.4.2),

$$Y_{24} = \frac{-1}{R'_{24} + jX'_{24}} = \frac{-1}{0.036 + j0.4} = -0.22319 + j2.47991 \text{ per unit}$$

$$= 2.48993\underline{/95.143°} \text{ per unit}$$

$$Y_{25} = \frac{-1}{R'_{25} + jX'_{25}} = \frac{-1}{0.018 + j0.2} = -0.44638 + j4.95983 \text{ per unit}$$

$$= 4.97988\underline{/95.143°} \text{ per unit}$$

$$Y_{22} = \frac{1}{R'_{24} + jX'_{24}} + \frac{1}{R'_{25} + jX'_{25}} + j\frac{B'_{24}}{2} + j\frac{B'_{25}}{2}$$

$$= (0.22319 - j\,2.47991) + (0.44638 - j4.95983) + j\frac{0.43}{2} + j\frac{0.22}{2}$$

$$= 0.66957 - j7.11474 = 7.14618\underline{/-84.624°} \text{ per unit}$$

where half of the shunt admittance of each line connected to bus 2 is included in Y_{22} (the other half is located at the other ends of these lines). ∎

Using Y_{bus}, we can write nodal equations for a power-system network, as follows:

$$I = Y_{bus} V \tag{7.4.3}$$

where I is the N vector of source currents injected into each bus and V is the N vector of bus voltages. For bus k, the kth equation in (7.4.3) is

$$I_k = \sum_{n=1}^{N} Y_{kn} V_n \tag{7.4.4}$$

The complex power delivered to bus k is

$$S_k = P_k + jQ_k = V_k I_k^* \tag{7.4.5}$$

Power-flow solutions by Gauss–Seidel are based on nodal equations, (7.4.4), where each current source I_k is calculated from (7.4.5). Using (7.4.4) in (7.4.5),

$$P_k + jQ_k = V_k \left[\sum_{n=1}^{N} Y_{kn} V_n \right]^* \qquad k = 1, 2, \dots, N \tag{7.4.6}$$

With the following notation,

$$V_n = \mathrm{V}_n e^{j\delta_n} \tag{7.4.7}$$

$$Y_{kn} = \mathrm{Y}_{kn} e^{j\theta_{kn}} \qquad k, n = 1, 2, \dots, N \tag{7.4.8}$$

(7.4.6) becomes

$$P_k + jQ_k = \mathrm{V}_k \sum_{n=1}^{N} \mathrm{Y}_{kn} \mathrm{V}_n e^{j(\delta_k - \delta_n - \theta_{kn})} \tag{7.4.9}$$

Taking the real and imaginary parts of (7.4.9)

$$P_k = \mathrm{V}_k \sum_{n=1}^{N} \mathrm{Y}_{kn} \mathrm{V}_n \cos(\delta_k - \delta_n - \theta_{kn}) \tag{7.4.10}$$

$$Q_k = \mathrm{V}_k \sum_{n=1}^{N} \mathrm{Y}_{kn} \mathrm{V}_n \sin(\delta_k - \delta_n - \theta_{kn}) \qquad k = 1, 2, \dots, N \tag{7.4.11}$$

Power-flow solutions by Newton–Raphson are based on the nonlinear power-flow equations given by (7.4.10) and (7.4.11).

SECTION 7.5

POWER-FLOW SOLUTION BY GAUSS–SEIDEL

Nodal equations $I = Y_{\mathrm{bus}} V$ are a set of linear equations analogous to $y = Ax$, solved in Section 7.2 using Gauss–Seidel. Since power-flow bus data consist of P_k and Q_k for load buses or P_k and V_k for voltage-controlled buses, nodal equations do not directly fit the linear equation format; the current source vector I is unknown and the equations are actually nonlinear. For each load bus, I_k can be calculated from (7.4.5), giving

$$I_k = \frac{P_k - jQ_k}{V_k^*} \tag{7.5.1}$$

Applying the Gauss–Seidel method, (7.2.9), to nodal equations, with I_k given above, we obtain

$$V_k(i+1) = \frac{1}{Y_{kk}} \left[\frac{P_k - jQ_k}{V_k^*(i)} - \sum_{n=1}^{k-1} Y_{kn} V_n(i+1) - \sum_{n=k+1}^{N} Y_{kn} V_n(i) \right] \tag{7.5.2}$$

Equation (7.5.2) can be applied twice during each iteration for load buses, first using $V_k^*(i)$, then replacing $V_k^*(i)$, by $V_k^*(i + 1)$ on the right side of (7.5.2).

For a voltage-controlled bus, Q_k is unknown, but can be calculated from (7.4.11), giving

$$Q_k = V_k(i) \sum_{n=1}^{N} Y_{kn} V_n(i) \sin[\delta_k(i) - \delta_n(i) - \theta_{kn}] \tag{7.5.3}$$

Also,

$$Q_{Gk} = Q_k + Q_{Lk}$$

If the calculated value of Q_{Gk} does not exceed its limits, then Q_k is used in (7.5.2) to calculate $V_k(i + 1) = V_k(i + 1)/\underline{\delta_k(i + 1)}$. Then the magnitude $V_k(i + 1)$ is changed to V_k, which is input data for the voltage-controlled bus. Thus we use (7.5.2) to compute only the angle $\delta_k(i + 1)$ for voltage-controlled buses.

If the calculated value exceeds its limit Q_{Gkmax} or Q_{Gkmin} during any iteration, then the bus type is changed from a voltage-controlled bus to a load bus, with Q_{Gk} set to its limit value. Under this condition, the voltage-controlling device (capacitor bank, static var system, and so on) is not capable of maintaining V_k as specified by the input data. The power-flow program then calculates a new value of V_k.

For the swing bus, denoted bus 1, V_1 and δ_1 are input data. As such, no iterations are required for bus 1. After the iteration process has converged, one pass through (7.4.10) and (7.4.11) can be made to compute P_1 and Q_1.

EXAMPLE 7.10	**Power-flow solution by Gauss–Seidel**

For the power system of Example 7.9, use Gauss–Seidel to calculate $V_2(1)$, the phasor voltage at bus 2 after the first iteration. Use zero initial phase angles and 1.0 per-unit initial voltage magnitudes (except at bus 3, where $V_3 = 1.05$) to start the iteration procedure.

Solution   Bus 2 is a load bus. Using the input data and bus admittance values from Example 7.9 in (7.5.2),

$$V_2(1) = \frac{1}{Y_{22}} \left\{ \frac{P_2 - jQ_2}{V_2^*(0)} - [Y_{21}V_1(1) + Y_{23}V_3(0) + Y_{24}V_4(0) + Y_{25}V_5(0)] \right\}$$

$$= \frac{1}{7.14618/\underline{-84.624°}} \left\{ \frac{-2 - j(-0.7)}{1.0/\underline{0°}} \right.$$

$$\left. - [(-0.44638 + j4.95983)(1.0) + (-0.22319 + j2.47991)(1.0)] \right\}$$

$$= \frac{(-2 + j0.7) - (-0.66957 + j7.43974)}{7.14618/\underline{-84.624°}}$$

$$= 0.96132/\underline{-16.543°} \quad \text{per unit}$$

Next, the above value is used in (7.5.2) to recalculate $V_2(1)$:

$$V_2(1) = \frac{1}{7.14618\underline{/-84.624°}} \left\{ \frac{-2+j0.7}{0.96132\underline{/16.543°}} - [-0.66957 + j7.43974] \right\}$$

$$= \frac{-1.11745 - j6.14933}{7.14618\underline{/-84.624°}} = 0.87460\underline{/-15.675°} \quad \text{per unit}$$

Computations are next performed at buses 3, 4, and 5 to complete the first Gauss–Seidel iteration. ∎

SECTION 7.6

POWER-FLOW SOLUTION BY NEWTON–RAPHSON

Equations (7.4.10) and (7.4.11) are analogous to the nonlinear equation $\mathbf{y} = \mathbf{f}(\mathbf{x})$, solved in Section 7.3 by Newton–Raphson. We define the $\mathbf{x}$, $\mathbf{y}$, and $\mathbf{f}$ vectors for the power-flow problem as

$$\mathbf{x} = \begin{bmatrix} \pmb{\delta} \\ \mathbf{V} \end{bmatrix} = \begin{bmatrix} \delta_2 \\ \vdots \\ \delta_N \\ V_2 \\ \vdots \\ V_N \end{bmatrix} ; \quad \mathbf{y} = \begin{bmatrix} \mathbf{P} \\ \mathbf{Q} \end{bmatrix} = \begin{bmatrix} P_2 \\ \vdots \\ P_N \\ Q_2 \\ \vdots \\ Q_N \end{bmatrix} ; \quad \mathbf{f}(\mathbf{x}) = \begin{bmatrix} \mathbf{P}(\mathbf{x}) \\ \mathbf{Q}(\mathbf{x}) \end{bmatrix} = \begin{bmatrix} P_2(\mathbf{x}) \\ \vdots \\ P_N(\mathbf{x}) \\ Q_2(\mathbf{x}) \\ \vdots \\ Q_N(\mathbf{x}) \end{bmatrix} \qquad (7.6.1)$$

where all V, P, and Q terms are in per-unit and δ terms are in radians. The swing bus variables δ_1 and V_1 are omitted from (7.6.1), since they are already known. Equations (7.4.10) and (7.4.11) then have the following form:

$$y_k = P_k = P_k(\mathbf{x}) = V_k \sum_{n=1}^{N} Y_{kn} V_n \cos(\delta_k - \delta_n - \theta_{kn}) \qquad (7.6.2)$$

$$y_{k+N} = Q_k = Q_k(\mathbf{x}) = V_k \sum_{n=1}^{N} Y_{kn} V_n \sin(\delta_k - \delta_n - \theta_{kn}) \qquad (7.6.3)$$

$$k = 2, 3, \dots, N$$

The Jacobian matrix of (7.3.10) has the form

$$
\mathbf{J} =
\begin{array}{cc}
\overbrace{\qquad\qquad\qquad\qquad}^{\mathbf{J1}} & \overbrace{\qquad\qquad\qquad\qquad}^{\mathbf{J2}} \\
\end{array}
$$

$$
\mathbf{J} =
\left[
\begin{array}{ccc|ccc}
\dfrac{\partial P_2}{\partial \delta_2} & \cdots & \dfrac{\partial P_2}{\partial \delta_N} & \dfrac{\partial P_2}{\partial V_2} & \cdots & \dfrac{\partial P_2}{\partial V_N} \\
\vdots & & & \vdots & & \\
\dfrac{\partial P_N}{\partial \delta_2} & \cdots & \dfrac{\partial P_N}{\partial \delta_N} & \dfrac{\partial P_N}{\partial V_2} & \cdots & \dfrac{\partial P_N}{\partial V_N} \\
\hline
\dfrac{\partial Q_2}{\partial \delta_2} & \cdots & \dfrac{\partial Q_2}{\partial \delta_N} & \dfrac{\partial Q_2}{\partial V_2} & \cdots & \dfrac{\partial Q_2}{\partial V_N} \\
\vdots & & & \vdots & & \\
\dfrac{\partial Q_N}{\partial \delta_2} & \cdots & \dfrac{\partial Q_N}{\partial \delta_N} & \dfrac{\partial Q_N}{\partial V_2} & \cdots & \dfrac{\partial Q_N}{\partial V_N} \\
\end{array}
\right]
\qquad (7.6.4)
$$

$$
\underbrace{\qquad\qquad\qquad}_{\mathbf{J3}} \qquad \underbrace{\qquad\qquad\qquad}_{\mathbf{J4}}
$$

Equation (7.6.4) is partitioned into four blocks. The partial derivatives in each block, derived from (7.6.2) and (7.6.3), are given in Table 7.5.

Table 7.5

Elements of the Jacobian matrix

$n \neq k$

$$
J1_{kn} = \frac{\partial P_k}{\partial \delta_n} = V_k Y_{kn} V_n \sin(\delta_k - \delta_n - \theta_{kn})
$$

$$
J2_{kn} = \frac{\partial P_k}{\partial V_n} = V_k Y_{kn} \cos(\delta_k - \delta_n - \theta_{kn})
$$

$$
J3_{kn} = \frac{\partial Q_k}{\partial \delta_n} = -V_k Y_{kn} V_n \cos(\delta_k - \delta_n - \theta_{kn})
$$

$$
J4_{kn} = \frac{\partial Q_k}{\partial V_n} = V_k Y_{kn} \sin(\delta_k - \delta_n - \theta_{kn})
$$

$n = k$

$$
J1_{kk} = \frac{\partial P_k}{\partial \delta_k} = -V_k \sum_{\substack{n=1 \\ n \neq k}}^{N} Y_{kn} V_n \sin(\delta_k - \delta_n - \theta_{kn})
$$

$$
J2_{kk} = \frac{\partial P_k}{\partial V_k} = V_k Y_{kk} \cos \theta_{kk} + \sum_{n=1}^{N} Y_{kn} V_n \cos(\delta_k - \delta_n - \theta_{kn})
$$

$$
J3_{kk} = \frac{\partial Q_k}{\partial \delta_k} = V_k \sum_{\substack{n=1 \\ n \neq k}}^{N} Y_{kn} V_n \cos(\delta_k - \delta_n - \theta_{kn})
$$

$$
J4_{kk} = \frac{\partial Q_k}{\partial V_k} = -V_k Y_{kk} \sin \theta_{kk} + \sum_{n=1}^{N} Y_{kn} V_n \sin(\delta_k - \delta_n - \theta_{kn})
$$

$$
k, n = 2, 3, \ldots, N
$$

We now apply to the power-flow problem the four Newton–Raphson steps outlined in Section 7.3, starting with $\mathbf{x}(i) = \begin{bmatrix} \boldsymbol{\delta}(i) \\ \mathbf{V}(i) \end{bmatrix}$ at the ith iteration.

Step 1 Use (7.6.2) and (7.6.3) to compute

$$\Delta \mathbf{y}(i) = \begin{bmatrix} \Delta \mathbf{P}(i) \\ \\ \Delta \mathbf{Q}(i) \end{bmatrix} = \begin{bmatrix} \mathbf{P} - \mathbf{P}[\mathbf{x}(i)] \\ \\ \mathbf{Q} - \mathbf{Q}[\mathbf{x}(i)] \end{bmatrix} \qquad (7.6.5)$$

Step 2 Use the equations in Table 7.5 to calculate the Jacobian matrix.

Step 3 Use Gauss elimination and back substitution to solve

$$\begin{bmatrix} \mathbf{J1}(i) & \mathbf{J2}(i) \\ \hline \mathbf{J3}(i) & \mathbf{J4}(i) \end{bmatrix} \begin{bmatrix} \Delta \boldsymbol{\delta}(i) \\ \Delta \mathbf{V}(i) \end{bmatrix} = \begin{bmatrix} \Delta \mathbf{P}(i) \\ \Delta \mathbf{Q}(i) \end{bmatrix} \qquad (7.6.6)$$

Step 4 Compute

$$\mathbf{x}(i+1) = \begin{bmatrix} \boldsymbol{\delta}(i+1) \\ \\ \mathbf{V}(i+1) \end{bmatrix} = \begin{bmatrix} \boldsymbol{\delta}(i) \\ \\ \mathbf{V}(i) \end{bmatrix} + \begin{bmatrix} \Delta \boldsymbol{\delta}(i) \\ \\ \Delta \mathbf{V}(i) \end{bmatrix} \qquad (7.6.7)$$

Starting with initial value $\mathbf{x}(0)$, the procedure continues until convergence is obtained or until the number of iterations exceeds a specified maximum. Convergence criteria are often based on $\Delta \mathbf{y}(i)$ (called *power mismatches*) rather than on $\Delta \mathbf{x}(i)$ (phase angle and voltage magnitude mismatches).

For each voltage-controlled bus, the magnitude V_k is already known, and the function $Q_k(\mathbf{x})$ is not needed. Therefore, we omit V_k from the $\mathbf{x}$ vector and Q_k from the $\mathbf{y}$ vector. We also omit from the Jacobian matrix the column corresponding to partial derivatives with respect to V_k and the row corresponding to partial derivatives of $Q_k(\mathbf{x})$.

At the end of each iteration, we compute $Q_k(\mathbf{x})$ from (7.6.3) and $Q_{Gk} = Q_k(\mathbf{x}) + Q_{Lk}$ for each voltage-controlled bus. If the computed value of Q_{Gk} exceeds its limits, then the bus type is changed to a load bus with Q_{Gk} set to its limit value. Also, the corresponding row and column are reinserted into the Jacobian matrix, and the power-flow program computes a new value for V_k.

EXAMPLE 7.11   **Jacobian matrix and power-flow solution by Newton–Raphson**

Determine the dimension of the Jacobian matrix for the power system in Example 7.9. Also calculate $\Delta P_2(0)$ in Step 1 and $J1_{24}(0)$ in Step 2 of the first Newton–Raphson iteration. Assume zero initial phase angles and 1.0 per-unit initial voltage magnitudes (except $V_3 = 1.05$).

Solution   Since there are $N = 5$ buses for Example 7.9, (7.6.2) and (7.6.3) constitute $2(N - 1) = 8$ equations, for which $\mathbf{J}(i)$ would have dimension 8×8. However, there is one voltage-controlled bus, bus 3. Therefore, V_3 and the equation for $Q_3(\mathbf{x})$ are eliminated, and $\mathbf{J}(i)$ is reduced to a 7×7 matrix.

From Step 1 and (7.6.2),

$$\Delta P_2(0) = P_2 - P_2(\mathbf{x}) = P_2 - V_2(0)\{Y_{21}V_1 \cos[\delta_2(0) - \delta_1(0) - \theta_{21}]$$
$$+ Y_{22}V_2 \cos[-\theta_{22}] + Y_{23}V_3 \cos[\delta_2(0) - \delta_3(0) - \theta_{23}]$$
$$+ Y_{24}V_4 \cos[\delta_2(0) - \delta_4(0) - \theta_{24}]$$
$$+ Y_{25}V_5 \cos[\delta_2(0) - \delta_5(0) - \theta_{25}]\}$$

$$\Delta P_2(0) = -2.0 - 1.0\{7.14618(1.0)\cos(84.624°)$$
$$+ 2.48993(1.0)\cos(-95.143°)$$
$$+ 4.97988(1.0)\cos(-95.143°)\}$$

$$= -2.0 - (-7.24 \times 10^{-5}) = -1.99993 \quad \text{per unit}$$

From Step 2 and J1 given in Table 7.5

$$J1_{24}(0) = V_2(0)Y_{24}V_4(0)\sin[\delta_2(0) - \delta_4(0) - \theta_{24}]$$
$$= (1.0)(2.48993)(1.0)\sin[-95.143°]$$
$$= -2.47991 \quad \text{per unit} \qquad \blacksquare$$

Voltage-controlled buses to which tap-changing or voltage-regulating transformers are connected can be handled by various methods. One method is to treat each of these buses as a load bus. The equivalent π circuit parameters (Figure 4.27) are first calculated with tap setting $c = 1.0$ for starting. During each iteration the computed bus voltage magnitude is compared with the desired value specified by the input data. If the computed voltage is low (or high), c is increased (or decreased) to its next setting, and the parameters of the equivalent π circuit as well as Y_{bus} are recalculated. The procedure continues until the computed bus voltage magnitude equals the desired value within a specified tolerance, or until the high or low tap-setting limit is reached. Phase-shifting transformers can be handled in a similar way by using a complex turns ratio $c = 1.0/\alpha$, and by varying the phase-shift angle α.

A method with faster convergence makes c a variable and includes it in the $\mathbf{x}$ vector of (7.6.1). An equation is then derived to enter into the Jacobian matrix [4].

SECTION 7.7

SPARSITY TECHNIQUES

A typical power system has an average of fewer than three lines connected to each bus. As such, each row of Y_{bus} has an average of fewer than four nonzero elements, one off-diagonal for each line and the diagonal. Such a matrix, which has only a few nonzero elements, is said to be *sparse*.

Newton–Raphson power-flow programs employ sparse matrix techniques to reduce computer storage and time requirements [2]. These techniques include compact storage of Y_{bus} and $J(i)$, and reordering of buses to avoid fill-in of $J(i)$ during Gauss elimination steps. Consider the following matrix:

$$S = \begin{bmatrix} 1.0 & -1.1 & -2.1 & -3.1 \\ -4.1 & 2.0 & 0 & -5.1 \\ -6.1 & 0 & 3.0 & 0 \\ -7.1 & 0 & 0 & 4.0 \end{bmatrix} \tag{7.7.1}$$

One method for compact storage of **S** consists of the following four vectors:

$$\textbf{DIAG} = [1.0 \quad 2.0 \quad 3.0 \quad 4.0] \tag{7.7.2}$$

$$\textbf{OFFDIAG} = [-1.1 \quad -2.1 \quad -3.1 \quad -4.1 \quad -5.1 \quad -6.1 \quad -7.1] \tag{7.7.3}$$

$$\textbf{COL} = [2 \quad 3 \quad 4 \quad 1 \quad 4 \quad 1 \quad 1] \tag{7.7.4}$$

$$\textbf{ROW} = [3 \quad 2 \quad 1 \quad 1] \tag{7.7.5}$$

DIAG contains the ordered diagonal elements and **OFFDIAG** contains the nonzero off-diagonal elements of **S**. **COL** contains the column number of each off-diagonal element. For example, the *fourth* element in **COL** is 1, indicating that the *fourth* element of **OFFDIAG**, -4.1, is located in column 1. **ROW** indicates the number of off-diagonal elements in each row of **S**. For example, the *first* element of **ROW** is 3, indicating the *first* three elements of **OFFDIAG**, -1.1, -2.1, and -3.1, are located in the *first* row. The *second* element of **ROW** is 2, indicating the next two elements of **OFFDIAG**, -4.1 and -5.1, are located in the *second* row. The **S** matrix can be completely reconstructed from these four vectors. Note that the dimension of **DIAG** and **ROW** equals the number of diagonal elements of **S**, whereas the dimension of **OFFDIAG** and **COL** equals the number of nonzero off-diagonals.

Now assume that computer storage requirements are 4 bytes to store each magnitude and 4 bytes to store each phase of Y_{bus} in an N-bus power system. Also assume Y_{bus} has an average of $3N$ nonzero off-diagonals (three lines per bus) along with its N diagonals. Using the preceding compact storage technique, we need $(4 + 4)3N = 24N$ bytes for **OFFDIAG** and $(4 + 4)N = 8N$ bytes for **DIAG**. Also, assuming 2 bytes to store each integer, we need $6N$ bytes for **COL** and $2N$ bytes for **ROW**. Total computer memory required is then $(24 + 8 + 6 + 2)N = 40N$ bytes with compact storage of Y_{bus}, compared to $8N^2$ bytes without compact storage. For a 1000-bus power system, this means 40 instead of 8000 kilobytes to store Y_{bus}. Further storage reduction could be obtained by storing only the upper triangular portion of the symmetric Y_{bus} matrix.

The Jacobian matrix is also sparse. From Table 7.5, whenever $Y_{kn} = 0$, $J1_{kn} = J2_{kn} = J3_{kn} = J4_{kn} = 0$. Compact storage of **J** for a 1000-bus power system requires less than 100 kilobytes with the above assumptions.

The other sparsity technique is to reorder buses. Suppose Gauss elimination is used to triangularize $\mathbf{S}$ in (7.7.1). After one Gauss elimination step, as described in Section 7.1, we have

$$\mathbf{S}^{(1)} = \begin{bmatrix} 1.0 & -1.1 & -2.1 & -3.1 \\ 0 & -2.51 & -8.61 & -7.61 \\ 0 & -6.71 & -9.81 & -18.91 \\ 0 & -7.81 & -14.91 & -18.01 \end{bmatrix} \qquad (7.7.6)$$

It can be seen that the zeros in columns 2, 3, and 4 of $\mathbf{S}$ are filled in with nonzero elements in $\mathbf{S}^{(1)}$. The original degree of sparsity is lost.

One simple reordering method is to start with those buses having the fewest connected branches and to end with those having the most connected branches. For example, $\mathbf{S}$ in (7.7.1) has three branches connected to bus 1 (three off-diagonals in row 1), two branches connected to bus 2, and one branch connected to buses 3 and 4. Reordering the buses 4, 3, 2, 1 instead of 1, 2, 3, 4 we have

$$\mathbf{S}_{\text{reordered}} = \begin{bmatrix} 4.0 & 0 & 0 & -7.1 \\ 0 & 3.0 & 0 & -6.1 \\ -5.1 & 0 & 2.0 & -4.1 \\ -3.1 & -2.1 & -1.1 & 1.0 \end{bmatrix} \qquad (7.7.7)$$

Now, after one Gauss elimination step,

$$\mathbf{S}^{(1)}_{\text{reordered}} = \begin{bmatrix} 4.0 & 0 & 0 & -7.1 \\ 0 & 3.0 & 0 & -6.1 \\ 0 & 0 & 2.0 & -13.15 \\ 0 & -2.1 & -1.1 & -4.5025 \end{bmatrix} \qquad (7.7.8)$$

Note that the original degree of sparsity is not lost in (7.7.8).

Reordering buses according to the fewest connected branches can be performed once, before the Gauss elimination process begins. Alternatively, buses can be renumbered during each Gauss elimination step in order to account for changes during the elimination process.

Sparsity techniques similar to those described in this section are a standard feature of today's Newton–Raphson power-flow programs. As a result of these techniques, typical 1000-bus power-flow solutions require less than 200 kilobytes of storage, less than 55 seconds per iteration of computer time, and less than 10 iterations to converge.

SECTION 7.8

PERSONAL COMPUTER PROGRAM: POWER FLOW

The software package that accompanies this text includes the program "POWER FLOW," which computes voltage magnitude and angle at each bus in a power system under balanced three-phase steady-state operation. Bus voltages are then used to compute generator, line, and transformer loadings.

Input data for the program include bus, line, and transformer data, as shown in Tables 7.1, 7.2, and 7.3. All input data are given in per-unit on a common MVA base.

The bus admittance matrix is calculated from the input data using (7.4.2). The program then uses the Newton–Raphson method described in Section 7.6.

For single runs, starting values of bus voltage magnitudes and angles are set equal to those of the swing bus, except for voltage-controlled buses, whose voltage magnitudes are known. For a series of runs with input data changes, the final values of each run are used as starting values for the next run.

The Newton–Raphson procedure is terminated when the magnitudes of all power mismatches $\Delta P(i)$ and $\Delta Q(i)$ are less than a tolerance level (typically 0.001 per unit) or when the number of iterations exceeds a maximum (typically 10). Both the tolerance level and maximum number of iterations are selected by the program user.

Output data consist of bus, line, and transformer data. Bus output data include voltage magnitude and angle, real and reactive power of each generator and load, and identification of buses with voltage magnitudes more than 5% above or below that of the swing bus. Line (and transformer) output data include real and reactive power flows entering the line terminals and identification of lines with MVA flows above their maximum ratings. Other useful output include generation, load, and line loss totals, number of iterations to converge, and total mismatches ΔP and ΔQ after convergence. If desired, reactive power delivered by line capacitances or other shunt reactive elements can also be obtained.

| EXAMPLE 7.12 | **POWER FLOW program** |

Run the POWER FLOW program for the power system given in Example 7.9.

Solution   Bus, line, and transformer output data are shown in Tables 7.6, 7.7, and 7.8. Note that the bus 2 voltage magnitude is low (0.834 per unit).

	VOLTAGE MAGNITUDE	PHASE ANGLE	GENERATION		LOAD		0.95 > V > 1.05
			PG	QG	PL	QL	
bus #	per unit	degrees	per unit	per unit	per unit	per unit	
1	1.000	0.000	0.987	0.286	0.000	0.000	
2	0.834	−22.407	0.000	0.000	2.000	0.700	****
3	1.050	−0.597	1.300	0.844	0.200	0.100	
4	1.019	−2.834	0.000	0.000	0.000	0.000	
5	0.974	−4.548	0.000	0.000	0.000	0.000	
		TOTAL	2.287	1.129	2.200	0.800	

Table 7.6   Example 7.12 bus output data

LINE #	BUS TO BUS		P	Q	S	RATING EXCEEDED
1	2	4	−0.730	−0.348	0.808	
	4	2	0.759	0.304	0.818	
2	2	5	−1.270	−0.352	1.318	
	5	2	1.314	0.658	1.469	
3	4	5	0.336	0.376	0.504	
	5	4	−0.333	−0.456	0.565	

Table 7.7   Example 7.12 line output data

TRAN. #	BUS TO BUS		P	Q	S	RATING EXCEEDED
1	1	5	0.987	0.286	1.028	
	5	1	−0.981	−0.201	1.001	
2	3	4	1.100	0.744	1.328	
	4	3	−1.095	−0.680	1.289	

Table 7.8   Example 7.12 transformer output data   ∎

SECTION 7.9

FAST DECOUPLED POWER FLOW

Contingencies are a major concern in power-system operations. For example, operating personnel need to know what power-flow changes will occur due to a particular generator outage or transmission-line outage. Contingency information, when obtained in real time, can be used to anticipate problems caused by such outages, and can be used to develop operating strategies to overcome the problems.

Fast power-flow algorithms have been developed to give power-flow solutions in seconds [8]. These algorithms are based on the following simplification of the Jacobian matrix. Neglecting $\mathbf{J}_2(i)$ and $\mathbf{J}_3(i)$, (7.6.6) reduces to two sets of decoupled equations:

$$\mathbf{J}_1(i)\,\Delta\delta(i) = \Delta\mathbf{P}(i) \tag{7.9.1}$$

$$\mathbf{J}_4(i)\,\Delta\mathbf{V}(i) = \Delta\mathbf{Q}(i) \tag{7.9.2}$$

The computer time required to solve (7.9.1) and (7.9.2) is significantly less than that required to solve (7.6.6). Further reduction in computer time can be obtained from additional simplification of the Jacobian matrix. For example, assume $V_k \approx V_n \approx 1.0$ per unit and $\delta_k \approx \delta_n$. Then $\mathbf{J}_1$ and $\mathbf{J}_4$ are constant matrices whose elements in Table 7.5 are the imaginary components of Y_{bus}. As such, $\mathbf{J}_1$ and $\mathbf{J}_4$ do not have to be recalculated during successive iterations.

Simplifications similar to these enable rapid power-flow solutions. For a fixed number of iterations, the fast decoupled algorithm given by (7.9.1) and (7.9.2) is not as accurate as the exact Newton–Raphson algorithm. But the savings in computer time is considered more important.

SECTION 7.10

CONTROL OF POWER FLOW

The following means are used to control system power flows:

1. Prime mover and excitation control of generators
2. Switching of shunt capacitor banks, shunt reactors, and static var systems
3. Control of tap-changing and regulating transformers

A simple model of a generator operating under balanced steady-state conditions is the Thévenin equivalent shown in Figure 7.3. V_t is the generator terminal voltage, E_g is the excitation voltage, δ is the power angle, and X_g is the positive-sequence synchronous reactance. From the figure, the generator current is

$$I = \frac{E_g e^{j\delta} - V_t}{jX_g} \tag{7.10.1}$$

Figure 7.3

Generator Thévenin equivalent

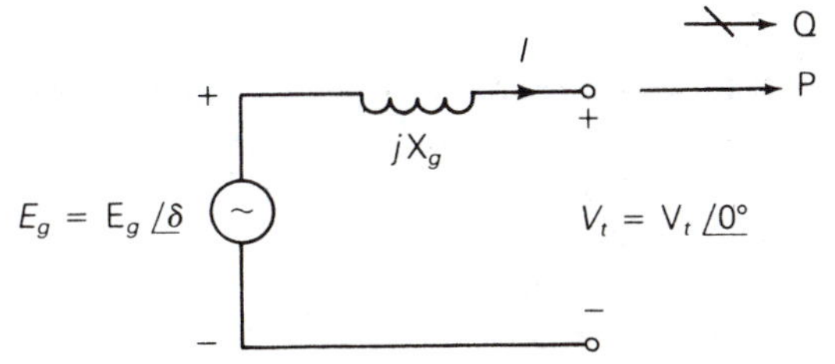

$E_g = E_g\,\underline{/\delta}$

jX_g

$V_t = V_t\,\underline{/0°}$

and the complex power delivered by the generator is

$$S = P + jQ = V_t I^* = V_t \left(\frac{E_g e^{-j\delta} - V_t}{-jX_g} \right)$$

$$= \frac{V_t E_g (j\cos\delta + \sin\delta) - jV_t^2}{X_g} \tag{7.10.2}$$

The real and reactive powers delivered are then

$$P = \text{Re } S = \frac{V_t E_g}{X_g} \sin\delta \tag{7.10.3}$$

$$Q = \text{Im } S = \frac{V_t}{X_g} (E_g \cos\delta - V_t) \tag{7.10.4}$$

Equation (7.10.3) shows that the real power P increases when the power angle δ increases. From an operational standpoint, when the prime mover increases the power input to the generator while the excitation voltage is held constant, the rotor speed increases. As the rotor speed increases, the power angle δ also increases, causing an increase in generator real power output P. There is also a decrease in reactive power output Q, given by (7.10.4). However, when δ is less than 15°, the increase in P is much larger than the decrease in Q. From the power-flow standpoint, an increase in prime-mover power corresponds to an increase in P at the constant-voltage bus to which the generator is connected. The power-flow program computes the increase in δ along with the small change in Q.

Equation (7.10.4) shows that reactive power output Q increases when the excitation voltage E_g increases. From the operational standpoint, when the generator exciter output increases while holding the prime-mover power constant, the rotor current increases. As the rotor current increases, the excitation voltage E_g also increases, causing an increase in generator reactive power output Q. There is also a small decrease in δ required to hold P constant in (7.10.3). From the power-flow standpoint, an increase in generator excitation corresponds to an increase in voltage magnitude at the constant-voltage bus to which the generator is connected. The power-flow program computes the increase in reactive power Q supplied by the generator along with the small change in δ.

Figure 7.4 shows the effect of adding a shunt capacitor bank to a power-system bus. The system is modeled by its Thévenin equivalent. Before the capacitor bank is connected, the switch SW is open and the bus voltage equals E_{Th}. After the bank is connected, SW is closed, and the capacitor current I_C leads the bus voltage V_t by 90°. The phasor diagram shows that V_t is larger than E_{Th} when SW is closed. From the power-flow standpoint, the addition of a shunt capacitor bank to a load bus corresponds to the addition of a negative reactive load, since a capacitor absorbs negative reactive power. The power-flow program computes the increase in bus voltage magnitude along with the small change in δ. Similarly, the addition of a shunt reactor

Figure 7.4

Effect of adding a shunt capacitor bank to a power-system bus

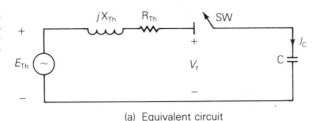

(a) Equivalent circuit

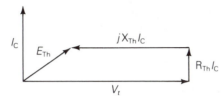

(b) Phasor diagram with switch SW closed

corresponds to the addition of a positive reactive load, wherein the power-flow program computes the decrease in voltage magnitude.

Tap-changing and voltage-magnitude–regulating transformers are used to control bus voltages as well as reactive power flows on lines to which they are connected. Similarly, phase-angle regulating transformers are used to control bus angles as well as real power flows on lines to which they are connected. Both tap-changing and regulating transformers are modeled by a transformer with an off-nominal turns ratio c (Figure 4.27). From the power-flow standpoint, a change in tap setting or voltage regulation corresponds to a change in c. The power-flow program computes the changes in Y_{bus}, bus voltage magnitudes and angles, and branch flows.

Besides the above controls, the power-flow program can be used to investigate the effect of switching in or out lines, transformers, loads, and generators. Proposed system changes to meet future load growth, including new transmission, new transformers, and new generation can also be investigated. Power-flow design studies are normally conducted by trial and error. Using engineering judgment, adjustments in generation levels and controls are made until the desired equipment loadings and voltage profile are obtained.

EXAMPLE 7.13 | **POWER FLOW program: effect of shunt capacitor banks**

Determine the effect of adding a 200-Mvar shunt capacitor bank at bus 2 on the power system in Example 7.9.

Solution The addition of a 200-Mvar (0.5 per-unit) shunt capacitor bank corresponds to adding -0.5 per unit of reactive load. Therefore, Q_L at bus 2 is changed from 0.7 to $(0.7 - 0.5) = 0.2$ per unit, and the POWER FLOW program is rerun. Bus, line, and transformer output data are shown in Tables 7.9, 7.10, and 7.11. As shown, the capacitor bank increases the voltage at bus 2 to an acceptable value of 0.968 per unit.

	VOLTAGE MAGNITUDE	PHASE ANGLE	GENERATION		LOAD		0.95 > V > 1.05
			PG	QG	PL	QL	
bus #	per unit	degrees	per unit	per unit	per unit	per unit	
1	1.000	0.000	0.963	−0.118	0.000	0.000	
2	0.968	−19.629	0.000	0.000	2.000	0.200	
3	1.050	−0.408	1.300	0.370	0.200	0.100	
4	1.037	−2.680	0.000	0.000	0.000	0.000	
5	1.007	−4.427	0.000	0.000	0.000	0.000	
		TOTAL	2.263	0.252	2.200	0.300	

Table 7.9   Example 7.13 bus output data

LINE #	BUS TO BUS		P	Q	S	RATING EXCEEDED
1	2	4	−0.731	−0.195	0.757	
	4	2	0.752	−0.010	0.752	
2	2	5	−1.269	−0.005	1.269	
	5	2	1.300	0.137	1.307	
3	4	5	0.345	0.234	0.416	
	5	4	−0.343	−0.329	0.476	

Table 7.10   Example 7.13 line output data

TRAN. #	BUS TO BUS		P	Q	S	RATING EXCEEDED
1	1	5	0.963	−0.118	0.970	
	5	1	−0.957	0.193	0.976	
2	3	4	1.100	0.270	1.133	
	4	3	−1.097	−0.224	1.119	

Table 7.11   Example 7.13 transformer output data   ∎

PROBLEMS

Section 7.1

7.1   Using Gauss elimination and back substitution, solve

$$\begin{bmatrix} 10 & 3 & 0 \\ 4 & 20 & 2 \\ 5 & 2 & 14 \end{bmatrix} \begin{bmatrix} x_1 \\ x_2 \\ x_3 \end{bmatrix} = \begin{bmatrix} 1 \\ 2 \\ 3 \end{bmatrix}$$

7.2   Rework Problem 7.1 with the value of A_{33} change from 14 to 0.14.

7.3   Show that the Gauss elimination method, which transforms the set of N linear equations $\mathbf{A}\mathbf{x} = \mathbf{y}$ to $\mathbf{A}^{(N-1)}\mathbf{x} = \mathbf{y}^{(N-1)}$, where $\mathbf{A}^{(N-1)}$ is triangular, requires $(N^3 - N)/3$

multiplications, $N(N - 1)/2$ divisions, and $(N^3 - N)/3$ subtractions. Assume that all the elements of $\mathbf{A}$ and $\mathbf{y}$ are nonzero and real. (*Hint:* Investigate (7.1.7). Note that during the first Gauss elimination step, *each* of the $(N - 1)$ rows that are changed requires *one* division, N multiplications, and N subtractions.)

7.4   Show that, after triangularizing $\mathbf{Ax} = \mathbf{y}$, the back substitution method of solving $\mathbf{A}^{(N-1)}\mathbf{x} = \mathbf{y}^{(N-1)}$ requires N divisions, $N(N - 1)/2$ multiplications, and $N(N - 1)/2$ subtractions. Assume that all the elements of $\mathbf{A}^{(N-1)}$ and $\mathbf{y}^{(N-1)}$ are nonzero and real.

7.5   For a digital computer with a 10^{-6}-s multiplication or division time and a 10^{-7}-s addition or subtraction time, determine how much computer time would be required to solve (7.1.1) for $N = 100$, using Gauss elimination and back substitution. Assume that all the elements of $\mathbf{A}$ and $\mathbf{y}$ are nonzero and real.

7.6   If 4 bytes are used to store each floating-point number in computer memory, how many kilobytes are required to store an N vector $\mathbf{y}$ and an $N \times N$ matrix $\mathbf{A}$, where $N = 1000$? Assume that all the elements of $\mathbf{y}$ and $\mathbf{A}$ are nonzero and real.

Section 7.2

7.7   Solve Problem 7.1 using the Jacobi iterative method. Start with $x_1(0) = x_2(0) = x_3(0) = 0$, and continue until (7.2.2) is satisfied with $\varepsilon = 0.01$.

7.8   Repeat Problem 7.7 using the Gauss–Seidel iterative method. Which method converges more rapidly?

7.9   Try to solve Problem 7.1 using the Jacobi and Gauss–Seidel iterative methods with the value of A_{33} changed from 14 to 0.14 and with $x_1(0) = x_2(0) = x_3(0) = 0$. Show that neither method converges to the unique solution.

7.10   Take the z-transform of (7.2.6) and show that $\mathbf{X}(z) = \mathbf{G}(z)\mathbf{Y}(z)$, where $\mathbf{G}(z) = (z\mathbf{U} - \mathbf{M})^{-1}\mathbf{D}^{-1}$ and $\mathbf{U}$ is the unit matrix.

$\mathbf{G}(z)$ is the matrix transfer function of a digital filter that represents the Jacobi or Gauss–Seidel methods. The filter poles are obtained by solving $\det(z\mathbf{U} - \mathbf{M}) = 0$. The filter is stable if and only if all the poles have magnitudes less than 1.

7.11   Using the results of Problem 7.10, determine the filter poles for Examples 7.3 and 7.5. Note that in Example 7.3 both poles have magnitudes less than 1, which means the filter is stable and Jacobi converges for this example. However, in Example 7.5, one pole has a magnitude greater than 1, which means the filter is unstable and Gauss–Seidel diverges for this example.

7.12   Determine the poles of the Jacobi and Gauss–Seidel digital filters for the general two-dimensional problem ($N = 2$):

$$
\begin{bmatrix} A_{11} & A_{12} \\ \hline A_{21} & A_{22} \end{bmatrix}
\begin{bmatrix} x_1 \\ x_2 \end{bmatrix}
=
\begin{bmatrix} y_1 \\ y_2 \end{bmatrix}
$$

Then determine a necessary and sufficient condition for convergence of these filters when $N = 2$.

7.13   For a digital computer with a 10^{-6}-s multiplication or division time and a 10^{-7}-s addition or subtraction time, determine how much computer time per iteration would be required to solve (7.1.1) with $N = 100$, using Gauss–Seidel. Assume that all the elements of $\mathbf{A}$ and $\mathbf{y}$ are nonzero and real.

Section 7.3

7.14   Use Newton–Raphson to find one solution to the polynomial equation $f(x) = y$, where $y = 0$ and $f(x) = 3x^3 + 4x^2 + 5x + 8$. Start with $x(0) = 1.0$ and continue until (7.2.2) is satisfied with $\varepsilon = 0.001$.

7.15   Use Newton–Raphson to find one solution to the polynomial equation $f(x) = y$, where $y = 0$ and $f(x) = x^4 + 12x^3 + 54x^2 + 108x + 81$. Start with $x(0) = -1$ and continue until (7.2.2) is satisfied with $\varepsilon = 0.001$.

7.16   Use Newton–Raphson to find a solution to

$$\begin{bmatrix} e^{x_1 x_2} \\ \\ \cos(x_1 + x_2) \end{bmatrix} = \begin{bmatrix} 1.2 \\ \\ 0.5 \end{bmatrix}$$

where x_1 and x_2 are in radians. (a) Start with $x_1(0) = 1.0$ and $x_2(0) = 0.5$ and continue until (7.2.2) is satisfied with $\varepsilon = 0.005$. (b) Show that Newton–Raphson diverges for this example if $x_1(0) = 1.0$ and $x_2(0) = 2.0$.

Section 7.4

7.17   Compute the elements of the third row of Y_{bus} for the power system in Example 7.9.

7.18   Figure 7.5 shows a single-line diagram of a three-bus power system. Power-flow input data are given in Tables 7.12 and 7.13. (a) Determine the 3×3 per-unit bus admittance matrix Y_{bus}. (b) For each bus $k = 1, 2, 3$ determine which of the variables V_k, ∂_k, P_k, and Q_k are input data and which are unknowns.

Figure 7.5

Single-line diagram for Problems 7.18, 7.20, 7.21, and 7.22 (per unit impedances and per unit real and reactive powers are shown)

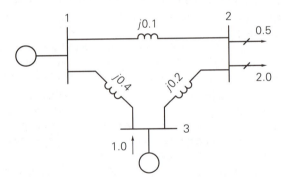

BUS	TYPE	V	∂	P_G	Q_G	P_L	Q_L	Q_{Gmin}	Q_{Gmax}
		per unit	degrees	per unit	per unit	per unit	per unit	per unit	per unit
1	Swing	1.0	0	—	—	0	0	—	—
2	Load	—	—	0	0	2.0	0.5	—	—
3	Constant voltage	1.0	—	1.0	—	0	0	−5.0	+5.0

Table 7.12   Bus input data for Problem 7.18

LINE	BUS-TO-BUS	R'	X'	G'	B'	MAXIMUM MVA
		per unit	per unit	per unit	per unit	per unit
1	1–2	0	0.1	0	0	3.0
2	2–3	0	0.2	0	0	3.0
3	1–3	0	0.4	0	0	3.0

Table 7.13

Line input data for Problem 7.18

(Note: There are no transformers)

Section 7.5

7.19 For the power system in Example 7.9, use Gauss–Seidel to calculate $V_3(1)$, the phasor voltage at bus 3 after the first iteration. Note that bus 3 is a voltage-controlled bus.

7.20 For the power system given in Problem 7.18, use Gauss–Seidel to compute $V_2(1)$ and $V_3(1)$, the phasor voltages at bus 2 and 3 after the first iteration. Use zero initial phase angles and 1.0 per-unit initial bus voltage magnitudes.

Section 7.6

7.21 For the power system in Example 7.9, calculate $\Delta P_4(0)$ in Step 1 and $J1_{44}(0)$ in Step 2 of the first Newton–Raphson iteration. Assume zero initial phase angles and 1.0 per-unit initial voltage magnitudes (except $V_3 = 1.05$).

7.22 For the power system given in Problem 7.18, use (7.6.2) and (7.6.3) to write the three power-flow equations to be solved by the Newton–Raphson method. Also, identify the three unknown variables to be solved. Do not solve the equations.

7.23 For the power system given in Problem 7.18, use Newton–Raphson to compute $V_2(1)$ and $V_3(1)$, the phasor voltages at bus 2 and 3 after the first iteration, as follows. (a) Step 1: use (7.6.2) and (7.6.3) to compute $\Delta y(0)$. (b) Step 2: compute the 3×3 Jacobian matrix $J(0)$ using the equations in Table 7.5. (c) Step 3: use Gauss elimination and back substitution to solve (7.6.6). (d) Step 4: compute $x(1)$ in (7.6.7). Also, use (7.5.3) to compute Q_{G3} and verify that it is within the limits shown in Table 7.12. In steps 1 and 2, use zero initial phase angles and 1.0 per-unit initial bus voltage magnitudes.

Section 7.7

7.24 Using the compact storage technique described in Section 7.7, determine the vectors **DIAG**, **OFFDIAG**, **COL**, and **ROW** for the following matrix:

$$
S = \begin{bmatrix}
17 & -9.1 & 0 & 0 & -2.1 & -7.1 \\
-9.1 & 25 & -8.1 & -1.1 & -6.1 & 0 \\
0 & -8.1 & 9 & 0 & 0 & 0 \\
0 & -1.1 & 0 & 2 & 0 & 0 \\
-2.1 & -6.1 & 0 & 0 & 14 & -5.1 \\
-7.1 & 0 & 0 & 0 & -5.1 & 15
\end{bmatrix}
$$

7.25 If 4 bytes of computer storage are used for each floating-point number and 2 bytes for each integer, determine the total number of bytes required to store the **S** matrix in Problem 7.24 (a) with compact storage and (b) without compact storage.

7.26 Reorder the rows of the matrix given in Problem 7.24 such that after one Gauss elimination step, $S^{(1)}_{reordered}$ has the same number of zeros in columns 2–6 as the original matrix.

⊟ **Sections 7.8–7.10**

7.27 For the power system given in Example 7.9, use the POWER FLOW program to find the Mvar rating of a shunt capacitor bank at bus 2 that increases V_2 to 1.00 per unit. Also determine the effect of this capacitor bank on line loadings and on total I^2R line losses.

7.28 For the power system given in Example 7.9, run the POWER FLOW program with an additional line installed from bus 2 to bus 4. The line parameters of the added line equal those of existing line 2–4. Determine the effect on V_2, on line loadings, and on total I^2R line losses.

7.29 For the power system given in Example 7.9, run the POWER FLOW program with V_3 decreased to 1.03 per unit. Determine the effect on V_2 and on the reactive power supplied by the generator at bus 3.

7.30 Assume that the transformer between buses 1 and 5 in Example 7.9 is a tap-changing transformer whose taps can be varied from 0.85 to 1.15 in increments of 0.05 per unit. Ue the POWER FLOW program to determine the tap setting required to increase V_5 to 1.05 per unit.

7.31 A new transformer is installed between buses 1 and 5, in parallel with the existing transformer in Example 7.9. The new transformer is identical to the existing transformer. The taps on the new transformer are set at 1.10 to give a 10% increase in voltage at bus 5, while the taps on the existing transformer are set at the nominal value 1.0 per unit. Use the POWER FLOW program to find the real power, reactive power, and MVA supplied by each of these transformers to bus 5.

7.32 Use the POWER FLOW program to solve Problem 7.23. Create the input data files shown in Tables 7.12 and 7.13. Then run the program with the maximum number of iterations set to 1. Select a flat start. Compare V_2 and V_3 in the bus output data with the values computed in Problem 7.23.

7.33 Using the input data files created in Problem 7.32, run the POWER FLOW program with the maximum number of iterations set to 20 and with a 0.0001 tolerance. (a) How many iterations are required for convergence? (b) Compare the converged values of V_2 and V_3 in the bus output data with those computed after only one iteration in Problem 7.32.

CASE STUDY QUESTIONS

A. Is voltage collapse more likely to occur in a power system serving an urban load, an industrial load, a residential-commercial load, or a mix of different load types?

B. The voltage collapse phenomenon is studied via power-flow programs. What other studies can be performed with power-flow programs?

References

1. W. F. Tinney and C. E. Hart, "Power Flow Solutions by Newton's Method," *IEEE Trans. PAS*, *86* (November 1967), p. 1449.

2. W. F. Tinney and J. W. Walker, "Direct Solution of Sparse Network Equations by Optimally Ordered Triangular Factorization," *Proc. IEEE*, *55* (November 1967); pp. 1801–1809.

3. Glenn W. Stagg and Ahmed H. El-Abiad, *Computer Methods in Power System Analysis* (New York: McGraw-Hill, 1968).

4. N. M. Peterson and W. S. Meyer, "Automatic Adjustment of Transformer and Phase Shifter Taps in Newton Power Flow," *IEEE Trans. PAS, 90* (January–February 1971), pp. 103–108.

5. W. D. Stevenson, Jr., *Elements of Power Systems Analysis*, 4th ed. (New York: McGraw-Hill, 1982).

6. A. Bramellar and R. N. Allan, *Sparsity* (London: Pitman, 1976).

7. C. A. Gross, *Power Systems Analysis* (New York: Wiley, 1979).

8. B. Stott, "Fast Decoupled Load Flow," *IEEE Trans. PAS*, Vol. PAS 91 (September–October 1972), pp. 1955–1959.

9. North American Electric Reliability Council (NERC), *Survey of the Voltage Collapse Phenomenon* (Princeton, NJ: NERC, 1991).

CHAPTER 8

SYMMETRICAL FAULTS

Three-phase 500 kV puffer-type circuit breaker
(Courtesy of Florida Power and Light Company)

Short circuits occur in power systems when equipment insulation fails, due to system overvoltages caused by lightning or switching surges, to insulation contamination (salt spray or pollution), or to other mechanical causes. The resulting short circuit or "fault" current is determined by the internal voltages of the synchronous machines and by the system impedances between the machine voltages and the fault. Short-circuit currents may be several orders of magnitude larger than normal operating currents and, if allowed to persist, may cause equipment thermal damage. Windings and busbars may also suffer mechanical damage due to high magnetic forces during faults. It is therefore necessary to remove from service, faulted sections of a power system, as soon as possible. Standard EHV protective equipment is designed to clear faults within 3 cycles (50 ms at 60 Hz). Lower voltage protective equipment operates more slowly (for example, 5 to 20 cycles).

We begin this chapter by reviewing series R–L circuit transients in Section 8.1, followed in Section 8.2 by a description of three-phase short-circuit currents at unloaded synchronous machines. Both the ac component, including subtransient, transient, and steady-state currents, and the dc component of fault current are analyzed. We then extend these results in Sections 8.3 and 8.4 to power-system three-phase short circuits by means of the superposition principle. We observe that the bus impedance matrix is the key to calculating fault currents. In Sections 8.5 and 8.6 we present a systematic procedure for computing the bus impedance matrix and a personal computer program based on this procedure to compute symmetrical short-circuit currents in power systems. This program may be utilized in power-system design to select, set, and coordinate protective equipment such as circuit breakers, fuses, relays, and instrument transformers. Circuit breaker and fuse selection is discussed in Section 8.7.

Balanced three-phase power systems are assumed throughout this chapter. Since three-phase fault currents in balanced systems are themselves balanced, only positive-sequence networks are used. We also work in per-unit.

CASE STUDY | Short circuits can cause severe damage when not interrupted promptly. In some cases, high-impedance fault currents may be insufficient to operate protective relays or blow fuses. Standard overcurrent protection schemes utilized on secondary distribution at some industrial, commercial, and large residential buildings may not detect high-impedance faults, commonly called arcing faults. In these cases, more careful design techniques, such as the use of ground fault circuit interruption, are required to detect arcing faults and prevent burndown. Examples of the destructive effects of arcing faults are given in the following case histories [11].

The Problem of Arcing Faults in Low-Voltage Power Distribution Systems

FRANCIS J. SHIELDS

Abstract

Many cases of electrical equipment burndown arising from low-level arcing-fault currents have occurred in recent years in low-voltage power distribution systems. Burndown, which is the severe damage or complete destruction of conductors, insulation systems and metallic enclosures, is caused by the concentrated release of energy in the fault arc. Both grounded and ungrounded electrical distribution systems have experienced burndown, and the reported incidents have involved both industrial and commercial building distribution equipment, without regard to manufacturer, geographical location, or operating environment.

Figure 1 Burndown damage caused by arcing fault. View shows low-voltage cable compartments of secondary unit substation.

Burndown case histories

The reported incidents of equipment burndown are many. One of the most publicized episodes involved a huge apartment building complex in New York City [Fig. 1], in which two main 480Y/277-volt switchboards were completely destroyed, and two 5000-ampere service entrance buses were burned-off right back to the utility vault. This arcing fault blazed and sputtered for over an hour, and inconvenienced some 10,000 residents of the development through loss of service to building water pumps, hall and stair lighting, elevators, appliances, and apartment lights. Several days elapsed before service resembling normal was restored through temporary hookups. Illustrations of equipment damage in this burndown are shown in Figs. 2 and 3.

Another example of burndown occurred in the

Reprinted with permission from Francis J. Shields, IEEE Transactions on Industry and General Applications, Vol. 1GA-3, No. 1, Jan./Feb. 1967, pp. 16–17. Copyright © 1967 IEEE.

Figure 2 Service entrance switch and current-limiting fuses completely destroyed by arcing fault in main low-voltage switchboard.

Figure 3 *Fused feeder switch consumed by arcing fault in high-rise apartment main switchboard. No intermediate segregating barriers had been used in construction.*

Midwest, and resulted in completely gutting a service entrance switchboard and burning up two 1000-kVA supply transformers. This burndown arc current flowed for about 15 minutes.

In still other reported incidents, a Maryland manufacturer experienced four separate burndowns of secondary unit substations in a little over a year; on the West Coast a unit substation at an industrial process plant burned for more than eight minutes, resulting in destruction of the low-voltage switchgear equipment; and this year* several burndowns have occurred in government office buildings at scattered locations throughout the country.

An example of the involvement of the latter type of equipment in arcing-fault burndowns is shown in Fig. 4. The arcing associated with this fault continued for over 20 minutes, and the fault was finally extinguished only when the relays on the primary system shut down the whole plant.

The electrical equipment destruction shown in the sample photographs is quite startling, but it is only one aspect of this type of fault. Other less graphic but no less serious effects of electrical equipment burndown may include personnel fatalities or serious injury, contingent fire damage, loss of vital services (lighting,

*1966

Figure 4 *Remains of main secondary circuit breaker burned down during arcing fault in low-voltage switchgear section of unit substation.*

elevators, ventilation, fire pumps, etc.), shutdown of critical loads, and loss of product revenue. It should be pointed out that the cases reported have involved both industrial and commercial building distribution equipment, without regard to manufacturer, geographical location, operating environment, or the presence or absence of electrical system neutral grounding. Also, the reported burndowns have included a variety of distribution equipment—load center unit substations, switchboards, busway, panelboards, service-entrance equipment, motor control centers, and cable in conduit, for example.

It is obvious, therefore, when all the possible effects of arcing-fault burndowns are taken into consideration, that engineers responsible for electrical power system layout and operation should be anxious both to minimize the probability of arcing faults in electrical systems and to alleviate or mitigate the destructive effects of such faults if they should inadvertently occur despite careful design and the use of quality equipment.

SECTION 8.1

SERIES R–L CIRCUIT TRANSIENTS

Consider the series R–L circuit shown in Figure 8.1. The closing of switch SW at $t = 0$ represents to a first approximation a three-phase short circuit at the terminals of an unloaded synchronous machine. For simplicity, assume zero fault impedance; that is, the short circuit is a solid or "bolted" fault. The current is assumed to be zero before SW closes, and the source angle α determines the source voltage at $t = 0$. Writing a KVL equation for the circuit,

$$\frac{L\,di(t)}{dt} + Ri(t) = \sqrt{2}V \sin(\omega t + \alpha) \qquad t \geqslant 0 \qquad (8.1.1)$$

The solution to (8.1.1) is

$$i(t) = i_{ac}(t) + i_{dc}(t) = \frac{\sqrt{2}V}{Z} \left[\sin(\omega t + \alpha - \theta) - \sin(\alpha - \theta)e^{-t/T}\right] \quad A$$

$$(8.1.2)$$

where

$$i_{ac}(t) = \frac{\sqrt{2}V}{Z} \sin(\omega t + \alpha - \theta) \quad A \qquad (8.1.3)$$

$$i_{dc}(t) = -\frac{\sqrt{2}V}{Z} \sin(\alpha - \theta)e^{-t/T} \quad A \qquad (8.1.4)$$

$$Z = \sqrt{R^2 + (\omega L)^2} = \sqrt{R^2 + X^2} \quad \Omega \qquad (8.1.5)$$

Figure 8.1

Current in a series R–L circuit with ac voltage source

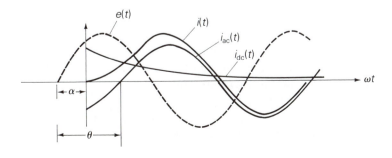

$$\theta = \tan^{-1} \frac{\omega L}{R} = \tan^{-1} \frac{X}{R} \tag{8.1.6}$$

$$T = \frac{L}{R} = \frac{X}{\omega R} = \frac{X}{2\pi f R} \quad s \tag{8.1.7}$$

The total fault current in (8.1.2), called the *asymmetrical fault current*, is plotted in Figure 8.1 along with its two components. The ac fault current (also called *symmetrical* or *steady-state fault current*), given by (8.1.3), is a sinusoid. The *dc offset current*, given by (8.1.4), decays exponentially with time constant $T = L/R$.

The rms ac fault current is $I_{ac} = V/Z$. The magnitude of the dc offset, which depends on α, varies from 0 when $\alpha = 0$ to $\sqrt{2} I_{ac}$ when $\alpha = (\theta \pm \pi/2)$. Note that a short circuit may occur at any instant during a cycle of the ac source; that is, α can have any value. Since we are primarily interested in the largest fault current, we choose $\alpha = (\theta - \pi/2)$. Then (8.1.2) becomes

$$i(t) = \sqrt{2} I_{ac} \left[\sin(\omega t - \pi/2) + e^{-t/T} \right] \quad A \tag{8.1.8}$$

where

$$I_{ac} = \frac{V}{Z} \quad A \tag{8.1.9}$$

The rms value of $i(t)$ is of interest. Since $i(t)$ in (8.1.8) is not strictly periodic, its rms value is not strictly defined. However, treating the exponential term as a constant, we stretch the rms concept to calculate the rms asymmetrical fault current with maximum dc offset, as follows:

$$
\begin{aligned}
I_{rms}(t) &= \sqrt{[I_{ac}]^2 + [I_{dc}(t)]^2} \\
&= \sqrt{[I_{ac}]^2 + [\sqrt{2} I_{ac} e^{-t/T}]^2} \\
&= I_{ac}\sqrt{1 + 2e^{-2t/T}} \quad A
\end{aligned}
\tag{8.1.10}
$$

It is convenient to use $T = X/(2\pi f R)$ and $t = \tau/f$, where τ is time in cycles, and write (8.1.10) as

$$I_{rms}(\tau) = K(\tau) I_{ac} \quad A \tag{8.1.11}$$

where

$$K(\tau) = \sqrt{1 + 2e^{-4\pi\tau/(X/R)}} \quad \text{per unit} \tag{8.1.12}$$

From (8.1.11) and (8.1.12), the rms asymmetrical fault current equals the rms ac fault current times an "asymmetry factor," $K(\tau)$. $I_{rms}(\tau)$ decreases from $\sqrt{3} I_{ac}$ when $\tau = 0$ to I_{ac} when τ is large. Also, higher X to R ratios (X/R) give higher values of $I_{rms}(\tau)$. The above series R–L short-circuit currents are summarized in Table 8.1.

Table 8.1

Short-circuit current—series R–L circuit*

COMPONENT / UNITS	INSTANTANEOUS CURRENT A	rms CURRENT A
Symmetrical (ac)	$i_{ac}(t) = \dfrac{\sqrt{2}V}{Z}\sin(\omega t + \alpha - \theta)$	$I_{ac} = \dfrac{V}{Z}$
dc offset	$i_{dc}(t) = \dfrac{-\sqrt{2}V}{Z}\sin(\alpha - \theta)e^{-t/T}$	
Asymmetrical (total)	$i(t) = i_{ac}(t) + i_{dc}(t)$	$I_{rms}(t) = \sqrt{I_{ac}^2 + i_{dc}(t)^2}$ with maximum dc offset: $I_{rms}(\tau) = K(\tau)I_{ac}$

*See Figure 8.1 and (8.1.1)–(8.1.12).

EXAMPLE 8.1

Fault currents: R–L circuit with ac source

A bolted short circuit occurs in the series R–L circuit of Figure 8.1 with $V = 20\,\text{kV}$, $X = 8\,\Omega$, $R = 0.8\,\Omega$, and with maximum dc offset. The circuit breaker opens 3 cycles after fault inception. Determine (a) the rms ac fault current, (b) the rms "momentary" current at $\tau = 0.5$ cycle, which passes through the breaker before it opens, and (c) the rms asymmetrical fault current which the breaker interrupts.

Solution   **a.** From (8.1.9),

$$I_{ac} = \frac{20 \times 10^3}{\sqrt{(8)^2 + (0.8)^2}} = \frac{20 \times 10^3}{8.040} = 2.488 \quad \text{kA}$$

b. From (8.1.11) and (8.1.12) with $(X/R) = 8/(0.8) = 10$ and $\tau = 0.5$ cycle,

$$K(0.5\ \text{cycle}) = \sqrt{1 + 2e^{-4\pi(0.5)/10}} = 1.438$$

$$I_{momentary} = K(0.5\ \text{cycle})I_{ac} = (1.438)(2.488) = 3.576 \quad \text{kA}$$

c. From (8.1.11) and (8.1.12) with $(X/R) = 10$ and $\tau = 3$ cycles,

$$K(3\ \text{cycles}) = \sqrt{1 + 2e^{-4\pi(3)/10}} = 1.023$$

$$I_{rms}(3\ \text{cycles}) = (1.023)(2.488) = 2.544 \quad \text{kA} \qquad \blacksquare$$

SECTION 8.2

THREE-PHASE SHORT CIRCUIT—UNLOADED SYNCHRONOUS MACHINE

One way to investigate a three-phase short circuit at the terminals of a synchronous machine is to perform a test on an actual machine. Figure 8.2 shows an oscillogram of the ac fault current in one phase of an unloaded

Figure 8.2

ac fault current in one phase of an unloaded synchronous machine during a three-phase short circuit (the dc offset current is removed)

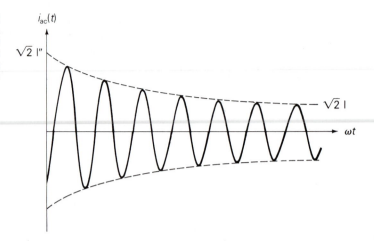

synchronous machine during such a test. The dc offset has been removed from the oscillogram. As shown, the amplitude of the sinusoidal waveform decreases from a high initial value to a lower steady-state value.

A physical explanation for this phenomenon is that the magnetic flux caused by the short-circuit armature currents (or by the resultant armature MMF) is initially forced to flow through high reluctance paths that do not link the field winding or damper circuits of the machine. This is a result of the theorem of constant flux linkages, which states that the flux linking a closed winding cannot change instantaneously. The armature inductance, which is inversely proportional to reluctance, is therefore initially low. As the flux then moves towards the lower reluctance paths, the armature inductance increases.

The ac fault current in a synchronous machine can be modeled by the series R–L circuit of Figure 8.1 if a time-varying inductance $L(t)$ or reactance $X(t) = \omega L(t)$ is employed. In standard machine theory texts [3, 4], the following reactances are defined:

X_d'' = direct axis subtransient reactance

X_d' = direct axis transient reactance

X_d = direct axis synchronous reactance

where $X_d'' < X_d' < X_d$. The subscript d refers to the direct axis. There are similar quadrature axis reactances X_q'', X_q', and X_q [3, 4]. However, if the armature resistance is small, the quadrature axis reactances do not significantly affect the short-circuit current. Using the above direct axis reactances, the instantaneous ac fault current can be written as

$$i_{\text{ac}}(t) = \sqrt{2}E_g \left[\left(\frac{1}{X_d''} - \frac{1}{X_d'} \right) e^{-t/T_d''} + \left(\frac{1}{X_d'} - \frac{1}{X_d} \right) e^{-t/T_d'} + \frac{1}{X_d} \right] \sin\left(\omega t + \alpha - \frac{\pi}{2} \right)$$

(8.2.1)

where E_g is the rms line-to-neutral prefault terminal voltage of the unloaded synchronous machine. Armature resistance is neglected in (8.2.1). Note that at $t = 0$, when the fault occurs, the rms value of $i_{ac}(t)$ in (8.2.1) is

$$I_{ac}(0) = \frac{E_g}{X_d''} = I'' \tag{8.2.2}$$

which is called the rms *subtransient fault current*, I''. The duration of I'' is determined by the time constant T_d'', called the *direct axis short-circuit subtransient time constant*.

At a later time, when t is large compared to T_d'' but small compared to the *direct axis short-circuit transient time constant* T_d', the first exponential term in (8.2.1) has decayed almost to zero, but the second exponential has not decayed significantly. The rms ac fault current then equals the rms *transient fault current*, given by

$$I' = \frac{E_g}{X_d'} \tag{8.2.3}$$

When t is much larger than T_d', the rms ac fault current approaches its steady-state value, given by

$$I_{ac}(\infty) = \frac{E_g}{X_d} = I \tag{8.2.4}$$

Since the three-phase no-load voltages are displaced 120° from each other, the three-phase ac fault currents are also displaced 120° from each other. In addition to the ac fault current, each phase has a different dc offset. The maximum dc offset in any one phase, which occurs when $\alpha = 0$ in (8.2.1), is

$$i_{dcmax}(t) = \frac{\sqrt{2}E_g}{X_d''} e^{-t/T_A} = \sqrt{2}I'' e^{-t/T_A} \tag{8.2.5}$$

where T_A is called the *armature time constant*. Note that the magnitude of the maximum dc offset depends only on the rms subtransient fault current I''. The above synchronous machine short-circuit currents are summarized in Table 8.2.

Machine reactances X_d'', X_d', and X_d as well as time constants T_d'', T_d', and T_A are usually provided by synchronous machine manufacturers. They can also be obtained from a three-phase short-circuit test, by analyzing an oscillogram such as that in Figure 8.2 [2]. Typical values of synchronous machine reactances and time constants are given in Appendix Table A.1.

Table 8.2

Short-circuit current—
unloaded synchronous
machine*

COMPONENT \ UNITS	INSTANTANEOUS CURRENT A	rms CURRENT A
Symmetrical (ac)	(8.2.1)	$I_{ac}(t) = E_g\left[\left(\dfrac{1}{X_d''}-\dfrac{1}{X_d'}\right)e^{-t/T_d''}\right.$ $\left.+\left(\dfrac{1}{X_d'}-\dfrac{1}{X_d}\right)e^{-t/T_d'}+\dfrac{1}{X_d}\right]$
Subtransient Transient Steady-state		$I'' = E_g/X_d''$ $I' = E_g/X_d'$ $I = E_g/X_d$
Maximum dc offset	$i_{dc}(t) = \sqrt{2}\,I''e^{-t/T_A}$	
Asymmetrical (total)	$i(t) = i_{ac}(t) + i_{dc}(t)$	$I_{rms}(t) = \sqrt{I_{ac}(t)^2 + i_{dc}(t)^2}$ with maximum dc offset: $I_{rms}(t) = \sqrt{I_{ac}(t)^2 + [\sqrt{2}\,I''e^{-t/T_A}]^2}$

*See Figure 8.2 and (8.2.1)–(8.2.5).

EXAMPLE 8.2   **Three-phase short-circuit currents, unloaded synchronous generator**

A 500-MVA 20-kV, 60-Hz synchronous generator with reactances $X_d'' = 0.15$, $X_d' = 0.24$, $X_d = 1.1$ per unit and time constants $T_d'' = 0.035$, $T_d' = 2.0$, $T_A = 0.20\,\text{s}$ is connected to a circuit breaker. The generator is operating at 5% above rated voltage and at no-load when a bolted three-phase short circuit occurs on the load side of the breaker. The breaker interrupts the fault 3 cycles after fault inception. Determine (a) the subtransient fault current in per-unit and kA rms; (b) maximum dc offset as a function of time; and (c) rms asymmetrical fault current, which the breaker interrupts, assuming maximum dc offset.

Solution   **a.** The no-load voltage before the fault occurs is $E_g = 1.05$ per unit. From (8.2.2), the subtransient fault current that occurs in each of the three phases is

$$I'' = \frac{1.05}{0.15} = 7.0 \quad \text{per unit}$$

The generator base current is

$$I_{base} = \frac{S_{rated}}{\sqrt{3}\,V_{rated}} = \frac{500}{(\sqrt{3})(20)} = 14.43 \quad \text{kA}$$

The rms subtransient fault current in kA is the per-unit value multiplied by the base current:

$$I'' = (7.0)(14.43) = 101.0 \quad \text{kA}$$

b. From (8.2.5), the maximum dc offset that may occur in any one phase is

$$i_{dcmax}(t) = \sqrt{2}(101.0)e^{-t/0.20} = 142.9\,e^{-t/0.20} \quad \text{kA}$$

c. From (8.2.1), the rms ac fault current at $t = 3$ cycles $= 0.05\,s$ is

$$I_{ac}(0.05\,s) = 1.05\left[\left(\frac{1}{0.15} - \frac{1}{0.24}\right)e^{-0.05/0.035}\right.$$

$$\left. + \left(\frac{1}{0.24} - \frac{1}{1.1}\right)e^{-0.05/2.0} + \frac{1}{1.1}\right]$$

$$= 4.920 \quad \text{per unit}$$

$$= (4.920)(14.43) = 71.01 \quad \text{kA}$$

Modifying (8.1.10) to account for the time-varying symmetrical component of fault current, we obtain

$$I_{rms}(0.05) = \sqrt{[I_{ac}(0.05)]^2 + [\sqrt{2}I''e^{-t/T_a}]^2}$$

$$= I_{ac}(0.05)\sqrt{1 + 2\left[\frac{I''}{I_{ac}(0.05)}\right]^2 e^{-2t/T_a}}$$

$$= (71.01)\sqrt{1 + 2\left[\frac{101}{71.01}\right]^2 e^{-2(0.05)/0.20}}$$

$$= (71.01)(1.8585)$$

$$= 132 \quad \text{kA} \qquad\qquad\qquad \blacksquare$$

SECTION 8.3

POWER-SYSTEM THREE-PHASE SHORT CIRCUITS

In order to calculate the subtransient fault current for a three-phase short circuit in a power system, we make the following assumptions:

1. Transformers are represented by their leakage reactances. Winding resistances, shunt admittances, and Δ–Y phase shifts are neglected.

2. Transmission lines are represented by their positive-sequence equivalent series reactances. Series resistances and shunt admittances are neglected.

3. Synchronous machines are represented by constant-voltage sources behind subtransient reactances. Armature resistance, saliency, and saturation are neglected.

4. All nonrotating impedance loads are neglected.

5. Induction motors are either neglected (especially for small motors rated less than 50 hp) or represented in the same manner as synchronous machines.

These assumptions are made for simplicity in this text, and in practice they should not be made for all cases. For example, in distribution systems,

Figure 8.3

Single-line diagram of a synchronous generator feeding a synchronous motor

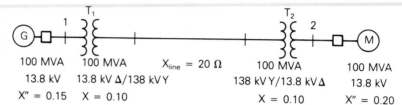

100 MVA	100 MVA	X_{line} = 20 Ω	100 MVA	100 MVA
13.8 kV	13.8 kV Δ/138 kV Y		138 kV Y/13.8 kV Δ	13.8 kV
X″ = 0.15	X = 0.10		X = 0.10	X″ = 0.20

Figure 8.4

Application of superposition to a power-system three-phase short circuit

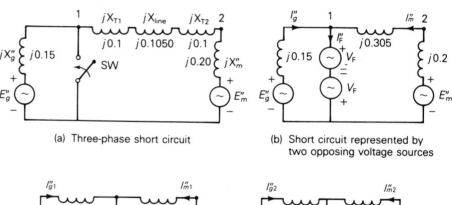

(a) Three-phase short circuit

(b) Short circuit represented by two opposing voltage sources

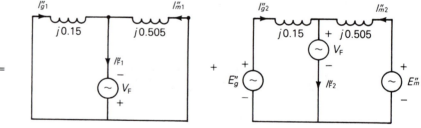

(c) Application of superposition

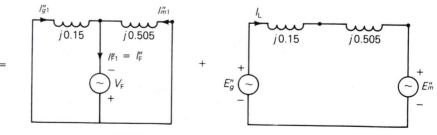

(d) V_F set equal to prefault voltage at fault

resistances of primary and secondary distribution lines may in some cases significantly reduce fault current magnitudes.

Figure 8.3 shows a single-line diagram consisting of a synchronous generator feeding a synchronous motor through two transformers and a transmission line. We shall consider a three-phase short circuit at bus 1. The positive-sequence equivalent circuit is shown in Figure 8.4(a), where the voltages E_g'' and E_m'' are the prefault internal voltages behind the subtransient reactances of the machines, and the closing of switch SW represents the fault. For purposes of calculating the subtransient fault current, E_g'' and E_m'' are assumed to be constant-voltage sources.

In Figure 8.4(b) the fault is represented by two opposing voltage sources with equal phasor values V_F. Using superposition, the fault current can then be calculated from the two circuits shown in Figure 8.4(c). However, if V_F equals the prefault voltage at the fault, then the second circuit in Figure 8.4(c) represents the system before the fault occurs. As such, $I_{F2}'' = 0$ and V_F, which has no effect, can be removed from the second circuit, as shown in Figure 8.4(d). The subtransient fault current is then determined from the first circuit in Figure 8.4(d), $I_F'' = I_{F1}''$. The contribution to the fault from the generator is $I_g'' = I_{g1}'' + I_{g2}'' = I_{g1}'' + I_L$, where I_L is the prefault generator current. Similarly, $I_m'' = I_{m1}'' - I_L$.

| EXAMPLE 8.3 | **Three-phase short-circuit currents, power system** |

The synchronous generator in Figure 8.3 is operating at rated MVA, 0.95 p.f. lagging and at 5% above rated voltage when a bolted three-phase short circuit occurs at bus 1. Calculate the per-unit values of (a) subtransient fault current; (b) subtransient generator and motor currents, neglecting prefault current; and (c) subtransient generator and motor currents including prefault current.

Solution   **a.** Using a 100-MVA base, the base impedance in the zone of the transmission line is

$$Z_{base,line} = \frac{(138)^2}{100} = 190.44 \quad \Omega$$

and

$$X_{line} = \frac{20}{190.44} = 0.1050 \quad \text{per unit}$$

The per-unit reactances are shown in Figure 8.4. From the first circuit in Figure 8.4(d), the Thévenin impedance as viewed from the fault is

$$Z_{Th} = jX_{Th} = j\frac{(0.15)(0.505)}{(0.15 + 0.505)} = j0.11565 \quad \text{per unit}$$

and the prefault voltage at the generator terminals is

$$V_F = 1.05\underline{/0^\circ} \quad \text{per unit}$$

The subtransient fault current is then

$$I_F'' = \frac{V_F}{Z_{Th}} = \frac{1.05\underline{/0^\circ}}{j0.11565} = -j9.079 \quad \text{per unit}$$

b. Using current division in the first circuit of Figure 8.4(d),

$$I_{g1}'' = \left(\frac{0.505}{0.505 + 0.15}\right) I_F'' = (0.7710)(-j9.079) = -j7.000 \quad \text{per unit}$$

$$I_{m1}'' = \left(\frac{0.15}{0.505 + 0.15}\right) I_F'' = (0.2290)(-j9.079) = -j2.079 \quad \text{per unit}$$

c. The generator base current is

$$I_{base,gen} = \frac{100}{(\sqrt{3})(13.8)} = 4.1837 \quad \text{kA}$$

and the prefault generator current is

$$I_L = \frac{100}{(\sqrt{3})(1.05 \times 13.8)} \underline{/-\cos^{-1}0.95} = 3.9845\underline{/-18.19^\circ} \quad \text{kA}$$

$$= \frac{3.9845\underline{/-18.19^\circ}}{4.1837} = 0.9524\underline{/-18.19^\circ}$$

$$= 0.9048 - j0.2974 \quad \text{per unit}$$

The subtransient generator and motor currents, including prefault current, are then

$$I_g'' = I_{g1}'' + I_L = -j7.000 + 0.9048 - j0.2974$$

$$= 0.9048 - j7.297 = 7.353\underline{/-82.9^\circ} \quad \text{per unit}$$

$$I_m'' = I_{m1}'' - I_L = -j2.079 - 0.9048 + j0.2974$$

$$= -0.9048 - j1.782 = 1.999\underline{/243.1^\circ} \quad \text{per unit}$$

An alternate method of solving Example 8.3 is to first calculate the internal voltages E_g'' and E_m'' using the prefault load current I_L. Then, instead of using superposition, the fault currents can be resolved directly from the circuit in Figure 8.4(a) (see Problem 8.6). However, in a system with many synchronous machines, the superposition method has the advantage that all machine voltage sources are shorted, and the prefault voltage is the only source required to calculate the fault current. Also, when calculating the contributions to fault current from each branch, prefault currents are usually small, and hence can be neglected. Otherwise, prefault load currents could be obtained from a power-flow program. ∎

SECTION 8.4

BUS IMPEDANCE MATRIX

We now extend the results of the previous section to calculate subtransient fault currents for three-phase faults in an N-bus power system. The system is modeled by its positive-sequence network, where lines and transformers are represented by series reactances and synchronous machines are represented by constant-voltage sources behind subtransient reactances. As before, all resistances, shunt admittances, and nonrotating impedance loads are neglected. For simplicity, we also neglect prefault load currents.

Consider a three-phase short circuit at any bus n. Using the superposition method described in Section 8.3, we analyze two separate circuits. (For example, see Figure 8.4[d].) In the first circuit all machine-voltage sources are short-circuited and the only source is due to the prefault voltage at the fault. Writing nodal equations for the first circuit,

$$Y_{bus} E^{(1)} = I^{(1)} \tag{8.4.1}$$

where Y_{bus} is the positive-sequence bus admittance matrix, $E^{(1)}$ is the vector of bus voltages, and $I^{(1)}$ is the vector of current sources. The superscript (1) denotes the first circuit. Solving (8.4.1),

$$Z_{bus} I^{(1)} = E^{(1)} \tag{8.4.2}$$

where

$$Z_{bus} = Y_{bus}^{-1} \tag{8.4.3}$$

Z_{bus}, the inverse of Y_{bus}, is called the positive-sequence *bus impedance matrix*. Both Z_{bus} and Y_{bus} are symmetric matrices.

Since the first circuit contains only one source, located at faulted bus n, the current source vector contains only one nonzero component, $I_n^{(1)} = -I_{Fn}''$. Also, the voltage at faulted bus n in the first circuit is $E_n^{(1)} = -V_F$. Rewriting (8.4.2),

$$
\begin{bmatrix}
Z_{11} & Z_{12} & \cdots & Z_{1n} & \cdots & Z_{1N} \\
Z_{21} & Z_{22} & \cdots & Z_{2n} & \cdots & Z_{2N} \\
\vdots & & & & & \\
Z_{n1} & Z_{n2} & \cdots & Z_{nn} & \cdots & Z_{nN} \\
\vdots & & & & & \\
Z_{N1} & Z_{N2} & \cdots & Z_{Nn} & \cdots & Z_{NN}
\end{bmatrix}
\begin{bmatrix}
0 \\
0 \\
\vdots \\
-I_{Fn}'' \\
\vdots \\
0
\end{bmatrix}
=
\begin{bmatrix}
E_1^{(1)} \\
E_2^{(1)} \\
\vdots \\
-V_F \\
\vdots \\
E_N^{(1)}
\end{bmatrix}
\tag{8.4.4}
$$

The minus sign associated with the current source in (8.4.4) indicates that the current injected into bus n is the negative of I_{Fn}'', since I_{Fn}'' flows away from bus n to the neutral. From (8.4.4), the subtransient fault current is

$$I_{Fn}'' = \frac{V_F}{Z_{nn}} \tag{8.4.5}$$

Also from (8.4.4) and (8.4.5), the voltage at any bus k in the first circuit is

$$E_k^{(1)} = Z_{kn}(-I_{Fn}'') = \frac{-Z_{kn}}{Z_{nn}} V_F \qquad (8.4.6)$$

The second circuit represents the prefault conditions. Neglecting prefault load current, all voltages throughout the second circuit are equal to the prefault voltage; that is, $E_k^{(2)} = V_F$ for each bus k. Applying superposition,

$$E_k = E_k^{(1)} + E_k^{(2)} = \frac{-Z_{kn}}{Z_{nn}} V_F + V_F = \left(1 - \frac{Z_{kn}}{Z_{nn}}\right) V_F \qquad k = 1, 2, \dots, N$$

$$(8.4.7)$$

EXAMPLE 8.4

Using Z_{bus} to compute three-phase short-circuit currents in a power system

Faults at bus 1 and 2 in Figure 8.3 are of interest. The prefault voltage is 1.05 per unit and prefault load current is neglected. (a) Determine the 2×2 positive-sequence bus impedance matrix. (b) For a bolted three-phase short circuit at bus 1, use Z_{bus} to calculate the subtransient fault current and the contribution to the fault current from the transmission line. (c) Repeat part (b) for a bolted three-phase short circuit at bus 2.

Figure 8.5

Circuit of Figure 8.4(a) showing per-unit admittance values

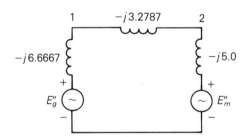

Solution   **a.** The circuit of Figure 8.4(a) is redrawn in Figure 8.5 showing per-unit admittance rather than per-unit impedance values. Neglecting prefault load current, $E_g'' = E_m'' = V_F = 1.05\underline{/0°}$ per unit. From Figure 8.5, the positive-sequence bus admittance matrix is

$$Y_{bus} = -j \begin{bmatrix} 9.9454 & -3.2787 \\ -3.2787 & 8.2787 \end{bmatrix} \text{ per unit}$$

Inverting Y_{bus},

$$Z_{bus} = Y_{bus}^{-1} = +j \begin{bmatrix} 0.11565 & 0.04580 \\ 0.04580 & 0.13893 \end{bmatrix} \text{ per unit}$$

b. Using (8.4.5) the subtransient fault current at bus 1 is

$$I_{F1}'' = \frac{V_F}{Z_{11}} = \frac{1.05\underline{/0°}}{j0.11565} = -j9.079 \quad \text{per unit}$$

which agrees with the result in Example 8.3, part (a). The voltages at buses 1 and 2 during the fault are, from (8.4.7),

$$E_1 = \left(1 - \frac{Z_{11}}{Z_{11}}\right) V_F = 0$$

$$E_2 = \left(1 - \frac{Z_{21}}{Z_{11}}\right) V_F = \left(1 - \frac{j0.04580}{j0.11565}\right) 1.05\underline{/0^\circ} = 0.6342\underline{/0^\circ}$$

The current to the fault from the transmission line is obtained from the voltage drop from bus 2 to 1 divided by the impedance of the line and transformers T_1 and T_2:

$$I_{21} = \frac{E_2 - E_1}{j(X_{\text{line}} + X_{T1} + X_{T2})} = \frac{0.6342 - 0}{j0.3050} = -j2.079 \quad \text{per unit}$$

which agrees with the motor current calculated in Example 8.3, part (b), where prefault load current is neglected.

c. Using (8.4.5), the subtransient fault current at bus 2 is

$$I''_{F2} = \frac{V_F}{Z_{22}} = \frac{1.05\underline{/0^\circ}}{j0.13893} = -j7.558 \quad \text{per unit}$$

and from (8.4.7),

$$E_1 = \left(1 - \frac{Z_{12}}{Z_{22}}\right) V_F = \left(1 - \frac{j0.04580}{j0.13893}\right) 1.05\underline{/0^\circ} = 0.7039\underline{/0^\circ}$$

$$E_2 = \left(1 - \frac{Z_{22}}{Z_{22}}\right) V_F = 0$$

The current to the fault from the transmission line is

$$I_{12} = \frac{E_1 - E_2}{j(X_{\text{line}} + X_{T1} + X_{T2})} = \frac{0.7039 - 0}{j0.3050} = -j2.308 \quad \text{per unit} \qquad \blacksquare$$

Figure 8.6 shows a bus impedance equivalent circuit that illustrates the short-circuit currents in an N-bus system. This circuit is given the name *rake equivalent* in Neuenswander [5] due to its shape, which is similar to a garden rake.

The diagonal elements $Z_{11}, Z_{22}, \ldots, Z_{NN}$ of the bus impedance matrix, which are the *self-impedances*, are shown in Figure 8.6. The off-diagonal elements, or the *mutual impedances*, are indicated by the brackets in the figure.

Neglecting prefault load currents, the internal voltage sources of all synchronous machines are equal both in magnitude and phase. As such, they can be connected, as shown in Figure 8.7, and replaced by one equivalent source V_F from neutral bus 0 to a references bus, denoted r. This equivalent source is also shown in the rake equivalent of Figure 8.6.

Figure 8.6

Bus impedance equivalent circuit (*rake equivalent*)

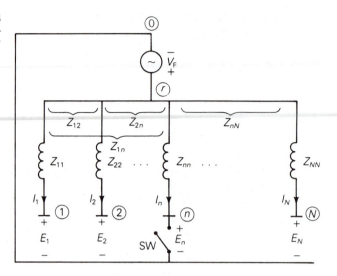

Figure 8.7

Parallel connection of unloaded synchronous machine internal-voltage sources

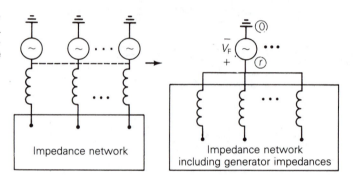

Impedance network

Impedance network including generator impedances

Using Z_{bus}, the fault currents in Figure 8.6 are given by

$$
\begin{bmatrix}
Z_{11} & Z_{12} & \cdots & Z_{1n} & \cdots & Z_{1N} \\
Z_{21} & Z_{22} & \cdots & Z_{2n} & \cdots & Z_{2N} \\
\vdots & & & & & \vdots \\
Z_{n1} & Z_{n2} & \cdots & Z_{nn} & \cdots & Z_{nN} \\
\vdots & & & & & \\
Z_{N1} & Z_{N2} & \cdots & Z_{Nn} & \cdots & Z_{NN}
\end{bmatrix}
\begin{bmatrix}
I_1 \\
I_2 \\
\vdots \\
I_n \\
\vdots \\
I_N
\end{bmatrix}
=
\begin{bmatrix}
V_F - E_1 \\
V_F - E_2 \\
\vdots \\
V_F - E_n \\
\vdots \\
V_F - E_N
\end{bmatrix}
\qquad (8.4.8)
$$

where $I_1, I_2, \ldots$ are the branch currents and $(V_F - E_1)$, $(V_F - E_2), \ldots$ are the voltages across the branches.

If switch SW in Figure 8.6 is open, all currents are zero and the voltage at each bus with respect to the neutral equals V_F. This corresponds to prefault conditions, neglecting prefault load currents. If switch SW is closed, corresponding to a short circuit at bus n, $E_n = 0$ and all currents except I_n remain zero. The fault current is $I''_{Fn} = I_n = V_F/Z_{nn}$, which agrees with (8.4.5). This fault current also induces a voltage drop $Z_{kn}I_n = (Z_{kn}/Z_{nn})V_F$ across each branch k. The voltage at bus k with respect to the neutral then equals V_F minus this voltage drop, which agrees with (8.4.7).

As shown by Figure 8.6 as well as (8.4.5), subtransient fault currents throughout an N-bus system can be determined from the bus impedance matrix and the prefault voltage. Once Z_{bus} has been obtained, these fault currents are easily computed.

SECTION 8.5

FORMATION OF Z_{bus} ONE STEP AT A TIME

In Section 8.4, Z_{bus} is obtained by inverting Y_{bus}. In this section we describe a method of building up Z_{bus} by adding branches one at a time until the complete network is formed [5, 6]. This method can be used to construct a new Z_{bus} or to modify an existing Z_{bus}. It is easily implemented on a digital computer and is much simpler than inverting Y_{bus}, especially for a large number of buses.

Consider the problem of adding a new branch with impedance Z_b to an existing n bus network whose $n \times n$ bus impedance matrix $Z_{bus(old)}$ is known. It is desired to construct the new bus impedance matrix $Z_{bus(new)}$. We recognize the following types of branch additions:

1. Add Z_b from the reference bus r to new bus p.

2. Add Z_b from old bus k to new bus p.

3. Add Z_b from the reference bus r to old bus k.

4. Add Z_b between two old buses k and m.

Since only three-phase short circuits in balanced three-phase systems are under consideration, we work with the positive-sequence network. For simplicity, we assume that there is no mutual impedance between the new branch and any existing branch. Also, all impedances are given in per-unit. The four types of branch additions are now discussed.

Type 1   Add Z_b from the reference bus r to new bus p. See Figure 8.8. The new equation set is:

$$
\underbrace{\left[
\begin{array}{ccc|c}
 & & & 0 \\
 & Z_{bus(old)} & & 0 \\
 & & & \vdots \\
 & & & 0 \\
\hline
0 & 0 \;\cdots\; 0 & & Z_b
\end{array}
\right]}_{Z_{bus(new)}}
\left[
\begin{array}{c}
I_1 \\
I_2 \\
\vdots \\
I_n \\
\hline
I_p
\end{array}
\right]
=
\left[
\begin{array}{c}
V_F - E_1 \\
V_F - E_2 \\
\vdots \\
V_F - E_n \\
\hline
V_F - E_p
\end{array}
\right]
\qquad (8.5.1)
$$

Figure 8.8

Type 1 addition

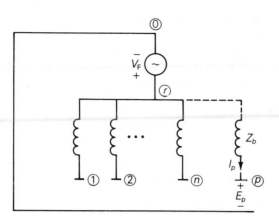

Since a new bus is added, a new row and column are added to $Z_{bus(old)}$ to form the $(n + 1) \times (n + 1)$ matrix $Z_{bus(new)}$. If a fault occurs at new bus p, then $E_p = 0$ and the fault current $I_p = V_F/Z_b$. Therefore, Z_b is the new diagonal element of $Z_{bus(new)}$. Also, I_p cannot change the existing voltages in the network, since Z_b is not mutually coupled to the existing branches. Therefore, the remaining elements in the new row and column are zero.

Type 2   Add Z_b from old bus k to new bus p. See Figure 8.9. The new equation set is

$$
\underbrace{\left[
\begin{array}{cccc|c}
 & & & & Z_{1k} \\
 & \multicolumn{3}{c}{Z_{bus(old)}} & Z_{2k} \\
 & & & & \vdots \\
 & & & & Z_{nk} \\
\hline
Z_{1k} & Z_{2k} & \cdots & Z_{nk} & Z_{kk} + Z_b
\end{array}
\right]}_{Z_{bus(new)}}
\left[
\begin{array}{c}
I_1 \\ I_2 \\ \vdots \\ I_n \\ \hline I_p
\end{array}
\right]
=
\left[
\begin{array}{c}
V_F - E_1 \\ V_F - E_2 \\ \vdots \\ V_F - E_n \\ \hline V_F - E_p
\end{array}
\right]
\qquad (8.5.2)
$$

Figure 8.9

Type 2 addition

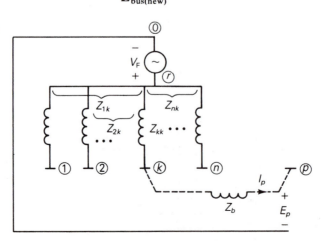

Again, since a new bus is added, a new row and column are created. If a fault occurs at new bus p, $E_p = 0$ and $I_p = V_F/(Z_{kk} + Z_b)$. Thus $(Z_{kk} + Z_b)$ is the new diagonal element. Also, the current I_p through branch k induces voltage drops $Z_{1k}I_p, Z_{2k}I_p, \ldots, Z_{nk}I_p$ in branches $1, 2, \ldots, n$, respectively. Therefore, the remaining elements of the new column of $Z_{bus(new)}$ are the elements of the kth column of $Z_{bus(old)}$. Also, since $Z_{bus(new)}$ is symmetric, the new row is identical to the new column.

Figure 8.10

Type 3 addition

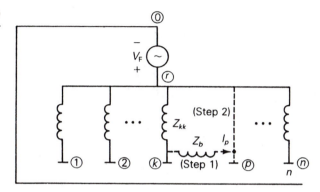

Type 3 Add Z_b from the reference bus r to old bus k. See Figure 8.10. $Z_{bus(new)}$ is formed in two steps. First, we temporarily create new bus p and connect Z_b from bus k to p, which is a type 2 addition. The equation set is given by (8.5.2). Second, we short-circuit new bus p to the reference. As such, $(V_F - E_p) = 0$ and I_p can be eliminated from (8.5.2). Writing (8.5.2) in partitioned form with $(V_F - E_p) = 0$,

$$Z_{bus(old)}I + Z_{col}I_p = V_F - E \tag{8.5.3}$$

$$Z_{col}^T I + (Z_{kk} + Z_b)I_p = 0 \tag{8.5.4}$$

where

$$Z_{col} = \begin{bmatrix} Z_{1k} \\ Z_{2k} \\ \vdots \\ Z_{nk} \end{bmatrix} \qquad I = \begin{bmatrix} I_1 \\ I_2 \\ \vdots \\ I_n \end{bmatrix} \qquad (V_F - E) = \begin{bmatrix} V_F - E_1 \\ V_F - E_2 \\ \vdots \\ V_F - E_n \end{bmatrix}$$

Solving (8.5.4) for I_p, (8.5.3) becomes

$$\left[\underbrace{Z_{bus(old)} - \frac{Z_{col}Z_{col}^T}{(Z_{kk} + Z_b)}}_{Z_{bus(new)}} \right] I = V_F - E \tag{8.5.5}$$

Therefore,

$$Z_{\text{bus(new)}} = Z_{\text{bus(old)}} - \frac{1}{(Z_{kk} + Z_b)} \begin{bmatrix} Z_{1k} \\ Z_{2k} \\ \vdots \\ Z_{nk} \end{bmatrix} [Z_{1k} \; Z_{2k} \; \cdots \; Z_{nk}] \qquad (8.5.6)$$

Note that $Z_{\text{bus(new)}}$ has the same dimension as the $n \times n$ matrix $Z_{\text{bus(old)}}$.

Figure 8.11

Type 4 addition

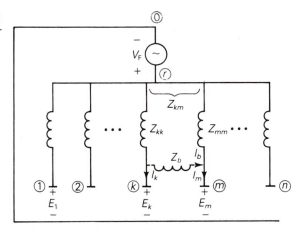

Type 4 Add Z_b between two old buses k and m. See Figure 8.11. Note that when Z_b is added, the current $(I_k + I_b)$ flows in branch k and $(I_m - I_b)$ flows in branch m. The voltage across branch 1 then becomes

$$Z_{11}I_1 + Z_{12}I_2 + \cdots + Z_{1k}(I_k + I_b) + \cdots + Z_{1m}(I_m - I_b) + \cdots + Z_{1n}I_n$$

$$= (V_F - E_1) \qquad (8.5.7)$$

Similarly,

$$Z_{1k}I_1 + Z_{2k}I_2 + \cdots + Z_{kk}(I_k + I_b) + \cdots + Z_{mk}(I_m - I_b)$$
$$+ \cdots + Z_{nk}I_n = (V_F - E_k) \qquad (8.5.8)$$

and

$$Z_{1m}I_1 + Z_{2m}I_2 + \cdots + Z_{km}(I_k + I_b) + \cdots + Z_{mm}(I_m - I_b)$$
$$+ \cdots + Z_{nm}I_n = (V_F - E_m) \qquad (8.5.9)$$

Also, the voltage drop across Z_b is:

$$Z_b I_b = E_k - E_m \qquad (8.5.10)$$

Using (8.5.8) and (8.5.9) in (8.5.10),

$$Z_b I_b = (Z_{1m} - Z_{1k})I_1 + (Z_{2m} - Z_{2k})I_2 + \cdots + (Z_{km} - Z_{kk})(I_k + I_b)$$
$$+ \cdots + (Z_{mm} - Z_{mk})(I_m - I_b) + \cdots + (Z_{nm} - Z_{nk})I_n \qquad (8.5.11)$$

or

$$(Z_{1k} - Z_{1m})I_1 + (Z_{2k} - Z_{2m})I_2 + \cdots + (Z_{kk} - Z_{km})I_k$$
$$+ \cdots + (Z_{mk} - Z_{mm})I_m + \cdots + (Z_{nk} - Z_{nm})I_n$$
$$+ (Z_b + Z_{kk} + Z_{mm} - 2Z_{km})I_b = 0 \qquad (8.5.12)$$

Equations (8.5.7)–(8.5.9) and (8.5.12) in matrix format are

$$
\left[
\begin{array}{c|c}
\textbf{Z}_{\text{bus(old)}} & \begin{matrix} (Z_{1k} - Z_{1m}) \\ (Z_{2k} - Z_{2m}) \\ \vdots \\ (Z_{nk} - Z_{nm}) \end{matrix} \\
\hline
(Z_{1k} - Z_{1m}) \cdots (Z_{nk} - Z_{nm}) & (Z_b + Z_{kk} + Z_{mm} - 2Z_{km})
\end{array}
\right]
\left[
\begin{array}{c}
I_1 \\ I_2 \\ \vdots \\ I_n \\ \hline I_b
\end{array}
\right]
=
\left[
\begin{array}{c}
V_F - E_1 \\ V_F - E_2 \\ \vdots \\ V_F - E_n \\ \hline 0
\end{array}
\right]
\qquad (8.5.13)
$$

Since the last element in the above voltage vector is zero, I_b can be eliminated from (8.5.13), just as I_p was eliminated from (8.5.3) and (8.5.4). The result is

$$\textbf{Z}_{\text{bus(new)}} = \textbf{Z}_{\text{bus(old)}} - \frac{1}{(Z_b + Z_{kk} + Z_{mm} - 2Z_{km})} \begin{bmatrix} Z_{1k} - Z_{1m} \\ Z_{2k} - Z_{2m} \\ \vdots \\ Z_{nk} - Z_{nm} \end{bmatrix} [(Z_{1k} - Z_{1m}) \cdots (Z_{nk} - Z_{nm})] \qquad (8.5.14)$$

When computing $\textbf{Z}_{\text{bus}}$ from scratch, or when modifying an existing $\textbf{Z}_{\text{bus}}$, we adopt the following ordering rules:

1. When computing $\textbf{Z}_{\text{bus}}$ from scratch, start with a generator impedance (a type 1 addition) at bus 1 so as to create the reference bus right from the beginning, and also to begin with bus 1. If there is no generator connected to bus 1, temporarily renumber a generator bus as bus 1.

2. When creating new buses, type 2 and 3 additions, proceed in the order of the buses. That is, create bus 2 after bus 1, then bus 3, 4, and so on. If this is not possible with the original bus numbers, then temporarily renumber the buses.

3. Avoid the addition of a branch between two new buses.

4. Subject to the above constraints, branches may be added in arbitrary order.

5. If renumbering of buses is necessary, reorder the elements of $\textbf{Z}_{\text{bus}}$ according to the original bus numbers after $\textbf{Z}_{\text{bus}}$ has been completely constructed.

EXAMPLE 8.5 | **Constructing $\textbf{Z}_{\text{bus}}$ one step at a time**

Use the one-step-at-a-time method to construct $\textbf{Z}_{\text{bus}}$ for the 2-bus network in Figure 8.3.

Solution   We construct Z_{bus} in the following three steps:

Step 1 Start by adding the generator impedance $jX''_g = j0.15$ per unit from new bus 1 to the reference bus (a type 1 addition):

$$Z_{bus} = j[0.15] \quad \text{per unit}$$

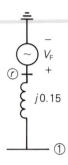

Step 2 Add $j(X_{T1} + X_{line} + X_{T2}) = j0.305$ per unit from old bus 1 to new bus 2, a type 2 addition:

$$Z_{bus} = j\begin{bmatrix} 0.15 & 0.15 \\ 0.15 & (0.305 + 0.15) \end{bmatrix} = j\begin{bmatrix} 0.15 & 0.15 \\ 0.15 & 0.455 \end{bmatrix} \quad \text{per unit}$$

Step 3 Add the synchronous motor impedance $jX''_m = j0.20$ per unit from the reference bus to old bus 2, a type 3 addition:

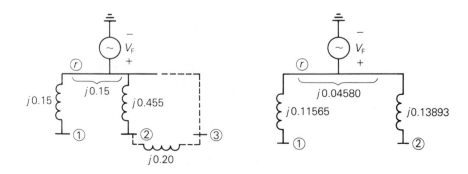

From (8.5.6),

$$Z_{bus} = j \begin{bmatrix} 0.15 & 0.15 \\ 0.15 & 0.455 \end{bmatrix} - \frac{j}{(0.455 + 0.2)} \begin{bmatrix} 0.15 \\ 0.455 \end{bmatrix} \begin{bmatrix} 0.15 & 0.455 \end{bmatrix}$$

$$= j \begin{bmatrix} \dfrac{0.15 - (0.15)^2}{0.655} & \dfrac{0.15 - (0.15)(0.455)}{0.655} \\[3mm] \dfrac{0.15 - (0.455)(0.15)}{0.655} & \dfrac{0.455 - (0.455)^2}{0.655} \end{bmatrix}$$

$$= j \begin{bmatrix} 0.11565 & 0.04580 \\ 0.04580 & 0.13893 \end{bmatrix} \quad \text{per unit}$$

which is the same as that obtained by inverting Y_{bus} in Example 8.4. Note that bus renumbering is unnecessary for this example. ∎

EXAMPLE 8.6 Z_{bus}: adding an impedance between two old buses

Add a new branch with $Z_b = j0.40$ per unit between buses 1 and 2 for the network in Figure 8.3. Construct the new bus impedance matrix assuming Z_b is not mutually coupled to the existing line.

Solution   This is a type 4 addition. From (8.5.14) with $k = 1$ and $m = 2$, and using $Z_{bus(old)}$ from the previous example:

$$Z_{bus(new)} = j \begin{bmatrix} 0.11565 & 0.04580 \\ 0.04580 & 0.13893 \end{bmatrix}$$

$$- j \frac{\begin{bmatrix} 0.11565 - 0.04580 \\ 0.04580 - 0.13893 \end{bmatrix} \begin{bmatrix} (0.11565 - 0.04580)(0.04580 - 0.13983) \end{bmatrix}}{(0.4 + 0.11565 + 0.13893 - 2 \times 0.04580)}$$

$$= j \begin{bmatrix} \dfrac{0.11565 - (0.06985)^2}{0.56299} & \dfrac{0.04580 - (0.06985)(-0.09313)}{0.56299} \\[3mm] \dfrac{0.04580 - (-0.09313)(0.06985)}{0.56299} & \dfrac{0.13893 - (-0.09313)^2}{0.56299} \end{bmatrix}$$

$$= j \begin{bmatrix} 0.10698 & 0.05735 \\ 0.05735 & 0.12352 \end{bmatrix} \quad \text{per unit}$$ ∎

SECTION 8.6

PERSONAL COMPUTER PROGRAM: SYMMETRICAL SHORT CIRCUITS

The software package that accompanies this text includes the program "SYMMETRICAL SHORT CIRCUITS," which computes the ac (symmetrical) fault current for a bolted three-phase short circuit at any bus in an N-bus power system. For each fault, the program also computes bus voltages and contributions to the fault current from transmission lines and transformers connected to the fault bus.

Input data for the program include machine, transmission line, and transformer data, as illustrated in Tables 8.3, 8.4, and 8.5, as well as the prefault voltage. When the machine reactance input data consist of direct axis subtransient reactances, the computed ac fault currents are subtransient fault currents. Alternatively, transient or steady-state ac fault currents are computed when these input data consist of direct axis transient or synchronous reactances. Transmission-line reactances are equivalent positive-sequence series reactances X' for long lines or positive-sequence series reactances X for medium and short lines. All machine, line, and transformer reactances are given in per-unit on a common MVA base. Resistances, shunt admittances, nonrotating impedance loads, and prefault load currents are neglected.

Table 8.3

Synchronous machine data for SYMMETRICAL SHORT CIRCUITS program*

BUS	MACHINE SUBTRANSIENT REACTANCE – X_d''
	per unit
1	0.18
3	0.09

*S_{base} = 400 MVA
V_{base} = 15 kV at buses 1, 3
= 345 kV at buses 2, 4, 5

The program computes the positive-sequence impedance matrix Z_{bus} using the one-step-at-a-time method described in Section 8.5. The program user can either compute Z_{bus} from scratch or begin with an existing Z_{bus} that has already been computed and stored. In both cases, type 1, 2, 3, and 4 modifications to Z_{bus} can be made. The program computes Z_{bus} in accordance with the ordering rules given in Section 8.5.

After Z_{bus} is computed, (8.4.5) and (8.4.7) are employed to compute the fault current and the voltages at each bus during the fault for a three-phase fault at bus 1. Contributions to this fault current from each line or trans-

Table 8.4

Line data for SYMMETRICAL SHORT CIRCUITS program

BUS-TO-BUS	EQUIVALENT POSITIVE-SEQUENCE SERIES REACTANCE
	per unit
2–4	0.4
2–5	0.2
4–5	0.1

	BUS-TO-BUS	LEAKAGE REACTANCE — X
Table 8.5		per unit
Transformer data for SYMMETRICAL SHORT CIRCUITS program	1–5	0.08
	3–4	0.04

former branch connected to the fault bus are also computed by dividing the voltage across the branch by the branch impedance. These computations are then repeated for a fault at bus 2, then bus 3, and so on to bus N.

For a fault at bus $1, 2, \ldots, N$, output data consist of: the fault current, contributions to the fault current from each branch connected to the fault bus, and bus voltages. Z_{bus} can also be included in the output, if desired.

EXAMPLE 8.7

SYMMETRICAL SHORT CIRCUITS program

Consider the 5-bus power system whose single-line diagram is shown in Figure 7.2. Machine, line, and transformer data are given in Tables 8.3, 8.4, and 8.5. The prefault voltage is 1.05 per unit. Run the SYMMETRICAL SHORT CIRCUITS program for this system.

Solution

The program is run with the input data from Tables 8.3, 8.4, and 8.5 and with a prefault voltage of 1.05 per unit. The 5×5 Z_{bus} matrix computed by the program is shown in Table 8.6, and the fault currents and bus voltages are given in Table 8.7. Note that these fault currents are subtransient fault currents, since the machine reactance input data consist of direct axis subtransient reactances.

Table 8.6

Z_{bus} for Example 8.7

$$j \begin{bmatrix} 0.11189 & 0.07081 & 0.03405 & 0.04919 & 0.08162 \\ 0.07081 & 0.22781 & 0.05459 & 0.07886 & 0.10228 \\ 0.03405 & 0.05459 & 0.07297 & 0.06541 & 0.04919 \\ 0.04919 & 0.07886 & 0.06541 & 0.09447 & 0.07105 \\ 0.08162 & 0.10228 & 0.04919 & 0.07105 & 0.11790 \end{bmatrix}$$

Table 8.7

Fault currents and bus voltages for Example 8.7

FAULT BUS	FAULT CURRENT	CONTRIBUTIONS TO FAULT CURRENT		
		GEN LINE OR TRSF	BUS-TO-BUS	CURRENT
FAULT BUS	per unit			per unit
1	9.384			
		G 1	GRND − 1	5.833
		T 1	5 − 1	3.551
2	4.609			
		L 1	4 − 2	1.716
		L 2	5 − 2	2.893
3	14.389			
		G 2	GRND − 3	11.667
		T 2	4 − 3	2.722
4	11.114			
		L 1	2 − 4	0.434
		L 3	5 − 4	2.603
		T 2	3 − 4	8.077
5	8.906			
		L 2	2 − 5	0.695
		L 3	4 − 5	4.172
		T 1	1 − 5	4.038

VF = 1.05	PER-UNIT BUS VOLTAGES DURING THE FAULT				
FAULT BUS	BUS1	BUS2	BUS3	BUS4	BUS5
1	0.0000	0.3855	0.7304	0.5884	0.2841
2	0.7236	0.0000	0.7984	0.6865	0.5786
3	0.5600	0.2644	0.0000	0.1089	0.3422
4	0.5033	0.1736	0.3231	0.0000	0.2603
5	0.3231	0.1391	0.6119	0.4172	0.0000

∎

EXAMPLE 8.8

SYMMETRICAL SHORT CIRCUITS program: effect of a new line installation on fault currents

For the power system given in Example 8.7, run the SYMMETRICAL SHORT CIRCUITS program with an additional line installed between buses 2 and 4. This line, whose reactance is 0.3 per unit, is not mutually coupled to any other line.

Solution

An additional line with reactance $X = 0.3$ per unit is added from bus 2 to bus 4 and the program is rerun. Z_{bus} along with the fault currents and bus voltages are shown in Tables 8.8 and 8.9.

Table 8.8

Z_{bus} for Example 8.8

$$
j\begin{bmatrix}
0.11089 & 0.06388 & 0.03456 & 0.04992 & 0.08017 \\
0.06388 & 0.18005 & 0.05806 & 0.08387 & 0.09227 \\
0.03456 & 0.05806 & 0.07272 & 0.06504 & 0.04992 \\
0.04992 & 0.08387 & 0.06504 & 0.09395 & 0.07210 \\
0.08017 & 0.09227 & 0.04992 & 0.07210 & 0.11580
\end{bmatrix}
$$

Table 8.9

Fault currents and bus voltages for Example 8.8

FAULT BUS	FAULT CURRENT	CONTRIBUTIONS TO FAULT CURRENT		
		GEN LINE OR TRSF	BUS-TO-BUS	CURRENT
	per unit			per unit
1	9.469			
		G 1	GRND − 1	5.833
		T 1	5 − 1	3.636
2	5.832			
		L 1	4 − 2	1.402
		L 2	5 − 2	2.560
		L 4	4 − 2	1.870
3	14.439			
		G 2	GRND − 3	11.667
		T 2	4 − 3	2.772
4	11.176			
		L 1	2 − 4	0.282
		L 3	5 − 4	2.442
		L 4	2 − 4	0.376
		T 2	3 − 4	8.077
5	9.067			
		L 2	2 − 5	1.067
		L 3	4 − 5	3.962
		T 1	1 − 5	4.038

VF = 1.05	PER-UNIT BUS VOLTAGES DURING THE FAULT				
FAULT BUS	BUS1	BUS2	BUS3	BUS4	BUS5
1	0.0000	0.4451	0.7228	0.5773	0.2909
2	0.6775	0.0000	0.7114	0.5609	0.5119
3	0.5510	0.2117	0.0000	0.1109	0.3293
4	0.4921	0.1127	0.3231	0.0000	0.2442
5	0.3231	0.2134	0.5974	0.3962	0.0000

SECTION 8.7

CIRCUIT BREAKER AND FUSE SELECTION

The computer programs SYMMETRICAL SHORT CIRCUITS (Section 8.6) and SHORT CIRCUITS (Section 9.6) may be utilized in power-system design to select, set, and coordinate protective equipment such as circuit breakers, fuses, relays, and instrument transformers. In this section we discuss basic principles of circuit breaker and fuse selection.

AC Circuit Breakers

A *circuit breaker* is a mechanical switch capable of interrupting fault currents and of reclosing. When circuit-breaker contacts separate while carrying current, an arc forms. The breaker is designed to extinguish the arc by elongating and cooling it. The fact that ac arc current naturally passes through zero twice during its 60-Hz cycle aids the arc extinction process.

Circuit breakers are classified as *power* circuit breakers when they are intended for service in ac circuits above 1500 V, and as *low-voltage* circuit breakers in ac circuits up to 1500 V. There are different types of circuit breakers depending on the medium—air, oil, SF_6 gas, or vacuum—in which the arc is elongated. Also, the arc can be elongated either by a magnetic force or by a blast of air.

Some circuit breakers are equipped with a high-speed automatic reclosing capability. Since most faults are temporary and self-clearing, reclosing is based on the idea that if a circuit is deenergized for a short time, it is likely that whatever caused the fault has disintegrated and the ionized arc in the fault has dissipated.

When reclosing breakers are employed in EHV systems, standard practice is to reclose only once, approximately 15 to 50 cycles (depending on operating voltage) after the breaker interrupts the fault. If the fault persists and the EHV breaker recloses into it, the breaker reinterrupts the fault current and then "locks out," requiring operator resetting. Multiple-shot reclosing in EHV systems is not standard practice because transient stability (Chapter 13) may be compromised. However, for distribution systems (2.4–46 kV) where customer outages are of concern, standard reclosers are equipped for two or more reclosures.

For low-voltage applications, molded case circuit breakers with dual trip capability are available. There is a magnetic instantaneous trip for large fault currents above a specified threshold, and a thermal trip with time delay for smaller fault currents.

Modern circuit-breaker standards are based on symmetrical interrupting current. It is usually necessary to calculate only symmetrical fault current at a system location, and then select a breaker with a symmetrical interrupting capability equal to or above the calculated current. The breaker has the additional capability to interrupt the asymmetrical (or total) fault current if the dc offset is not too large.

Recall from Section 8.1 that the maximum asymmetry factor K ($\tau = 0$) is $\sqrt{3}$, which occurs at fault inception ($\tau = 0$). After fault inception, the dc fault current decays exponentially with time constant $T = (L/R) = (X/\omega R)$, and the asymmetry factor decreases. Power circuit breakers with a 2-cycle rated interruption time are designed for an asymmetrical interrupting capability up to 1.4 times their symmetrical interrupting capability, whereas slower circuit breakers have a lower asymmetrical interrupting capability.

A simplified method for breaker selection is called the "E/X simplified method" [1, 7]. The maximum symmetrical short-circuit current at the system location in question is calculated from the prefault voltage and system reactance characteristics, using, for example, computer programs similar to those described in Sections 8.6 and 9.6. Resistances, shunt admittances, nonrotating impedance loads, and prefault load currents are neglected. Then, if the X/R ratio at the system location is less than 15, a breaker with a symmetrical interrupting capability equal to or above the calculated current at the given operating voltage is satisfactory. However, if X/R is greater than 15, the dc offset may not have decayed to a sufficiently low value. In this case, a method for correcting the calculated fault current to account for dc and ac time constants as well as breaker speed can be used [10]. If X/R is unknown, the calculated fault current should not be greater than 80% of the breaker interrupting capability.

When selecting circuit breakers for generators, two cycle breakers are employed in practice, and the subtransient fault current is calculated; therefore subtransient machine reactances X_d'' are used in fault calculations. For synchronous motors, subtransient reactances X_d'' or transient reactances X_d' are used, depending on breaker speed. Also, induction motors can momentarily contribute to fault current. Large induction motors are usually modeled as sources in series with X_d'' or X_d', depending on breaker speed. Smaller induction motors (below 50 hp) are often neglected entirely.

Table 8.10 shows a schedule of preferred ratings for outdoor power circuit breakers. Some of the more important ratings shown are described next.

Voltage ratings

Rated maximum voltage: Designated the maximum rms line-to-line operating voltage. The breaker should be used in systems with an operating voltage less than or equal to this rating.

Rated low frequency withstand voltage: The maximum 60-Hz rms line-to-line voltage that the circuit breaker can withstand without insulation damage.

Rated impulse withstand voltage: The maximum crest voltage of a voltage pulse with standard rise and delay times that the breaker insulation can withstand.

Rated voltage range factor K: The range of voltage for which the symmetrical interrupting capability times the operating voltage is constant.

Current ratings

Rated continuous current: The maximum 60-Hz rms current that the breaker can carry continuously while it is in the closed position without overheating.

Rated short-circuit current: The maximum rms symmetrical current that the breaker can safely interrupt at rated maximum voltage.

Rated momentary current: The maximum rms asymmetrical current that the breaker can withstand while in the closed position without damage. Rated momentary current for standard breakers is 1.6 times the symmetrical interrupting capability.

Rated interrupting time: The time in cycles on a 60-Hz basis from the

Table 8.10 Preferred ratings for outdoor circuit breakers (symmetrical current basis of rating) [10]

IDENTIFICATION		RATED VALUES					
		VOLTAGE		INSULATION LEVEL		CURRENT	
				RATED WITHSTAND TEST VOLTAGE			
NOMINAL VOLTAGE CLASS kV, rms	NOMINAL 3-PHASE MVA CLASS	RATED MAX VOLTAGE kV, rms	RATED VOLTAGE RANGE FACTOR K	LOW FREQUENCY kV, rms	IMPULSE kV, CREST	RATED CONTINU-OUS CURRENT AT 60 Hz AMPERES, rms	RATED SHORT-CIRCUIT CURRENT (AT RATED MAX kV) kA, rms
COL 1	COL 2	COL 3	COL 4	COL 5	COL 6	COL 7	COL 8
14.4	250	15.5	2.67			600	8.9
14.4	500	15.5	1.29			1200	18
23	500	25.8	2.15			1200	11
34.5	1500	38	1.65			1200	22
46	1500	48.3	1.21			1200	17
69	2500	72.5	1.21			1200	19
115		121	1.0			1200	20
115		121	1.0			1600	40
115		121	1.0			2000	40
115		121	1.0			2000	63
115		121	1.0			3000	40
115		121	1.0			3000	63
138		145	1.0			1200	20
138	Not	145	1.0			1600	40
138		145	1.0			2000	40
138		145	1.0			2000	63
138		145	1.0			2000	80
138	Applica-	145	1.0			3000	40
138		145	1.0			3000	63
138		145	1.0			3000	80
161	ble	169	1.0			1200	16
161		169	1.0			1600	31.5
161		169	1.0			2000	40
161		169	1.0			2000	50
230		242	1.0			1600	31.5
230		242	1.0			2000	31.5
230		242	1.0			3000	31.5
230		242	1.0			2000	40
230		242	1.0			3000	40
230		242	1.0			3000	63
345		362	1.0			2000	40
345		362	1.0			3000	40
500		550	1.0			2000	40
500		550	1.0			3000	40
700		765	1.0			2000	40
700		765	1.0			3000	40

Table 8.10 (continued)

RATED VALUES			RELATED REQUIRED CAPABILITIES		
			CURRENT VALUES		
RATED INTER-RUPTING TIME CYCLES	RATED PER-MISSIBLE TRIPPING DELAY SECONDS	RATED MAX VOLTAGE DIVIDED BY *K* kV, rms	MAX SYMMET-RICAL INTER-RUPTING CA-PABILITY *K* TIMES RATED SHORT-CIRCUIT CURRENT kA, rms	3-SECOND SHORT-TIME CURRENT CARRY-ING CA-PABILITY *K* TIMES RATED SHORT-CIRCUIT CURRENT kA, rms	CLOSING AND LATCHING CA-PABILITY 1.6*K* TIMES RATED SHORT-CIRCUIT CURRENT kA, rms
COL 9	COL 10	COL 11	COL 12	COL 13	COL 14
5	2	5.8	24	24	38
5	2	12	23	23	37
5	2	12	24	24	38
5	2	23	36	36	58
5	2	40	21	21	33
5	2	60	23	23	37
3	1	121	20	20	32
3	1	121	40	40	64
3	1	121	40	40	64
3	1	121	63	63	101
3	1	121	40	40	64
3	1	121	63	63	101
3	1	145	20	20	32
3	1	145	40	40	64
3	1	145	40	40	64
3	1	145	63	63	101
3	1	145	80	80	128
3	1	145	40	40	64
3	1	145	63	63	101
3	1	145	80	80	128
3	1	169	16	16	26
3	1	169	31.5	31.5	50
3	1	169	40	40	64
3	1	169	50	50	80
3	1	242	31.5	31.5	50
3	1	242	31.5	31.5	50
3	1	242	31.5	31.5	50
3	1	242	40	40	64
3	1	242	40	40	64
3	1	242	63	63	101
3	1	362	40	40	64
3	1	362	40	40	64
2	1	550	40	40	64
2	1	550	40	40	64
2	1	765	40	40	64
2	1	765	40	40	64

instant the trip coil is energized to the instant the fault current is cleared.

Rated interrupting MVA: For a three-phase circuit breaker, this is $\sqrt{3}$ times the rated maximum voltage in kV times the rated short-circuit current in kA. It is more common to work with current and voltage ratings than with MVA rating.

As an example, the symmetrical interrupting capability of the 69-kV class breaker listed in Table 8.10 is plotted versus operating voltage in Figure 8.12. As shown, the symmetrical interrupting capability increases from its rated short-circuit current $I = 19$ kA at rated maximum voltage $V_{max} = 72.5$ kV up to $I_{max} = KI = (1.21)\,(19) = 23$ kA at an operating voltage $V_{min} = V_{max}/K = 72.5/1.21 = 60$ kV. At operating voltages V between V_{min} and V_{max}, the symmetrical interrupting capability is $I \times V_{max}/V = 1378/V$ kA. At operating voltages below V_{min}, the symmetrical interrupting capability remains at $I_{max} = 23$ kA.

Breakers of the 115-kV class and higher have a voltage range factor $K = 1.0$; that is, their symmetrical interrupting current capability remains constant.

Figure 8.12

Symmetrical interrupting capability of a 69-kV class breaker

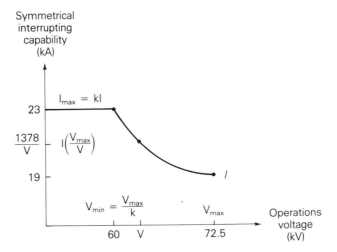

EXAMPLE 8.9   **Circuit breaker selection**

The calculated symmetrical fault current is 17 kA at a three-phase bus where the operating voltage is 64 kV. The X/R ratio at the bus is unknown. Select a circuit breaker from Table 8.10 for this bus.

Solution   The 69-kV-class breaker has a symmetrical interrupting capability

$$I\left(\frac{V_{max}}{V}\right) = 19\left(\frac{72.5}{64}\right) = 21.5\ \text{kA}$$

at the operating voltage $V = 64$ kV. The calculated symmetrical fault current, 17 kA, is less than 80% of this capability (less than $0.80 \times 21.5 = 17.2$ kA), which is a requirement when X/R is unknown. Therefore, we select the 69-kV-class breaker from Table 8.10.   ∎

Fuses

Figure 8.13(a) shows a cutaway view of a fuse, which is one of the simplest overcurrent devices. The fuse consists of a metal "fusible" link or links encapsulated in a tube, packed in filler material, and connected to contact terminals. Silver is a typical link metal, and sand is a typical filler material.

Figure 8.13

Typical fuse

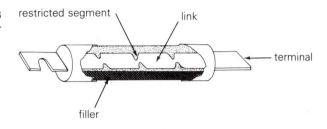

(a) Cutaway view

(b) The link melts and an arc is established under sustained overload current

(c) The "open" link after clearing the overload current.

During normal operation, when the fuse is operating below its continuous current rating, the electrical resistance of the link is so low that it simply acts as a conductor. If an overload current from one to about six times its continuous current rating occurs and persists for more than a short interval of time, the temperature of the link eventually reaches a level that causes a restricted segment of the link to melt. As shown in Figure 8.13(b), a gap is then formed and an electric arc is established. As the arc causes the link metal to burn back, the gap width increases. The resistance of the arc eventually reaches such a high level that the arc cannot be sustained and it is extinguished, as in Figure 8.13(c). The current flow within the fuse is then completely cut off.

If the fuse is subjected to fault currents higher than about six times its continuous current rating, several restricted segments melt simultaneously, resulting in rapid arc suppression and fault clearing. Arc suppression is accelerated by the filler material in the fuse.

Many modern fuses are current limiting. As shown in Figure 8.14, a current-limiting fuse has such a high speed of response that it cuts off a high fault current in less than a half cycle—before it can build up to its full peak value. By limiting fault currents, these fuses permit the use of motors, transformers, conductors, and bus structures that could not otherwise withstand the destructive forces of high fault currents.

Figure 8.14

Operation of a current-limiting fuse

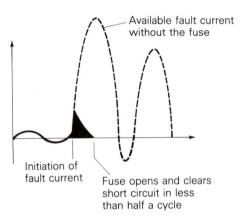

Fuse specification is normally based on the following four factors.

1. *Voltage rating.* This rms voltage determines the ability of a fuse to suppress the internal arc that occurs after the fuse link melts. A blown fuse should be able to withstand its voltage rating. Most low-voltage fuses have 250- or 600-V ratings. Ratings of medium-voltage fuses range from 2.4 to 34.5 kV.

2. *Continuous current rating.* The fuse should carry this rms current indefinitely, without melting and clearing.

3. *Interrupting current rating.* This is the largest rms asymmetrical current that the fuse can safely interrupt. Most modern, low-voltage current-limiting fuses have a 200-kA interrupting rating. Standard interrupting ratings for medium-voltage current-limiting fuses include 65, 80, and 100 kA.

4. *Time response.* The melting and clearing time of a fuse depends on the magnitude of the overcurrent or fault current, and is usually specified by a "time–current" curve. Figure 8.15 shows the time–current curve of a 15.5-kV, 100-A (continuous) current-limiting fuse. As shown, the fuse link melts within 2 s and clears within 5 s for a 500-A current. For a 5-kA current, the fuse link melts in less than 0.01 s and clears within 0.015 s.

It is usually a simple matter to coordinate fuses in a power circuit such that

Figure 8.15

Time–current curves for a 15.5-kV, 100-A current-limiting fuse

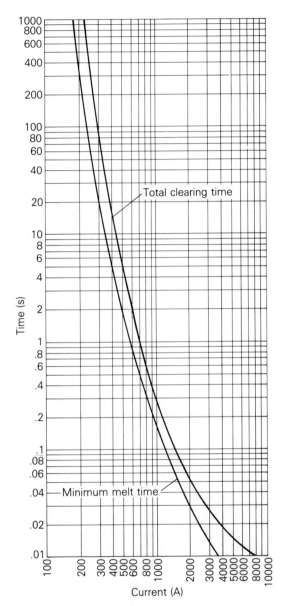

only the fuse closest to the fault opens the circuit. In a radial circuit, fuses with larger continuous current ratings are located closer to the source, such that the fuse closest to the fault clears before other, upstream fuses melt.

Fuses are inexpensive, fast operating, easily coordinated, and reliable, and they do not require protective relays or instrument transformers. Their chief disadvantage is that the fuse or the fuse link must be manually replaced after it melts. They are basically one-shot devices that are, for example, incapable of high-speed reclosing.

PROBLEMS

Section 8.1

8.1 In the circuit of Figure 8.1, $V = 277$ volts, $L = 3\,mH$, $R = 0.5\,\Omega$, and $\omega = 2\pi60\,rad/s$. Determine (a) the rms symmetrical fault current; (b) the rms asymmetrical fault current at the instant the switch closes, assuming maximum dc offset; (c) the rms asymmetrical fault current 5 cycles after the switch closes, assuming maximum dc offset; (d) the dc offset as a function of time if the switch closes when the instantaneous source voltage is 300 volts.

8.2 Repeat Example 8.1 with $V = 4\,kV$, $X = 3\,\Omega$, and $R = 1\,\Omega$.

Section 8.2

8.3 A 1300-MVA 20-kV, 60-Hz three-phase generator is connected through a 1300-MVA 20-kV Δ/500-kV Y transformer to a 500-kV circuit breaker and a 500-kV transmission line. The generator reactances are $X_d'' = 0.17$, $X_d' = 0.30$, and $X_d = 1.5$ per unit, and its time constants are $T_d'' = 0.05$, $T_d' = 1.0$, and $T_A = 0.10\,s$. The transformer series reactance is 0.10 per unit; transformer losses and exciting current are neglected. A three-phase short-circuit occurs on the line side of the circuit breaker when the generator is operated at rated terminal voltage and at no-load. The breaker interrupts the fault 3 cycles after fault inception. Determine (a) the subtransient current through the breaker in per-unit and in kA rms; and (b) the rms asymmetrical fault current the breaker interrupts, assuming maximum dc offset. Neglect the effect of the transformer on the time constants.

8.4 For Problem 8.3, determine (a) the instantaneous symmetrical fault current in kA in phase a of the generator as a function of time, assuming maximum dc offset occurs in this generator phase; and (b) the maximum dc offset current in kA as a function of time that can occur in any one generator phase.

Section 8.3

8.5 Recalculate the subtransient current through the breaker in Problem 8.3 if the generator is initially delivering rated MVA at 0.80 p.f. lagging and at rated terminal voltage.

8.6 Solve Example 8.3, parts (a) and (c) without using the superposition principle. First calculate the internal machine voltages E_g'' and E_m'', using the prefault load current. Then determine the subtransient fault, generator, and motor currents directly from Figure 8.4(a). Compare your answers with those of Example 8.3.

8.7 Equipment ratings for the four-bus power system shown in Figure 8.16 are as follows:

> Generator G1:   500 MVA, 13.8 kV, $X'' = 0.20$ per unit
> Generator G2:   750 MVA, 18 kV, $X'' = 0.18$ per unit
> Generator G3:   1000 MVA, 20 kV, $X'' = 0.17$ per unit
> Transformer T1:   500 MVA, 13.8 Δ/500 Y kV, $X = 0.12$ per unit
> Transformer T2:   750 MVA, 18 Δ/500 Y kV, $X = 0.10$ per unit
> Transformer T3:   1000 MVA, 20 Δ/500 Y kV, $X = 0.10$ per unit
> Each 500-kV line: $X_1 = 50\,\Omega$

A three-phase short circuit occurs at bus 1, where the prefault voltage is 525 kV. Prefault load current is neglected. Draw the positive-sequence reactance diagram in per-unit on a 1000-MVA, 20-kV base in the zone of generator G3. Determine (a) the

Figure 8.16

Problems 8.7, 8.8, 8.12, 8.18

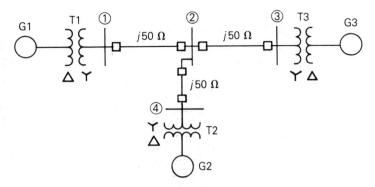

Thévenin reactance in per-unit at the fault, (b) the subtransient fault current in per-unit and in kA rms, and (c) contributions to the fault current from generator G1 and from line 1–2.

8.8   For the power system given in Problem 8.7, a three-phase short circuit occurs at bus 2, where the prefault voltage is 525 kV. Prefault load current is neglected. Determine the (a) Thévenin equivalent at the fault, (b) subtransient fault current in per-unit and in kA rms, and (c) contributions to the fault from lines 1–2, 2–3, and 2–4.

8.9   Equipment ratings for the five-bus power system shown in Figure 8.17 are as follows:

> Generator G1:   50 MVA, 12 kV, $X'' = 0.2$ per unit
> Generator G2:   100 MVA, 15 kV, $X'' = 0.2$ per unit
> Transformer T1:   50 MVA, 10 kV Y/138 kV Y, $X = 0.10$ per unit
> Transformer T2:   100 MVA, 15 kV Δ/138 kV Y, $X = 0.10$ per unit
> Each 138-kV line: $X_1 = 40\,\Omega$

Figure 8.17

Problems 8.9, 8.10, 8.13, 8.19

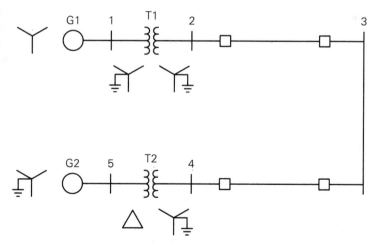

A three-phase short circuit occurs at bus 5, where the prefault voltage is 15 kV. Prefault load current is neglected. (a) Draw the positive-sequence reactance diagram in per-unit on a 100-MVA, 15-kV base in the zone of generator G2. Determine: (b) the Thévenin equivalent at the fault, (c) the subtransient fault current in per-unit and in kA rms, and (d) contributions to the fault from generator G2 and from transformer T2.

8.10   For the power system given in Problem 8.9, a three-phase short circuit occurs at bus 4, where the prefault voltage is 138 kV. Prefault load current is neglected. Determine: (a) the Thévenin equivalent at the fault, (b) the subtransient fault current in per-unit and in kA rms, and (c) contributions to the fault from transformer T2 and from line 3–4.

Section 8.4

8.11   The bus impedance matrix for a three-bus power system is

$$Z_{bus} = j \begin{bmatrix} 0.12 & 0.08 & 0.04 \\ 0.08 & 0.12 & 0.06 \\ 0.04 & 0.06 & 0.08 \end{bmatrix} \text{ per unit}$$

where subtransient reactances were used to compute Z_{bus}. Prefault voltage is 1.0 per unit and prefault current is neglected. (a) Draw the bus impedance matrix equivalent circuit (rake equivalent). Identify the per-unit self- and mutual impedances as well as the prefault voltage in the circuit. (b) A three-phase short circuit occurs at bus 2. Determine the subtransient fault current and the voltages at buses 1, 2, and 3 during the fault.

8.12   Determine Y_{bus} in per-unit for the circuit in Problem 8.7. Then invert Y_{bus} to obtain Z_{bus}.

8.13   Determine Y_{bus} in per-unit for the circuit in Problem 8.9. Then invert Y_{bus} to obtain Z_{bus}.

Section 8.5

8.14   For the three-bus power system whose Z_{bus} is given in Problem 8.11, a new impedance $Z_b = j0.7$ per unit is added from old bus 2 to new bus 4. The new impedance element is not mutually coupled to any of the existing impedance elements. Determine the new bus impedance matrix.

8.15   Repeat Problem 8.14 if Z_b is added from old bus 2 to the reference bus.

8.16   Repeat Problem 8.14 if Z_b is added from old bus 2 to old bus 3.

8.17   Repeat Problem 8.14 if Z_b is added from new bus 4 to the reference bus.

8.18   Determine Z_{bus} for the circuit given in Problem 8.7, using the one-step-at-a-time method.

8.19   Determine Z_{bus} for the circuit given in Problem 8.9, using the one-step-at-a-time method.

8.20   Repeat Problem 8.19 if a new impedance $Z_b = j0.4$ per unit is added from bus 2 to bus 4.

Section 8.6

8.21   Using the SYMMETRICAL SHORT CIRCUITS program for the circuit given in Problem 8.7, compute: (a) Z_{bus} and (b) the fault current and bus voltages during the fault for a three-phase short circuit at each bus.

8.22   Rework Problems 8.14, 8.15, 8.16, and 8.17 using the SYMMETRICAL SHORT CIRCUITS program. Refer to SAMPLE RUN 8.2 in the software manual to construct Z_{bus}.

8.23   Using the SYMMETRICAL SHORT CIRCUITS program for the circuit given in

Problem 8.9, compute: (a) Z_{bus} and (b) the fault current and bus voltages during the fault for a three-phase short circuit at each bus.

8.24 Rework Problem 8.20 using the SYMMETRICAL SHORT CIRCUITS program.

8.25 Run the SYMMETRICAL SHORT CIRCUITS program for the power system given in Example 8.7 with a new line installed between buses 4 and 5. This line, whose series impedance is $Z_b = j0.4$ per unit, is not mutually coupled to any other line.

8.26 Starting with the Z_{bus} matrix computed in Problem 8.25, remove the new line by adding an impedance $Z_b = -j0.4$ per unit between buses 4 and 5. Verify that the computed Z_{bus} is the same as that shown in Table 8.6.

Section 8.7

8.27 A three-phase circuit breaker has a 15.5-kV rated maximum voltage, 9.0-kA rated short-circuit current, and a 2.67-rated voltage range factor. (a) Determine the symmetrical interrupting capability at 10-kV and at 5-kV operating voltages. (b) Can this breaker be safely installed at a three-phase bus where the symmetrical fault current is 10 kA, the operating voltage is 13.8 kV, and the (X/R) ratio is 12?

8.28 A 500-kV three-phase transmission line has a 2.2 kA continuous current rating and a 2.5 kA maximum short-time overload rating, with a 525 kV maximum operating voltage. Maximum symmetrical fault current on the line is 30 kA. Select a circuit breaker for this line from Table 8.10.

CASE STUDY QUESTIONS

A. Why are arcing (high-impedance) faults more difficult to detect than low-impedance faults?

B. What methods are available to prevent the destructive effects of arcing faults from occurring?

References

1. Westinghouse Electric Corporation, *Electrical Transmission and Distribution Reference Book*, 4th ed. (East Pittsburgh, PA, 1964).

2. E. W. Kimbark, *Power System Stability, Synchronous Machines*, vol. 3 (New York: Wiley, 1956).

3. A. E. Fitzgerald, C. Kingsley, and S. Umans, *Electric Machinery*, 4th ed. (New York: McGraw-Hill, 1985).

4. M. S. Sarma, *Electric Machines* (Dubuque, IA: Brown, 1985).

5. J. R. Neuenswander, *Modern Power Systems* (New York: Intext Educational Publishers, 1971).

6. H. E. Brown, *Solution of Large Networks by Matrix Methods* (New York: Wiley, (1975).

7. G. N. Lester, "High Voltage Circuit Breaker Standards in the USA—Past, Present and Future," *IEEE Transactions PAS*, vol. PAS-93 (1974): pp. 590-600.

8. W. D. Stevenson, Jr., *Elements of Power System Analysis*, 4th ed. (New York: McGraw-Hill, 1982).

9. C. A. Gross, *Power System Analysis* (New York: Wiley, 1979).

10. *Application Guide for AC High-Voltage Circuit Breakers Rated on a Symmetrical Current Basis*, ANSI C 37.010 (New York: American National Standards Institute, 1972).

11. F. Shields, "The Problem of Arcing Faults in Low-Voltage Power Distribution Systems," *IEEE Transactions on Industry and General Applications*, vol. IGA-3, no. 1, (January/February 1967), pp. 15-25.

CHAPTER 9

UNSYMMETRICAL FAULTS

500/220 kV substation with SF6 gas-insulated switchgear and conventional oil-filled transformers
(Courtesy of Southern California Edison)

Short circuits occur in three-phase power systems as follows, in order of frequency of occurrence: single line-to-ground, line-to-line, double line-to-ground, and balanced three-phase faults. The path of the fault current may have either either zero impedance, which is called a *bolted* short circuit, or nonzero impedance. Other types of faults include one-conductor-open and two-conductors-open, which can occur when conductors break or when one or two phases of a circuit breaker inadvertently open.

Although the three-phase short circuit occurs the least, it was considered first, in Chapter 8, because of its simplicity. When a balanced three-phase fault occurs in a balanced three-phase system, there is only positive-sequence fault current; the zero-, positive-, and negative-sequence networks are completely uncoupled.

When an unsymmetrical fault occurs in an otherwise balanced system, the sequence networks are interconnected only at the fault location. As such, the computation of fault currents is greatly simplified by the use of sequence networks. Readers may wish to briefly review Chapter 3, "Symmetrical Components," before proceeding with unsymmetrical faults.

As in the case of balanced three-phase faults, unsymmetrical faults have two components of fault current; an ac or symmetrical component, including subtransient, transient, and steady-state currents; and a dc component. The simplified E/X method for breaker selection described in Section 8.7 is also applicable to unsymmetrical faults. The dc offset current need not be considered unless it is too large—for example, when the X/R ratio is too large.

We begin this chapter by using the per-unit zero-, positive-, and negative-sequence networks to represent a three-phase system. Also, certain assumptions are made to simplify fault-current calculations, and the balanced three-phase fault is briefly reviewed. Single line-to-ground, line-to-line, and double line-to-ground faults are presented in Sections 9.2, 9.3, and 9.4. The use of the positive-sequence bus impedance matrix for three-phase fault calculations in Section 8.4 is extended in Section 9.5 to unsymmetrical fault calculations by considering a bus impedance matrix for each sequence network. In Section 9.6 a personal computer program, SHORT CIRCUITS, which is based on the use of bus impedance matrices, is presented; it computes symmetrical fault currents for both three-phase and unsymmetrical faults. This program may be used in power-system design to select, set, and coordinate protective equipment.

CASE STUDY When short circuits are not interrupted promptly, electrical fires and explosions can occur. In order to minimize the probability of electrical fire and explosion, the following are recommended:

Careful design of electric power-system layouts

Quality equipment installation

Power-system protection that provides rapid detection and isolation of faults (see Chapter 10)

Automatic fire-suppression systems

Formal maintenance programs and inspection intervals

Repair or retirement of damaged or decrepit equipment

The following article describes recent incidents at three U.S. utilities [8].

Fires at U.S. Utilities

GLENN ZORPETTE

Electrical fires in substations were the cause of three major midsummer power outages in the United States, two on Chicago's West Side and one in New York City's downtown financial district. In Chicago, the trouble began Saturday night, July 28, with a fire in switch house No. 1 at the Commonwealth Edison Co.'s Crawford substation, according to spokesman Gary Wald.

Some 40,000 residents of Chicago's West Side lost electricity. About 25,000 had service restored within a day or so and the rest, within three days. However, as part of the restoration, Commonwealth Edison installed a temporary line configuration around the Crawford substation. But when a second fire broke out on Aug. 5 in a different, nearby substation, some of the protective systems that would have isolated that fire were inoperable because of that configuration. Thus, what would have been a minor mishap resulted in a one-day loss of power to 25,000 customers—the same 25,000 whose electricity was restored first after the Crawford fire.

The New York outage began around midday on

Reprinted with permission. Glenn Zorpette, IEEE Spectrum, 28, *1 (Jan. 1991), p. 64. Copyright © 1991 IEEE.*

Aug. 13, after an electrical fire broke out in switching equipment at Consolidated Edison's Seaport substation, a point of entry into Manhattan for five 138-kilovolt transmission lines. To interrupt the flow of energy to the fire, Edison had to disconnect the five lines, which cut power to four networks in downtown Manhattan, according to Con Ed spokeswoman Martha Liipfert.

Power was restored to three of the networks within about five hours, but the fourth network, Fulton—which carried electricity to about 2400 separate residences and 815 businesses—was out until Aug. 21. Liipfert said much of the equipment in the Seaport substation will have to be replaced, at an estimated cost of about $25 million.

Mounting concern about underground electrical vaults in some areas was tragically validated by an explosion in Pasadena, Calif., that killed three city workers in a vault. Partly in response to the explosion, the California Public Utilities Commission adopted new regulations last Nov. 21 requiring that utilities in the state set up formal maintenance programs, inspection intervals, and guidelines for rejecting decrepit or inferior equipment. "They have to maintain a paper trail, and we as a commission will do inspections of underground vaults and review their records to make sure they're maintaining their vaults and equipment in good order," said Russ Copeland, head of the commission's utility safety branch.

SECTION 9.1

SYSTEM REPRESENTATION

A three-phase power system is represented by its sequence networks in this chapter. The zero-, positive-, and negative-sequence networks of system components—generators, motors, transformers, and transmission lines—as developed in Chapters 3, 4, and 5 can be used to construct system zero-, positive-, and negative-sequence networks. We make the following assumptions:

1. The power system operates under balanced steady-state conditions before the fault occurs. Thus the zero-, positive-, and negative-

sequence networks are uncoupled before the fault occurs. During unsymmetrical faults they are interconnected only at the fault location.

2. Prefault load current is neglected. Because of this, the positive-sequence internal voltages of all machines are equal to the prefault voltage V_F. Therefore, the prefault voltage at each bus in the positive-sequence network equals V_F.

3. Transformer winding resistances and shunt admittances are neglected.

4. Transmission-line series resistances and shunt admittances are neglected.

5. Synchronous machine armature resistance, saliency, and saturation are neglected.

6. All nonrotating impedance loads are neglected.

7. Induction motors are either neglected (especially for motors rated 50 hp or less) or represented in the same manner as synchronous machines.

Note that these assumptions are made for simplicity in this text, and in practice should not be made for all cases. For example, in primary and secondary distribution systems, prefault currents may be in some cases comparable to short-circuit currents, and in other cases line resistances may significantly reduce fault currents.

Although fault currents as well as contributions to fault currents on the fault side of Δ–Y transformers are not affected by Δ–Y phase shifts, contributions to the fault from the other side of such transformers are affected by Δ–Y phase shifts for unsymmetrical faults. Therefore, we include Δ–Y phase-shift effects in this chapter.

We consider faults at the general three-phase bus shown in Figure 9.1. Terminals *abc*, denoted the *fault terminals*, are brought out in order to make external connections that represent faults. Before a fault occurs, the currents I_a, I_b, and I_c are zero.

Figure 9.2(a) shows general sequence networks as viewed from the fault

Figure 9.1   Three-phase power system

General three-phase bus

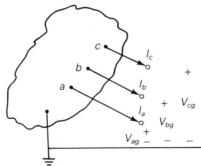

Figure 9.2

Sequence networks at a general three-phase bus in a balanced system

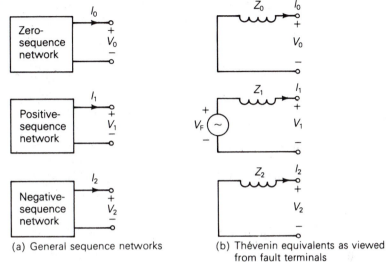

(a) General sequence networks   (b) Thévenin equivalents as viewed from fault terminals

terminals. Since the prefault system is balanced, these zero-, positive-, and negative-sequence networks are uncoupled. Also, the sequence components of the fault currents, I_0, I_1, and I_2, are zero before a fault occurs. The general sequence networks in Figure 9.2(a) are reduced to their Thévenin equivalents as viewed from the fault terminals in Figure 9.2(b). Each sequence network has a Thévenin equivalent impedance. Also, the positive-sequence network has a Thévenin equivalent voltage source, which equals the prefault voltage V_F.

EXAMPLE 9.1   **Power-system sequence networks and their Thévenin equivalents**

A single-line diagram of the power system considered in Example 8.3 is shown in Figure 9.3, where negative- and zero-sequence reactances are also given. The neutrals of the generator and Δ–Y transformers are solidly grounded. The motor neutral is grounded through a reactance $X_n = 0.05$ per unit on the motor base. (a) Draw the per-unit zero-, positive-, and negative-sequence networks on a 100-MVA, 13.8-kV base in the zone of the generator. (b) Reduce the sequence networks to their Thévenin equivalents, as viewed from

Figure 9.3

Single-line diagram for Example 9.1

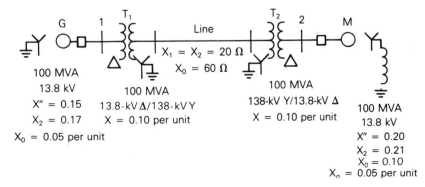

bus 2. Prefault voltage is $V_F = 1.05\underline{/0°}$ per unit. Prefault load current and $\Delta-Y$ transformer phase shift are neglected.

Solution **a.** The sequence networks are shown in Figure 9.4. The positive-sequence network is the same as that shown in Figure 8.4(a). The negative-sequence network is similar to the positive-sequence network, except that there are no sources, and negative-sequence machine reactances are shown. $\Delta-Y$ phase shifts are omitted from the positive- and negative-sequence networks for this example. In the zero-sequence network the zero-sequence generator, motor, and transmission-line reactances are shown. Since the motor neutral is grounded through a neutral reactance X_n, $3X_n$ is included in the zero-sequence motor circuit. Also, the zero-sequence $\Delta-Y$ transformer models are taken from Figure 4.17.

Figure 9.4

Sequence networks for
Example 9.1

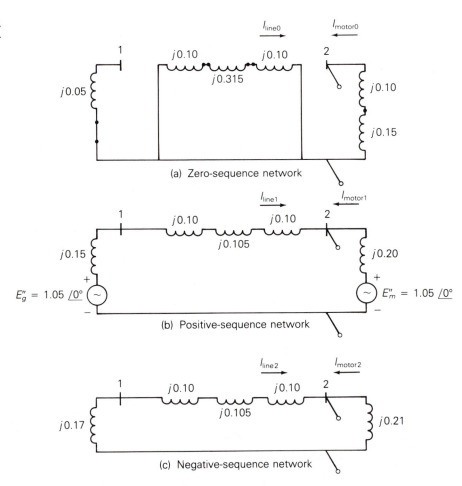

(a) Zero-sequence network

(b) Positive-sequence network

(c) Negative-sequence network

b. Figure 9.5 shows the sequence networks reduced to their Thévenin equivalents, as viewed from bus 2. For the positive-sequence equivalent, the Thévenin voltage source is the prefault voltage $V_F = 1.05\underline{/0°}$ per unit. From

Figure 9.5

Thévenin equivalents of sequence networks for Example 9.1

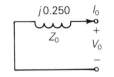

(a) Zero-sequence network

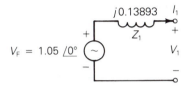

(b) Positive-sequence network

(c) Negative-sequence network

Figure 9.4, the positive-sequence Thévenin impedance at bus 2 is the motor impedance $j0.20$, as seen to the right of bus 2, in parallel with $j(0.15 + 0.10 + 0.105 + 0.10) = j0.455$, as seen to the left; the parallel combination is $j0.20//j0.455 = j0.13893$ per unit. Similarly, the negative-sequence Thévenin impedance is $j0.21//j(0.17 + 0.10 + 0.105 + 0.10) = j0.21//j0.475 = j0.14562$ per unit. In the zero-sequence network of Figure 9.4, the Thévenin impedance at bus 2 consists only of $j(0.10 + 0.15) = j0.25$ per unit, as seen to the right of bus 2; due to the Δ connection of transformer T_2, the zero-sequence network looking to the left of bus 2 is open. ∎

Recall that for three-phase faults, as considered in Chapter 8, the fault currents are balanced and have only a positive-sequence component. Therefore we work only with the positive-sequence network when calculating three-phase fault currents.

EXAMPLE 9.2 **Three-phase short-circuit calculations using sequence networks**

Calculate the per-unit subtransient fault currents in phases a, b, and c for a bolted three-phase-to-ground short circuit at bus 2 in Example 9.1.

Solution The terminals of the positive-sequence network in Figure 9.5(b) are shorted, as shown in Figure 9.6. The positive-sequence fault current is

$$I_1 = \frac{V_F}{Z_1} = \frac{1.05/\underline{0°}}{j0.13893} = -j7.558 \quad \text{per unit}$$

which is the same result as obtained in part (c) of Example 8.4. Note that since

Figure 9.6

Example 9.2: Bolted three-phase-to-ground fault at bus 2

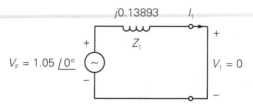

subtransient machine reactances are used in Figures 9.4 to 9.6, the current calculated above is the positive-sequence subtransient fault current at bus 2. Also, the zero-sequence current I_0 and negative-sequence current I_2 are both zero. Therefore, the subtransient fault currents in each phase are, from (3.1.16),

$$
\begin{bmatrix} I_a'' \\ I_b'' \\ I_c'' \end{bmatrix} = \begin{bmatrix} 1 & 1 & 1 \\ 1 & a^2 & a \\ 1 & a & a^2 \end{bmatrix} \begin{bmatrix} 0 \\ -j7.558 \\ 0 \end{bmatrix} = \begin{bmatrix} 7.558\underline{/-90^\circ} \\ 7.558\underline{/150^\circ} \\ 7.558\underline{/30^\circ} \end{bmatrix} \quad \text{per unit}
$$ ∎

The sequence components of the line-to-ground voltages at the fault terminals are, from Figure 9.2(b),

$$
\begin{bmatrix} V_0 \\ V_1 \\ V_2 \end{bmatrix} = \begin{bmatrix} 0 \\ V_F \\ 0 \end{bmatrix} - \begin{bmatrix} Z_0 & 0 & 0 \\ 0 & Z_1 & 0 \\ 0 & 0 & Z_2 \end{bmatrix} \begin{bmatrix} I_0 \\ I_1 \\ I_2 \end{bmatrix} \tag{9.1.1}
$$

During a bolted three-phase fault, the sequence fault currents are $I_0 = I_2 = 0$ and $I_1 = V_F/Z_1$; therefore, from (9.1.1), the sequence fault voltages are $V_0 = V_1 = V_2 = 0$, which must be true since $V_{ag} = V_{bg} = V_{cg} = 0$. However, fault voltages need not be zero during unsymmetrical faults, which are considered next.

SECTION 9.2

SINGLE LINE-TO-GROUND FAULT

Consider a single line-to-ground fault from phase a to ground at the general three-phase bus shown in Figure 9.7(a). For generality, we include a fault impedance Z_F. In the case of a bolted fault, $Z_F = 0$, whereas for an arcing fault, Z_F is the arc impedance. In the case of a transmission-line insulator flashover, Z_F includes the total fault impedance between the line and ground,

Figure 9.7

Single line-to-ground fault

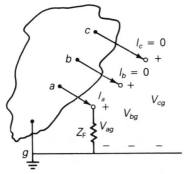

(a) General three-phase bus

Fault conditions
in phase domain:

$V_{ag} = Z_F I_a$

$I_b = I_c = 0$

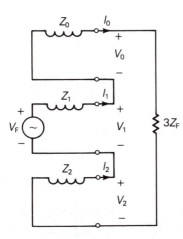

(b) Interconnected sequence networks

Fault conditions
in sequence domain:

$I_0 = I_1 = I_2$

$(V_0 + V_1 + V_2) = 3Z_F I_1$

including the impedances of the arc and the transmission tower, as well as the tower footing if there are no neutral wires.

The relations to be derived here apply only to a single line-to-ground fault on phase *a*. However, since any of the three phases can be arbitrarily labeled phase *a*, we do not consider single line-to-ground faults on other phases.

From Figure 9.7(a):

$$\begin{array}{ll} \text{Fault conditions in phase domain} \\ \text{Single line-to-ground fault} \end{array} \left\} \begin{array}{l} I_b = I_c = 0 \qquad\qquad (9.2.1) \\ V_{ag} = Z_F I_a \qquad\qquad (9.2.2) \end{array} \right.$$

We now transform (9.2.1) and (9.2.2) to the sequence domain. Using (9.2.1) in (3.1.19),

$$\begin{bmatrix} I_0 \\ I_1 \\ I_2 \end{bmatrix} = \frac{1}{3} \begin{bmatrix} 1 & 1 & 1 \\ 1 & a & a^2 \\ 1 & a^2 & a \end{bmatrix} \begin{bmatrix} I_a \\ 0 \\ 0 \end{bmatrix} = \frac{1}{3} \begin{bmatrix} I_a \\ I_a \\ I_a \end{bmatrix} \qquad (9.2.3)$$

Also, using (3.1.3) and (3.1.20) in (9.2.2),

$$(V_0 + V_1 + V_2) = Z_F(I_0 + I_1 + I_2) \tag{9.2.4}$$

From (9.2.3) and (9.2.4):

Fault conditions in sequence domain $\left.\begin{array}{l} I_0 = I_1 = I_2 \hspace{2cm} (9.2.5) \\ (V_0+V_1+V_2)=(3Z_F)I_1 \hspace{0.5cm} (9.2.6) \end{array}\right\}$
Single line-to-ground fault

Equations (9.2.5) and (9.2.6) can be satisfied by interconnecting the sequence networks in series at the fault terminals through the impedance $(3Z_F)$, as shown in Figure 9.7(b). From this figure, the sequence components of the fault currents are:

$$I_0 = I_1 = I_2 = \frac{V_F}{Z_0 + Z_1 + Z_2 + (3Z_F)} \tag{9.2.7}$$

Transforming (9.2.7) to the phase domain via (3.1.20),

$$I_a = I_0 + I_1 + I_2 = 3I_1 = \frac{3V_F}{Z_0 + Z_1 + Z_2 + (3Z_F)} \tag{9.2.8}$$

Note also from (3.1.21) and (3.1.22),

$$I_b = (I_0 + a^2 I_1 + a I_2) = (1 + a^2 + a)I_1 = 0 \tag{9.2.9}$$

$$I_c = (I_0 + a I_1 + a^2 I_2) = (1 + a + a^2)I_1 = 0 \tag{9.2.10}$$

These are obvious, since the single line-to-ground fault is on phase a, not phase b or c.

The sequence components of the line-to-ground voltages at the fault are determined from (9.1.1). The line-to-ground voltages at the fault can then be obtained by transforming the sequence voltages to the phase domain.

| EXAMPLE 9.3 | **Single line-to-ground short-circuit calculations using sequence networks** |

Calculate the subtransient fault current in per-unit and in kA for a bolted single line-to-ground short circuit from phase a to ground at bus 2 in Example 9.1. Also calculate the per-unit line-to-ground voltages at faulted bus 2.

Solution The zero-, positive-, and negative-sequence networks in Figure 9.5 are connected in series at the fault terminals, as shown in Figure 9.8. Since the short circuit is bolted, $Z_F = 0$. From (9.2.7), the sequence currents are:

$$I_0 = I_1 = I_2 = \frac{1.05\underline{/0°}}{j(0.25 + 0.13893 + 0.14562)}$$

$$= \frac{1.05}{j0.53455} = -j1.96427 \quad \text{per unit}$$

From (9.2.8), the subtransient fault current is

$$I_a'' = 3(-j1.96427) = -j5.8928 \quad \text{per unit}$$

Figure 9.8

Example 9.3: Single line-to-ground fault at bus 2

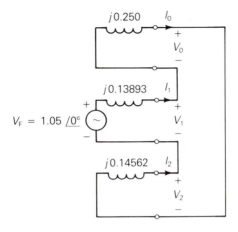

The base current at bus 2 is $100/(13.8\sqrt{3}) = 4.1837\,\text{kA}$. Therefore,

$$I_a'' = (-j5.8928)(4.1837) = 24.65\underline{/-90°}\quad \text{kA}$$

From (9.1.1), the sequence components of the voltages at the fault are

$$\begin{bmatrix} V_0 \\ V_1 \\ V_2 \end{bmatrix} = \begin{bmatrix} 0 \\ 1.05\underline{/0°} \\ 0 \end{bmatrix} - \begin{bmatrix} j0.25 & 0 & 0 \\ 0 & j0.13893 & 0 \\ 0 & 0 & j0.14562 \end{bmatrix}\begin{bmatrix} -j1.96427 \\ -j1.96427 \\ -j1.96427 \end{bmatrix}$$

$$= \begin{bmatrix} -0.49107 \\ 0.77710 \\ -0.28604 \end{bmatrix}\quad \text{per unit}$$

Transforming to the phase domain, the line-to-ground voltages at faulted bus 2 are

$$\begin{bmatrix} V_{ag} \\ V_{bg} \\ V_{cg} \end{bmatrix} = \begin{bmatrix} 1 & 1 & 1 \\ 1 & a^2 & a \\ 1 & a & a^2 \end{bmatrix}\begin{bmatrix} -0.49107 \\ 0.77710 \\ -0.28604 \end{bmatrix} = \begin{bmatrix} 0 \\ 1.179\underline{/231.3°} \\ 1.179\underline{/128.7°} \end{bmatrix}\quad \text{per unit}$$

Note that $V_{ag} = 0$, as specified by the fault conditions. Also $I_b'' = I_c'' = 0$. ∎

SECTION 9.3

LINE-TO-LINE FAULT

Consider a line-to-line fault from phase b to c, shown in Figure 9.9(a). Again, we include a fault impedance Z_F for generality. From Figure 9.9(a):

Fault conditions in phase domain } $I_a = 0$ (9.3.1)
Line-to-line fault   } $I_c = -I_b$ (9.3.2)
                                        $V_{bg} - V_{cg} = Z_F I_b$ (9.3.3)

Figure 9.9

Line-to-line fault

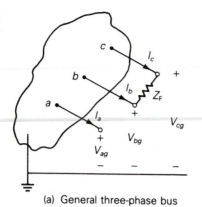

(a) General three-phase bus

Fault conditions
in phase domain:

$I_a = 0$

$I_c = -I_b$

$(V_{bg} - V_{cg}) = Z_F I_b$

Fault conditions
in sequence domain:

$I_0 = 0$

$I_2 = -I_1$

$(V_1 - V_2) = Z_F I_1$

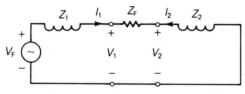

(b) Interconnected sequence networks

We transform (9.3.1)–(9.3.3) to the sequence domain. Using (9.3.1) and (9.3.2) in (3.1.19),

$$\begin{bmatrix} I_0 \\ I_1 \\ I_2 \end{bmatrix} = \frac{1}{3}\begin{bmatrix} 1 & 1 & 1 \\ 1 & a & a^2 \\ 1 & a^2 & a \end{bmatrix}\begin{bmatrix} 0 \\ I_b \\ -I_b \end{bmatrix} = \begin{bmatrix} 0 \\ \frac{1}{3}(a - a^2)I_b \\ \frac{1}{3}(a^2 - a)I_b \end{bmatrix} \tag{9.3.4}$$

Using (3.1.4), (3.1.5), and (3.1.21) in (9.3.3),

$$(V_0 + a^2 V_1 + a V_2) - (V_0 + a V_1 + a^2 V_2) = Z_F(I_0 + a^2 I_1 + a I_2) \tag{9.3.5}$$

Noting from (9.3.4) that $I_0 = 0$ and $I_2 = -I_1$, (9.3.5) simplifies to

$$(a^2 - a)V_1 - (a^2 - a)V_2 = Z_F(a^2 - a)I_1$$

or

$$V_1 - V_2 = Z_F I_1 \tag{9.3.6}$$

Therefore, from (9.3.4) and (9.3.6):

Fault conditions in sequence domain $\Big\}$ $I_0 = 0$ (9.3.7)

Line-to-line fault $I_2 = -I_1$ (9.3.8)

$$V_1 - V_2 = Z_F I_1 \qquad (9.3.9)$$

Equations (9.3.7)–(9.3.9) are satisfied by connecting the positive- and negative-sequence networks in parallel at the fault terminals through the fault impedance Z_F, as shown in Figure 9.9(b). From this figure, the fault currents are:

$$I_1 = -I_2 = \frac{V_F}{(Z_1 + Z_2 + Z_F)} \qquad I_0 = 0 \qquad (9.3.10)$$

Transforming (9.3.10) to the phase domain and using the identity $(a^2 - a) = -j\sqrt{3}$, the fault current in phase b is

$$I_b = I_0 + a^2 I_1 + a I_2 = (a^2 - a) I_1$$

$$= -j\sqrt{3} I_1 = \frac{-j\sqrt{3} V_F}{(Z_1 + Z_2 + Z_F)} \qquad (9.3.11)$$

Note also from (3.1.20) and (3.1.22) that

$$I_a = I_0 + I_1 + I_2 = 0 \qquad (9.3.12)$$

and

$$I_c = I_0 + a I_1 + a^2 I_2 = (a - a^2) I_1 = -I_b \qquad (9.3.13)$$

which verify the fault conditions given by (9.3.1) and (9.3.2). The sequence components of the line-to-ground voltages at the fault are given by (9.1.1).

EXAMPLE 9.4	**Line-to-line short circuit calculations using sequence networks**

Calculate the subtransient fault current in per-unit and in kA for a bolted line-to-line fault from phase b to c at bus 2 in Example 9.1.

Solution The positive- and negative-sequence networks in Figure 9.5 are connected in parallel at the fault terminals, as shown in Figure 9.10. From (9.3.10) with $Z_F = 0$, the sequence fault currents are

$$I_1 = -I_2 = \frac{1.05\underline{/0°}}{j(0.13893 + 0.14562)} = 3.690\underline{/-90°}$$

$$I_0 = 0$$

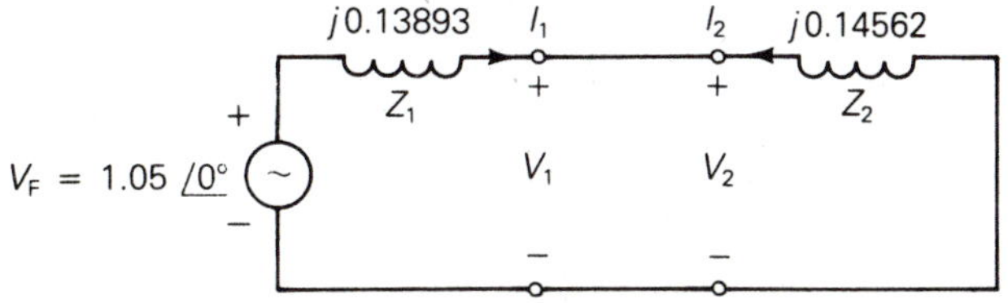

Figure 9.10

Example 9.4: Line-to-line fault at bus 2

From (9.3.11), the subtransient fault current in phase b is

$$I_b'' = (-j\sqrt{3})(3.690\underline{/-90°}) = -6.391 = 6.391\underline{/180°} \quad \text{per unit}$$

Using 4.1837 kA as the base current at bus 2,

$$I_b'' = (6.391\underline{/180°})(4.1837) = 26.74\underline{/180°} \quad \text{kA}$$

Also, from (9.3.12) and (9.3.13),

$$I_a'' = 0 \qquad I_c'' = 26.74\underline{/0°} \quad \text{kA} \qquad \blacksquare$$

DOUBLE LINE-TO-GROUND FAULT

A double line-to-ground fault from phase b to phase c through fault impedance Z_F to ground is shown in Figure 9.11(a). From this figure:

Fault conditions in the phase domain $\left.\right\}$ $I_a = 0$ (9.4.1)

Double line-to-ground fault $\left.\right\}$ $V_{cg} = V_{bg}$ (9.4.2)

$V_{bg} = Z_F(I_b + I_c)$ (9.4.3)

Figure 9.11

Double line-to-ground fault

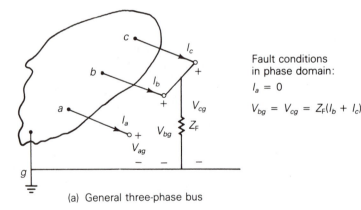

Fault conditions
in phase domain:

$I_a = 0$

$V_{bg} = V_{cg} = Z_F(I_b + I_c)$

(a) General three-phase bus

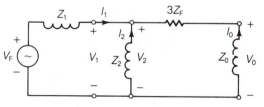

Fault conditions
in sequence domain:

$I_0 + I_1 + I_2 = 0$

$V_0 - V_1 = (3Z_F)I_0$

$V_1 = V_2$

(b) Interconnected sequence networks

Transforming (9.4.1) to the sequence domain via (3.1.20),

$$I_0 + I_1 + I_2 = 0 \tag{9.4.4}$$

Also, using (3.1.4) and (3.1.5) in (9.4.2),

$$(V_0 + aV_1 + a^2V_2) = (V_0 + a^2V_1 + aV_2)$$

Simplifying:

$$(a^2 - a)V_2 = (a^2 - a)V_1$$

or

$$V_2 = V_1 \tag{9.4.5}$$

Now, using (3.1.4), (3.1.21), and (3.1.22) in (9.4.3),

$$(V_0 + a^2V_1 + aV_2) = Z_F(I_0 + a^2I_1 + aI_2 + I_0 + aI_1 + a^2I_2) \tag{9.4.6}$$

Using (9.4.5) and the identity $a^2 + a = -1$ in (9.4.6),

$$(V_0 - V_1) = Z_F(2I_0 - I_1 - I_2) \tag{9.4.7}$$

From (9.4.4), $I_0 = -(I_1 + I_2)$; therefore, (9.4.7) becomes

$$V_0 - V_1 = (3Z_F)I_0 \tag{9.4.8}$$

From (9.4.4), (9.4.5), and (9.4.8), we summarize:

$$
\left.
\begin{array}{l}
\text{Fault conditions in the sequence domain} \\
\text{Double line-to-ground fault}
\end{array}
\right\}
\begin{array}{ll}
I_0 + I_1 + I_2 = 0 & (9.4.9) \\
V_2 = V_1 & (9.4.10) \\
V_0 - V_1 = (3Z_F)I_0 & (9.4.11)
\end{array}
$$

Equations (9.4.9)–(9.4.11) are satisfied by connecting the zero-, positive-, and negative-sequence networks in parallel at the fault terminal; additionally, $(3Z_F)$ is included in series with the zero-sequence network. This connection is shown in Figure 9.11(b). From this figure the positive-sequence fault current is

$$I_1 = \frac{V_F}{Z_1 + [Z_2//(Z_0 + 3Z_F)]} = \frac{V_F}{Z_1 + \left[\dfrac{Z_2(Z_0 + 3Z_F)}{Z_2 + Z_0 + 3Z_F} \right]} \tag{9.4.12}$$

Using current division in Figure 9.11(b), the negative- and zero-sequence fault currents are

$$I_2 = (-I_1)\left(\frac{Z_0 + 3Z_F}{Z_0 + 3Z_F + Z_2} \right) \tag{9.4.13}$$

$$I_0 = (-I_1)\left(\frac{Z_2}{Z_0 + 3Z_F + Z_2} \right) \tag{9.4.14}$$

These sequence fault currents can be transformed to the phase domain via (3.1.16). Also, the sequence components of the line-to-ground voltages at the fault are given by (9.1.1).

| **EXAMPLE 9.5** | **Double line-to-ground short-circuit calculations using sequence networks** |

Calculate (a) the subtransient fault current in each phase, (b) neutral fault current, and (c) contributions to the fault current from the motor and from the transmission line, for a bolted double line-to-ground fault from phase b to c to ground at bus 2 in Example 9.1. Neglect the Δ–Y transformer phase shifts.

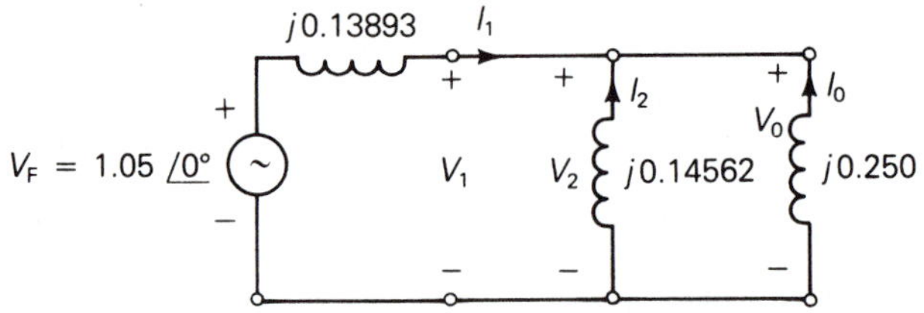

Figure 9.12

Example 9.5: Double line-to-ground fault at bus 2

Solution **a.** The zero-, positive-, and negative-sequence networks in Figure 9.5 are connected in parallel at the fault terminals in Figure 9.12. From (9.4.12) with $Z_F = 0$,

$$I_1 = \frac{1.05\underline{/0^\circ}}{j\left[0.13893 + \dfrac{(0.14562)(0.25)}{0.14562 + 0.25}\right]} = \frac{1.05\underline{/0^\circ}}{j0.23095}$$

$$= -j4.5464 \quad \text{per unit}$$

From (9.4.13) and (9.4.14),

$$I_2 = (+j4.5464)\left(\frac{0.25}{0.25 + 0.14562}\right) = j2.8730 \quad \text{per unit}$$

$$I_0 = (+j4.5464)\left(\frac{0.14562}{0.25 + 0.14562}\right) = j1.6734 \quad \text{per unit}$$

Transforming to the phase domain, the subtransient fault currents are:

$$\begin{bmatrix} I_a'' \\ I_b'' \\ I_c'' \end{bmatrix} = \begin{bmatrix} 1 & 1 & 1 \\ 1 & a^2 & a \\ 1 & a & a^2 \end{bmatrix} \begin{bmatrix} +j1.6734 \\ -j4.5464 \\ +j2.8730 \end{bmatrix} = \begin{bmatrix} 0 \\ 6.8983\underline{/158.66^\circ} \\ 6.8983\underline{/21.34^\circ} \end{bmatrix} \quad \text{per unit}$$

Using the base current of 4.1837 kA at bus 2,

$$\begin{bmatrix} I_a'' \\ I_b'' \\ I_c'' \end{bmatrix} = \begin{bmatrix} 0 \\ 6.8983\underline{/158.66^\circ} \\ 6.8983\underline{/21.34^\circ} \end{bmatrix}(4.1837) = \begin{bmatrix} 0 \\ 28.86\underline{/158.66^\circ} \\ 28.86\underline{/21.34^\circ} \end{bmatrix} \quad \text{kA}$$

b. The neutral fault current is

$$I_n = (I_b'' + I_c'') = 3I_0 = j5.0202 \quad \text{per unit}$$

$$= (j5.0202)(4.1837) = 21.00\underline{/90^\circ} \quad \text{kA}$$

c. Neglecting Δ–Y transformer phase shifts, the contributions to the fault current from the motor and transmission line can be obtained from Figure 9.4. From the zero-sequence network, Figure 9.4(a), the contribution to the zero-sequence fault current from the line is zero, due to the transformer connection. That is,

$$I_{\text{line}0} = 0$$

$$I_{\text{motor}0} = I_0 = j1.6734 \quad \text{per unit}$$

From the positive-sequence network, Figure 9.4(b), the positive terminals of the internal machine voltages can be connected, since $E_g'' = E_m''$. Then, by current division,

$$
\begin{aligned}
I_{\text{line 1}} &= \frac{X_m''}{X_m'' + (X_g'' + X_{T1} + X_{\text{line 1}} + X_{T2})} I_1 \\[2mm]
&= \frac{0.20}{0.20 + (0.455)} (-j4.5464) = -j1.3882 \quad \text{per unit}
\end{aligned}
$$

$$
I_{\text{motor 1}} = \frac{0.455}{0.20 + 0.455} (-j4.5464) = -j3.1582 \quad \text{per unit}
$$

From the negative-sequence network, Figure 9.4(c), using current division,

$$
I_{\text{line 2}} = \frac{0.21}{0.21 + 0.475} (j2.8730) = j0.8808 \quad \text{per unit}
$$

$$
I_{\text{motor 2}} = \frac{0.475}{0.21 + 0.475} (j2.8730) = j1.9922 \quad \text{per unit}
$$

Transforming to the phase domain with base currents of 0.41837 kA for the line and 4.1837 kA for the motor,

$$
\begin{bmatrix} I_{\text{line }a}'' \\ I_{\text{line }b}'' \\ I_{\text{line }c}'' \end{bmatrix}
=
\begin{bmatrix} 1 & 1 & 1 \\ 1 & a^2 & a \\ 1 & a & a^2 \end{bmatrix}
\begin{bmatrix} 0 \\ -j1.3882 \\ j0.8808 \end{bmatrix}
=
\begin{bmatrix} 0.5074\underline{/-90^\circ} \\ 1.9813\underline{/172.643^\circ} \\ 1.9813\underline{/7.357^\circ} \end{bmatrix} \quad \text{per unit}
$$

$$
=
\begin{bmatrix} 0.2123\underline{/-90^\circ} \\ 0.8289\underline{/172.643^\circ} \\ 0.8289\underline{/7.357^\circ} \end{bmatrix} \quad \text{kA}
$$

$$
\begin{bmatrix} I_{\text{motor }a}'' \\ I_{\text{motor }b}'' \\ I_{\text{motor }c}'' \end{bmatrix}
=
\begin{bmatrix} 1 & 1 & 1 \\ 1 & a^2 & a \\ 1 & a & a^2 \end{bmatrix}
\begin{bmatrix} j1.6734 \\ -j3.1582 \\ j1.9922 \end{bmatrix}
=
\begin{bmatrix} 0.5074\underline{/90^\circ} \\ 4.9986\underline{/153.17^\circ} \\ 4.9986\underline{/26.83^\circ} \end{bmatrix} \quad \text{per unit}
$$

$$
=
\begin{bmatrix} 2.123\underline{/90^\circ} \\ 20.91\underline{/153.17^\circ} \\ 20.91\underline{/26.83^\circ} \end{bmatrix} \quad \text{kA}
$$

∎

EXAMPLE 9.6   **Effect of Δ–Y transformer phase shift on fault currents**

Rework Example 9.5, with the Δ–Y transformer phase shifts included. Assume American standard phase shift.

Solution   The sequence networks of Figure 9.4 are redrawn in Figure 9.13 with ideal phase-shifting transformers representing Δ–Y phase shifts. In accordance with the American standard, positive-sequence quantities on the high-voltage side of the transformers lead their corresponding quantities on the low-voltage side by 30°. Also, the negative-sequence phase shifts are the reverse of the positive-sequence phase shifts.

a. Recall from Section 4.1 and (4.1.26) that per-unit impedance is unchanged when it is referred from one side of an ideal phase-shifting transformer to

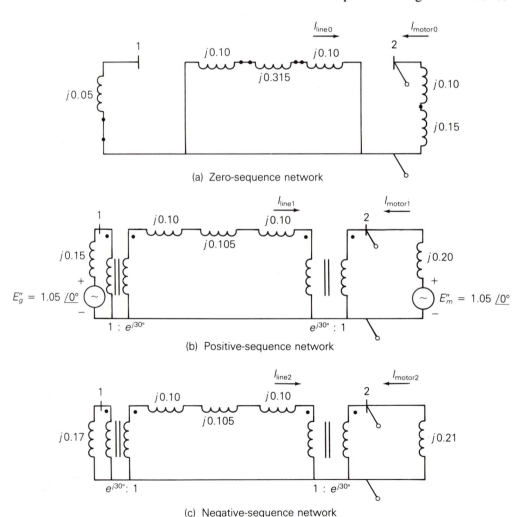

(a) Zero-sequence network

(b) Positive-sequence network

(c) Negative-sequence network

Figure 9.13   Sequence networks for Example 9.6

the other. Accordingly, the Thévenin equivalents of the sequence networks in Figure 9.13, as viewed from fault bus 2, are the same as those given in Figure 9.5. Therefore, the sequence components as well as the phase components of the fault currents are the same as those given in Example 9.5(a).

b. The neutral fault current is the same as that given in Example 9.5(b).

c. The zero-sequence network, Figure 9.13(a), is the same as that given in Figure 9.4(a). Therefore, the contributions to the zero-sequence fault current from the line and motor are the same as those given in Example 9.5(c).

$$I_{\text{line }0} = 0 \qquad I_{\text{motor }0} = I_0 = j1.6734 \quad \text{per unit}$$

The contribution to the positive-sequence fault current from the line in Figure 9.13(b) leads that in Figure 9.4(b) by 30°. That is,

$$I_{\text{line }1} = (-j1.3882)(1\underline{/30°}) = 1.3882\underline{/-60°} \quad \text{per unit}$$

$$I_{\text{motor }1} = -j3.1582 \quad \text{per unit}$$

Similarly, the contribution to the negative-sequence fault current from the line in Figure 9.13(c) lags that in Figure 9.4(c) by 30°. That is,

$$I_{\text{line }2} = (j0.8808)(1\underline{/-30°}) = 0.8808\underline{/60°} \quad \text{per unit}$$

$$I_{\text{motor }2} = j1.9922 \quad \text{per unit}$$

Thus, the sequence currents as well as the phase currents from the motor are the same as those given in Example 9.5(c). Also, the sequence currents from the line have the same magnitudes as those given in Example 9.5(c), but the positive- and negative-sequence line currents are shifted by $+30°$ and $-30°$, respectively. Transforming the line currents to the phase domain:

$$\begin{bmatrix} I''_{\text{line }a} \\ I''_{\text{line }b} \\ I''_{\text{line }c} \end{bmatrix} = \begin{bmatrix} 1 & 1 & 1 \\ 1 & a^2 & a \\ 1 & a & a^2 \end{bmatrix} \begin{bmatrix} 0 \\ 1.3882\underline{/-60°} \\ 0.8808\underline{/60°} \end{bmatrix}$$

$$= \begin{bmatrix} 1.2166\underline{/-21.17°} \\ 2.2690\underline{/180°} \\ 1.2166\underline{/21.17°} \end{bmatrix} \quad \text{per unit} = \begin{bmatrix} 0.5090\underline{/-21.17°} \\ 0.9492\underline{/180°} \\ 0.5090\underline{/21.17°} \end{bmatrix} \quad \text{kA}$$

In conclusion, Δ–Y transformer phase shifts have no effect on the fault currents and no effect on the contribution to the fault currents on the fault side of the Δ–Y transformers. However, on the other side of the Δ–Y transformers, the positive- and negative-sequence components of the contributions to the fault currents are shifted by $\pm 30°$, which affects both the magnitude as well as the angle of the phase components of these fault contributions for unsymmetrical faults.   ∎

Figure 9.14 summarizes the sequence network connections for both the

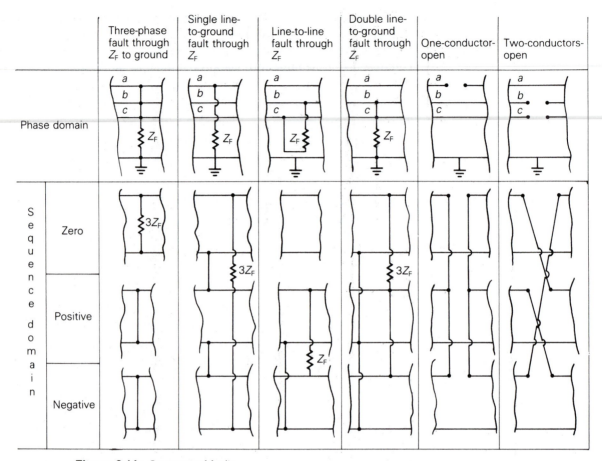

		Three-phase fault through Z_F to ground	Single line-to-ground fault through Z_F	Line-to-line fault through Z_F	Double line-to-ground fault through Z_F	One-conductor-open	Two-conductors-open
Phase domain							
Sequence domain	Zero						
	Positive						
	Negative						

Figure 9.14 Summary of faults

balanced three-phase fault and the unsymmetrical faults that we have considered. Sequence network connections for two additional faults, one-conductor-open and two-conductors-open, are also shown in Figure 9.14 and are left as an exercise for the reader to verify (see Problems 9.23 and 9.24).

SECTION 9.5

SEQUENCE BUS IMPEDANCE MATRICES

We use the positive-sequence bus impedance matrix in Section 8.4 for calculating currents and voltages during balanced three-phase faults. This method is extended here to unsymmetrical faults by representing each sequence network as a bus impedance equivalent circuit (or as a rake equivalent). A bus impedance matrix can be computed for each sequence network using the one-step-at-a-time method given in Section 8.5. For simplicity, resistances, shunt admittances, nonrotating impedance loads, and prefault load currents are neglected.

Figure 9.15 shows the connection of sequence rake equivalents for both symmetrical and unsymmetrical faults at bus n of an N-bus three-phase power system. Each bus impedance element has an additional subscript, 0, 1, or 2, that identifies the sequence rake equivalent in which it is located. Mutual impedances are not shown in the figure. The prefault voltage V_F is included in the positive-sequence rake equivalent. From the figure the sequence components of the fault current for each type of fault at bus n are as follows:

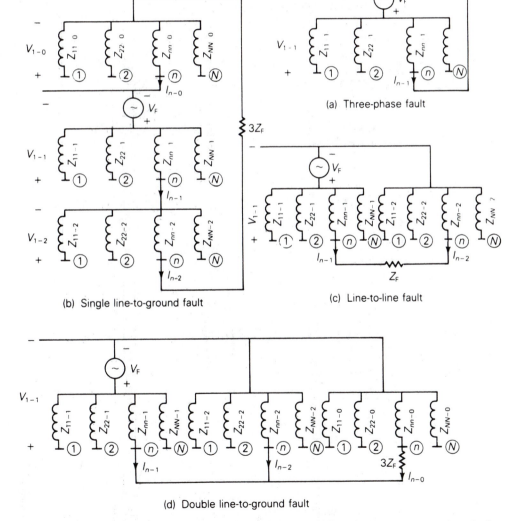

(a) Three-phase fault

(b) Single line-to-ground fault

(c) Line-to-line fault

(d) Double line-to-ground fault

Figure 9.15   Connection of rake equivalent sequence networks for three-phase system faults (mutual impedances not shown)

Balanced three-phase fault:

$$I_{n-1} = \frac{V_F}{Z_{nn-1}}$$ (9.5.1)

$$I_{n-0} = I_{n-2} = 0$$ (9.5.2)

Single line-to-ground fault (phase *a* to ground):

$$I_{n-0} = I_{n-1} = I_{n-2} = \frac{V_F}{Z_{nn-0} + Z_{nn-1} + Z_{nn-2} + 3Z_F}$$ (9.5.3)

Line-to-line fault (phase *b* to *c*):

$$I_{n-1} = -I_{n-2} = \frac{V_F}{Z_{nn-1} + Z_{nn-2} + Z_F}$$ (9.5.4)

$$I_{n-0} = 0$$ (9.5.5)

Double line-to-ground fault (phase *b* to *c* to ground):

$$I_{n-1} = \frac{V_F}{Z_{nn-1} + \left[\dfrac{Z_{nn-2}(Z_{nn-0} + 3Z_F)}{Z_{nn-2} + Z_{nn-0} + 3Z_F} \right]}$$ (9.5.6)

$$I_{n-2} = (-I_{n-1})\left(\frac{Z_{nn-0} + 3Z_F}{Z_{nn-0} + 3Z_F + Z_{nn-2}} \right)$$ (9.5.7)

$$I_{n-0} = (-I_{n-1})\left(\frac{Z_{nn-2}}{Z_{nn-0} + 3Z_F + Z_{nn-2}} \right)$$ (9.5.8)

Also from Figure 9.15, the sequence components of the line-to-ground voltages at any bus *k* during a fault at bus *n* are:

$$\begin{bmatrix} V_{k-0} \\ V_{k-1} \\ V_{k-2} \end{bmatrix} = \begin{bmatrix} 0 \\ V_F \\ 0 \end{bmatrix} - \begin{bmatrix} Z_{kn-0} & 0 & 0 \\ 0 & Z_{kn-1} & 0 \\ 0 & 0 & Z_{kn-2} \end{bmatrix} \begin{bmatrix} I_{n-0} \\ I_{n-1} \\ I_{n-2} \end{bmatrix}$$ (9.5.9)

If bus *k* is on the unfaulted side of a Δ–Y transformer, then the phase angles of V_{k-1} and V_{k-2} in (9.5.9) are modified to account for Δ–Y phase shifts. Also, the above sequence fault currents and sequence voltages can be transformed to the phase domain via (3.1.16) and (3.1.9).

EXAMPLE 9.7	**Single line-to-ground short-circuit calculations using Z_{bus0}, $Z_{bus\,1}$, and Z_{bus2}**

Faults at buses 1 and 2 for the three-phase power system given in Example 9.1 are of interest. The prefault voltage is 1.05 per unit. Prefault load current is neglected. (a) Determine the per-unit zero-, positive-, and negative-sequence bus impedance matrices using the one-step-at-a-time method given in Section 8.5. Find the subtransient fault current in per-unit for a bolted single line-to-ground fault current from phase *a* to ground (b) at bus 1 and (c) at bus 2. Find the per-unit line-to-ground voltages at (d) bus 1 and (e) bus 2 during the single line-to-ground fault at bus 1.

Solution   **a.** Referring to Figure 9.4(a), the zero-sequence bus impedance matrix is determined in two steps as follows:

Step 1
Add $Z_b = j0.05$
from new bus 1
to the reference bus,
a type 1 addition

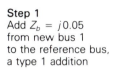
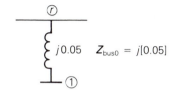

$j0.05$ $Z_{bus0} = j[0.05]$

Step 2
Add $Z_b = j0.25$
from new bus 2
to the reference bus,
a type 1 addition

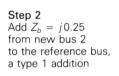

$j0.05$ $j0.25$ $Z_{bus0} = j\begin{bmatrix} 0.05 & 0 \\ 0 & 0.25 \end{bmatrix}$ per unit

Note that the transformer leakage reactances and the zero-sequence transmission-line reactance in Figure 9.4(a) have no effect on $Z_{bus\,0}$. The transformer Δ connections block the flow of zero-sequence current from the transformers to bus 1 and 2.

The positive-sequence bus impedance matrix, determined in Example 8.5, is

$$Z_{bus\,1} = j\begin{bmatrix} 0.11565 & 0.04580 \\ 0.04580 & 0.13893 \end{bmatrix} \text{ per unit}$$

Referring to Figure 9.4(c), the negative-sequence bus impedance matrix is determined in three steps, as follows:

Step 1
Add $Z_b = j0.17$
from new bus 1
to the reference bus,
a type 1 addition

$j0.17$ $Z_{bus2} = j[0.17]$

Step 2
Add $Z_b = j(0.10 + 0.105 + 0.10) = j0.305$
from old bus 1 to new bus 2,
a type 2 addition

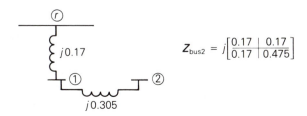

$Z_{bus2} = j\begin{bmatrix} 0.17 & 0.17 \\ 0.17 & 0.475 \end{bmatrix}$

$j0.17$

$j0.305$

Step 3
Add $Z_b = j0.21$
from old bus 2
to the reference bus,
a type 3 addition

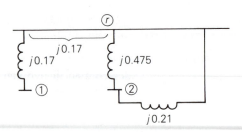

$$Z_{\text{bus 2}} = j \left\{ \begin{bmatrix} 0.17 & 0.17 \\ 0.17 & 0.475 \end{bmatrix} - \frac{1}{(0.475 + 0.21)} \begin{bmatrix} 0.17 \\ 0.475 \end{bmatrix} [0.17 \quad 0.475] \right\}$$

$$= j \left[\begin{array}{c|c} 0.12781 & 0.05212 \\ \hline 0.05212 & 0.14562 \end{array} \right] \quad \text{per unit}$$

b. From (9.5.3), with $n = 1$ and $Z_F = 0$, the sequence fault currents are

$$I_{1-0} = I_{1-1} = I_{1-2} = \frac{V_F}{Z_{11-0} + Z_{11-1} + Z_{11-2}}$$

$$= \frac{1.05\underline{/0°}}{j(0.05 + 0.11565 + 0.12781)} = \frac{1.05}{j0.29346} = -j3.578 \quad \text{per unit}$$

The subtransient fault currents at bus 1 are, from (3.1.16),

$$\begin{bmatrix} I''_{1a} \\ I''_{1b} \\ I''_{1c} \end{bmatrix} = \begin{bmatrix} 1 & 1 & 1 \\ 1 & a^2 & a \\ 1 & a & a^2 \end{bmatrix} \begin{bmatrix} -j3.578 \\ -j3.578 \\ -j3.578 \end{bmatrix} = \begin{bmatrix} -j10.73 \\ 0 \\ 0 \end{bmatrix} \quad \text{per unit}$$

c. Again from (9.5.3), with $n = 2$ and $Z_F = 0$,

$$I_{2-0} = I_{2-1} = I_{2-2} = \frac{V_F}{Z_{22-0} + Z_{22-1} + Z_{22-2}}$$

$$= \frac{1.05\underline{/0°}}{j(0.25 + 0.13893 + 0.14562)} = \frac{1.05}{j0.53455}$$

$$= -j1.96427 \quad \text{per unit}$$

and

$$\begin{bmatrix} I''_{2a} \\ I''_{2b} \\ I''_{2c} \end{bmatrix} = \begin{bmatrix} 1 & 1 & 1 \\ 1 & a^2 & a \\ 1 & a & a^2 \end{bmatrix} \begin{bmatrix} -j1.96427 \\ -j1.96427 \\ -j1.96427 \end{bmatrix} = \begin{bmatrix} -j5.8928 \\ 0 \\ 0 \end{bmatrix}$$

This is the same result as obtained in Example 9.3.

d. The sequence components of the line-to-ground voltages at bus 1 during the fault at bus 1 are, from (9.5.9), with $k = 1$ and $n = 1$,

$$
\begin{bmatrix} V_{1-0} \\ V_{1-1} \\ V_{1-2} \end{bmatrix} = \begin{bmatrix} 0 \\ 1.05\underline{/0°} \\ 0 \end{bmatrix} - \begin{bmatrix} j0.05 & 0 & 0 \\ 0 & j0.11565 & 0 \\ 0 & 0 & j0.12781 \end{bmatrix} \begin{bmatrix} -j3.578 \\ -j3.578 \\ -j3.578 \end{bmatrix}
$$

$$
= \begin{bmatrix} -0.1789 \\ 0.6362 \\ -0.4573 \end{bmatrix} \quad \text{per unit}
$$

and the line-to-ground voltages at bus 1 during the fault at bus 1 are

$$
\begin{bmatrix} V_{1-ag} \\ V_{1-bg} \\ V_{1-cg} \end{bmatrix} = \begin{bmatrix} 1 & 1 & 1 \\ 1 & a^2 & a \\ 1 & a & a^2 \end{bmatrix} \begin{bmatrix} -0.1789 \\ +0.6362 \\ -0.4573 \end{bmatrix} = \begin{bmatrix} 0 \\ 0.9843\underline{/254.2°} \\ 0.9843\underline{/105.8°} \end{bmatrix} \quad \text{per unit}
$$

e. The sequence components of the line-to-ground voltages at bus 2 during the fault at bus 1 are, from (9.5.9), with $k = 2$ and $n = 1$,

$$
\begin{bmatrix} V_{2-0} \\ V_{2-1} \\ V_{2-2} \end{bmatrix} = \begin{bmatrix} 0 \\ 1.05\underline{/0°} \\ 0 \end{bmatrix} - \begin{bmatrix} 0 & 0 & 0 \\ 0 & j0.04580 & 0 \\ 0 & 0 & j0.05212 \end{bmatrix} \begin{bmatrix} -j3.578 \\ -j3.578 \\ -j3.578 \end{bmatrix}
$$

$$
= \begin{bmatrix} 0 \\ 0.8861 \\ -0.18649 \end{bmatrix} \quad \text{per unit}
$$

Note that since both bus 1 and 2 are on the low-voltage side of the Δ–Y transformers in Figure 9.3, there is no shift in the phase angles of these sequence voltages. From the above, the line-to-ground voltages at bus 2 during the fault at bus 1 are

$$
\begin{bmatrix} V_{2-ag} \\ V_{2-bg} \\ V_{2-cg} \end{bmatrix} = \begin{bmatrix} 1 & 1 & 1 \\ 1 & a^2 & a \\ 1 & a & a^2 \end{bmatrix} \begin{bmatrix} 0 \\ 0.8861 \\ -0.18649 \end{bmatrix} = \begin{bmatrix} 0.70 \\ 0.9926\underline{/249.4°} \\ 0.9926\underline{/110.6°} \end{bmatrix} \quad \text{per unit}
$$

■

SECTION 9.6

PERSONAL COMPUTER PROGRAM: SHORT CIRCUITS

The software package that accompanies this text includes the program "SHORT CIRCUITS," which computes the symmetrical fault current for each of the following faults at any bus in an N-bus power system: balanced three-phase fault, single line-to-ground fault, line-to-line fault, or double line-to-ground fault. For each fault, the program also computes bus voltages and contributions to the fault current from transmission lines and transformers connected to the fault bus.

Input data for the program include machine, transmission-line, and transformer data, as illustrated in Tables 9.1, 9.2, and 9.3 as well as the prefault voltage V_F and fault impedance Z_F. When the machine positive-sequence reactance input data consist of direct axis subtransient reactances, the computed symmetrical fault currents are subtransient fault currents. Alternatively, transient or steady-state fault currents are computed when this input data consist of direct axis transient or synchronous reactances. Transmission-line positive- and zero-sequence series reactances are those of the equivalent π circuits for long lines or of the nominal π circuit for medium or short lines. Also, recall that the negative-sequence transmission-line reactance equals the positive-sequence transmission-line reactance. All machine, line, and transformer reactances are given in per-unit on a common MVA base. Resistances, shunt admittances, nonrotating impedance loads, and prefault load currents are neglected.

Table 9.1

Synchronous machine data for Example 9.8

BUS	X_0	$X_1 = X_d''$	X_2	NEUTRAL REACTANCE X_n
	per unit	per unit	per unit	per unit
1	0.05	0.18	0.18	0
3	0.02	0.09	0.09	0.01

Table 9.2

Line data for Example 9.8

BUS-TO-BUS	X_0	X_1
	per unit	per unit
2–4	1.2	0.4
2–5	0.6	0.2
4–5	0.3	0.1

Table 9.3

Transformer data for Example 9.8

LOW-VOLTAGE (connection) bus	HIGH-VOLTAGE (connection) bus	LEAKAGE REACTANCE	NEUTRAL REACTANCE
		per unit	per unit
1 (Δ)	5 (Y)	0.08	0
3 (Δ)	4 (Y)	0.04	0

$S_{base} = 400$ MVA

$V_{base} = \begin{cases} 15\text{kV at buses } 1, 3 \\ 345\text{kV at buses } 2, 4, 5 \end{cases}$

The program computes the zero-, positive-, and negative-sequence bus impedance matrices $Z_{bus\,0}$, $Z_{bus\,1}$, and $Z_{bus\,2}$, using the one-step-at-a-time method described in Section 8.5. The program user can either compute these impedance matrices from scratch or begin with existing matrices that have already been computed and stored. In both cases, type 1, 2, 3, and 4 modifications, as described in Section 8.5, can be made. The program computes the matrices in accordance with the ordering rules given in Section 8.5.

After $Z_{bus\,0}$, $Z_{bus\,1}$, and $Z_{bus\,2}$ are computed, (9.5.1)–(9.5.9) are used to compute the sequence fault currents and the sequence voltages at each bus during a fault at bus 1 for the fault type selected by the program user (for example, three-phase fault, or single line-to-ground fault, and so on). Contributions to the sequence fault currents from each line or transformer branch connected to the fault bus are computed by dividing the sequence voltage across the branch by the branch sequence impedance. The phase angles of positive- and negative-sequence voltages are also modified to account for Δ–Y transformer phase shifts. The sequence currents and sequence voltages are then transformed to the phase domain via (3.1.16) and (3.1.9). All these computations are then repeated for a fault at bus 2, then bus 3, and so on to bus N.

For large power systems with hundreds or thousands of buses, computer storage requirements can be reduced by about one-third if $Z_{bus\,2}$ is set equal to $Z_{bus\,1}$. This is a good approximation when calculating subtransient fault currents because the negative-sequence reactance of a synchronous machine is approximately equal to its direct axis subtransient reactance.

Output data for the fault type and fault impedance selected by the user consist of: the fault current in each phase, contributions to the fault current from each branch connected to the fault bus for each phase, and the line-to-ground voltages at each bus—for a fault at bus 1, then bus 2, and so on to bus N. $Z_{bus\,0}$, $Z_{bus\,1}$, and $Z_{bus\,2}$ can also be included in the output if desired.

| EXAMPLE 9.8 | **SHORT CIRCUITS program** |

Consider the five-bus power system whose single-line diagram is shown in Figure 7.2. Machine, line, and transformer data are given in Tables 9.1, 9.2, and 9.3. Note that the neutrals of both transformers and generator 1 are solidly grounded, as indicated by a neutral reactance of zero for these equipments. However, a neutral reactance = 0.01 per unit is connected to the generator 2 neutral. The prefault voltage is 1.05 per unit. Run the SHORT CIRCUITS program for a bolted single line-to-ground fault at bus 1, then bus 2, and so on to bus 5.

Solution The program was run with the input data from Tables 9.1, 9.2, and 9.3, with $V_F = 1.05$, and with $Z_F = 0$ for a single line-to-ground fault. The 5×5 bus impedance matrices $Z_{bus\,0}$, $Z_{bus\,1}$, and $Z_{bus\,2}$ are shown in Table 9.4 and the fault currents and bus voltages are given in Tables 9.5 and 9.6. Note that these fault currents are subtransient currents, since the machine positive-sequence reactance input consists of direct axis subtransient reactances.

Table 9.4

Z_{bus} impedance matrices for Example 9.8

$$Z_F = 0 + j0$$

$$Z_{bus0} = j \begin{bmatrix} 0.05000 & 0.00000 & 0.00000 & 0.00000 & 0.00000 \\ 0.00000 & 0.43576 & 0.00000 & 0.01758 & 0.04485 \\ 0.00000 & 0.00000 & 0.05000 & 0.00000 & 0.00000 \\ 0.00000 & 0.01758 & 0.00000 & 0.03576 & 0.00848 \\ 0.00000 & 0.04485 & 0.00000 & 0.00848 & 0.06303 \end{bmatrix}$$

$$Z_{bus1} = j \begin{bmatrix} 0.11189 & 0.07081 & 0.03405 & 0.04919 & 0.08162 \\ 0.07081 & 0.22781 & 0.05459 & 0.07886 & 0.10228 \\ 0.03405 & 0.05459 & 0.07297 & 0.06541 & 0.04919 \\ 0.04919 & 0.07886 & 0.06541 & 0.09447 & 0.07105 \\ 0.08162 & 0.10228 & 0.04919 & 0.07105 & 0.11790 \end{bmatrix}$$

$$Z_{bus2} = j \begin{bmatrix} 0.11189 & 0.07081 & 0.03405 & 0.04919 & 0.08162 \\ 0.07081 & 0.22781 & 0.05459 & 0.07886 & 0.10228 \\ 0.03405 & 0.05459 & 0.07297 & 0.06541 & 0.04919 \\ 0.04919 & 0.07886 & 0.06541 & 0.09447 & 0.07105 \\ 0.08162 & 0.10228 & 0.04919 & 0.07105 & 0.11790 \end{bmatrix}$$

FAULT BUS	SINGLE LINE-TO-GROUND FAULT CURRENT (PHASE A) per unit/degrees	CONTRIBUTIONS TO FAULT CURRENT				
		GEN LINE OR TRSF	BUS-TO-BUS	PHASE A	CURRENT PHASE B	PHASE C
				per unit/degrees		
1	11.505/−90.00	G1	GRND−1	8.6032/ −90.00	1.4511/ −90.00	1.4511/ −90.00
		T1	5−1	2.9023/ −90.00	1.4511/ 90.00	1.4511/ 90.00
2	3.534/−90.00	L1	4−2	1.2878/ −90.00	0.0281/ 90.00	0.0281/ 90.00
		L2	5−2	2.2461/ −90.00	0.0281/ −90.00	0.0281/ −90.00
3	16.076/−90.00	G2	GRND−3	14.0483/ −90.00	1.0138/ −90.00	1.0138/ −90.00
		T2	4−3	2.0276/ −90.00	1.0138/ 90.00	1.0138/ 90.00
4	14.018/−90.00	L1	2−4	0.4356/ −90.00	0.1116/ 90.00	0.1116/ 90.00
		L3	5−4	2.6138/ −90.00	0.6697/ 90.00	0.6697/ 90.00
		T2	3−4	10.9688/ −90.00	0.7813/ −90.00	0.7813/ −90.00
5	10.541/−90.00	L2	2−5	0.6552/ −90.00	0.1679/ 90.00	0.1679/ 90.00
		L3	4−5	3.9310/ −90.00	1.0072/ 90.00	1.0072/ 90.00
		T1	1−5	5.9551/ −90.00	1.1751/ −90.00	1.1751/ −90.00

Table 9.5 Fault currents for Example 9.8

$V_F = 1.05/0$		BUS VOLTAGES DURING FAULT		
FAULT BUS	BUS #	PHASE A	PHASE B per unit/degrees	PHASE C
1	1	0.0000/0.00	0.9537/252.45	0.9537/107.55
	2	0.6843/50.10	1.0500/−90.00	0.6843/129.90
	3	0.7888/0.00	0.9912/246.55	0.9912/113.45
	4	0.7842/42.02	1.0500/−90.00	0.7842/137.98
	5	0.6406/55.03	1.0500/−90.00	0.6406/124.97
2	1	0.9277/−34.47	0.9277/214.47	1.0500/90.00
	2	0.0000/0.00	1.1915/229.74	1.1915/130.26
	3	0.9552/−33.34	0.9552/213.34	1.0500/90.00
	4	0.8435/0.00	1.0158/243.53	1.0158/116.47
	5	0.7562/0.00	1.0179/243.30	1.0179/116.70
3	1	0.6850/0.00	0.9717/249.36	0.9717/110.64
	2	0.6616/52.52	1.0500/−90.00	0.6616/127.48
	3	0.0000/0.00	0.9942/246.16	0.9942/113.84
	4	0.6058/60.07	1.0500/−90.00	0.6058/119.93
	5	0.6933/49.22	1.0500/−90.00	0.6933/130.78
4	1	0.7328/−45.76	0.7328/225.76	1.0500/90.00
	2	0.2309/0.00	0.9401/255.30	0.9401/104.70
	3	0.6481/−54.10	0.6481/234.10	1.0500/90.00
	4	0.0000/0.00	0.9432/254.59	0.9432/105.41
	5	0.3463/0.00	0.9386/255.65	0.9386/104.35
5	1	0.6677/−51.84	0.6677/231.84	1.0500/90.00
	2	0.1736/0.00	0.9651/250.43	0.9651/109.57
	3	0.8048/−40.72	0.8048/220.72	1.0500/90.00
	4	0.5209/0.00	0.9592/251.45	0.9592/108.55
	5	0.0000/0.00	0.9681/249.93	0.9681/110.07

Table 9.6 Bus voltages for Example 9.8   ■

PROBLEMS

Section 9.1

9.1 The single-line diagram of a three-phase power system is shown in Figure 9.16. Equipment ratings are given as follows:

Synchronous generators:

G1   1000 MVA   15 kV $X''_d = X_2 = 0.18, X_0 = 0.07$ per unit

G2   1000 MVA   15 kV $X''_d = X_2 = 0.20, X_0 = 0.10$ per unit

G3   500 MVA   13.8 kV $X''_d = X_2 = 0.15, X_0 = 0.05$ per unit

G4   750 MVA   13.8 kV $X''_d = 0.30, X_2 = 0.40, X_0 = 0.10$ per unit

Figure 9.16

Problem 9.1

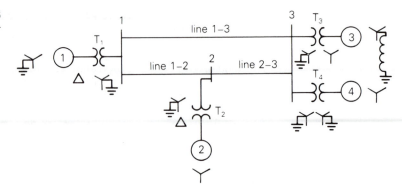

Transformers:

T1	1000 MVA	15 kV Δ/765 kVY	X = 0.10 per unit
T2	1000 MVA	15 kV Δ/765 kVY	X = 0.10 per unit
T3	500 MVA	15 kVY/765 kVY	X = 0.12 per unit
T4	750 MVA	15 kVY/765 kVY	X = 0.11 per unit

Transmission lines:

1–2	765 kV	$X_1 = 50\,\Omega, X_0 = 150\,\Omega$
1–3	765 kV	$X_1 = 40\,\Omega, X_0 = 100\,\Omega$
2–3	765 kV	$X_1 = 40\,\Omega, X_0 = 100\,\Omega$

The inductor connected to generator 3 neutral has a reactance of 0.05 per unit using generator 3 ratings as a base. Draw the zero-, positive-, and negative-sequence reactance diagrams using a 1000-MVA, 765-kV base in the zone of line 1–2. Neglect the Δ–Y transformer phase shifts.

9.2 Faults at bus n in Problem 9.1 are of interest (the instructor selects $n = 1$, 2, or 3). Determine the Thévenin equivalent of each sequence network as viewed from the fault bus. Prefault voltage is 1.0 per unit. Prefault load currents and Δ–Y transformer phase shifts are neglected.
Hint: Use the Y–Δ conversion in Figure 2.23.

9.3 Determine the subtransient fault current in per-unit and in kA during a bolted three-phase fault at the fault bus selected in Problem 9.2.

9.4 Equipment ratings for the four-bus power system shown in Figure 8.16 are given as follows:

Generator G1:	500 MVA, 13.8 kV, $X_d'' = X_2 = 0.20$, $X_0 = 0.10$ per unit
Generator G2:	750 MVA, 18 kV, $X_d'' = X_2 = 0.18$, $X_0 = 0.09$ per unit
Generator G3:	1000 MVA, 20 kV, $X_d'' = 0.17$, $X_2 = 0.20$, $X_0 = 0.09$ per unit
Transformer T1:	500 MVA, 13.8 kV Δ/500 kV Y, X = 0.12 per unit
Transformer T2:	750 MVA, 18 kV Δ/500 kV Y, X = 0.10 per unit
Transformer T3:	1000 MVA, 20 kV Δ/500 kV Y, X = 0.10 per unit
Each line:	$X_1 = 50$ ohms, $X_0 = 150$ ohms

The inductor connected to generator G3 neutral has a reactance of $0.028\,\Omega$. Draw the zero-, positive-, and negative-sequence reactance diagrams using a 1000-MVA, 20-kV base in the zone of generator G3. Neglect $\Delta-Y$ transformer phase shifts.

9.5 Faults at bus n in Problem 9.4 are of interest (the instructor selects $n = 1, 2, 3,$ or 4). Determine the Thévenin equivalent of each sequence network as viewed from the fault bus. Prefault voltage is 1.0 per unit. Prefault load currents and $\Delta-Y$ phase shifts are neglected.

9.6 Determine the subtransient fault current in per-unit and in kA during a bolted three-phase fault at the fault bus selected in Problem 9.5.

9.7 Equipment ratings for the five-bus power system shown in Figure 8.17 are given as follows:

> Generator G1:   50 MVA, 12 kV, $X_d'' = X_2 = 0.20$, $X_0 = 0.10$ per unit
>
> Generator G2:   100 MVA, 15 kV, $X_d'' = 0.2$, $X_2 = 0.23$, $X_0 = 0.1$ per unit
>
> Transformer T1:   50 MVA, 10 kV Y/138 kV Y, $X = 0.10$ per unit
>
> Transformer T2:   100 MVA, 15 kV Δ/138 kV Y, $X = 0.10$ per unit
>
> Each 138-kV line: $X_1 = 40$ ohms, $X_0 = 100$ ohms

Draw the zero-, positive-, and negative-sequence reactance diagrams using a 100-MVA, 15-kV base in the zone of generator G2. Neglect $\Delta-Y$ transformer phase shifts.

9.8 Faults at bus n in Problem 9.7 are of interest (the instructor selects $n = 1, 2, 3, 4$ or 5). Determine the Thévenin equivalent of each sequence network as viewed from the fault bus. Prefault voltage is 1.0 per unit. Prefault load currents and $\Delta-Y$ phase shifts are neglected.

9.9 Determine the subtransient fault current in per-unit and in kA during a bolted three-phase fault at the fault bus selected in Problem 9.8.

Section 9.2–9.4

9.10 Determine the subtransient fault current in per-unit and in kA, as well as the per-unit line-to-ground voltages at the fault bus for a bolted single line-to-ground fault at the fault bus selected in Problem 9.2.

9.11 Repeat Problem 9.10 for a single line-to-ground arcing fault with arc impedance $Z_F = 30 + j0\,\Omega$.

9.12 Repeat Problem 9.10 for a bolted line-to-line fault.

9.13 Repeat Problem 9.10 for a bolted double line-to-ground fault.

9.14 Repeat Problems 9.1 and 9.10 including $\Delta-Y$ transformer phase shifts. Assume American standard phase shift. Also calculate the sequence components and phase components of the contribution to the fault current from generator n ($n = 1, 2,$ or 3 as specified by the instructor in Problem 9.2).

9.15 A 500-MVA, 13.8-kV synchronous generator with $X_d'' = X_2 = 0.20$ and $X_0 = 0.05$ per unit is connected to a 500-MVA, 13.8-kV Δ/500-kV Y transformer with 0.10 per-unit leakage reactance. The generator and transformer neutrals are solidly grounded. The generator is operated at no-load and rated voltage, and the high-voltage side of the transformer is disconnected from the power system. Compare the subtransient fault currents for the following bolted faults at the transformer high-voltage terminals: three-phase fault, single line-to-ground fault, line-to-line fault, and double line-to-ground fault.

9.16 Determine the subtransient fault current in per-unit and in kA, as well as contributions to the fault current from each line and transformer connected to the fault bus for a bolted single line-to-ground fault at the fault bus selected in Problem 9.5.

9.17 Repeat Problem 9.16 for a bolted line-to-line fault.

9.18 Repeat Problem 9.16 for a bolted double line-to-ground fault.

9.19 Determine the subtransient fault current in per-unit and in kA, as well as contributions to the fault current from each line, transformer, and generator connected to the fault bus for a bolted single line-to-ground fault at the fault bus selected in Problem 9.8.

9.20 Repeat Problem 9.19 for a single line-to-ground arcing fault with arc impedance $Z_F = 0.05 + j0$ per unit.

9.21 Repeat Problem 9.19 for a bolted line-to-line fault.

9.22 Repeat Problem 9.19 for a bolted double line-to-ground fault.

9.23 As shown in Figure 9.17(a), two three-phase buses abc and $a'b'c'$ are interconnected by short circuits between phases b and b' and between c and c', with an open circuit between phases a and a'. The fault conditions in the phase domain are $I_a = I_{a'} = 0$ and $V_{bb'} = V_{cc'} = 0$. Determine the fault conditions in the sequence domain and verify the interconnection of the sequence networks as shown in Figure 9.14 for this one-conductor-open fault.

Figure 9.17

Problems 9.23 and 9.24: open conductor faults

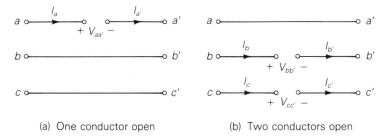

(a) One conductor open   (b) Two conductors open

9.24 Repeat Problem 9.23 for the two-conductors-open fault shown in Figure 9.17(b). The fault conditions in the phase domain are

$$I_b = I_{b'} = I_c = I_{c'} = 0 \qquad \text{and} \qquad V_{aa'} = 0$$

Section 9.5

9.25 The zero-, positive-, and negative-sequence bus impedance matrices for a three-bus three-phase power system are

$$Z_{\text{bus }0} = j \begin{bmatrix} 0.10 & 0 & 0 \\ 0 & 0.20 & 0 \\ 0 & 0 & 0.10 \end{bmatrix} \text{ per unit}$$

$$Z_{\text{bus }1} = Z_{\text{bus }2} = j \begin{bmatrix} 0.12 & 0.08 & 0.04 \\ 0.08 & 0.12 & 0.06 \\ 0.04 & 0.06 & 0.08 \end{bmatrix}$$

Determine the per-unit fault current and per-unit voltage at bus 2 for a bolted three-phase fault at bus 1. The prefault voltage is 1.0 per unit.

9.26 Repeat Problem 9.25 for a bolted single line-to-ground fault at bus 1.

9.27 Repeat Problem 9.25 for a bolted line-to-line fault at bus 1.

9.28 Repeat Problem 9.25 for a bolted double line-to-ground fault at bus 1.

9.29 Use the one-step-at-a-time method to find the 3×3 per-unit zero-, positive-, and negative-sequence bus impedance matrices for the power system given in Problem 9.1. Use a base of 1000 MVA and 765 kV in the zone of line 1–2.

9.30 Using the bus impedance matrices determined in Problem 9.29, verify the fault currents for the faults given in Problems 9.3, 9.10, 9.11, 9.12, and 9.13.

9.31 Use the one-step-at-a-time method to find the 4×4 per-unit zero-, positive-, and negative-sequence bus impedance matrices for the power system given in Problem 9.4. Use a base of 1000 MVA and 20 kV in the zone of generator G3.

9.32 Using the bus impedance matrices determined in Problem 9.31, verify the fault currents for the faults given in Problems 9.6, 9.16, 9.17, and 9.18.

9.33 Use the one-step-at-a-time method to find the 5×5 per-unit zero-, positive-, and negative-sequence bus impedance matrices for the power system given in Problem 9.7. Use a base of 100 MVA and 15 kV in the zone of generator G2.

9.34 Using the bus impedance matrices determined in Problem 9.33, verify the fault currents for the faults given in Problems 9.9, 9.19, 9.20, 9.21, and 9.22.

🖬 **Section 9.6**

9.35 Rework Problem 9.29 using the SHORT CIRCUITS program.

9.36 Rework Problem 9.30 using the SHORT CIRCUITS program.

9.37 Run the SHORT CIRCUITS program for the power system given in Example 9.8 for a bolted double line-to-ground fault at bus 1, then bus 2, and so on to bus 5. Prefault voltage is 1.05 per unit. Compare these fault currents with the single line-to-ground faults computed in Example 9.8.

9.38 Rework Problem 9.37 with a new line installed between buses 4 and 5. This line has a positive-sequence series impedance $Z_{b1} = j0.4$ per unit and a zero-sequence series impedance $Z_{b0} = j1.2$ per unit. The new line is not mutually coupled to any other line. Are the fault currents larger or smaller than those of Problem 9.37?

9.39 Repeat Problem 9.31 using the SHORT CIRCUITS program.

9.40 Repeat Problem 9.32 using the SHORT CIRCUITS program.

9.41 Repeat Problem 9.33 using the SHORT CIRCUITS program.

9.42 Repeat Problem 9.34 using the SHORT CIRCUITS program.

CASE STUDY QUESTIONS

A. Are safety hazards associated with generation, transmission, and distribution of electric power by the electric utility industry greater than or less than safety hazards associated with the transportation industry? The chemical products industry? The medical services industry? The agriculture industry?

B. What is the public's perception of the electric utility industry's safety record?

References

1.   Westinghouse Electric Corporation, *Electrical Transmission and Distribution Reference Book*, 4th ed. (East Pittsburgh, PA, 1964).

2.   Westinghouse Electric Corporation, *Applied Protective Relaying* (Newark, NJ, 1976).

3.   P. M. Anderson, *Analysis of Faulted Power Systems* (Ames: Iowa State University Press, 1973).

4.   J. R. Neuenswander, *Modern Power Systems* (New York: Intext Educational Publishers, 1971).

5.   H. E. Brown, *Solution of Large Networks by Matrix Methods* (New York: Wiley, (1975).

6.   W. D. Stevenson, Jr., *Elements of Power System Analysis*, 4th ed. (New York: McGraw-Hill, 1982).

7.   C. A. Gross, *Power System Analysis* (New York: Wiley, 1979).

8.   Glenn Zorpette, "Fires at U.S. Utilities," *IEEE Spectrum, 28*, 1 (January, 1991), p. 64.

CHAPTER 10

SYSTEM PROTECTION

Lightning slices through rainy skies above a city, in a time-exposure view
(Courtesy of Uniphoto, Inc.)

Short circuits occur in power systems when equipment insulation fails, due to system overvoltages caused by lightning or switching surges, to insulation contamination, or to other mechanical and natural causes. Careful design, operation, and maintenance can minimize the occurrence of short circuits but cannot eliminate them. Methods for calculating short-circuit currents for balanced and unbalanced faults were discussed in Chapters 8 and 9. Such currents can be several orders of magnitude larger than normal operating currents and, if allowed to persist, may cause insulation damage, conductor melting, fire, and explosion. Windings and busbars may also suffer mechanical damage due to high magnetic forces during faults. Clearly, faults must be quickly removed from a power system. Standard EHV protective equipment is designed to clear faults within 3 cycles, whereas lower-voltage protective equipment typically operates within 5 to 20 cycles.

This chapter provides an introduction to power-system protection. Blackburn defines protection as "the science, skill, and art of applying and setting relays and/or fuses to provide maximum sensitivity to faults and undesirable conditions, but to avoid their operation on all permissible or tolerable conditions" [1]. The basic idea is to define the undesirable conditions and look for differences between the undesirable and permissible conditions that relays or fuses can sense. It is also important to remove only the faulted equipment from the system while maintaining as much of the unfaulted system as possible in service, in order to continue to supply as much of the load as possible.

Although fuses and reclosers (circuit breakers with built-in instrument transformers and relays) are widely used to protect primary distribution systems (with voltages in the 2.4 to 46 kV range), we focus primarily in this chapter on circuit breakers and relays, which are used to protect HV (115 to 230 kV) and EHV (345 to 765 kV) power systems. The IEEE defines a relay as "a device whose function is to detect defective lines or apparatus or other power system conditions of an abnormal or dangerous nature and to initiate appropriate control action" [1]. In practice, a relay is a device that closes or opens a contact when energized. Relays are also used in low-voltage (600-V and below) power systems and almost anywhere that electricity is used. For example, they are used in heating, air conditioning, stoves, clothes washers and dryers, refrigerators, dishwashers, telephone networks, traffic controls, airplane and other transportation systems, and robotics, as well as many other applications.

Problems with the protection equipment itself can occur. A second line of defense, called *backup* relays, may be used to protect the first line of defense, called *primary* relays. In HV and EHV systems, separate current- or voltage-measuring devices, separate trip coils on the circuit breakers, and separate batteries for the trip coils may be used. Also, the various protective devices must be properly coordinated such that primary relays assigned to protect equipment in a particular zone operate first. If the primary relays fail, then backup relays should operate after a specified time delay.

This chapter begins with a discussion of the basic system-protection components.

CASE STUDY Three relay technologies are presently used in power-system protection: (1) electromechanical relays, introduced in the early 1900s; (2) solid-state relays, introduced in the late 1950s; and (3) computer (microprocessor) relays, introduced in the 1980s. The numerous benefits of computer relays include flexibility, reliability with self-checking capability, and fault location/event-recording capabilities with local and remote reporting. Experts describe the rapid changes in computer relaying with constantly expanding product development as an industry revolution. The following article reviews applications of computer relays, their benefits and impacts [14].

The Impact of Microprocessor Protective Relays in the Electric Utility Industry

KEITH W. JEFFERS

With increasing awareness of the possibilities and benefits of the microprocessor protective relay, the past decade has been a critical time in the development of microprocessor technology.

Use of the microprocessor relay in numerous Burns & McDonnell projects over the past 7 yr has progressed from experimental to frontline complex protective schemes, such as piloted schemes. Microprocessor relays are now applied in both primary and secondary schemes; the first and second lines of defense. Every application reveals some new value inherent to its usage.

Although the industry is changing, the potential selectors of the microprocessor relays remain fairly constant: utility managers, in-house engineers, outside consultants, operations personnel and technical staff. These people will be affected by the change from electro-mechanical relay systems to microprocessor methods. While implementation of this new technology may appear to be overwhelming to many old-school operators, cases exist where experienced personnel readily grasp the concepts and advantages of the new technology and its expanded features.

The resistance to change is understandable, but the microprocessor technology is here to stay as

Reprinted with permission from Transmission and Distribution.

demonstrated in the personal-computer business and countless home and industrial applications. In order to meet the needs of the protective-relay applications, individuals must become accepting, knowledgeable and accommodating to the new methodology.

Manufacturing is competitive

Diversification within the manufacturing realm has been most accentuated. The electro-mechanical relay world held two major manufacturers in the United States for 90 yr: General Electric and Westinghouse. Three other domestic manufacturers have crept into the business in the past 20 yr with solid-state discrete relays. The conversion process to the microprocessor relay is altered now by the influence of European manufacturers merging in the United States and Europe.

The introduction of the microprocessor relay witnessed an influx of new competitors, such as Schweitzer Engineering Laboratories, Inc., Pullman, WA, whose founder essentially incurred a revolution in the business. The others, large and small, in the business have been introducing more discrete microprocessor relays in rapid numbers. Competition is fierce in this rapidly growing field.

Standards are still needed

To date, the industry has not established standards. A serious examination of the application of the microprocessor relay is necessary to comprehend the major features involved in its implementation. This examination encompasses the why and how questions, and the benefits of the innovative method of protective relaying.

More and more benefits are being realized through microprocessor experimentation. The infiltration and success of this new technology over the past few years proves without a doubt that it is the future of the industry. Numerous case studies emphasize the benefits of the microprocessor relay, which features:

Lower costs than electro-mechanical relays.

Fault-location capability with local and remote reporting.

Space saving (switchboard and control building).

Event-recording capability with local, remote reporting.

Reliable, with self-checking capability.

Flexible applications, both in program and protection capability.

Short delivery time.

Simple panel layout.

Simple wiring.

Simple setting process.

Reduced engineering and drafting requirements.

First application

In 1984, Burns & McDonnell first experimented with the microprocessor relay at Blue Ridge Electric Membership Corp., Lenior, NC. The microprocessor relay was recommended as an alternative to an electro-magnetic backup. It was employed as a 230-kV step-distance backup to an electro-mechanical DCB (Directional Comparison Blocking) scheme, and also as the primary step-distance relay on the 95-kV transmission line system. At the time of startup, the two technicians learned the operation and maintenance of these schemes in 2 weeks. It appeared in this case that the engineering level involvement in the procurement process was reinforced in the rapid adaption by the relay technicians.

Expanded technology benefits Imperial Irrigation District

When major developers of geothermal power plants in the Imperial Valley contracted to deliver power to Southern California Edison Co., the Imperial Irrigation District (IID), Imperial, CA, had to provide the transmission facilities to deliver that power. The 230-kV transmission line project was protected by nine terminals of primary POTT (Permissive Overreaching Transfer Trip) schemes using microprocessor relays. The six radial 92-kV transmission line terminals used microprocessor-based primary step-distance and directional ground-overcurrent relays. All 230- and 92-kV lines used conventional electro-mechanical back-up relaying. Irby Construction Co., Jackson, MS, reported that the change requiring electro-mechanical backup relaying versus backup microprocessor relaying added more than 20 ft to the length of one control building.

Compatibility was proven as two microprocessors were designed and implemented within one operating substation. The G.E.C. Quadramho was installed on the 230-kV line as the primary relay in a POTT scheme, with an electro-mechanical step-distance backup. A Schweitzer design SEL-121G also was implemented for its fault-location capability on the same 230-kV line. The fault-location feature is an intrinsic benefit, and over the past 2 yr has shown fault locations accurately.

Remote communication by modem, RS 232C switching ports, telephone and IID's new system-wide microwave system allowed access to its relay engineer's desk in El Centro, CA, and to the consulting engineer's desk in Kansas City, MO. Gulf Western Electric, Baton Rouge, LA, a testing firm, performed microprocessor-based relay testing in the field, converting past electro-mechanical testing methods to microprocessor methodology within 2 weeks.

The metering, event-reporting and fault-analysis tools of the microprocessor relays proved invaluable at startup time. The relay engineer working closely with the testing personnel was able to confirm power flows, phase angles and trends, then check this data against switchboard meters, phase-angle meter readings and SCADA (Supervisory Control and Data Acquisition) transducer outputs. In this application, any discrepancy in readings proved later that the microprocessor was more accurate and correct than any other data available. A hard copy generated by the printer in the substation control building aided the energization process.

IID relay application engineers immediately began using the microprocessor-based relays on new transmission applications and old-relay replacement programs. IID, like other utilities, learned that replacement of old troublesome balanced-beam electromechanical impedance relays could be done on existing relay switchboards with less cost than two new electro-mechanical impedance relays. But with the third-zone impedance, timing, directional ground overcurrent features of the microprocessor-based relay, the entire obsolete line-relaying package could be replaced easily and economically. This activity also has been reported by the Board of Public Utilities of Kansas City, KS, and the Central Electric Power Corp., Jefferson City, MO. In these cases, the microprocessor relay was placed in one of the panel holes vacated by four previous electro-mechanical relays.

More recently, the microprocessor was used to fit a 92-kV transmission and 13.2-kV distribution system for IID. The 92-kV line predominantly was protected by a primary electro-mechanical pilot-wire method. The Schweitzer SEL-121G was employed as a back-up step-distance with directional ground overcurrent (Fig. 1). The microprocessor technology was capable of providing remote fault locations and event-recording features not available in the primary scheme.

Once IID gained experience in the 230-kV levels with microprocessor relaying, similar technology was used in transformer and 13.2-kV distribution protection. Asea Brown Boveri's (ABB) microprocessor Distribution Protective Unit (DPU) was applied as 13.2-kV feeder relays in conjunction with the Schweitzer SEL-167. In this case, both the DPU and the SEL-167 offer combined strengths. Either serves as primary/secondary relaying, reclosing and event recording. One provides fault location while the other offers

Figure 1 *Minimum panel space is required for this microprocessor relay*

front-keyboard access, spring charge and trip-coil monitoring. In this case, each relay system complemented the other.

Since the most critical and most expensive item in the substation is the station transformer, three different microprocessor schemes were applied at various IID sites. In all three cases, the low-voltage side was equipped with either an ABB RACID, ABB IMPRS, or a Schweitzer SEL-167. The ascending levels of monitoring not only provided event recording for analysis of fault magnitudes, but served as the low-side backup or main-breaker relay-protection scheme. Yet, conservative practices applied conventional electro-mechanical transformer-differential and high-side backup overcurrent relays.

Versatility is evident

A 1988 project for Lincoln Electric System, Lincoln, NE, provided the opportunity to test the microprocessor's flexibility and integrity in yet another manner. In this case, the utility had successful previous experience with the microprocessor relays at 115-kV level. The critical point in this instance is twofold:

There was a much higher voltage of 345 kV.

The microprocessor was established at the primary, secondary and tertiary levels.

The primary relay was installed as a DCUB (Directional Comparison Unblocking) system, the secondary as a POTT scheme, and the third as a step-distance backup. At the same location, a microprocessor relay was implemented on a 115-kV line as as a primary POTT scheme and DCB scheme. An electromechanical step-distance relay remains as backup. Use of the microprocessor in four variant applications (DCUB, DCB, POTT and stepped-distance) at two voltage levels, all at one site, exemplifies its versatility and substantiates its reliability.

The low cost and wide range of applications of the microprocessor relay encompasses the lower-voltage application as well. Willmar Municipal Utilities, Willmar, MN, operates a 69-kV transmission system. Since the Engineering Manager had a strong interest in the personal-computer-driven systems such as computer-aided drafting and SCADA, the conversion to microprocessor relays did not appear to be insurmountable. Evaluation of the looped 69-kV system determined their best application to be a microprocessor relay-POTT application. The same microprocessor relay was used as a backup stepped-distance scheme. A small control room required a miniature relay switchboard that was possible with the use of microprocessor relays. A small staff required minimal help with learning the techniques. The adjoining utility, United Power Association, Elk River, MN, which uses the same microprocessor relay, easily accepted the adjacent terminal because of their experience with microprocessor relays.

Spacing saving is an important benefit

Comparative panel area utilization is shown in Fig. 2. Note that the four primary devices (one being the microprocessor relay) require only 18% (480 sq inches) of the panel area while a typical electromechanical backup package (seven devices) requires 45% (1210 sq inches) of the panel total area of 2700 sq inches. Examination shows that a vertical-mounted microprocessor relay would fit in the space of one conventional electro-mechanical impedance relay in the case of a replacement installation.

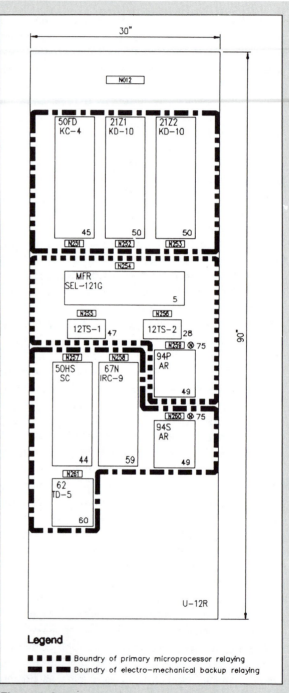

Legend

■ ■ ■ ■ ■ Boundry of primary microprocessor relaying
■ ■ ■ ■ ■ Boundry of electro-mechanical backup relaying

Figure 2 Space-saving advantages of the microprocessor relay are shown in this panel-utilization drawing

A 1989 study for Rochester Public Utility Dept., Rochester, MN, demonstrated the impact of space savings. Replacing a 115-kV with a new 161-kV substation on the same site required three new 161-kV line-backup relay schemes. The electro-mechanical panel layout required a costly 20-ft brick control-building extension that would have blocked the required 161-kV ring-bus construction space. An application of the Schweitzer SEL-121F relay and associated DTA metering display unit saved space. In addition, its reclosing, metering and synchronizing capability made it possible to install three line terminals in a 6-ft space, eliminating the necessity for the control-building extension.

The savings of the control-building expansion not only met the site space restriction, but funded the new relay board purchases. The owner also gained replacement of 40-yr old electro-mechanical relays and gained fault locating, event recording and combined metering-transducer package; all in one microprocessor relay application.

At IID, a skid-mounted, prewired control building (14 by 25 ft) manufactured by Electrical Power Products, Des Moines, IA, was possible for Carreon Substation, by the microprocessor relay space savings.

Fundamentally, the microprocessor relay is superlative to any previous methodology. The system usually operates on a single power supply. If the supply is down, the whole system is down. Similarly, nonintegrated electro-mechanical units may also trip upon the failure of one of the other components. Yet, the drawback to the relay can be overlooked because:

1. There is a backup, an electro-mechanical or microprocessor.

2. The microprocessor's intrinsic self-diagnosis reporting alerts the utility that a malfunction must be resolved.

Certainly, the microprocessor is not cost prohibitive. As is usually the case with technology, the price stabilizes or decreases after development, then further as equipment is refined, improved, tested, implemented and marketed competitively. The prices range from $3500 to $14,000, and considering the versatility of the units, it is cost-effective when the equivalent electro-mechanical system can cost from 50 to 200% more.

The computer era is making a marked impression on the microprocessor relay business. Experts concur that eventually the microprocessor will evolve into a central computer system controlling an entire substation. The manufacturers currently producing microprocessors will be constantly expanding product development, and the competition for customer acceptance will intensify. A revolution is changing the face of the industry and many are fortunate to participate and witness its evolution.

Keith W. Jeffers is the Transmission and Distribution Consultant for Burns & McDonnell Engineering Co., Kansas City, MO. He has the BSEE degree from Kansas State University and the MSEE degree from the University of Kansas. His current responsibilities include design and project management of conventional and turnkey substations. Mr. Jeffers is a member of IEEE and is a Registered Professional Engineer in the States of Missouri and Arizona.

SECTION 10.1

SYSTEM-PROTECTION COMPONENTS

Protection systems have three basic components:

1. Instrument transformers
2. Relays
3. Circuit breakers

Figure 10.1 shows a simple overcurrent protection schematic with: (1) one type of instrument transformer—the current transformer (CT), (2) an overcurrent relay (OC), and (3) a circuit breaker (CB) for a single-phase line. The function of the CT is to reproduce in its secondary winding a current I' that is proportional to the primary current I. The CT converts primary currents in the kiloamp range to secondary currents in the 0–5 ampere range for convenience of measurement, with the following advantages.

Figure 10.1

Overcurrent protection schematic

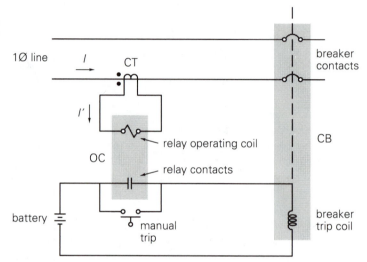

Safety: Instrument transformers provide electrical isolation from the power system so that personnel working with relays will work in a safer environment.

Economy: Lower-level relay inputs enable relays to be smaller, simpler, and less expensive.

Accuracy: Instrument transformers accurately reproduce power-system currents and voltages over wide operating ranges.

The function of the relay is to discriminate between normal operation and fault conditions. The OC relay in Figure 10.1 has an operating coil, which is connected to the CT secondary winding, and a set of contacts. When $|I'|$ exceeds a specified "pickup" value, the operating coil causes the normally open contacts to close. When the relay contacts close, the trip coil of the circuit breaker is energized, which then causes the circuit breaker to open.

Note that the circuit breaker does not open until its operating coil is energized, either manually or by relay operation. Based on information from instrument transformers, a decision is made and "relayed" to the trip coil of the breaker, which actually opens the power circuit—hence the name *relay*.

System-protection components have the following design criteria [2]:

Reliability: Operate dependably when fault conditions occur, even after remaining idle for months or years. Failure to do so may result in costly damages.

Selectivity: Avoid unnecessary, false trips.

Speed: Operate rapidly to minimize fault duration and equipment damage. Any intentional time delays should be precise.

Economy: Provide maximum protection at minimum cost.

Simplicity: Minimize protection equipment and circuitry.

Since it is impossible to satisfy all these criteria simultaneously, compromises must be made in system protection.

SECTION 10.2

INSTRUMENT TRANSFORMERS

There are two basic types of instrument transformers: voltage transformers (VTs), formerly called potential transformers (PTs), and current transformers (CTs). Figure 10.2 shows a schematic representation for the VT and CT. The transformer primary is connected to or into the power system and is insulated for the power-system voltage. The VT reduces the primary voltage and the CT reduces the primary current to much lower, standardized levels suitable for operation of relays. Photos of VTs and CTs are shown in Figures 10.3–10.6.

Figure 10.2

VT and CT schematic

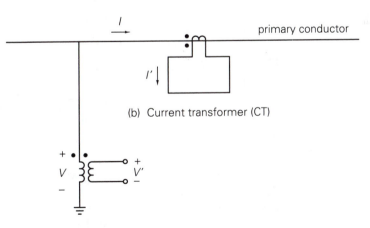

(b) Current transformer (CT)

(a) Voltage transformer (VT)

For system-protection purposes, VTs are generally considered to be sufficiently accurate. Therefore, the VT is usually modeled as an ideal transformer, where

$$V' = (1/n)V \qquad (10.2.1)$$

Figure 10.3

Three 34.5-kV voltage transformers with 34.5 kV : 115/67 volt VT ratios, at Lisle substation, Lisle, IL (Courtesy of Commonwealth Edison Company)

Figure 10.4

Three 500-kV coupling capacitor voltage transformers on the right with 303.1 kV : 115/67 volt VT ratios, with wave traps on the left, at Westwing 500-kV Switching Substation (Courtesy of Arizona Public Service Company)

Figure 10.5

Three 15-kV class current transformers with 2000:5 CT ratios, at Mountain View 69/12.5 kV substation (Courtesy of Arizona Public Service Company)

Figure 10.6

Three 500-kV class current transformers with 2000:5 CT ratios in front of 500-kV SF6 circuit breakers, Westwing 500-kV Switching Substation (Courtesy of Arizona Public Service Company)

V' is a scaled-down representation of V and is in phase with V. A standard VT secondary voltage rating is 115 V (line-to-line). Standard VT ratios are given in Table 10.1.

Ideally, the VT secondary is connected to a voltage-sensing device with infinite impedance, such that the entire VT secondary voltage is across the sensing device. In practice, the secondary voltage divides across the high-

Table 10.1

Standard VT ratios

VOLTAGE RATIOS						
1:1	2:1	2.5:1	4:1	5:1	20:1	40:1
60:1	100:1	200:1	300:1	400:1	600:1	800:1
1000:1	2000:1	3000:1	4500:1			

impedance sensing device and the VT series leakage impedances. VT leakage impedances are kept low in order to minimize voltage drops and phase-angle differences from primary to secondary.

The primary winding of a current transformer usually consists of a single turn, obtained by running the power system's primary conductor through the CT core. The normal current rating of CT secondaries is standardized at 5 A in the United States, whereas 1 A is standard in Europe and some other regions. Currents of 10 to 20 times (or greater) normal rating often occur in CT windings for a few cycles during short circuits. Standard CT ratios are given in Table 10.2.

Table 10.2

Standard CT ratios

CURRENT RATIOS						
50:5	100:5	150:5	200:5	250:5	300:5	400:5
450:5	500:5	600:5	800:5	900:5	1000:5	1200:5
1500:5	1600:5	2000:5	2400:5	2500:5	3000:5	3200:5
4000:5	5000:5	6000:5				

Ideally, the CT secondary is connected to a current-sensing device with zero impedance, such that the entire CT secondary current flows through the sensing device. In practice, the secondary current divides, with most flowing through the low-impedance sensing device and some flowing through the CT shunt excitation impedance. CT excitation impedance is kept high in order to minimize excitation current.

An approximate equivalent circuit of a CT is shown in Figure 10.7, where

Z' = CT secondary leakage impedance

X_e = (Saturable) CT excitation reactance

Z_B = Impedance of terminating device (relay, including leads)

Figure 10.7

CT equivalent circuit

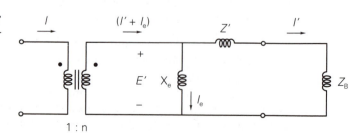

The total impedance Z_B of the terminating device is called the *burden* and is typically expressed in values of less than an ohm. The burden on a CT may also be expressed as volt-amperes at a specified current.

Associated with the CT equivalent circuit is an excitation curve that determines the relationship between the CT secondary voltage E' and excitation current I_e. Excitation curves for a multiratio bushing CT with ANSI classification C100 are shown in Figure 10.8.

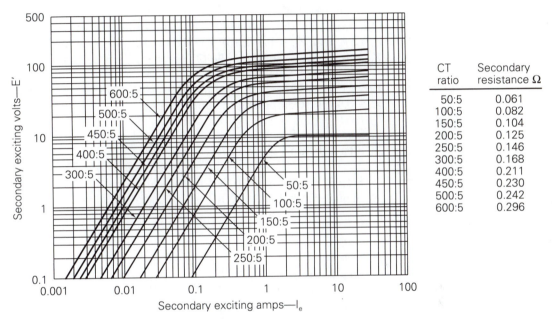

CT ratio	Secondary resistance Ω
50:5	0.061
100:5	0.082
150:5	0.104
200:5	0.125
250:5	0.146
300:5	0.168
400:5	0.211
450:5	0.230
500:5	0.242
600:5	0.296

Figure 10.8 Excitation curves for a multiratio bushing CT with a C100 ANSI accuracy classification [3]

Current = transformer performance is based on the ability to deliver a secondary output current I' that accurately reproduces the primary current I. Performance is determined by the highest current that can be reproduced without saturation to cause large errors. Using the CT equivalent circuit and excitation curves, the following procedure can be used to determine CT performance.

Step 1 Assume a CT secondary output current I'.

Step 2 Compute $E' = (Z' + Z_B)I'$.

Step 3 Using E', find I_e from the excitation curve.

Step 4 Compute $I = n(I' + I_e)$.

Step 5 Repeat Steps 1–4 for different values of I', then plot I' versus I.

For simplicity, approximate computations are made with magnitudes rather than with phasors. Also, the CT error is the percent difference between

$(I' + I_e)$ and I', given by:

$$\text{CT error} = \frac{I_e}{I' + I_e} \times 100\% \qquad (10.2.2)$$

The following examples illustrate the procedure.

| **EXAMPLE 10.1** | **Current transformer (CT) performance** |

Evaluate the performance of the multiratio CT in Figure 10.8 with a 100:5 CT ratio, for the following secondary output currents and burdens: (a) $I' = 5\,\text{A}$ and $Z_B = 0.5\,\Omega$; (b) $I' = 8\,\text{A}$ and $Z_B = 0.8\,\Omega$; and (c) $I' = 15\,\text{A}$ and $Z_B = 1.5\,\Omega$. Also, compute the CT error for each output current.

Solution From Figure 10.8, the CT with a 100:5 CT ratio has a secondary resistance $Z' = 0.082\,\Omega$. Completing the above steps:

a. Step 1 $I' = 5\,\text{A}$
   Step 2 From Figure 10.7,
   $$E' = (Z' + Z_B)I' = (0.082 + 0.5)(5) = 2.91 \text{ V}$$
   Step 3 From Figure 10.8, $I_e = 0.25\,\text{A}$
   Step 4 From Figure 10.7, $I = (100/5)(5 + 0.25) = 105\,\text{A}$
   $$\text{CT error} = \frac{0.25}{5.25} \times 100 = 4.8\%$$

b. Step 1 $I' = 8\,\text{A}$
   Step 2 From Figure 10.7,
   $$E' = (Z' + Z_B)I' = (0.082 + 0.8)(8) = 7.06 \text{ V}$$
   Step 3 From Figure 10.8, $I_e = 0.4\,\text{A}$
   Step 4 From Figure 10.7, $I = (100/5)(8 + 0.4) = 168\,\text{A}$
   $$\text{CT error} = \frac{0.4}{8.4} \times 100 = 4.8\%$$

c. Step 1 $I' = 15\,\text{A}$
   Step 2 From Figure 10.7,
   $$E' = (Z' + Z_B)I' = (0.082 + 1.5)(15) = 23.73 \text{ V}$$
   Step 3 From Figure 10.8, $I_e = 20\,\text{A}$
   Step 4 From Figure 10.7, $I = (100/5)(15 + 20) = 700\,\text{A}$
   $$\text{CT error} = \frac{20}{35} \times 100 = 57.1\%$$

Note that for the 15-A secondary current in (c), high CT saturation causes a large CT error of 57.1%. Standard practice is to select a CT ratio to give a little less than 5-A secondary output current at maximum normal load. From (a), the 100:5 CT ratio and 0.5 Ω burden are suitable for a maximum primary load current of about 100 A. This example is extended in Problem 10.2 to obtain a plot of I' versus I. ∎

| EXAMPLE 10.2 | **Relay operation versus fault current and CT burden** |

An overcurrent relay set to operate at 8 A is connected to the multiratio CT in Figure 10.8 with a 100:5 CT ratio. Will the relay detect a 200-A primary fault current if the burden Z_B is (a) .8 Ω, (b) 3.0 Ω?

Solution   Note that if an ideal CT is assumed, $(100/5) \times 8 = 160$-A primary current would cause the relay to operate.

a. From Example 10.1(b), a 168-A primary current with $Z_B = 0.8 \, \Omega$ produces a secondary output current of 8 A, which would cause the relay to operate. Therefore, the higher 200-A fault current will also cause the relay to operate.

b. **Step 1** $I' = 8 \, A$
   Step 2 From Figure 10.7,
   $$E' = (Z' + Z_B)I' = (0.05 + 3.0)(8) = 24.4 \, V$$
   Step 3 From Figure 10.8, $I_e = 30 \, A$
   Step 4 From Figure 10.7, $I = (100/5)(8 + 30) = 760 \, A$

With a 3.0-Ω burden, 760 A is the lowest primary current that causes the relay to operate. Therefore, the relay will not operate for the 200-A fault current.

■

SECTION 10.3

OVERCURRENT RELAYS

As shown in Figure 10.1, the CT secondary current I' is the input to the overcurrent relay operating coil. Instantaneous overcurrent relays respond to the magnitude of their input current, as shown by the trip and block regions in Figure 10.9. If the current magnitude $I' = |I'|$ exceeds a specified adjustable current magnitude I_p, called the *pickup* current, then the relay contacts close "instantaneously" to energize the circuit breaker trip coil. If I' is less than the pickup current I_p, then the relay contacts remain open, blocking the trip coil.

Time-delay overcurrent relays also respond to the magnitude of their input current, but with an intentional time delay. As shown in Figure 10.10, the time delay depends on the magnitude of the relay input current. If I' is a large multiple of the pickup current I_p, then the relay operates (or trips) after a small time delay. For smaller multiples of pickup, the relay trips after a longer time delay. And if $I' < I_p$, the relay remains in the blocking position.

The ABB-Westinghouse series of CO relays, one of which is shown in Figure 10.11, is a typical time-delay overcurrent relay product line. Characteristic curves of the CO-8 relay are shown in Figure 10.12. These relays have two settings:

Figure 10.9

Instantaneous overcurrent relay block and trip regions

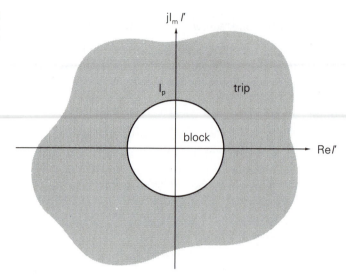

Figure 10.10

Time-delay overcurrent relay block and trip regions

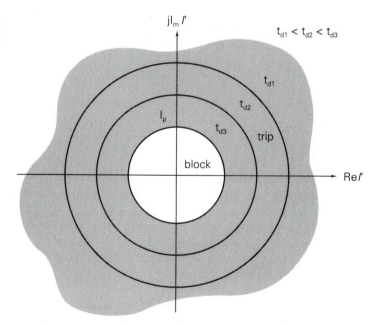

Current tap setting: The pickup current in amperes

Time-dial setting: The adjustable amount of time delay

The characteristic curves are usually shown with operating time in seconds versus relay input current as a multiple of the pickup current. The curves are asymptotic to the vertical axis and decrease with some inverse power of current magnitude for values exceeding the pickup current. This

Figure 10.11

Electromechanical time-delay overcurrent relay (Courtesy of ABB-Westinghouse)

inverse time characteristic can be shifted up or down by adjustment of the time-dial setting. Although discrete time-dial settings are shown in Figure 10.12 on page 372, intermediate values can be obtained by interpolating between the discrete curves.

EXAMPLE 10.3   **Operating time for a CO-8 time-delay overcurrent relay**

The CO-8 relay with a current tap setting of 6 amperes and a time-dial setting of 1 is used with the 100:5 CT in Example 10.1. Determine the relay operating time for each case.

Solution   **a.** From Example 10.1(a)

$$I' = 5 \, \text{A} \qquad \frac{I'}{I_p} = \frac{5}{6} = 0.83$$

The relay does not operate. It remains in the blocking position.

b. $I' = 8 \, \text{A} \qquad \dfrac{I'}{I_p} = \dfrac{8}{6} = 1.33$

Using curve 1 in Figure 10.12, $t_{operating} = 6$ seconds.

c. $I' = 15\,A$ $\qquad \dfrac{I'}{I_p} = \dfrac{15}{6} = 2.5$

From curve 1, $t_{operating} = 1.2$ seconds. ∎

Figure 10.13 shows the time-current characteristics of five CO time-delay overcurrent relays used in transmission and distribution lines. The time-dial settings are selected in the figure so that all relays operate in 0.2 seconds at 20 times the pickup current. The choice of relay time-current characteristic depends on the sources, lines, and loads. The definite (CO-6) and moderately inverse (CO-7) relays maintain a relatively constant operating time above 10 times pickup. The inverse (CO-8), very inverse (CO-9), and extremely inverse (CO-11) relays operate respectively faster on higher fault currents.

Figure 10.14 illustrates the operating principle of an electromechanical time-delay overcurrent relay. The ac input current to the relay operating coil sets up a magnetic field that is perpendicular to a conducting aluminum disc.

Figure 10.12

CO-8 time-delay overcurrent relay characteristics (Courtesy of Westinghouse Electric Corporation)

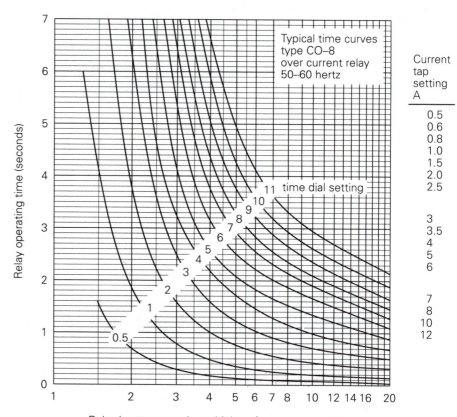

Relay input current in multiples of current tap setting

Figure10.13

Comparison of CO relay characteristics (Courtesy of ABB-Westinghouse)

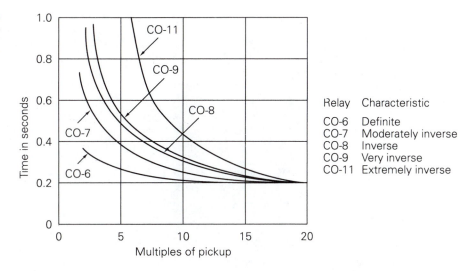

Relay	Characteristic
CO-6	Definite
CO-7	Moderately inverse
CO-8	Inverse
CO-9	Very inverse
CO-11	Extremely inverse

Figure 10.14

Electromechanical time-delay overcurrent relay—induction disc type

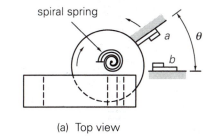

(a) Top view

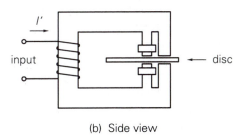

(b) Side view

The disc can rotate and is restrained by a spiral spring. Current is induced in the disc, interacts with the magnetic field, and produces a torque. If the input current exceeds the pickup current, the disc rotates through an angle ϕ to close the relay contacts. The larger the input current, the larger the torque and the faster the contact closing. After the current is removed or reduced below the pickup, the spring provides reset of the contacts.

A solid state relay panel between older-style electromechanical relays is shown in Figure 10.15.

Figure 10.15

Solid-state relay panel (center) for a 345-kV transmission line, with electromechanical relays on each side, at Electric Junction Substation, Naperville, IL (Courtesy of Commonwealth Edison Company)

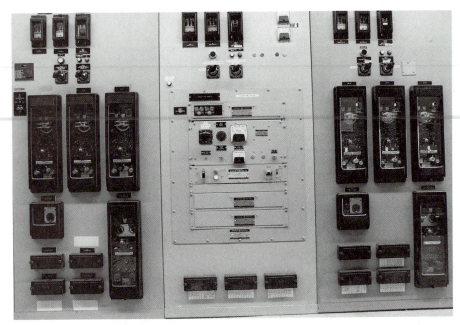

SECTION 10.4

RADIAL SYSTEM PROTECTION

Many radial systems are protected by time-delay overcurrent relays. Adjustable time delays can be selected such that the breaker closest to the fault opens, while other upstream breakers with larger time delays remain closed. That is, the relays can be coordinated to operate in sequence so as to interrupt minimum load during faults. Successful relay coordination is obtained when fault currents are much larger than normal load currents. Also, coordination of overcurrent relays usually limits the maximum number of breakers in a radial system to five or less, otherwise the relay closest to the source may have an excessive time delay.

Consider a fault at P_1 to the right of breaker B3 for the radial system of Figure 10.16. For this fault we want breaker B3 to open while B2 (and B1) remains closed. Under these conditions, only load L3 is interrupted. We could select a longer time delay for the relay at B2, so that B3 operates first. Thus, for any fault to the right of B3, B3 provides primary protection. Only if B3 fails to open will B2 open, after time delay, thus providing backup protection.

Similarly, consider a fault at P_2 between B2 and B3. We want B2 to open while B1 remains closed. Under these conditions, loads L2 and L3 are interrupted. Since the fault is closer to the source, the fault current will be larger than for the previous fault considered. B2, set to open for the previous, smaller fault current after time delay, will open more rapidly for this fault. We

Figure 10.16

Single-line diagram of a 34.5-kV radial system

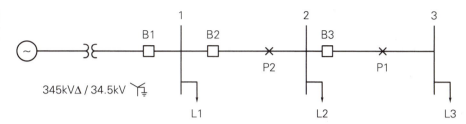

345kVΔ / 34.5kV

also select the B1 relay with a longer time delay than B2, so that B2 opens first. Thus, B2 provides primary protection for faults between B2 and B3, as well as backup protection for faults to the right of B3. Similarly, B1 provides primary protection for faults between B1 and B2, as well as backup protection for further downstream faults.

The *coordination time interval* is the time interval between the primary and remote backup protective devices. It is the difference between the time that the backup relaying operates and the time that circuit breakers clear the fault under primary relaying. Precise determination of relay operating times is complicated by several factors, including CT error, dc offset component of fault current, and relay overtravel. Therefore, typical coordination time intervals from 0.2 to 0.5 seconds are selected to account for these factors in most practical applications.

EXAMPLE 10.4 | **Coordinating time-delay overcurrent relays in a radial system**

Data for the 60-Hz radial system of Figure 10.16 are given in Tables 10.3, 10.4, and 10.5. Select current tap settings (TSs) and time-dial settings (TDSs) to protect the system from faults. Assume three CO-8 relays for each breaker, one for each phase, with a 0.3-second coordination time interval. The relays for each breaker are connected as shown in Figure 10.17, so that all three phases of the breaker open when a fault is detected on any one phase. Assume a 34.5-kV (line-to-line) voltage at all buses during normal operation. Also,

Figure 10.3

Maximum loads—Example 10.4

BUS	S	LAGGING p.f.
	MVA	
1	11.0	0.95
2	4.0	0.95
3	6.0	0.95

Table 10.4

Symmetrical fault currents—Example 10.4

BUS	MAXIMUM FAULT CURRENT (Bolted Three-Phase)	MINIMUM FAULT CURRENT (L–G or L–L)
	A	A
1	3000	2200
2	2000	1500
3	1000	700

Figure 10.17

Relay connections to trip all
three phases

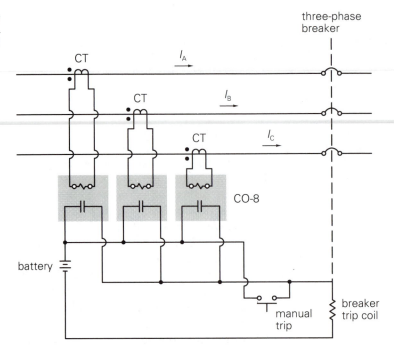

Table 10.5

Breaker, CT, and relay data—
Example 10.4

BREAKER	BREAKER OPERATING TIME	CT RATIO	RELAY
B1	5 cycles	400:5	CO-8
B2	5 cycles	200:5	CO-8
B3	5 cycles	200:5	CO-8

future load growth is included in Table 10.3, such that maximum loads over
the operating life of the radial system are given in this table.

Solution First select TSs such that the relays do not operate for maximum load
currents. Starting at B3, the primary and secondary CT currents for maximum load L3 are

$$I_{L3} = \frac{S_{L3}}{V_3\sqrt{3}} = \frac{6 \times 10^6}{(34.5 \times 10^3)\sqrt{3}} = 100.4 \, A$$

$$I'_{L3} = \frac{100.4}{(200/5)} = 2.51 \, A$$

From Figure 10.12, we select for the B3 relay a 3-A TS, which is the lowest
TS above 2.51 A.

Note that $|S_{L2} + S_{L3}| = |S_{L2}| + |S_{L3}|$ because the load power factors are
identical. Thus, at B2, the primary and secondary CT currents for maximum

load are

$$I_{L2} = \frac{S_{L2} + S_{L3}}{V_2\sqrt{3}} = \frac{(4 + 6) \times 10^6}{(34.5 \times 10^3)\sqrt{3}} = 167.3 \text{ A}$$

$$I'_{L2} = \frac{167.3}{(200/5)} = 4.18 \text{ A}$$

From Figure 10.12, select for the B2 relay a 5-A TS, the lowest TS above 4.18 A. At B1,

$$I_{L1} = \frac{S_{L1} + S_{L2} + S_{L3}}{V_1\sqrt{3}} = \frac{(11 + 4 + 6) \times 10^6}{(34.5 \times 10^3)\sqrt{3}} = 351.4 \text{ A}$$

$$I'_{L1} = \frac{351.4}{(400/5)} = 4.39 \text{ A}$$

Select a 5-A TS for the B1 relay.

Next select the TDSs. We first coordinate for the maximum fault currents in Table 10.4, checking coordination for minimum fault currents later. Starting at B3, the largest fault current through B3 is 2000 A, which occurs for the three-phase fault at bus 2 (just to the right of B3). Neglecting CT saturation, the fault-to-pickup current ratio at B3 for this fault is

$$\frac{I'_{3\text{Fault}}}{\text{TS3}} = \frac{2000/(200/5)}{3} = 16.7$$

Since we want to clear faults as rapidly as possible, select a 1/2 TDS for the B3 relay. Then, from the 1/2 TDS curve in Figure 10.12, the relay operating time is T3 = 0.05 seconds. Adding the breaker operating time (5 cycles = 0.083 s), primary protection clears this fault in T3 + T_{breaker} = 0.05 + 0.083 = 0.133 seconds.

For this same fault, the fault-to-pickup current ratio at B2 is

$$\frac{I'_{2\text{Fault}}}{\text{TS2}} = \frac{2000/(200/5)}{5} = 10.0$$

Adding the B3 relay operating time (T3 = 0.05 s), breaker operating time (0.083 s), and 0.3 s coordination time interval, we want a B2 relay operating time

$$T2 = T3 + T_{\text{breaker}} + T_{\text{coordination}} = 0.05 + 0.083 + 0.3 \approx 0.43 \text{ s}$$

From Figure 10.12, select TDS2 = 2.

Next select the TDS at B1. The largest fault current through B2 is 3000 A, for a three-phase fault at bus 1 (just to the right of B2). The fault-to-pickup current ratio at B2 for this fault is

$$\frac{I'_{2\text{Fault}}}{\text{TS2}} = \frac{3000/(200/5)}{5} = 15.0$$

From the 2 TDS curve in Figure 10.12, T2 = 0.38 s. For this same fault,

$$\frac{I'_{1\text{Fault}}}{\text{TS1}} = \frac{3000/(400/5)}{5} = 7.5$$

$$\text{T1} = \text{T2} + \text{T}_{\text{breaker}} + \text{T}_{\text{coordination}} = 0.38 + 0.083 + 0.3 \approx 0.76\,\text{s}$$

From Figure 10.12, select TDS1 = 3. The relay settings are shown in Table 10.6. Note that for reliable relay operation the fault-to-pickup current ratios with minimum fault currents should be greater than 2. Coordination for minimum fault currents listed in Table 10.4 is evaluated in Problem 10.6.

Table 10.6

Solution—Example 10.4

BREAKER	RELAY	TS	TDS
B1	CO-8	5	3
B2	CO-8	5	2
B3	CO-8	3	1/2

■

Note that separate relays are used for each phase in Example 10.4, and therefore these relays will operate for three-phase as well as line-to-line, single line-to-ground, and double line-to-ground faults. However, in many cases single line-to-ground fault currents are much lower than three-phase fault currents, especially for distribution feeders with high zero-sequence impedances. In these cases a separate ground relay with a lower current tap setting than the phase relays is used. The ground relay is connected to operate on zero-sequence current from three of the phase CTs connected in parallel or from a CT in the grounded neutral.

SECTION 10.5

RECLOSERS AND FUSES

Automatic circuit reclosers are commonly used for distribution circuit protection. A *recloser* is a self-controlled device for automatically interrupting and reclosing an ac circuit with a preset sequence of openings and reclosures. Unlike circuit breakers, which have separate relays to control breaker opening and reclosing, reclosers have built-in controls. More than 80% of faults on overhead distribution circuits are temporary, caused by tree limb contact, by animal interference, by wind bringing bare conductors in contact, or by lightning. The automatic tripping-reclosing sequence of reclosers clears these temporary faults and restores service with only momentary outages, thereby significantly improving customer service. A disadvantage of reclosers is the increased hazard when a circuit is physically contacted by people, for example in the case of a broken conductor at ground level that remains

Figure 10.18

Single-line diagram of a 13.8-kV radial distribution feeder with fuse/recloser/relay protection

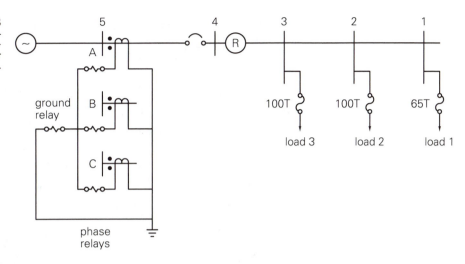

Table 10.7

Data for Figure 10.18

Bus	Maximum load current A	3φ fault current A	IL-G fault current A
1	60	1000	850
2	95	1500	1300
3	95	2000	1700
4	250	3000	2600
5	250	4000	4050

energized. Also, reclosing should be locked out during live-line maintenance by utility personnel.

Figure 10.18 shows a common protection scheme for radial distribution circuits utilizing fuses, reclosers, and time-delay overcurrent relays. Data for the 13.8-kV feeder in this figure is given in Table 10.7. There are three load taps protected by fuses. The recloser ahead of the fuses is set to open and reclose for faults up to and beyond the fuses. For temporary faults the recloser can be set for one or more instantaneous or time-delayed trips and reclosures in order to clear the faults and restore service. If faults persist, the fuses operate for faults to their right (downstream), or the recloser opens after time delay and locks out for faults between the recloser and fuses. Separate time-delay overcurrent phase and ground relays open the substation breaker after multiple reclosures of the recloser.

Coordination of the fuses, recloser, and time-delay overcurrent relays is shown via the time-current curves in Figure 10.19. Type T (slow) fuses are selected because their time-current characteristics coordinate well with reclosers. The fuses are selected on the basis of maximum loads served from the taps. A 65T fuse is selected for the bus 1 tap, which has a 60-A maximum load current, and 100T is selected for the bus 2 and 3 taps, which have 95-A

Figure 10.19

Time-current curves for the radial distribution circuit of Figure 10.18

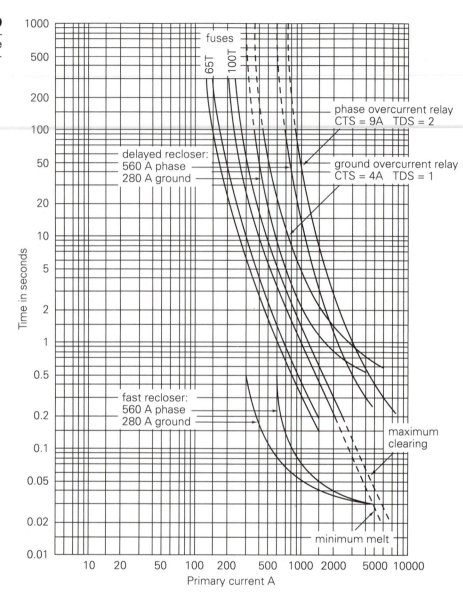

maximum load currents. The fuses should also have a rated voltage larger than the maximum bus voltage and an interrupting current rating larger than the maximum asymmetrical fault current at the fuse location. Type T fuses with voltage ratings of 15 kV and interrupting current ratings of 10 kA and higher are standard.

Standard reclosers have minimum trip ratings of 50, 70, 100, 140, 200, 280, 400, 560, 800, 1120, and 1600 A, with voltage ratings up to 38 kV and maximum interrupting currents up to 16 kA. A minimum trip rating of 200–250% of maximum load current is typically selected for the phases, in order to override cold load pickup with a safety factor. The minimum trip rating of the ground unit is typically set at maximum load and should be higher than the maximum allowable load unbalance. For the recloser in Figure 10.18, which carries a 250-A maximum load, minimum trip ratings of 560 A for each phase and 280 A for the ground unit are selected.

A popular operation sequence for reclosers is two fast operations, without intentional time delay, followed by two delayed operations. The fast operations allow temporary faults to self-clear, while the delayed operations allow downstream fuses to clear permanent faults. Note that the time-current curves of the fast recloser lie below the fuse curves in Figure 10.19, such that the recloser opens before the fuses melt. The fuse curves lie below the delayed recloser curves, such that the fuses clear before the recloser opens. The recloser is typically programmed to reclose $\frac{1}{2}$ s after the first fast trip, 2 s after the second fast trip, and 5–10 s after a delayed trip.

Time-delay overcurrent relays with an extremely inverse characteristic coordinate with both reclosers and type T fuses. A 300 : 5 CT ratio is selected to give a secondary current of $250 \times (5/300) = 4.17$ A at maximum load. Relay settings are selected to allow the recloser to operate effectively to clear faults before relay operation. A current tap setting of 9 A is selected for the CO-11 phase relays so that minimum pickup exceeds twice the maximum load. A time-dial setting of 2 is selected so that the delayed recloser trips at least 0.2 s before the relay. The ground relay is set with a current tap setting of 4 A and a time-dial setting of 1.

| EXAMPLE 10.5 | **Fuse/recloser coordination**

For the system of Figure 10.18, describe the operating sequence of the protective devices for the following faults: (a) a self-clearing, temporary, three-phase fault on the load side of tap 2; and (b) a permanent three-phase fault on the load side of tap 2.

Solution   **a.** From Table 10.7, the three-phase fault current at bus 2 is 1500 A. From Figure 10.19, the 560-A fast recloser opens 0.05 s after the 1500-A fault current occurs, and then recloses $\frac{1}{2}$ s later. Assuming the fault has self-cleared, normal service is restored. During the 0.05-s fault duration, the 100T fuse does not melt.

b. For a permanent fault the fast recloser opens after 0.05 s, recloses $\frac{1}{2}$ s later into the permanent fault, opens again after 0.05 s, and recloses into the fault a second time after a 2-s delay. Then the 560-A delayed recloser opens 3 seconds later. During this interval the 100T fuse clears the fault. The delayed recloser then recloses 5 to 10 s later, restoring service to loads 1 and 3.   ■

DIRECTIONAL RELAYS

Directional relays are designed to operate for fault currents in only one direction. Consider the directional relay D in Figure 10.20, which is required to operate only for faults to the right of the CT. Since the line impedance is mostly reactive, a fault at P_1 to the right of the CT will have a fault current I from bus 1 to bus 2 that lags the bus voltage V by an angle of almost 90°. This fault current is said to be in the forward direction. On the other hand, a fault at P_2, to the left of the CT, will have a fault current I that leads V by almost 90°. This fault current is said to be in the reverse direction.

Figure 10.20

Directional relay in series with overcurrent relay (only phase A is shown)

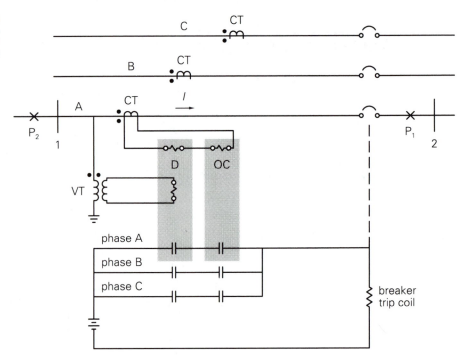

The directional relay has two inputs: the reference voltage $V = V\underline{/0°}$, and current $I = I\underline{/\phi}$. The relay trip and block regions, shown in Figure 10.21, can be described by

$$-180° < (\phi - \phi_1) < 0° \quad \text{(Trip)}$$

$$\text{Otherwise} \quad \text{(Block)} \tag{10.6.1}$$

where ϕ is the angle of the current with respect to the voltage and ϕ_1, typically 2° to 8°, defines the boundary between the trip and block regions.

Figure 10.21

Directional relay block and trip regions in the complex plane

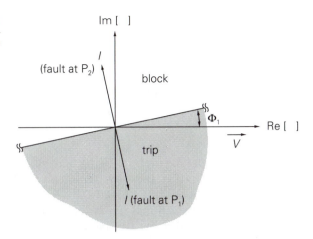

The contacts of the overcurrent relay OC and the directional relay D are connected in series in Figure 10.20, so that the breaker trip coil is energized only when the CT secondary current (1) exceeds the OC relay pickup value, and (2) is in the forward tripping direction.

Although construction details differ, the operating principle of an electromechanical directional relay is similar to that of a watt-hour meter. There are two input coils, a voltage coil and a current coil, both located on a stator, and there is a rotating disc element. Suppose that the reference voltage is passed through a phase-shifting element to obtain $V_1 = V\underline{/\phi_1 - 90°}$. If V_1 and $I = I\underline{/\phi}$ are applied to a watt-hour meter, the torque on the rotating element is

$$T = kVI \cos(\phi_1 - \phi - 90°) = kVI \sin(\phi_1 - \phi) \qquad (10.6.2)$$

Note that for faults in the forward direction the current lags the voltage, and the angle $(\phi_1 - \phi)$ in (10.6.2) is close to 90°. This results in maximum positive torque on the rotating disc, which would cause the relay contacts to close. On the other hand, for faults in the reverse direction the current leads the voltage, and $(\phi_1 - \phi)$ is close to $-90°$. This results in maximum negative torque tending to rotate the disc element in the backward direction. Backward motion can be restrained by mechanical stops.

SECTION 10.7

PROTECTION OF TWO-SOURCE SYSTEM WITH DIRECTIONAL RELAYS

It becomes difficult and in some cases impossible to coordinate overcurrent relays when there are two or more sources at different locations. Consider the system with two sources shown in Figure 10.22. Suppose there is a fault at P_1. We want B23 and B32 to clear the fault so that service to the three loads

Figure 10.22

System with two sources

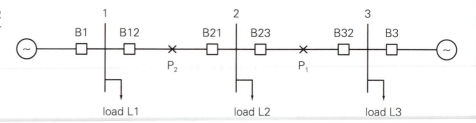

continues without interruption. Using time-delay overcurrent relays, we could set B23 faster than B21. But now consider a fault at P_2 instead. Breaker B23 will open faster than B21, and load L2 will be disconnected. When a fault can be fed from both the left and right, overcurrent relays cannot be coordinated. However, directional relays can be used to overcome this problem.

EXAMPLE 10.6   **Two-source system protection with directional and time-delay overcurrent relays**

Explain how directional and time-delay overcurrent relays can be used to protect the system in Figure 10.22. Which relays should be coordinated for a fault (a) at P_1, (b) at P_2? (c) Is the system also protected against bus faults?

Solution   Breakers B12, B21, B23, and B32 should respond only to faults on their "forward" or "line" sides. Directional overcurrent relays connected as shown in Figure 10.20 can be used for these breakers. Overcurrent relays alone can be used for breakers B1 and B3, which do not need to be directional.

   a. For a fault at P_1, the B21 relay would not operate; B12 should coordinate with B23 so that B23 trips before B12 (and B1). Also, B3 should coordinate with B32.

   b. For a fault at P_2, B23 would not operate; B32 should coordinate with B21 so that B21 trips before B32 (and B3). Also, B1 should coordinate with B12.

   c. Yes, the directional overcurrent relays also protect the system against bus faults. If the fault is at bus 2, relays at B21 and B23 will not operate, but B12 and B32 will operate to clear the fault. B1 and B21 will operate to clear a fault at bus 1. B3 and B23 will clear a fault at bus 3. ■

SECTION 10.8

ZONES OF PROTECTION

Protection of simple systems has been discussed so far. For more general power-system configurations, a fundamental concept is the division of a system into protective zones [1]. If a fault occurs anywhere within a zone, action will be taken to isolate that zone from the rest of the system. Zones are defined for:

Generators

Transformers

Buses

Transmission and distribution lines

Motors

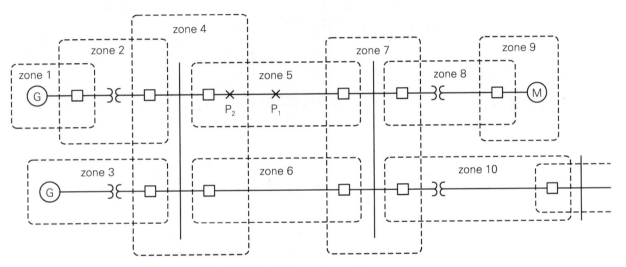

Figure 10.23   Power-system protective zones

Figure 10.23 illustrates the protective zone concept. Each zone is defined by a closed, dashed line. Zone 1, for example, contains a generator and connecting leads to a transformer. In some cases a zone may contain more than one component. For example, zone 3 contains a generator-transformer unit and connecting leads to a bus, and zone 10 contains a transformer and a line.

Protective zones have the following characteristics:

Zones are overlapped.

Circuit breakers are located in the overlap regions.

For a fault anywhere in a zone, all circuit breakers in that zone open to isolate the fault.

Neighboring zones are overlapped to avoid the possibility of unprotected areas. Without overlap the small area between two neighboring zones would not be located in any zone and thus would not be protected.

Since isolation during faults is done by circuit breakers, they should be inserted between equipment in a zone and each connection to the system. That is, breakers should be inserted in each overlap region. As such, they identify the boundaries of protective zones. For example, zone 5 in Figure 10.23 is connected to zones 4 and 7. Therefore, a circuit breaker is located in the overlap region between zones 5 and 4, as well as between zones 5 and 7.

If a fault occurs anywhere within a zone, action is taken to open all breakers in that zone. For example, if a fault occurs at P_1 on the line in zone 5, then the two breakers in zone 5 should open. If a fault occurs at P_2 within the overlap region of zones 4 and 5, then all five breakers in zones 4 and 5 should open. Clearly, if a fault occurs within an overlap region, two zones will be isolated and a larger part of the system will be lost from service. In order to minimize this possibility, overlap regions are kept as small as possible.

Overlap is accomplished by having two sets of instrument transformers and relays for each circuit breaker. For example, the breaker in Figure 10.24 shows two CTs, one for zone 1 and one for zone 2. Overlap is achieved by the order of the arrangement: first the equipment in the zone, second the breaker, and then the CT for that zone.

Figure 10.24

Overlapping protection around a circuit breaker

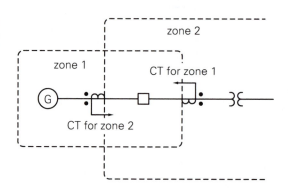

Figure 10.25

Power system for Example 10.6

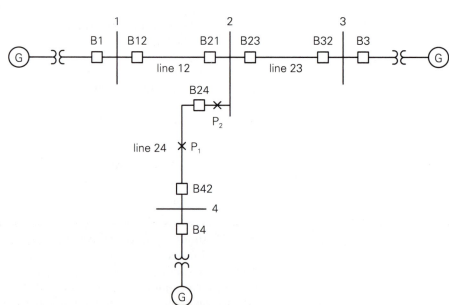

EXAMPLE 10.7	**Zones of protection**

Draw the protective zones for the power system shown in Figure 10.25. Which circuit breakers should open for a fault at P_1? at P_2?

Solution   Noting that circuit breakers identify zone boundaries, protective zones are drawn with dashed lines as shown in Figure 10.26. For a fault at P_1, located in zone 5, breakers B24 and B42 should open. For a fault at P_2, located in the overlap region of zones 4 and 5, breakers B24, B42, B21, and B23 should open.

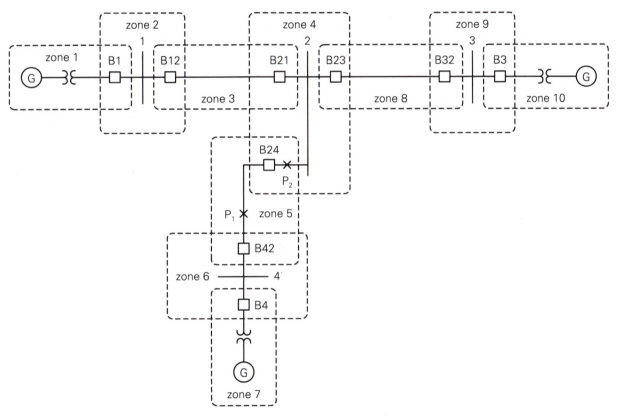

Figure 10.26   Protective zones for Example 10.7   ∎

LINE PROTECTION WITH IMPEDANCE (DISTANCE) RELAYS

Coordinating time-delay overcurrent relays can also be difficult for some radial systems. If there are too many radial lines and buses, the time delay for the breaker closest to the source becomes excessive.

Also, directional overcurrent relays are difficult to coordinate in transmission loops with multiple sources. Consider the use of these relays for the transmission loop shown in Figure 10.27. For a fault at P_1, we want the B21 relay to operate faster than the B32 relay. For a fault at P_2, we want B32 faster than B13. And for a fault at P_3, we want B13 faster than B21. Proper coordination, which depends on the magnitudes of the fault currents, becomes a tedious process. Furthermore, when consideration is given to various lines or sources out of service, coordination becomes extremely difficult.

Figure 10.27

345-kV transmission loop

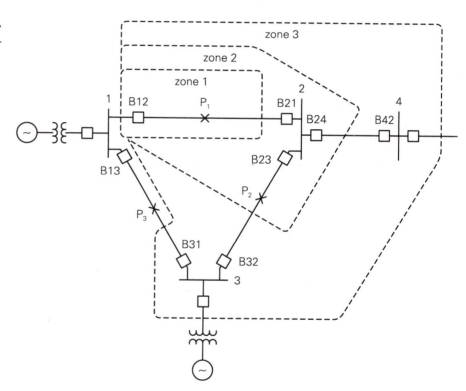

To overcome these problems, relays that respond to a voltage-to-current ratio can be used. Note that during a three-phase fault, current increases while bus voltages close to the fault decrease. If, for example, current increases by a factor of 5 while voltage decreases by a factor of 2, then the voltage-to-current ratio decreases by a factor of 10. That is, the voltage-to-current ratio is more sensitive to faults than current alone. A relay that operates on the basis of voltage-to-current ratio is called an *impedance* relay. It is also called a *distance* relay or a *ratio* relay.

Impedance relay block and trip regions are shown in Figure 10.28, where the impedance Z is defined as the voltage-to-current ratio at the relay location. The relay trips for $|Z| < |Z_r|$, where Z_r is an adjustable relay setting.

Figure 10.28

Impedance relay block and trip regions

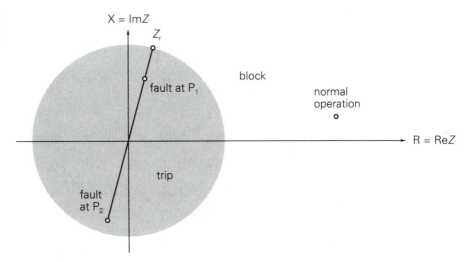

The impedance circle that defines the border between the block and trip regions passes through Z_r.

A straight line called the *line impedance locus* is shown for the impedance relay in Figure 10.28. This locus is a plot of positive sequence line impedances, predominantly reactive, as viewed between the relay location and various points along the line. The relay setting Z_r is a point in the R-X plane through which the impedance circle that defines the trip-block boundary must pass.

Consider an impedance relay for breaker B12 in Figure 10.27, for which $Z = V_1/I_{12}$. During normal operation, load currents are usually much smaller than fault currents, and the ratio Z has a large magnitude (and some arbitrary phase angle). Therefore, Z will lie outside the circle of Figure 10.28, and the relay will not trip during normal operation.

During a three-phase fault at P_1, however, Z appears to relay B12 to be the line impedance from the B12 relay to the fault. If $|Z_r|$ in Figure 10.28 is set to be larger than the magnitude of this impedance, then the B12 relay will trip. Also, during a three-phase fault at P_3, Z appears to relay B12 to be the negative of the line impedance from the relay to the fault. If $|Z_r|$ is larger than the magnitude of this impedance, the B12 relay will trip. Thus, the impedance relay of Figure 10.28 is not directional; a fault to the left or right of the relay can cause a trip.

Two ways to include directional capability with an impedance relay are shown in Figure 10.29. In Figure 10.29(a), an impedance relay with directional restraint is obtained by including a directional relay in series with an impedance relay, just as was done previously with an overcurrent relay. In Figure 10.29(b), a modified impedance relay is obtained by offsetting the center of the impedance circle from the origin. This modified impedance relay is sometimes called a *Mho* relay. If either of these relays is used at B12 in

Figure 10.29

Impedance relays with directional capability

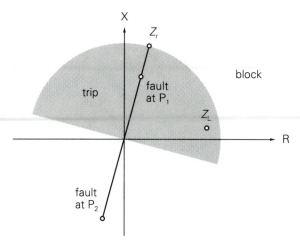

(a) Impedance relay with directional restraint

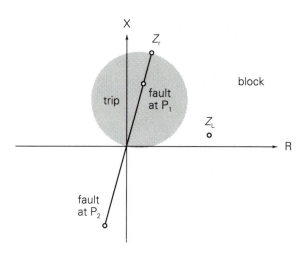

(b) Modified impedance relay (mho relay)

Figure 10.27, a fault at P_1 will result in a trip decision, but a fault at P_3 will result in a block decision.

Note that the radius of the impedance circle for the modified impedance relay is half of the corresponding radius for the impedance relay with directional restraint. The modified impedance relay has the advantage of better selectivity for high power factor loads. For example, the high power factor load Z_L lies outside the trip region of Figure 10.29(b) but inside the trip region of Figure 10.29(a).

The *reach* of an impedance relay denotes how far down the line the relay detects faults. For example, an 80% reach means that the relay will detect any (solid three-phase) fault between the relay and 80% of the line length. This explains the term *distance* relay.

It is common practice to use three directional impedance relays per phase, with increasing reaches and longer time delays. For example, Figure 10.27 shows three protection zones for B12. The zone 1 relay is typically set for an 80% reach and instantaneous operation, in order to provide primary protection for line 1–2. The zone 2 relay is set for about 120% reach, extending beyond bus 2, with a typical time delay of 0.2 to 0.3 seconds. The zone 2 relay provides backup protection for faults on line 1–2 as well as remote backup for faults on line 2–3 or 2–4 in zone 2.

Note that in the case of a fault on line 2–3 we want the B23 relay to trip, not the B12 relay. Since the impedance seen by B12 for faults near bus 2, either on line 1–2 or line 2–3, is essentially the same, we cannot set the B12 zone 1 relay for 100% reach. Instead, an 80% reach is selected to avoid instantaneous operation of B12 for a fault on line 2–3 near bus 2. For example, if there is a fault at P_2 on line 2–3, B23 should trip instantaneously; if it fails, B12 will trip after time delay. Other faults at or near bus 2 also cause tripping of the B12 zone 2 relay after time delay.

Reach for the zone 3 B12 relay is typically set to extend beyond buses 3 and 4 in Figure 10.27, in order to provide remote backup for neighboring lines. As such, the zone 3 reach is set for 100% of line 1–2 plus 120% of either line 2–3 or 2–4, whichever is longer, with an even larger time delay, typically one second.

Typical block and trip regions are shown in Figure 10.30 for both types of three-zone, directional impedance relays. Relay connections for a three-zone impedance relay with directional restraint are shown in Figure 10.31.

EXAMPLE 10.8   **Three-zone impedance relay settings**

Table 10.8 gives positive-sequence line impedances as well as CT and VT ratios at B12 for the 345-kV system shown in Figure 10.27. (a) Determine the settings Z_{r1}, Z_{r2}, and Z_{r3} for the B12 three-zone, directional impedance relays connected as shown in Figure 10.31. Consider only solid, three-phase faults. (b) Maximum current for line 1–2 during emergency loading conditions is 1500 A at a power factor of 0.95 lagging. Verify that B12 does not trip during normal and emergency loadings.

Solution   a. Denoting V_{LN} as the line-to-neutral voltage at bus 1 and I_L as the line current through B12, the primary impedance Z viewed at B12 is

$$Z = \frac{V_{LN}}{I_L} \quad \Omega$$

Using the CT and VT ratios given in Table 10.8, the secondary impedance viewed by the B12 impedance relays is

$$Z' = \frac{V_{LN} \left/ \left(\frac{3000}{1}\right)\right.}{I_L \left/ \left(\frac{1500}{5}\right)\right.} = \frac{Z}{10}$$

Figure 10.30

Three-zone, directional impedance relay

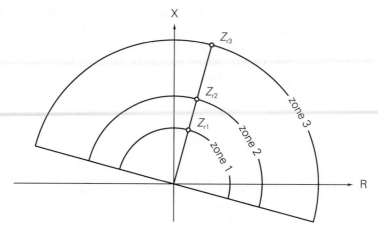

(a) Impedance relay with directional restraint

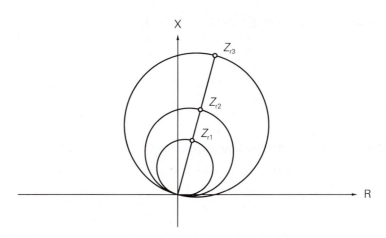

(b) Modified impedance relay (mho relay)

Table 10.8

Data for Example 10.8

LINE	POSITIVE-SEQUENCE IMPEDANCE		
	Ω		
1–2	$8 + j50$		
2–3	$8 + j50$		
2–4	$5.3 + j33$		
1–3	$4.3 + j27$		
BREAKER	CT RATIO		VT RATIO
B12	1500:5		3000:1

Figure 10.31

Relay connections for a three-zone directional impedance relay (only phase A is shown)

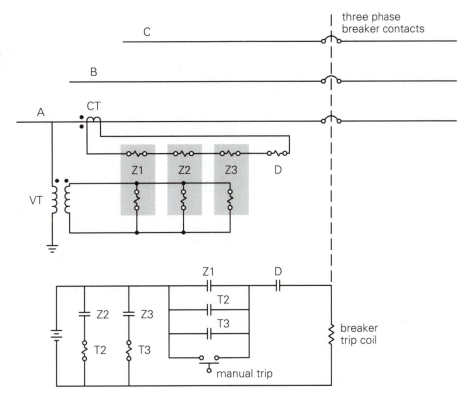

T2 : zone 2 timing relay

T3 : zone 3 timing relay

We set the B12 zone 1 relay for 80% reach, that is, 80% of the line 1–2 (secondary) impedance:

$$Z_{r1} = 0.80(8 + j50)/10 = 0.64 + j4 = 4.05\underline{/80.9°}\,\Omega \text{ secondary}$$

Setting the B12 zone 2 relay for 120% reach:

$$Z_{r2} = 1.2(8 + j50)/10 = 0.96 + j6 = 6.08\underline{/80.9°}\,\Omega \quad \text{secondary}$$

From Table 10.8, line 2–4 has a larger impedance than line 2–3. Therefore, we set the B12 zone 3 relay for 100% reach of line 1–2 plus 120% reach of line 2–4.

$$Z_{r3} = 1.0(8 + j50)/10 + 1.2(5.3 + j33)/10 = 1.44 + j8.96 = 9.07\underline{/80.9°}\,\Omega$$

secondary

b. The secondary impedance viewed by B12 during emergency loading, using $V_{LN} = 345/\sqrt{3}\underline{/0°} = 199.2\underline{/0°}\,kV$ and $I_L = 1500\underline{/-\cos^{-1}(0.95)} = 1500\underline{/-18.19°}\,A$, is

$$Z' = Z/10 = \left(\frac{199.2 \times 10^3}{1500\underline{/-18.19°}}\right)\Big/ 10 = 13.28\underline{/18.19°}\,\Omega \quad \text{secondary}$$

Since this impedance exceeds the zone 3 setting of $9.07\underline{/80.9°}\,\Omega$, the impedance during emergency loading lies outside the trip regions of the three-zone, directional impedance relay. Also, lower line loadings during normal operation will result in even larger impedances farther away from the trip regions. B12 will trip during faults but not during normal and emergency loadings. ∎

Remote backup protection of adjacent lines using zone 3 of an impedance relay may be ineffective. In practice, buses have multiple lines of different lengths with sources at their remote ends. Contributions to fault currents from the multiple lines may cause the zone 3 relay to underreach. This "infeed effect" is illustrated in Problem 10.14.

The impedance relays considered so far use line-to-neutral voltages and line currents and are called *ground fault relays*. They respond to three-phase, single line-to-ground, and double line-to-ground faults very effectively. The impedance seen by the relay during unbalanced faults will generally not be the same as seen during three-phase faults and will not be truly proportional to the distance to the fault location. But the relay can be accurately set for any fault location after computing impedance to the fault using fault currents and voltages. For other fault locations farther away (or closer), the impedance to the fault will increase (or decrease).

Ground fault relays are relatively insensitive to line-to-line faults. Impedance relays that use line-to-line voltages V_{ab}, V_{bc}, V_{ca} and line-current differences $I_a - I_b$, $I_b - I_c$, $I_c - I_a$ are called *phase relays*. Phase relays respond effectively to line-to-line faults and double line-to-ground faults but are relatively insensitive to single line-to-ground faults. Therefore, both phase and ground fault relays need to be used.

SECTION 10.10

DIFFERENTIAL RELAYS

Differential relays are commonly used to protect generators, buses, and transformers. Figure 10.32 illustrates the basic method of differential relaying for generator protection. The protection of only one phase is shown. The method is repeated for the other two phases. When the relay in any one phase operates, all three phases of the main circuit breaker will open, as well as the generator neutral and field breakers (not shown).

Figure 10.32

Differential relaying for generator protection (protection for one phase shown)

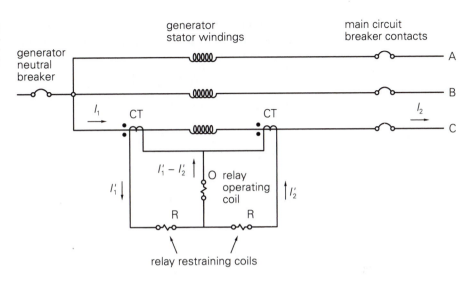

For the case of no internal fault within the generator windings, $I_1 = I_2$, and, assuming identical CTs, $I'_1 = I'_2$. For this case the current in the relay operating coil is zero, and the relay does not operate. On the other hand, for an internal fault such as a phase-to-ground or phase-to-phase short within the generator winding, $I_1 \neq I_2$, and $I'_1 \neq I'_2$. Therefore, a difference current $I'_1 - I'_2$ flows in the relay operating coil, which may cause the relay to operate. Since this relay operation depends on a *difference* current, it is called a *differential* relay.

Figure 10.33

Balance beam differential relay

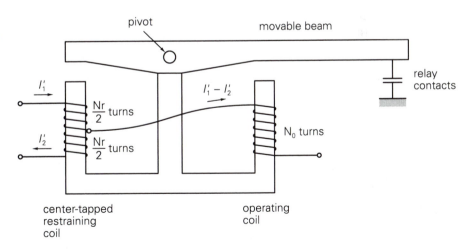

An electromechanical differential relay called a *balance beam* relay is shown in Figure 10.33. The relay contacts close if the downward force on the right side exceeds the downward force on the left side. The electromagnetic force on the right, operating coil is proportional to the square of the operating coil mmf, that is, to $[N_0(I'_1 - I'_2)]^2$. Similarly, the electromagnetic force on the left, restraining coil is proportional to $[N_r(I'_1 + I'_2)/2]^2$. The condition for relay operation is then

$$[N_0(I'_1 - I'_2)]^2 > [N_r(I'_1 + I'_2)/2]^2 \qquad (10.10.1)$$

Taking the square root:

$$|I'_1 - I'_2| > k|(I'_1 + I'_2)/2| \qquad (10.10.2)$$

where

$$k = N_r/N_0 \qquad (10.10.3)$$

Assuming I'_1 and I'_2 are in phase, (10.10.2) is solved to obtain

$$I'_2 > \frac{2 + k}{2 - k} I'_1 \quad \text{for } I'_2 > I'_1$$

$$I'_2 < \frac{2 - k}{2 + k} I'_1 \quad \text{for } I'_2 < I'_1 \qquad (10.10.4)$$

Equation (10.10.4) is plotted in Figure 10.34 to obtain the block and trip regions of the differential relay for $k = 0.1$. Note that as k increases, the block region becomes larger; that is, the relay becomes less sensitive. In practice, no two CTs are identical, and the differential relay current $I'_1 - I'_2$ can become appreciable during external faults, even though $I_1 = I_2$. The balanced beam relay solves this problem without sacrificing sensitivity during normal currents, since the block region increases as the currents increase, as shown in Figure 10.34. Also, the relay can be easily modified to enlarge the block region for very small currents near the origin, in order to avoid false trips at low currents.

Note that differential relaying provides primary zone protection without backup. Coordination with protection in adjacent zones is eliminated, which permits high speed tripping. Precise relay settings are unnecessary. Also, the need to calculate system fault currents and voltages is avoided.

Figure 10.34

Differential relay block and trip regions

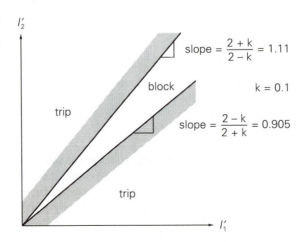

SECTION 10.11

BUS PROTECTION WITH DIFFERENTIAL RELAYS

Differential bus protection is illustrated by the single-line diagram of Figure 10.35. In practice, three differential relays are required, one for each phase. Operation of any one relay would cause all of the three-phase circuit breakers connected to the bus to open, thereby isolating the three-phase bus from service.

Figure 10.35

Single-line diagram of differential bus protection

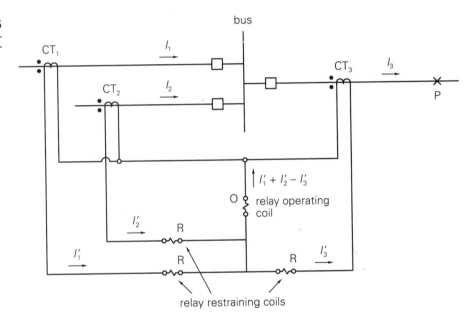

For the case of no internal fault between the CTs, that is, no bus fault, $I_1 + I_2 = I_3$. Assuming identical CTs, the differential relay current $I'_1 + I'_2 - I'_3$ equals zero, and the relay does not operate. However, if there is a bus fault, the differential current $I'_1 + I'_2 - I'_3$, which is not zero, flows in the operating coil to operate the relay. Use of the restraining coils overcomes the problem of nonidentical CTs.

A problem with differential bus protection can result from different levels of fault currents and varying amounts of CT saturation. For example, consider an external fault at point P in Figure 10.35. Each of the CT_1 and CT_2 primaries carries part of the fault current, but the CT_3 primary carries the sum $I_3 = I_1 + I_2$. CT_3, energized at a higher level, will have more saturation, such that $I'_3 \neq I'_1 + I'_2$. If the saturation is too high, the differential current in the relay operating coil could result in a false trip. This problem becomes more difficult when there are large numbers of circuits connected to the bus. Various schemes have been developed to overcome this problem [1].

TRANSFORMER PROTECTION WITH DIFFERENTIAL RELAYS

The protection method used for power transformers depends on the transformer MVA rating. Fuses are often used to protect transformers with small MVA ratings, whereas differential relays are commonly used to protect transformers with ratings larger than 10 MVA.

Figure 10.36

Differential protection of a single-phase, two-winding transformer

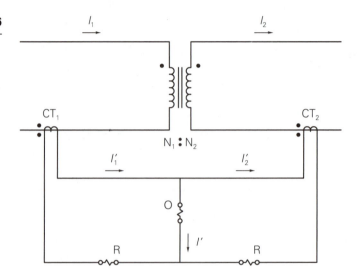

The differential protection method is illustrated in Figure 10.36 for a single-phase, two-winding transformer. Denoting the turns ratio of the primary and secondary CTs by $1/n_1$ and $1/n_2$, respectively (a CT with 1 primary turn and n secondary turns has a turns ratio a = 1/n), the CT secondary currents are

$$I_1' = \frac{I_1}{n_1} \qquad I_2' = \frac{I_2}{n_2} \tag{10.12.1}$$

and the current in the relay operating coil is

$$I' = I_1' - I_2' = \frac{I_1}{n_1} - \frac{I_2}{n_2} \tag{10.12.2}$$

For the case of no fault between the CTs, that is, no internal transformer fault, the primary and secondary currents for an ideal transformer are related by

$$I_2 = \frac{N_1 I_1}{N_2} \tag{10.12.3}$$

Using (10.12.3) in (10.12.2),

$$I' = \frac{I_1}{n_1}\left(1 - \frac{N_1/N_2}{n_2/n_1}\right) \tag{10.12.4}$$

To prevent the relay from tripping for the case of no internal transformer fault, where (10.12.3) and (10.12.4) are satisfied, the differential relay current I' must be zero. Therefore, from (10.12.4), we select

$$\frac{n_2}{n_1} = \frac{N_1}{N_2} \tag{10.12.5}$$

If an internal transformer fault between the CTs does occur, (10.12.3) is not satisfied and the differential relay current $I' = I'_1 - I'_2$ is not zero. The relay will trip if the operating condition given by (10.10.4) is satisfied. Also, the value of k in (10.10.4) can be selected to control the size of the block region shown in Figure 10.34, thereby controlling relay sensitivity.

| EXAMPLE 10.9 | **Differential relay protection for a single-phase transformer** |

A single-phase two-winding, 10-MVA, 80 kV/20 kV transformer has differential relay protection. Select suitable CT ratios. Also, select k such that the relay blocks for up to 25% mismatch between I'_1 and I'_2.

Solution   The transformer-rated primary current is

$$I_{1\,rated} = \frac{10 \times 10^6}{80 \times 10^3} = 125\,A$$

From Table 10.2, select a 150:5 primary CT ratio to give $I'_1 = 125(5/150) = 4.17\,A$ at rated conditions. Similarly, $I_{2\,rated} = 500\,A$. Select a 600:5 secondary CT ratio to give $I'_2 = 500(5/600) = 4.17\,A$ and a differential current $I' = I'_1 - I'_2 = 0$ (neglecting magnetizing current) at rated conditions. Also, for a 25% mismatch between I'_1 and I'_2, select a 1.25 upper slope in Figure 10.34. That is,

$$\frac{2 + k}{2 - k} = 1.25 \qquad k = 0.2222$$ ∎

A common problem in differential transformer protection is the mismatch of relay currents that occurs when standard CT ratios are used. If the primary winding in Example 10.9 has a 138-kV instead of 80-kV rating, then $I_{1\,rated} = 10 \times 10^6/138 \times 10^3 = 72.46\,A$, and a 100:5 primary CT would give $I'_1 = 72.46(5/100) = 3.62\,A$ at rated conditions. This current does not balance $I'_2 = 4.17\,A$ using a 5:600 secondary CT, nor $I'_2 = 3.13\,A$ using a 5:800 secondary CT. The mismatch is about 15%.

One solution to this problem is to use auxiliary CTs, which provide a wide range of turns ratios. A 5:5.76 auxiliary CT connected to the 5:600 secondary CT in the above example would reduce I'_2 to $4.17(5/5.76) = 3.62\,A$, which does balance I'_1. Unfortunately, auxiliary CTs add their own burden to the main CTs and also increase transformation errors. A better solution is to use tap settings on the relays themselves, which have the same effect as auxiliary CTs. Most transformer differential relays have taps that provide for differences in restraining windings in the order of 2 or 3 to 1.

When a transformer is initially energized, it can draw a large "inrush" current, a transient current that flows in the shunt magnetizing branch and decays after a few cycles to a small steady-state value. Inrush current appears as a differential current since it flows only in the primary winding. If a large inrush current does occur upon transformer energization, a differential relay will see a large differential current and trip out the transformer unless the protection method is modified to detect inrush current.

One method to prevent tripping during transformer inrush is based on the fact that inrush current is nonsinusoidal with a large second-harmonic component. A filter can be used to pass fundamental and block harmonic components of the differential current I' to the relay operating coil. Another method is based on the fact that inrush current has a large dc component, which can be used to desensitize the relay. Time-delay relays may also be used to temporarily desensitize the differential relay until the inrush current has decayed to a low value.

Figure 10.37 illustrates differential protection of a three-phase Y–Δ two-winding transformer. Note that a Y–Δ transformer produces 30° phase shifts in the line currents. The CTs must be connected to compensate for the 30° phase shifts, such that the CT secondary currents as seen by the relays are in phase. The correct phase-angle relationship is obtained by connecting CTs on the Y side of the transformer in Δ, and CTs on the Δ side in Y.

EXAMPLE 10.10   **Differential relay protection for a three-phase transformer**

A 30-MVA, 34.5 kV Y/138 kV Δ transformer is protected by differential relays with taps. Select CT ratios, CT connections, and relay tap settings. Also determine currents in the transformer and in the CTs at rated conditions. Assume that the available relay tap settings are 5:5, 5:5.5, 5:6.6, 5:7.3, 5:8, 5:9, and 5:10, giving relay tap ratios of 1.00, 1.10, 1.32, 1.46, 1.60, 1.80, and 2.00.

Solution   As shown in Figure 10.37, CTs are connected in Δ on the (34.5-kV) Y side of the transformer, and CTs are connected in Y on the (138-kV) Δ side, in order to obtain the correct phasing of the relay currents.

Rated current on the 138-kV side of the transformer is

$$I_{A\,rated} = \frac{30 \times 10^6}{\sqrt{3}(138 \times 10^3)} = 125.51\,A$$

Select a 150:5 CT on the 138-kV side to give $I_A' = 125.51(5/150) = 4.184\,A$ in the 138-kV CT secondaries and in the righthand restraining windings of Figure 10.37.

Figure 10.37

Differential protection of a three-phase, Y–Δ, two-winding transformer

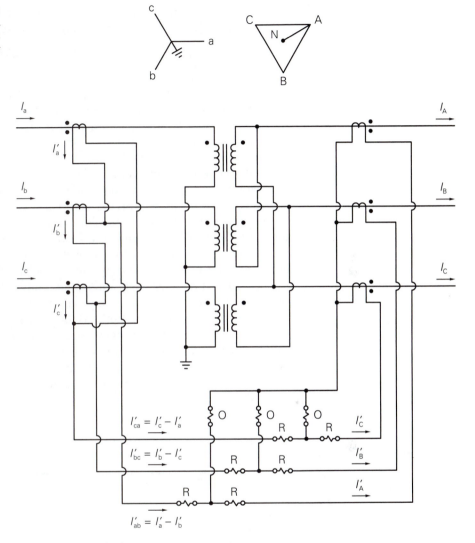

Next, rated current on the 34.5-kV side of the transformer is

$$I_{a\,rated} = \frac{30 \times 10^6}{\sqrt{3}(34.5 \times 10^3)} = 502.04 \text{ A}$$

Select a 500:5 CT on the 34.5-kV side to give $I'_a = 502.0(5/500) = 5.02$ A in the 34.5-kV CT secondaries and $I'_{ab} = 5.02\sqrt{3} = 8.696$ A in the lefthand restraining windings of Figure 10.37.

Finally, select relay taps to balance the currents in the restraining windings. The ratio of the currents in the left- to righthand restraining windings is

$$\frac{I'_{ab}}{I'_A} = \frac{8.696}{4.184} = 2.078$$

The closest relay tap ratio is $T'_{AB}/T'_A = 2.0$, corresponding to a relay tap setting of $T'_A : T'_{ab} = 5 : 10$. The percentage mismatch for this tap setting is

$$\left| \frac{(I'_A/T'_A) - (I'_{ab}/T'_{ab})}{(I'_{ab}/T'_{ab})} \right| \times 100 = \left| \frac{(4.184/5) - (8.696/10)}{(8.696/10)} \right| \times 100 = 3.77\%$$

This is a good mismatch; since transformer differential relays typically have their block regions adjusted between 20% and 60% (by adjusting k in Figure 10.34), a 3.77% mismatch gives an ample safety margin in the event of CT and relay differences.   ∎

For three-phase transformers (Y–Y, Y–Δ, Δ–Y, Δ–Δ), the general rule is to connect CTs on the Y side in Δ and CTs on the Δ side in Y. This arrangement compensates for the 30° phase shifts in Y–Δ or Δ–Y banks. Note also that zero-sequence current cannot enter a Δ side of a transformer nor the CTs on that side, and zero-sequence current on a grounded Y side cannot enter the Δ-connected CTs on that side. Therefore, this arrangement also blocks zero-sequence currents in the differential relays during external ground faults. For internal ground faults, however, the relays can operate from the positive- and negative-sequence currents involved in these faults.

Differential protection methods have been modified to handle multi-winding transformers, voltage-regulating transformers, phase-angle regulating transformers, power-rectifier transformers, transformers with special connections (such as zig-zag), and other, special-purpose transformers. Also, other types of relays such as gas-pressure detectors for liquid-filled transformers are used.

SECTION 10.13

PILOT RELAYING

Pilot relaying refers to a type of differential protection that compares the quantities at the terminals via a communication channel rather than by a direct wire interconnection of the relays. Differential protection of generators, buses, and transformers considered in previous sections does not require pilot relaying because each of these devices is at one geographical location where CTs and relays can be directly interconnected. However, differential relaying of transmission lines requires pilot relaying because the terminals are widely separated (often by many kilometers). In actual practice, pilot relaying is typically applied to short transmission lines (up to 80 km) with 69 to 115 kV ratings.

Four types of communication channels are used for pilot relaying:

1. *Pilot wires:* Separate electrical circuits operating at dc, 50 to 60 Hz, or audio frequencies. These could be owned by the power company or leased from the telephone company.

2. *Power-line carrier:* The transmission line itself is used as the communication circuit, with frequencies between 30 and 300 kHz being transmitted. The communication signals are applied to all three phases using an L–C voltage divider and are confined to the line under protection by blocking filters called *line traps* at each end.

3. *Microwave:* A 2 to 12 GHz signal transmitted by line-of-sight paths between terminals using dish antennas.

4. *Fiber optic cable:* Signals transmitted by light modulation through electrically nonconducting cable. This cable eliminates problems due to electrical insulation, inductive coupling from other circuits, and atmospheric disturbances.

Two common fault detection methods are *directional comparison*, where the power flows at the line terminals are compared, and *phase comparison*, where the relative phase angles of the currents at the terminals are compared. Also, the communication channel can either be required for trip operations, which is known as a *transfer trip system*, or not be required for trip operations, known as a *blocking system*. A particular pilot-relaying method is usually identified by specifying the fault-detection method and the channel use. The four basic combinations are directional comparison blocking, directional comparison transfer trip, phase comparison blocking, and phase comparison transfer trip.

Like differential relays, pilot relays provide primary zone protection without backup. Thus, coordination with protection in adjacent zones is eliminated, resulting in high-speed tripping. Precise relay settings are unnecessary. Also, the need to calculate system fault currents and voltages is eliminated.

SECTION 10.14

COMPUTER RELAYING

Previous sections describe the operating principle of relays built with electromechanical components, including the induction disc time-delay overcurrent relay, Figure 10.14; the directional relay, similar in operation to a watt-hour meter; and the balance-beam differential relay, Figure 10.33. These electromechanical relays, introduced in the early 1900s, have performed well over the years and continue in relatively maintenance-free operation today. Solid-state relays using analog circuits and logic gates, with block-trip regions similar to those of electromechanical relays and with newer types of block/trip regions, have been available since the late 1950s. Such relays, widely used in HV and EHV systems, offer the reliability and ruggedness of their electromechanical counterparts at a competitive price. Beyond solid-state analog relays, a new generation of relays based on digital computer technology has been under development since the 1980s.

Benefits of computer relays include accuracy, improved sensitivity to faults, better selectivity, flexibility, user-friendliness, easy testing, and self-monitoring capabilities. Computer relaying also has the advantage that modifications to tripping characteristics, either changes in conventional settings or shaping of entirely new block/trip regions, can be made by updating software from a remote computer terminal. For example, the relay engineer could reprogram tripping characteristics of field-installed, in-service relays without leaving the engineering office. Alternatively, relay software could be updated in real time, based on operating conditions, from a central computer.

An important feature of power-system protection is the decentralized, local nature of relays. Except for pilot relaying, each relay receives information from nearby, local CTs and VTs and trips only local breakers. Interest in computer relaying is not directed at replacing local relays by a central computer. Instead, each electromechanical or solid-state analog relay would be replaced by a dedicated, local computer relay with a similar operating principle, such as time-delay overcurrent, impedance, or differential relaying. The central computer would interact with local computer relays in a supervisory role.

As one computer relay example, the Houston Power and Lighting Company installed a microprocessor-based protection system on a radial distribution line in 1983 for field testing and subsequent operation [4]. In addition to conventional overcurrent protection, the system provides arcing fault detection, line monitoring, fault data acquisition, and communications that permit changing relay settings from a remote computer terminal. Many other examples are appearing in the literature [5, 13] as computer relay development progresses.

PROBLEMS

Section 10.2

10.1 The primary conductor in Figure 10.2 is one phase of a three-phase transmission line operating at 345 kV, 600 MVA, 0.95 power factor lagging. The CT ratio is 1200:5 and the VT ratio is 3000:1. Determine the CT secondary current I' and the VT secondary voltage V'. Assume zero CT error.

10.2 A Westinghouse CO-8 relay with a current tap setting of 5 amperes is used with the 100:5 CT in Example 10.1. The CT secondary current I' is the input to the relay operating coil. The CO-8 relay burden is shown in the following table for various relay input currents.

CO-8 relay input current I', A	5	8	10	13	15
CO-8 relay burden Z_B, Ω	0.5	0.8	1.0	1.3	1.5

Primary current and CT error are computed in Example 10.1 for the 5-, 8-, and 15-A relay input currents. Compute the primary current and CT error for (a) $I' = 10$ A and $Z_B = 1.0\,\Omega$, and for (b) $I' = 13$ A and $Z_B = 1.3\,\Omega$. (c) Plot I' versus I for the above five values of I'. (d) For reliable relay operation, the fault-to-pickup current ratio with minimum fault current should be greater than two. Determine the minimum fault current for application of this CT and relay with 5-A tap setting.

10.3 An overcurrent relay set to operate at 10 A is connected to the CT in Figure 10.8 with a 200:5 CT ratio. Determine the minimum primary fault current that the relay will detect if the burden Z_B is (a) 1.0 Ω, (b) 4.0 Ω, and (c) 5.0 Ω.

Section 10.3

10.4 The input current to a Westinghouse CO-8 relay is 10 A. Determine the relay operating time for the following current tap settings (TS) and time dial settings (TDS): (a) TS = 1.0, TDS = 1/2; (b) TS = 2.0, TDS = 1.5; (c) TS = 2.0, TDS = 7; (d) TS = 3.0, TDS = 7; and (e) TS = 12.0, TDS = 1.

10.5 The relay in Problem 10.2 has a time-dial setting of 4. Determine the relay operating time if the primary fault current is 500 A.

Section 10.4

10.6 Evaluate relay coordination for the minimum fault currents in Example 10.4. For the selected current tap settings and time-dial settings, (a) determine the operating time of relays at B2, and B3 for the 700-A fault current. (b) Determine the operating time of relays at B1 and B2 for the 1500-A fault current. Are the fault-to-pickup current ratios ⩾ 2.0 (a requirement for reliable relay operation) in all cases? Are the coordination time intervals ⩾ 0.3 seconds in all cases?

10.7 Repeat Example 10.4 for the following system data. Coordinate the relays for the maximum fault currents.

BUS	MAXIMUM LOAD		SYMMETRICAL FAULT CURRENT	
	MVA	LAGGING p.f.	MAXIMUM A	MINIMUM A
1	9.0	0.95	5000	3750
2	9.0	0.95	3000	2250
3	9.0	0.95	2000	1500

BREAKER	BREAKER OPERATING TIME	CT RATIO	RELAY
B1	5 cycles	600:5	CO-8
B2	5 cycles	400:5	CO-8
B3	5 cycles	200:5	CO-8

10.8 Using the current tap settings and time-dial settings that you have selected in Problem 10.7, evaluate relay coordination for the minimum fault currents. Are the fault-to-pickup current ratios ⩾ 2.0, and are the coordination time delays ⩾ 0.3 seconds in all cases?

Section 10.5

10.9 Rework Example 10.5 for the following faults: (a) a three-phase, permanent fault on the load side of tap 3; (b) a single line-to-ground, permanent fault at bus 4 on the load side of the recloser; and (c) a three-phase, permanent fault at bus 4 on the source side of the recloser.

Section 10.7

10.10 For the system shown in Figure 10.38, directional overcurrent relays are used at breakers B12, B21, B23, B32, B34, and B43. Overcurrent relays alone are used at B1

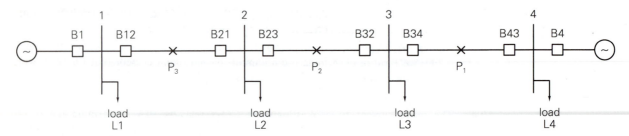

Figure 10.38   Problem 10.10

and B4. (a) For a fault at P_1, which breakers do not operate? Which breakers should be coordinated? Repeat (a) for a fault at (b) P_2, (c) P_3. (d) Explain how the system is protected against bus faults.

Section 10.8

10.11   (a) Draw the protective zones for the power system shown in Figure 10.39. Which circuit breakers should open for a fault at (a) P_1, (b) P_2, and (c) P_3?

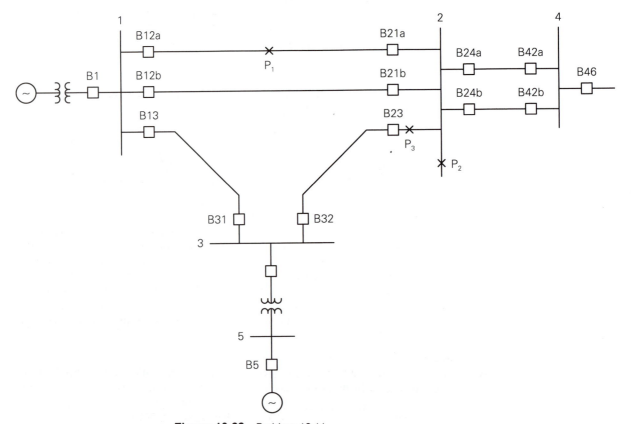

Figure 10.39   Problem 10.11

10.12   Figure 10.40 shows three typical bus arrangements. Although the number of lines connected to each arrangement varies widely in practice, four lines are shown for

Figure 10.40

Problem 10.12—typical bus arrangements

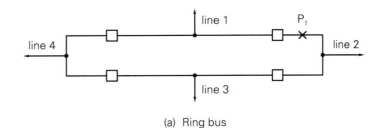

(a) Ring bus

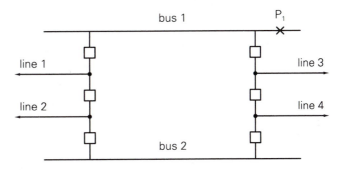

(b) Breaker-and-a-half double bus

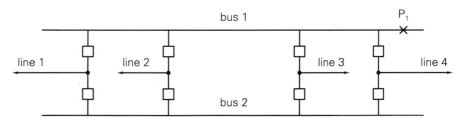

(c) Double-breaker double bus

convenience and comparison. Note that the required number of circuit breakers per line is 1 for the ring bus, $1\frac{1}{2}$ for the breaker-and-a-half double-bus, and 2 for the double-breaker double-bus arrangement. For each arrangement: (a) Draw the protective zones. (b) Identify the breakers that open under primary protection for a fault on line 1. (c) Identify the lines that are removed from service under primary protection during a bus fault at P_1. (d) Identify the breakers that open under backup protection in the event a breaker fails to clear a fault on line 1 (that is, a stuck breaker during a fault on line 1).

Section 10.9

10.13 Three-zone mho relays are used for transmission line protection of the power system shown in Figure 10.25. Positive-sequence line impedances are given as follows.

LINE	POSITIVE-SEQUENCE IMPEDANCE, Ω
1–2	$6 + j60$
1–3	$4 + j40$
2–3	$5 + j50$

Rated voltage for the high-voltage buses is 500 kV. Assume a 1500:5 CT ratio and a 4500:1 VT ratio at B12. (a) Determine the settings Z_{t1}, Z_{t2}, and Z_{t3} for the mho relay at B12. (b) Maximum current for line 1–2 under emergency loading conditions is 1400 A at 0.90 power factor lagging. Verify that B12 does not trip during emergency loading conditions.

10.14   Line impedances for the power system shown in Figure 10.41 are $Z_{12} = Z_{23} = 3.0 + j40.0\,\Omega$, and $Z_{24} = 6.0 + j80.0\,\Omega$. Reach for the zone 3 B12 impedance relays is set for 100% of line 1–2 plus 120% of line 2–4. (a) For a bolted three-phase fault at bus 4, show that the apparent primary impedance "seen" by the B12 relays is

$$Z_{\text{apparent}} = Z_{12} + Z_{24} + (I_{32}/I_{12})Z_{24}$$

where (I_{32}/I_{12}) is the line 2–3 to line 1–2 fault current ratio. (b) If $|I_{32}/I_{12}| > 0.20$, does the B12 relay see the fault at bus 4?

   Note: This problem illustrates the "infeed effect." Fault currents from line 2–3 can cause the zone 3 B12 relay to underreach. As such, remote backup of line 2–4 at B12 is ineffective.

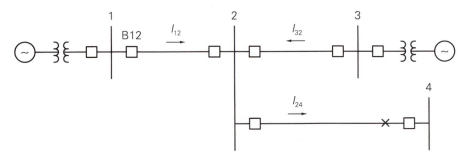

Section 10.10

10.15   Select k such that the differential relay characteristic shown in Figure 10.34 blocks for up to 20% mismatch between I'_1 and I'_2.

Section 10.12

10.16   A single-phase, 5-MVA, 20/8.66-kV transformer is protected by a differential relay with taps. Available relay tap settings are 5:5, 5:5.5, 5:6.6, 5:7.3, 5:8, 5:9, and 5:10, giving tap ratios of 1.00, 1.10, 1.32, 1.46, 1.60, 1.80, and 2.00. Select CT ratios and relay tap settings. Also, determine the percentage mismatch for the selected tap setting.

10.17   A three-phase, 500-MVA, 345 kV Δ/500 kV Y transformer is protected by differential relays with taps. Select CT ratios, CT connections, and relay tap settings. Determine the currents in the transformer and in the CTs at rated conditions. Also determine the percentage mismatch for the selected relay tap settings. Available relay tap settings are given in Problem 10.16.

CASE STUDY QUESTIONS

   A.   Do computer (microprocessor) relays provide better protection than electromechanical relays? If so, in what ways?

   B.   Do computer relays impose any new risks?

   C.   Is computer relaying an art or a science?

References

1.   J. L. Blackburn, *Protective Relaying* (New York: Dekker, 1987).

2.   J. L. Blackburn et al., *Applied Protective Relaying* (Newark, NJ: Westinghouse Electric Corporation, 1976).

3.   *Westinghouse Relay Manual, A New Silent Sentinels Publication* (Newark, NJ: Westinghouse Electric Corporation, 1972).

4.   M. Narendorf, B. D. Russel, and M. Aucoin, "Microcomputer Based Feeder Protection and Monitoring System—Utility Experience," *IEEE Transactions on Power Delivery*, vol. PWRD-2, no. 4, pp. 1046–1052 (October 1987).

5.   J. W. Ingleson et al., "Bibliography of Relay Literature. 1986–1987. IEEE Committee Report," *IEEE Transactions on Power Delivery*, *4*, 3, pp. 1649–1658 (July 1989).

6.   *IEEE Recommended Practice for Protection and Coordination of Commercial Power Systems* (New York: Wiley Interscience, 1975).

7.   *Distribution Manual* (New York: Ebasco/Electrical World, 1990).

8.   C. Russel Mason, *The Art and Science of Protective Relaying* (New York: Wiley, 1956).

9.   C. A. Gross, *Power System Analysis* (New York: Wiley, 1979).

10.   W. D. Stevenson, Jr., *Elements of Power System Analysis*, 4th ed. (New York: McGraw-Hill, 1982).

11.   A. R. Bergen, *Power System Analysis*, (Englewood Cliffs, NJ: Prentice-Hall, 1986).

12.   S. H. Horowitz and A. G. Phadke, *Power System Relaying* (New York: Research Studies Press, 1992).

13.   A. G. Phadke and J. S. Thorpe, *Computer Relaying for Power Systems* (New York: Wiley, 1988).

14.   K. Jeffers, "The Impact of Microprocessor Protective Relays in The Electric Utility Industry," *Transmission and Distribution*, *43*, 8 (August 1991), pp. 152–157.

CHAPTER 11

POWER-SYSTEM CONTROLS

New England Power Exchange (NEPEX) control room
(Courtesy of New England Electric)

Automatic control systems are used extensively in power systems. Local controls are employed at turbine-generator units and at selected voltage-controlled buses. Central controls are employed at area control centers.

Figure 11.1 shows two basic controls of a steam turbine-generator: the voltage regulator and turbine-governor. The voltage regulator adjusts the power output of the generator exciter in order to control the magnitude of generator terminal voltage V_t. When a reference voltage V_{ref} is raised (or lowered), the output voltage V_r of the regulator increases (or decreases) the exciter voltage E_{fd} applied to the generator field winding, which in turn acts to increase (or decrease) V_t. Also a voltage transformer and rectifier monitor V_t, which is used as a feedback signal in the voltage regulator. If V_t decreases, the voltage regulator increases V_r to increase E_{fd}, which in turn acts to increase V_t.

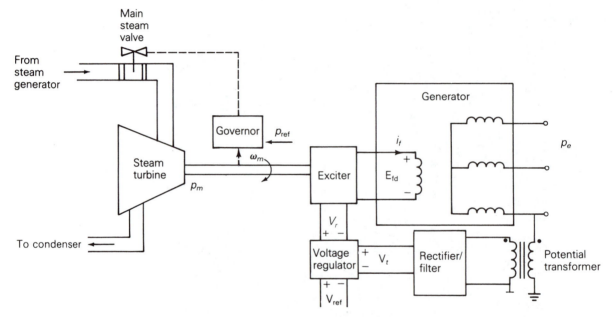

Figure 11.1   Voltage regulator and turbine-governor controls for a steam-turbine generator

The turbine-governor shown in Figure 11.1 adjusts the steam valve position to control the mechanical power output p_m of the turbine. When a reference power level p_{ref} is raised (or lowered), the governor moves the steam valve in the open (or close) direction to increase (or decrease) p_m. The governor also monitors rotor speed ω_m, which is used as a feedback signal to control the balance between p_m and the electrical power output p_e of the generator. Neglecting losses, if p_m is greater than p_e, ω_m increases, and the governor moves the steam valve in the close direction to reduce p_m. Similarly, if p_m is less than p_e, ω_m decreases, and the governor moves the valve in the open direction.

In addition to voltage regulators at generator buses, equipment is used to control voltage magnitudes at other selected buses. Tap-changing transformers, switched capacitor banks, and static var systems can be automatically regulated for rapid voltage control.

Central controls also play an important role in modern power systems. Today's systems are composed of interconnected areas, where each area has its own control center. There are many advantages to interconnections. For example, interconnected areas can share their reserve power to handle anticipated load peaks and unanticipated generator outages. Interconnected areas can also tolerate larger load changes with smaller frequency deviations than an isolated area.

Figure 11.2

Daily load cycle

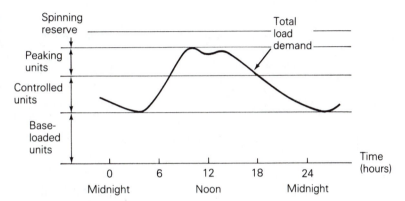

Figure 11.2 shows how a typical area meets its daily load cycle. The base load is carried by base-loaded generators running at 100% of their rating for 24 hours. Nuclear units and large fossil-fuel units are typically base-loaded. The variable part of the load is carried by units that are controlled from the central control center. Medium-sized fossil-fuel units and hydro units are used for control. During peak load hours, smaller, less efficient units such as gas-turbine or diesel-generating units are employed. In addition, generators operating at partial output (with *spinning reserve*) and standby generators provide a reserve margin.

The central control center monitors information including area frequency, generating unit outputs, and tie-line power flows to interconnected areas. This information is used by automatic *load-frequency control* (LFC) in order to maintain area frequency at its scheduled value (60 Hz) and net tie-line power flow out of the area at its scheduled value. Raise and lower reference power signals are dispatched to the turbine-governors of controlled units.

Operating costs vary widely among controlled units. Larger units tend to be more efficient, but the varying cost of different fuels such as coal, oil, and gas is an important factor. *Economic dispatch* determines the megawatt outputs of the controlled units that minimize the total operating cost for a given load demand. Economic dispatch is coordinated with LFC such that reference power signals dispatched to controlled units move the units toward their economic loadings and satisfy LFC objectives.

In this chapter we investigate automatic controls employed in power systems under normal operation. Sections 11.1 and 11.2 describe the operation of the two generator controls: voltage and turbine-governor. Load-frequency control is discussed in Section 11.3 and economic dispatch is presented in Section 11.4.

CASE STUDY A central control center within an area of an interconnected power system serves many functions, all under the general heading "Energy Management System" (EMS). EMS functions include:

- Supervisory control and data acquisition (Scada)
- Automatic generation control (AGC)
- Load forecasting
- State estimation
- Security analysis
- Voltage reduction
- Load shedding

Recent technology applications at control centers include:

- Fiberoptic communications networks that tie monitored power flows, voltages, and circuit breaker status from generators and substations into the Scada computer
- Updated hardware including newer generations of computers and work stations
- Real-time software with full graphic presentations and windows

The following articles describe recent upgrades at two power-system control centers, one by the California Department of Water Resources and the other by Kansas City Gas and Electric [12, 13].

Scada Upgraded with AGC Using Different Vendor's Computer

JOSEPH KIMBRIEL, LARRY DOUGLAS, AND JOE BODMANN

The power needed to pump water throughout the state of California has a far greater effect on operating costs of the California Dept of Water Resources (CDWR) than the water itself. In fact, CDWR is the largest purchaser of power in the state and is also an important power generator. In 1985, CDWR dis-

covered that it could not reach a satisfactory contract with the utilities that supply it, unless it controlled all of its hydro generation and pumping loads from one control center as a single control area.

Today, the department operates as one control area, using all the power it generates for its own pumping operations and paying neighboring utilities a wheeling charge to transmit the energy from CDWR powerplants to its pumping stations. CDWR continues to be a net purchaser of power from Pacific Gas & Electric Co (PSE&G) and Southern California Edison Co (SCE).

Most significant aspect of the system upgrading was that the department's existing Scada system was left intact and operating. The automatic generation

Reprinted with permission from Electrical World.

control (AGC) needed to implement the single control area is processed in an entirely different computer from the Scada system and the necessary data to establish the generation setpoints are exchanged between the two computers.

Before 1988, power generated in the department's three hydroelectric plants and 17 reversible pumping/generating plants was sold to neighboring utilities. Power for the pumping stations was repurchased from those same utilities. The department operates 24 major pumping stations (Fig. 1), including the largest in the world, which pushes water up over the 2000-ft Thachpee Mountains to Los Angeles. Most of this pumping is done at night to take advantage of off-peak rates. Unfortunately, when the department has excess power to sell, neighboring

utilities also have ample hydroelectric capacity. Occasionally, the department has to give away excess power.

Utility requires control area

In 1988, a new agreement with PG&E was only one year away and the department's Control System Engineering Branch was notified that the single control area must be in place for the next contract period. At the time, the department's operating center contained a 1100/62 computer, supplied by Univac Corp (now Unisys Corp, Bluebell, PA), which performed Scada operations for the state water project (map), interrogating the tie-line metering points and issuing manual generation setpoints upon direction from the power

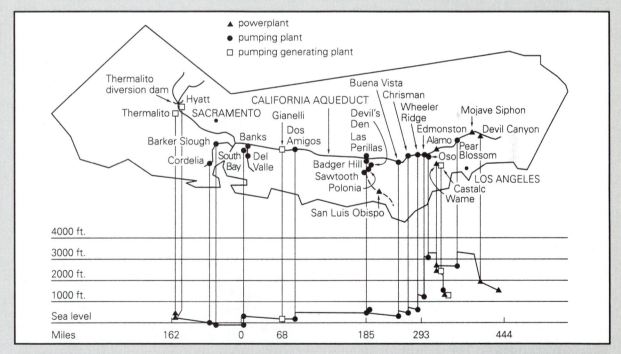

Figure 1 *Primary purpose of California Dept of Water Resources is to operate the California Water Project to meet agricultural, urban, and industrial water needs, and to maintain stream flows for recreation, fish, wildlife, esthetic values, water quality, and navigation. It operates 18 reservoirs, 24 major pumping plants, three hydro plants, 550 miles of aqueducts and pipelines, and maintains about 1100 dams. At full capacity, it delivers 4.2-million acre-ft of water per year. CDWR is also a substantial user and generator of electric power. In addition to its hydro capacity, it owns geothermal and thermal powerplants, including a thermal plant in Nevada not included in the control area*

dispatcher. This computer also performed dispatch water scheduling, power scheduling, accounting, and engineering analysis. Although the system was working well, there was insufficient excess capacity to easily handle the AGC software.

The department owned two unused MicroVAX II computers, supplied by Digital Equipment Corp, Maynard, MA, that were left over from a previous system configuration. Engineers discovered that it was both feasible and cost-effective to operate the AGC processing in the DEC computers and exchange necessary data between the DEC computers and the existing Univac computer. The contract for hardware integration and software rewriting was awarded to Johnson Yokogawa Corp, Carrollton, TX.

Secret to having the two different computers communicate with each other is TCP/IP protocol, which evolved from a Dept of Defense protocol. The Univac 1100/62 computer was not built to communicate with other machines, but Unisys engineers were able to develop an interface for TCP/IP protocol. The

DEC computers can communicate with several different protocols. Johnson Yokogawa developed the AGC system using its JC-6000 platform and off-the-shelf software.

Full power monitoring needed

First step in the conversion process was to monitor power flow from each plant and interconnect point and tie them into the Scada computer. To avoid the need to use dedicated telephone lines, the department uses a fiberoptic communications network. The network, including nodes and multiplexers, was designed and installed by DCA Co, Atlanta, GA. Fiberoptic communications lines were installed by MCI Telecommunications Corp, Washington, DC.

Generating and pumping plants in the State Water project are separated by distances of up to 600 miles and all of the department's energy is wheeled on PG&E and SCE lines. An important function of the new Scada/AGC system is to observe limits on utility

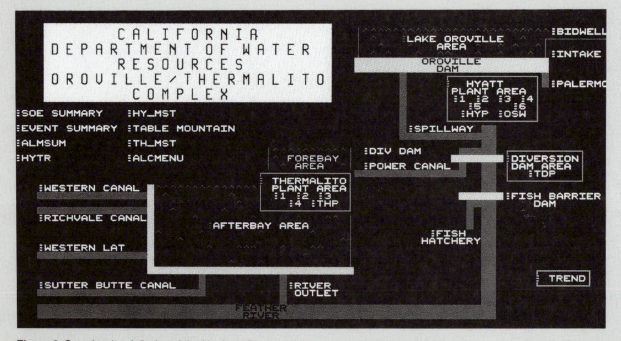

Figure 2 Overview-level display of the Northern Div—one of the department's five divisions

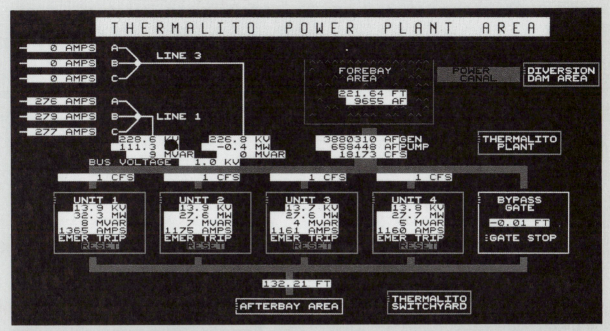

Figure 3 Summary display of the Thermalito 119-MW, four-unit pumping/generating plant located in northern California

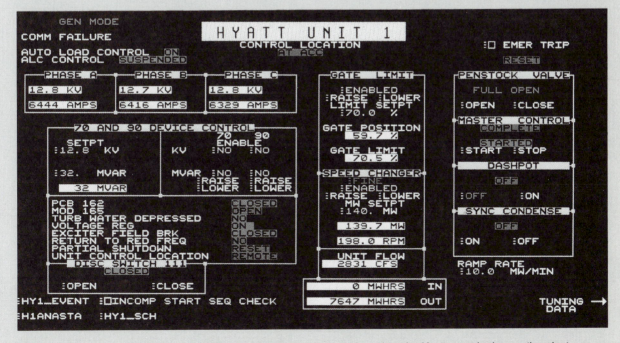

Figure 4 Unit control display showing all the controllable parameters of a unit at the Hyatt pumping/generating plant

tie lines according to the power that is already flowing on them.

The real-time generation software is the chief component of the AGC system. It allows the dispatcher to monitor, analyze, and control generation within the control area. Specifically, the dispatcher can:

Schedule generation for all generating units in the control area.

Allocate and distribute generation to meet power demand in the PG&E and SCE areas.

Define unit schedules, set base points, and any necessary limitation on the output of individual units.

Handle combinations of generator load and required spinning reserve capacity.

The dispatcher is also able to determine generation demand and schedules, study general field conditions, and coordinate power exchanges. A unique advantage of the Scada/AGC system is that the department can now control its pumping loads as well as its generation. This enables it to minimize power costs by reducing load whenever possible. The software automatically observes frequency and individual generating unit constraints—such as ramp rates, wheeling constraints, and time-error correction.

Following the success of this system, the department is now using a JC-6000 platform for a new Scada system to replace all of the control systems in the Oroville field division, which is the department's primary generation area. The installation is about 75% complete.

Edited by John Reason, Senior Editor. Joseph Kimbriel is with the CDWR, Sacramento, Calif, and Larry Douglas and Joe Bodmann are with Johnson Yokogawa Corp, Carrollton, Tex.

Full Graphics Boosts Dispatcher's Capacity

JOHN REASON

Early this year, Kansas Gas & Electric Co (KG&E) vastly increased the capabilities of its dispatchers by moving directly from two aging character/graphics Scada systems, purchased in the early 1970s, to a full-graphics energy management system (EMS). Using the old equipment, the three shifts of three operators were able to perform Scada and automatic generation control, plus a few off-line operations. Today, the same staff can also handle a wide range of real-time functions: from transaction evaluation and energy accounting to state estimator and security analysis to voltage reduction and load shedding.

Jerry Young, KG&E's supervisor of operations services, adds that the new system also drives the map boards and can double as a training simulator.

To streamline the move across several generations of Scada equipment, KG&E used the services of Advanced Scada Consulting Engineering (ASCE), Coto De Caza, CA. ASCE's Tom Green put together a bid package, helped evaluate five bids, and assisted in the final selection of the Telegyr 8500 full-graphics EMS, supplied by Landis & Gyr, San Jose, CA.

The sample displays shown here demonstrate a few of the full-graphics capabilities that enable the dispatch staff to handle the significant increase in its work load.

The weather-adaptive load-forecast program (Fig. 1) predicts hour-by-hour demand for the week. Using this program, dispatchers can estimate hourly electrical demand by region and for the total system. By using historical load patterns and seasonal weather data for the base-load component, the program enables the utility to fine-tune its resource optimization and evaluation functions.

The three-dimensional bar-chart display of Fig. 1 shows a seven-day load forecast for one region. Each bar represents the load for a particular hour. A simple point-and-click operation can bring any day's hourly load forecast to the front of the display. Full-graphics presentations allow the system operator to quickly recognize daily load patterns, a task that was cumbersome with the tabular representation of data used in the old character/graphics system.

Representation of large displays is made possible by the use of a world coordinate system. The first display (Fig. 2) is a geographical overview of the KG&E system showing counties and 345-kV lines.

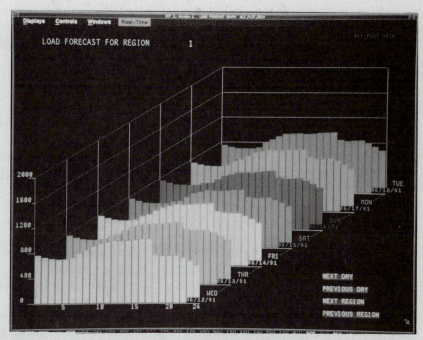

Figure 1 Load forecast program represents one week's load forecast for a region or the entire KG&E system

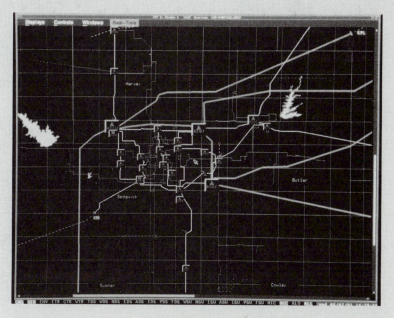

Figure 2 System overview shows KG&E's entire system. User can zoom in on geographical areas, transmission lines, or individual substations

The user has the ability to zoom in on selected areas of interest and more detail is added at each display level. In addition, each transmission line can be selected as a one-line diagram and each substation symbol can be selected to display the corresponding one-line diagram. KG&E finds the overview particularly valuable as a training tool.

The substation analog tabular displays (Fig. 3) show the current value of each station's analog values, as well as the high and low warning and alarm limits. Colors adjust dynamically based on user-defined limits and color schemes. The information on this display is shown numerically and graphically. Graphical representation allows a quick scan for abnormal conditions, avoiding the need to read the actual values.

To determine the most economical distribution of generation among dispatchable generating units,

an economic-dispatch control program is used. Fig. 4 is the summary display, which illustrates, in a normalized framework, the base point, actual generation, regulating limits, economic limits, and regulating margins for each unit. Different colors show regulating margins and economic dispatch ranges in the positive and negative directions. The diamond shape in each bar indicates the ideal generation point.

To monitor system conditions and to observe alarm conditions that extend over a period of time, KG&E operators use the data trending function (Fig. 5). The recent history of up to four analog values is displayed. The time line is shown at the bottom; current values and scale factors are shown at the top. The user has the option of filling portions of the curve above an alarm limit (as shown) or above a base line value. The screen may be split horizontally and vertically to show up to four sets of trend data at the same

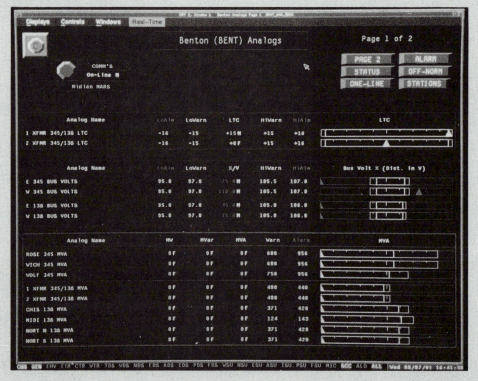

Figure 3 *Substation analogs are shown numerically and graphically so that abnormal conditions can be seen with a quick scan*

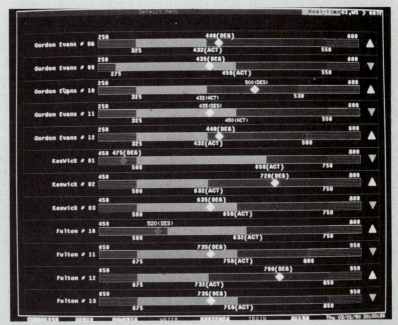

Figure 4 *Economic-dispatch summary display shows actual generating level of each unit in the system, together with limits and margin*

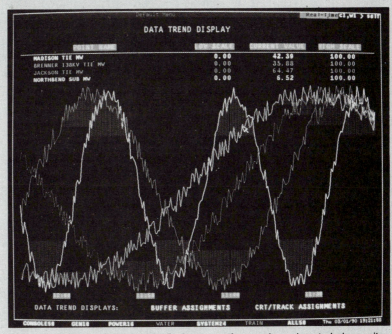

Figure 5 *Data trending function displays up to four analog values to help monitor system conditions and observe long-term alarm conditions*

time. Other windows may display a different set of trending data either horizontally or vertically.

Windowing opens up many possibilities for improving operator efficiency. Fig. 6 shows a workstation screen that has been segmented into three windows. The top two windows display interconnected substations, while the bottom window displays the transmission line that connects them. This window configuration allows operators to perform operations on the transmission line and observe the effects on the substations without having to call up new displays or overwrite existing ones.

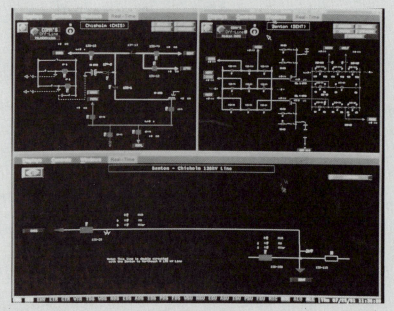

Figure 6 *Windows are particularly valuable when operators need to observe the effect of changes on interconnected subsystems*

SECTION 11.1

GENERATOR-VOLTAGE CONTROL

The *exciter* delivers dc power to the field winding on the rotor of a synchronous generator. For older generators, the exciter consists of a dc generator driven by the rotor. The dc power is transferred to the rotor via slip rings and brushes. For newer generators, *static* or *brushless* exciters are often employed.

For static exciters, ac power is obtained directly from the generator terminals or a nearby station service bus. The ac power is then rectified via thyristors and transferred to the rotor of the synchronous generator via slip rings and brushes.

For brushless exciters, ac power is obtained from an "inverted" synchronous generator whose three-phase armature windings are located on the main generator rotor and whose field winding is located on the stator. The ac power from the armature windings is rectified via diodes mounted on the rotor and is transferred directly to the field winding. For this design, slip rings and brushes are eliminated.

Block diagrams of several standard types of generator-voltage control systems have been developed by the IEEE Working Group on Exciters [1]. A simplified block diagram of generator-voltage control, similar to those given in [1], is shown in Figure 11.3. Nonlinearities due to exciter saturation and limits on exciter output are not shown in this figure.

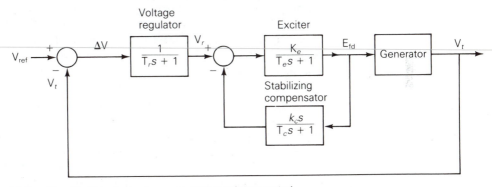

Figure 11.3   Simplified block diagram—generator-voltage control

The generator terminal voltage V_t in Figure 11.3 is compared with a voltage reference V_{ref} to obtain a voltage error signal ΔV, which in turn is applied to the voltage regulator. The $1/(T_r s + 1)$ block accounts for voltage-regulator time delay, where s is the Laplace operator and T_r is the voltage-regulator time constant. Note that if a unit step is applied to a $1/(T_r s + 1)$ block, the output rises exponentially to unity with time constant T_r.

Neglecting the stabilizing compensator in Figure 11.3, the output V_r of the voltage regulator is applied to the exciter, which is represented by a $K_e/(T_e s + 1)$ block. The output of this exciter block is the field voltage E_{fd}, which is applied to the generator field winding and acts to adjust the generator terminal voltage. The generator block, which relates the effect of changes in E_{fd} to V_t, can be developed from synchronous-machine equations [2].

The stabilizing compensator shown in Figure 11.3 is used to improve the dynamic response of the exciter by reducing excessive overshoot. The compensator is represented by a $(K_c s)/(T_c s + 1)$ block, which provides a filtered first derivative. The input to this block is the exciter voltage E_{fd}, and the output is a stabilizing feedback signal that is subtracted from the regulator voltage V_r.

Block diagrams such as those shown in Figure 11.3 are used for computer representation of generator-voltage control in transient stability computer programs (see Chapter 13). In practice, high-gain, fast-responding exciters provide large, rapid increases in field voltage E_{fd} during short circuits at the generator terminals in order to improve transient stability after fault clearing. Equations represented in the block diagram can be used to compute the transient response of generator-voltage control.

SECTION 11.2

TURBINE-GOVERNOR CONTROL

Turbine-generator units operating in a power system contain stored kinetic energy due to their rotating masses. If the system load suddenly increases, stored kinetic energy is released to initially supply the load increase. Also, the electrical torque T_e of each turbine-generating unit increases to supply the load increase, while the mechanical torque T_m of the turbine initially remains constant. From Newton's second law, $J\alpha = T_m - T_e$, the acceleration α is therefore negative. That is, each turbine-generator decelerates and the rotor speed drops as kinetic energy is released to supply the load increase. The electrical frequency of each generator, which is proportional to rotor speed for synchronous machines, also drops.

From this, we conclude that either rotor speed or generator frequency indicates a balance or imbalance of generator electrical torque T_e and turbine mechanical torque T_m. If speed or frequency is decreasing, then T_e is greater than T_m (neglecting generator losses). Similarly, if speed or frequency is increasing, T_e is less than T_m. Accordingly, generator frequency is an appropriate control signal for governing the mechanical output power of the turbine.

The steady-state frequency–power relation for turbine-governor control is

$$\Delta p_m = \Delta p_{ref} - \frac{1}{R} \Delta f \tag{11.2.1}$$

where Δf is the change in frequency, Δp_m is the change in turbine mechanical power output, and Δp_{ref} is the change in a reference power setting. R is called the *regulation constant*. The equation is plotted in Figure 11.4 as a family of curves, with Δp_{ref} as a parameter. Note that when Δp_{ref} is fixed, Δp_m is directly proportional to the drop in frequency.

Figure 11.4 illustrates a steady-state frequency–power relation. When an electrical load change occurs, the turbine-generator rotor accelerates or decelerates, and frequency undergoes a transient disturbance. Under normal operating conditions, the rotor acceleration eventually becomes zero, and the frequency reaches a new steady-state, shown in the figure.

The regulation constant R in (11.2.1) is the negative of the slope of the Δf versus Δp_m curves shown in Figure 11.4. The units of R are Hz/MW when

Figure 11.4

Steady-state frequency–power relation for a turbine-governor

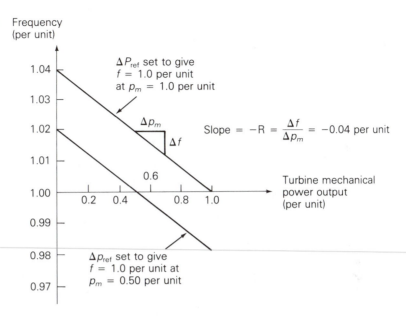

Frequency (per unit)

ΔP_{ref} set to give $f = 1.0$ per unit at $p_m = 1.0$ per unit

Δp_m

Slope $= -R = \dfrac{\Delta f}{\Delta p_m} = -0.04$ per unit

Δf

Turbine mechanical power output (per unit)

Δp_{ref} set to give $f = 1.0$ per unit at $p_m = 0.50$ per unit

Δf is in Hz and Δp_m is in MW. When Δf and Δp_m are given in per-unit, however, R is also in per-unit.

EXAMPLE 11.1 | **Turbine-governor response to frequency change at a generating unit**

A 500-MVA, 60-Hz turbine-generator has a regulation constant R = 0.05 per unit based on its own rating. If the generator frequency increases by 0.01 Hz in steady-state, what is the decrease in turbine mechanical power output? Assume a fixed reference power setting.

Solution   The per-unit change in frequency is

$$\Delta f_{\text{p.u.}} = \frac{\Delta f}{f_{\text{base}}} = \frac{0.01}{60} = 1.6667 \times 10^{-4} \quad \text{per unit}$$

Then, from (11.2.1), with $\Delta p_{\text{ref}} = 0$,

$$\Delta p_{m\text{p.u.}} = \left(\frac{-1}{0.05}\right)(1.6667 \times 10^{-4}) = -3.3333 \times 10^{-4} \quad \text{per unit}$$

$$\Delta p_m = (\Delta p_{m\text{p.u.}})S_{\text{base}} = (-3.3333 \times 10^{-4})(500) = -1.6667 \quad \text{MW}$$

The turbine mechanical power output decreases by 1.67 MW.   ∎

The steady-state frequency–power relation for one area of an interconnected power system can be determined by summing (11.2.1) for each turbine-generating unit in the area. Noting that Δf is the same for each unit,

$$\Delta p_m = \Delta p_{m1} + \Delta p_{m2} + \Delta p_{m3} + \cdots$$

$$= (\Delta p_{ref1} + \Delta p_{ref2} + \cdots) - \left(\frac{1}{R_1} + \frac{1}{R_2} + \cdots \right) \Delta f$$

$$= \Delta p_{ref} - \left(\frac{1}{R_1} + \frac{1}{R_2} + \cdots \right) \Delta f \tag{11.2.2}$$

where Δp_m is the total change in turbine mechanical powers and Δp_{ref} is the total change in reference power settings within the area. We define the *area frequency response characteristic* β as

$$\beta = \left(\frac{1}{R_1} + \frac{1}{R_2} + \cdots \right) \tag{11.2.3}$$

Using (11.2.3) in (11.2.2),

$$\Delta p_m = \Delta p_{ref} - \beta \Delta f \tag{11.2.4}$$

Equation (11.2.4) is the area steady-state frequency–power relation. The units of β are MW/Hz when Δf is in Hz and Δp_m is in MW. β can also be given in per-unit. In practice, β is somewhat higher than that given by (11.2.3) due to system losses and the frequency dependence of loads.

A standard figure for the regulation constant is R = 0.05 per unit. When all turbine-generating units have the same per-unit value of R based on their own ratings, then each unit shares total power changes in proportion to its own ratings. This desirable feature is illustrated by the following example.

EXAMPLE 11.2 | **Response of turbine-governors to a load change in an interconnected power system**

One area of an interconnected 60-Hz power system has three turbine-generator units rated 1000, 750, and 500 MVA, respectively. The regulation constant of each unit is R = 0.05 per unit based on its own rating. Each unit is initially operating at one-half of its own rating, when the system load suddenly increases by 200 MW. Determine (a) the per-unit area frequency response characteristic β on a 1000 MVA system base; (b) the steady-state drop in area frequency, and (c) the increase in turbine mechanical power output of each unit. Assume that the reference power setting of each turbine-generator remains constant. Neglect losses and the dependence of load on frequency.

Solution   **a.** The regulation constants are converted to per-unit on the system base using

$$R_{p.u.new} = R_{p.u.old} \frac{S_{base(new)}}{S_{base(old)}}$$

We obtain

$$R_{1p.u.new} = R_{1p.u.old} = 0.05$$

$$R_{2p.u.new} = (0.05)\left(\frac{1000}{750}\right) = 0.06667$$

$$R_{3p.u.new} = (0.05)\left(\frac{1000}{500}\right) = 0.10 \quad \text{per unit}$$

Using (11.2.3),

$$\beta = \frac{1}{R_1} + \frac{1}{R_2} + \frac{1}{R_3} = \frac{1}{0.05} + \frac{1}{0.06667} + \frac{1}{0.10} = 45.0 \quad \text{per unit}$$

b. Neglecting losses and dependence of load on frequency, the steady-state increase in total turbine mechanical power equals the load increase, 200 MW or 0.20 per unit. Using (11.2.4) with $\Delta p_{ref} = 0$,

$$\Delta f = \left(\frac{-1}{\beta}\right)\Delta p_m = \left(\frac{-1}{45}\right)(0.20) = -4.444 \times 10^{-3} \quad \text{per unit}$$

$$= (-4.444 \times 10^{-3})(60) = -0.2667 \quad \text{Hz}$$

The steady-state frequency drop is 0.2667 Hz.

c. From (11.2.1), using $\Delta f = -4.444 \times 10^{-3}$ per unit,

$$\Delta p_{m1} = \left(\frac{-1}{0.05}\right)(-4.444 \times 10^{-3}) = 0.08888 \quad \text{per unit}$$

$$= 88.88 \quad \text{MW}$$

$$\Delta p_{m2} = \left(\frac{-1}{0.06667}\right)(-4.444 \times 10^{-3}) = 0.06666 \quad \text{per unit}$$

$$= 66.66 \quad \text{MW}$$

$$\Delta p_{m3} = \left(\frac{-1}{0.10}\right)(-4.444 \times 10^{-3}) = 0.04444 \quad \text{per unit}$$

$$= 44.44 \quad \text{MW} \qquad \blacksquare$$

Note that unit 1, whose MVA rating is $33\frac{1}{3}\%$ larger than that of unit 2 and 100% larger than that of unit 3, picks up $33\frac{1}{3}\%$ more load than unit 2 and 100% more load than unit 3. That is, each unit shares the total load change in proportion to its own rating.

Figure 11.5 shows a block diagram of a nonreheat steam turbine-governor, which includes nonlinearities and time delays that were not included in (11.2.1). The deadband block in this figure accounts for the fact that speed governors do not respond to changes in frequency or to reference power settings that are smaller than a specified value. The limiter block

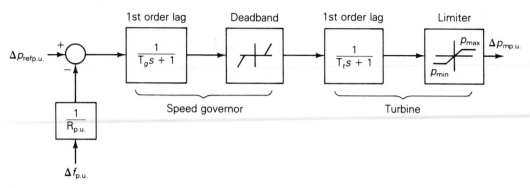

Figure 11.5 Turbine-governor block diagram

accounts for the fact that turbines have minimum and maximum outputs. The $1/(Ts + 1)$ blocks account for time delays, where s is the Laplace operator and T is a time constant. Typical values are $T_g = 0.10$ and $T_t = 1.0$ seconds. Block diagrams for steam turbine-governors with reheat and hydro turbine-governors are also available [3].

SECTION 11.3

LOAD-FREQUENCY CONTROL

As shown in Section 11.2, turbine-governor control eliminates rotor accelerations and decelerations following load changes during normal operation. However, there is a steady-state frequency error Δf when the change in turbine-governor reference setting Δp_{ref} is zero. One of the objectives of load-frequency control (LFC), therefore, is to return Δf to zero.

In a power system consisting of interconnected areas, each area agrees to export or import a scheduled amount of power through transmission-line interconnections, or tie-lines, to its neighboring areas. Thus, a second LFC objective is to have each area absorb its own load changes during normal operation. This objective is achieved by maintaining the net tie-line power flow out of each area at its scheduled value.

The following summarizes the two basic LFC objectives for an interconnected power system:

1. Following a load change, each area should assist in returning the steady-state frequency error Δf to zero.

2. Each area should maintain the net tie-line power flow out of the area at its scheduled value, in order for the area to absorb its own load changes.

The following control strategy developed by N. Cohn [4] meets these LFC objectives. We first define the *area control error* (ACE) as follows:

$$\text{ACE} = (p_{\text{tie}} - p_{\text{tie,sched}}) + B_f(f - 60)$$

$$= \Delta p_{\text{tie}} + B_f \Delta f \tag{11.3.1}$$

where Δp_{tie} is the deviation in net tie-line power flow out of the area from its scheduled value $p_{\text{tie,sched}}$, and Δf is the deviation of area frequency from its scheduled value (60 Hz). Thus, the ACE for each area consists of a linear combination of tie-line error Δp_{tie} and frequency error Δf. The constant B_f is called a *frequency bias constant*.

The change in reference power setting Δp_{refi} of each turbine-governor operating under LFC is proportional to the integral of the area control error. That is,

$$\Delta p_{\text{refi}} = -K_i \int \text{ACE} \, dt \tag{11.3.2}$$

Each area monitors its own tie-line power flows and frequency at the area control center. The ACE given by (11.3.1) is computed and a percentage of the ACE is allocated to each controlled turbine-generator unit. Raise or lower commands are dispatched to the turbine-governors at discrete time intervals of two or more seconds in order to adjust the reference power settings. As the commands accumulate, the integral action in (11.3.2) is achieved.

The constant K_i in (11.3.2) is an integrator gain. The minus sign in (11.3.2) indicates that if either the net tie-line power flow out of the area or the area frequency is low—that is, if the ACE is negative—then the area should increase its generation.

When a load change occurs in any area, a new steady-state operation can be obtained only after the power output of every turbine-generating unit in the interconnected system reaches a constant value. This occurs only when all reference power settings are zero, which in turn occurs only when the ACE of every area is zero. Furthermore, the ACE is zero in every area only when both Δp_{tie} and Δf are zero. Therefore, in steady-state, both LFC objectives are satisfied.

EXAMPLE 11.3 | **Response of LFC to a load change in an interconnected power system**

As shown in Figure 11.6, a 60-Hz power system consists of two interconnected areas. Area 1 has 2000 MW of total generation and an area frequency response characteristic $\beta_1 = 700$ MW/Hz. Area 2 has 4000 MW of total generation and $\beta_2 = 1400$ MW/Hz. Each area is initially generating one-half of its total generation, at $\Delta p_{\text{tie}1} = \Delta p_{\text{tie}2} = 0$ and at 60 Hz when the load in area 1 suddenly increases by 100 MW. Determine the steady-state frequency error Δf and the steady-state tie-line error Δp_{tie} of each area for the following two cases: (a) without LFC, and (b) with LFC given by (11.3.1) and (11.3.2). Neglect losses and the dependence of load on frequency.

Figure 11.6

Example 11.3

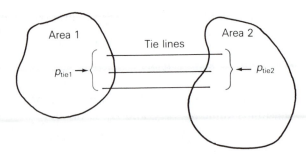

Solution   **a.** Since the two areas are interconnected, the steady-state frequency error Δf is the same for both areas. Adding (11.2.4) for each area,

$$(\Delta p_{m1} + \Delta p_{m2}) = (\Delta p_{ref\,1} + \Delta p_{ref\,2}) - (\beta_1 + \beta_2)\Delta f$$

Neglecting losses and the dependence of load on frequency, the steady-state increase in total mechanical power of both areas equals the load increase, 100 MW. Also, without LFC, $\Delta p_{ref\,1}$ and $\Delta p_{ref\,2}$ are both zero. The above equation then becomes

$$100 = -(\beta_1 + \beta_2)\Delta f = -(700 + 1400)\Delta f$$

$$\Delta f = -100/2100 = -0.0476 \quad \text{Hz}$$

Next, using (11.2.4) for each area, with $\Delta p_{ref} = 0$,

$$\Delta p_{m1} = -\beta_1 \Delta f = -(700)(-0.0476) = 33.33 \quad \text{MW}$$

$$\Delta p_{m2} = -\beta_2 \Delta f = -(1400)(-0.0476) = 66.67 \quad \text{MW}$$

In response to the 100-MW load increase in area 1, area 1 picks up 33.33 MW and area 2 picks up 66.67 MW of generation. The 66.67-MW increase in area 2 generation is transferred to area 1 through the tie-lines. Therefore, the change in net tie-line power flow out of each area is

$$\Delta p_{tie2} = +66.67 \quad \text{MW}$$

$$\Delta p_{tie1} = -66.67 \quad \text{MW}$$

b. From (11.3.1), the area control error for each area is

$$\text{ACE}_1 = \Delta p_{tie1} + B_1 \Delta f_1$$

$$\text{ACE}_2 = \Delta p_{tie2} + B_2 \Delta f_2$$

Neglecting losses, the sum of the net tie-line flows must be zero; that is, $\Delta p_{tie1} + \Delta p_{tie2} = 0$ or $\Delta p_{tie2} = -\Delta p_{tie1}$. Also, in steady-state $\Delta f_1 = \Delta f_2 = \Delta f$. Using these relations in the above equations,

$$\text{ACE}_1 = \Delta p_{tie1} + B_1 \Delta f$$

$$\text{ACE}_2 = -\Delta p_{tie1} + B_2 \Delta f$$

In steady-state, $\text{ACE}_1 = \text{ACE}_2 = 0$; otherwise, the LFC given by (11.3.2) would be changing the reference power settings of turbine-governors on

LFC. Adding the above two equations,

$$\text{ACE}_1 + \text{ACE}_2 = 0 = (\text{B}_1 + \text{B}_2)\Delta f$$

Therefore, $\Delta f = 0$ and $\Delta p_{\text{tie}1} = \Delta p_{\text{tie}2} = 0$. That is, in steady-state the frequency error is returned to zero, area 1 picks up its own 100-MW load increase, and area 2 returns to its original operating condition—that is, the condition before the load increase occurred.

We note that turbine-governor controls act almost instantaneously, subject only to the time delays shown in Figure 11.5. However, LFC acts more slowly. LFC raise and lower signals are dispatched from the area control center to turbine-governors at discrete-time intervals of 2 or more seconds. Also, it takes time for the raise or lower signals to accumulate. Thus, case (a) represents the first action. Turbine-governors in both areas rapidly respond to the load increase in area 1 in order to stabilize the frequency drop. Case (b) represents the second action. As LFC signals are dispatched to turbine-governors, Δf and Δp_{tie} are slowly returned to zero. ∎

The choice of the B_f and K_i constants in (11.3.1) and (11.3.2) affects the transient response to load changes—for example, the speed and stability of the response. The frequency bias B_f should be high enough such that each area adequately contributes to frequency control. Cohn [4] has shown that choosing B_f equal to the area frequency response characteristic, $\text{B}_f = \beta$, gives satisfactory performance of the interconnected system. The integrator gain K_i should not be too high; otherwise, instability may result. Also, the time interval at which LFC signals are dispatched, 2 or more seconds, should be long enough so that LFC does not attempt to follow random or spurious load changes. A detailed investigation of the effect of B_f, K_i and LFC time interval on the transient response of LFC and turbine-governor controls is beyond the scope of this text.

Two additional LFC objectives are to return the integral of frequency error and the integral of net tie-line error to zero in steady-state. By meeting these objectives, LFC controls both the time of clocks that are driven by 60-Hz motors and energy transfers out of each area. These two objectives are achieved by making temporary changes in the frequency schedule and tie-line schedule in (11.3.1).

Finally, note that LFC maintains control during normal changes in load and frequency—that is, changes that are not too large. During emergencies, when large imbalances between generation and load occur, LFC is bypassed and other, emergency controls are applied.

SECTION 11.4

ECONOMIC DISPATCH

Section 11.3 describes how LFC adjusts the reference power settings of turbine-governors in an area to control frequency and net tie-line power flow out of the area. This section describes how the real power output of each controlled generating unit in an area is selected to meet a given load and to minimize the total operating costs in the area. This is the *economic dispatch* problem [5].

We begin this section by considering an area with only fossil-fuel generating units, with no constraints on maximum and minimum generator outputs, and with no transmission losses. The economic dispatch problem is first solved for this idealized case. Then we include inequality constraints on generator outputs; then we include transmission losses. Next we discuss the coordination of economic dispatch with LFC. Finally, we briefly discuss the dispatch of other types of units including nuclear, pumped-storage hydro, and hydro units.

Fossil-Fuel Units, No Inequality Constraints, No Transmission Losses

Figure 11.7 shows the operating cost C_i of a fossil-fuel generating unit versus real power output P_i. Fuel cost is the major portion of the variable cost of operation, although other variable costs, such as maintenance, could have been included in the figure. Fixed costs, such as the capital cost of installing the unit, are not included. Only those costs that are a function of unit power output—that is, those costs that can be controlled by operating strategy—enter into the economic dispatch formulation.

In practice, C_i is constructed of piecewise continuous functions valid for ranges of output P_i, based on empirical data. The discontinuities in Figure 11.7 may be due to the firing of equipment such as additional boilers or condensers as power output is increased. It is often convenient to express C_i

Figure 11.7

Unit operating cost versus real power output—fossil-fuel generating unit

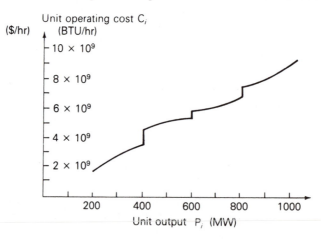

in terms of BTU/hr, which is relatively constant over the lifetime of the unit, rather than $/hr, which can change monthly or daily. C_i can be converted to $/hr by multiplying the fuel input in BTU/hr by the cost of fuel in $/BTU.

Figure 11.8 shows the unit incremental operating cost dC_i/dP_i versus unit output P_i, which is the slope or derivative of the C_i versus P_i curve in Figure 11.7. When C_i consists of only fuel costs, dC_i/dP_i is the ratio of the incremental fuel energy input in BTU to incremental energy output in kWhr, which is called incremental *heat rate*. Note that the reciprocal of the heat rate, which is the ratio of output energy to input energy, gives a measure of fuel efficiency for the unit. For the unit shown in Figure 11.7, maximum efficiency occurs at $P_i = 600$ MW, where the heat rate is $\mathbf{C/P} = 5.4 \times 10^9/600 \times 10^3 = 9000$ BTU/kWhr. The efficiency at this output is

$$\text{percent efficiency} = \left(\frac{1}{9000}\,\frac{\text{kWhr}}{\text{BTU}}\right)\left(3413\,\frac{\text{BTU}}{\text{kWhr}}\right) \times 100 = 37.92\%$$

The dC_i/dP_i curve in Figure 11.8 is also represented by piecewise continuous functions valid for ranges of output P_i. For analytical work, the actual curves are often approximated by straight lines. The ratio dC_i/dP_i can also be converted to $/kWhr by multiplying the incremental heat rate in BTU/kWhr by the cost of fuel in $/BTU.

Figure 11.8

Unit incremental operating cost versus real power output—fossil-fuel generating unit

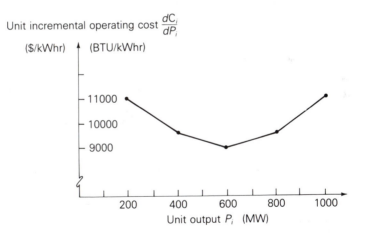

For the area of an interconnected power system consisting of N units operating on economic dispatch, the total variable cost C_T of operating these units is

$$C_T = \sum_{i=i}^{N} C_i$$
$$= C_1(P_1) + C_2(P_2) + \cdots + C_N(P_N) \quad \text{\$/hr} \tag{11.4.1}$$

where C_i, expressed in $/hr, includes fuel cost as well as any other variable costs of unit i. Let P_T equal the total load demand in the area. Neglecting transmission losses,

$$P_1 + P_2 + \cdots + P_N = P_T \qquad (11.4.2)$$

Due to relatively slow changes in load demand, P_T may be considered constant for periods of 2 to 10 minutes. The economic dispatch problem can be stated as follows:

> Find the values of unit outputs $P_1, P_2, \ldots, P_N$ that minimize C_T given by (11.4.1), subject to the equality constraint given by (11.4.2).

A criterion for the solution to this problem is: All units on economic dispatch should operate at equal incremental operating cost. That is,

$$\frac{dC_1}{dP_1} = \frac{dC_2}{dP_2} = \cdots = \frac{dC_N}{dP_N} \qquad (11.4.3)$$

An intuitive explanation of this criterion is the following. Suppose one unit is operating at a higher incremental operating cost than the other units. If the output power of that unit is reduced and transferred to units with lower incremental operating costs, then the total operating cost C_T decreases. That is, reducing the output of the unit with the *higher* incremental cost results in a *greater cost decrease* than the cost increase of adding that same output reduction to units with lower incremental costs. Therefore, all units must operate at the same incremental operating cost (the economic dispatch criterion).

A mathematical solution to the economic dispatch problem can also be given. The minimum value of C_T occurs when the total differential dC_T is zero. That is,

$$dC_T = \frac{\partial C_T}{\partial P_1} dP_1 + \frac{\partial C_T}{\partial P_2} dP_2 + \cdots + \frac{\partial C_T}{\partial P_N} dP_N = 0 \qquad (11.4.4)$$

Using (11.4.1), (11.4.4) becomes

$$dC_T = \frac{dC_1}{dP_1} dP_1 + \frac{dC_2}{dP_2} dP_2 + \cdots + \frac{dC_N}{dP_N} dP_N = 0 \qquad (11.4.5)$$

Also, assuming P_T is constant, the differential of (11.4.2) is

$$dP_1 + dP_2 + \cdots + dP_N = 0 \qquad (11.4.6)$$

Multiplying (11.4.6) by λ and subtracting the resulting equation from (11.4.5),

$$\left(\frac{dC_1}{dP_1} - \lambda\right) dP_1 + \left(\frac{dC_2}{dP_2} - \lambda\right) dP_2 + \cdots + \left(\frac{dC_N}{dP_N} - \lambda\right) dP_N = 0 \qquad (11.4.7)$$

Equation (11.4.7) is satisfied when each term in parentheses equals zero. That is,

$$\frac{dC_1}{dP_1} = \frac{dC_2}{dP_2} = \cdots = \frac{dC_N}{dP_N} = \lambda \qquad (11.4.8)$$

Therefore, all units have the same incremental operating cost, denoted here by λ, in order to minimize the total operating cost C_T.

EXAMPLE 11.4 | **Economic dispatch solution neglecting generator limits and line losses**

An area of an interconnected power system has two fossil-fuel units operating on economic dispatch. The variable operating costs of these units are given by

$$C_1 = 10P_1 + 8 \times 10^{-3}P_1^2 \quad \$/hr$$

$$C_2 = 8P_2 + 9 \times 10^{-3}P_2^2 \quad \$/hr$$

where P_1 and P_2 are in megawatts. Determine the power output of each unit, the incremental operating cost, and the total operating cost C_T that minimizes C_T as the total load demand P_T varies from 500 to 1500 MW. Generating unit inequality constraints and transmission losses are neglected.

Solution The incremental operating costs of the units are

$$\frac{dC_1}{dP_1} = 10 + 16 \times 10^{-3}P_1 \quad \$/MWhr$$

$$\frac{dC_2}{dP_2} = 8 + 18 \times 10^{-3}P_2 \quad \$/MWhr$$

Using (11.4.8), the minimum total operating cost occurs when

$$\frac{dC_1}{dP_1} = 10 + 16 \times 10^{-3}P_1 = \frac{dC_2}{dP_2} = 8 + 18 \times 10^{-3}P_2$$

Using $P_2 = P_T - P_1$, the preceding equation becomes

$$10 + 16 \times 10^{-3}P_1 = 8 + 18 \times 10^{-3}(P_T - P_1)$$

Solving for P_1,

$$P_1 = \frac{18 \times 10^{-3}P_T - 2}{34 \times 10^{-3}} = 0.5294P_T - 58.82 \quad MW$$

Also, the incremental operating cost when C_T is minimized is

$$\frac{dC_2}{dP_2} = \frac{dC_1}{dP_1} = 10 + 16 \times 10^{-3}P_1 = 10 + 16 \times 10^{-3}(0.5294P_T - 58.82)$$

$$= 9.0589 + 8.4704 \times 10^{-3}P_T \quad \$/MWhr$$

and the minimum total operating cost is

$$C_T = C_1 + C_2 = (10P_1 + 8 \times 10^{-3}P_1^2) + (8P_2 + 9 \times 10^{-3}P_2^2) \quad \$/hr$$

The economic dispatch solution is shown in Table 11.1 for values of P_T from 500 to 1500 MW.

Table 11.1

Economic dispatch solution for
Example 11.4

P_T	P_1	P_2	dC_1/dP_1	C_T
MW	MW	MW	$/MWhr	$/hr
500	206	294	13.29	5529
600	259	341	14.14	6901
700	312	388	14.99	8358
800	365	435	15.84	9899
900	418	482	16.68	11525
1000	471	529	17.53	13235
1100	524	576	18.38	15030
1200	576	624	19.22	16910
1300	629	671	20.07	18875
1400	682	718	20.92	20924
1500	735	765	21.76	23058

◾

Effect of Inequality Constraints

Each generating unit must not operate above its rating or below some minimum value. That is,

$$P_{i\min} < P_i < P_{i\max} \qquad i = 1, 2, \ldots, N \tag{11.4.9}$$

Other inequality constraints may also be included in the economic dispatch problem. For example, some unit outputs may be restricted so that certain transmission lines or other equipments are not overloaded. Also, under adverse weather conditions, generation at some units may be limited to reduce emissions.

When inequality constraints are included, we modify the economic dispatch solution as follows. If one or more units reach their limit values, then these units are held at their limits, and the remaining units operate at equal incremental operating cost λ. The incremental operating cost of the area equals the common λ for the units that are not at their limits.

EXAMPLE 11.5 | Economic dispatch solution including generator limits

Rework Example 11.4 if the units are subject to the following inequality constraints:

$$100 \leq P_1 \leq 600 \quad \text{MW}$$

$$400 \leq P_2 \leq 1000 \quad \text{MW}$$

Solution   At light loads, unit 2 operates at its lower limit of 400 MW, where its incremental operating cost is $dC_2/dP_2 = 15.2$ $/MWhr. Additional load comes from unit 1 until $dC_1/dP_1 = 15.2$ $/MWhr, or

$$\frac{dC_1}{dP_1} = 10 + 16 \times 10^{-3}P_1 = 15.2$$

$$P_1 = 325 \quad \text{MW}$$

For P_T less than 725 MW, where P_1 is less than 325 MW, the incremental operating cost of the area is determined by unit 1 alone.

At heavy loads, unit 1 operates at its upper limit of 600 MW, where its incremental operating cost is $dC_1/dP_1 = 19.60$ \$/MWhr. Additional load comes from unit 2 for all values of dC_2/dP_2 greater than 19.60 \$/MWhr. At $dC_2/dP_2 = 19.60$ \$/MWhr,

$$\frac{dC_2}{dP_2} = 8 + 18 \times 10^{-3}P_2 = 19.60$$

$$P_2 = 644 \quad \text{MW}$$

For P_T greater than 1244 MW, where P_2 is greater than 644 MW, the incremental operating cost of the area is determined by unit 2 alone.

For $725 < P_T < 1244$ MW, neither unit has reached a limit value, and the economic dispatch solution is the same as that given in Table 11.1.

The solution to this example is summarized in Table 11.2 for values of P_T from 500 to 1500 MW.

Table 11.2

Economic dispatch solution for Example 11.5

P_T	P_1	P_2	dC/dP	C_T
MW	MW	MW	\$/MWhr	\$/hr
500	100	400	$\dfrac{dC_1}{dP_1}\begin{cases}11.60\\13.20\\14.80\\15.20\end{cases}$	5720
600	200	400		6960
700	300	400		8360
725	325	400		8735
800	365	435	15.84	9899
900	418	482	16.68	11525
1000	471	529	17.53	13235
1100	524	576	18.38	15030
1200	576	624	19.22	16910
1244	600	644	$\dfrac{dC_2}{dP_2}\begin{cases}19.60\\20.60\\22.40\\24.20\end{cases}$	17765
1300	600	700		18890
1400	600	800		21040
1500	600	900		23370

∎

Effect of Transmission Losses

Although one unit may be very efficient with a low incremental operating cost, it may also be located far from the load center. The transmission losses associated with this unit may be so high that the economic dispatch solution requires the unit to decrease its output, while other units with higher incremental operating costs but lower transmission losses increase their outputs.

When transmission losses are included in the economic dispatch problem, (11.4.2) becomes

$$P_1 + P_2 + \cdots + P_N - P_L = P_T \tag{11.4.10}$$

where P_T is the total load demand and P_L is the total transmission loss in the area. In general, P_L is not constant, but depends on the unit outputs $P_1, P_2, \ldots, P_N$. The total differential of (11.4.10) is

$$(dP_1 + dP_2 + \cdots + dP_N) - \left(\frac{\partial P_L}{\partial P_1} dP_1 + \frac{\partial P_L}{\partial P_2} dP_2 + \cdots + \frac{\partial P_L}{\partial P_N} dP_N \right) = 0$$

(11.4.11)

Multiplying (11.4.11) by λ and subtracting the resulting equation from (11.4.5),

$$\left(\frac{dC_1}{dP_1} + \lambda \frac{\partial P_L}{\partial P_1} - \lambda \right) dP_1 + \left(\frac{dC_2}{dP_2} + \lambda \frac{\partial P_L}{\partial P_2} - \lambda \right) dP_2 + \cdots + \left(\frac{dC_N}{dP_N} + \lambda \frac{\partial P_L}{\partial P_N} - \lambda \right) dP_N = 0 \qquad (11.4.12)$$

Equation (11.4.12) is satisfied when each term in parentheses equals zero. That is,

$$\frac{dC_i}{dP_i} + \lambda \frac{\partial P_L}{\partial P_i} - \lambda = 0$$

or

$$\lambda = \frac{dC_i}{dP_i}(L_i) = \frac{dC_i}{dP_i} \left(\frac{1}{1 - \dfrac{\partial P_L}{\partial P_i}} \right) \qquad i = 1, 2, \ldots, N \qquad (11.4.13)$$

Equation (11.4.13) gives the economic dispatch criterion, including transmission losses. Each unit that is not at a limit value operates such that its incremental operating cost dC_i/dP_i multiplied by the *penalty factor* L_i is the same. Note that when transmission losses are negligible, $\partial P_L/\partial P_i = 0$, $L_i = 1$, and (11.4.13) reduces to (11.4.8).

| EXAMPLE 11.6 | **Economic dispatch solution including generator limits and line losses** |

Total transmission losses for the power-system area given in Example 11.5 are given by

$$P_L = 1.5 \times 10^{-4} P_1^2 + 2 \times 10^{-5} P_1 P_2 + 3 \times 10^{-5} P_2^2 \quad \text{MW}$$

where P_1 and P_2 are given in megawatts. Determine the output of each unit, total transmission losses, total load demand, and total operating cost C_T when the area $\lambda = 16.00$ \$/MWhr.

Solution   Using the incremental operating costs from Example 11.4 in (11.4.13),

$$\frac{dC_1}{dP_1} \left(\frac{1}{1 - \dfrac{\partial P_L}{\partial P_1}} \right) = \frac{10 + 16 \times 10^{-3} P_1}{1 - (3 \times 10^{-4} P_1 + 2 \times 10^{-5} P_2)} = 16.00$$

$$\frac{dC_2}{dP_2} \left(\frac{1}{1 - \dfrac{\partial P_L}{\partial P_2}} \right) = \frac{8 + 18 \times 10^{-3} P_2}{1 - (6 \times 10^{-5} P_2 + 2 \times 10^{-5} P_1)} = 16.00$$

Rearranging the above two equations,

$$20.8 \times 10^{-3}P_1 + 32 \times 10^{-5}P_2 = 6.00$$

$$32 \times 10^{-5}P_1 + 18.96 \times 10^{-3}P_2 = 8.00$$

Solving,

$$P_1 = 282 \quad MW \qquad P_2 = 417 \quad MW$$

Using the equation for total transmission losses,

$$P_L = 1.5 \times 10^{-4}(282)^2 + 2 \times 10^{-5}(282)(417) + 3 \times 10^{-5}(417)^2$$

$$= 19.5 \quad MW$$

From (11.4.10), the total load demand is

$$P_T = P_1 + P_2 - P_L = 282 + 417 - 19.5 = 679.5 \quad MW$$

Also, using the cost formulas given in Example 11.4, the total operating cost is

$$C_T = C_1 + C_2 = 10(282) + 8 \times 10^{-3}(282)^2 + 8(417) + 9 \times 10^{-3}(417)^2$$

$$= 8357 \quad \$/hr$$

Note that when transmission losses are included, λ given by (11.4.13) is no longer the incremental operating cost of the area. Instead, λ is the unit incremental operating cost dC_i/dP_i multiplied by the unit penalty factor L_i.

∎

In Example 11.6, total transmission losses are expressed as a quadratic function of unit output powers. For an area with N units, this formula generalizes to

$$P_L = \sum_{i=1}^{N} \sum_{j=1}^{N} P_i B_{ij} P_j \tag{11.4.14}$$

where the B_{ij} terms are called *loss coefficients* or **B** *coefficients*. The **B** coefficients are not truly constant, but vary with unit loadings. However, the **B** coefficients are often assumed constant in practice since the calculation of $\partial P_L/\partial P_i$ is thereby simplified. Using (11.4.14),

$$\frac{\partial P_L}{\partial P_i} = 2 \sum_{j=1}^{N} B_{ij} P_j \tag{11.4.15}$$

This equation can be used to compute the penalty factor L_i in (11.4.13).

Various methods of evaluating **B** coefficients from power-flow studies are available [6]. In practice, more than one set of **B** coefficients may be used during the daily load cycle.

When the unit incremental cost curves are linear, an analytic solution to the economic dispatch problem is possible, as illustrated by Examples 11.4–11.6. However, in practice, the incremental cost curves are nonlinear and contain discontinuities. In this case, an iterative solution by digital computer can be obtained. Given the load demand P_T, the unit incremental cost curves, generator limits, and B coefficients, such an iterative solution can be obtained by the following nine steps. Assume that the incremental cost curves are stored in tabular form, such that a unique value of P_i can be read for each dC_i/dP_i.

Step 1 Set iteration index $m = 1$.

Step 2 Estimate mth value of λ.

Step 3 Skip this step for all $m > 1$. Determine initial unit outputs P_i $(i = 1, 2, \ldots, N)$. Use $dC_i/dP_i = \lambda$ and read P_i from each incremental operating cost table. Transmission losses are neglected here.

Step 4 Compute $\partial P_L/\partial P_i$ from (11.4.15) $(i = 1, 2, \ldots, N)$.

Step 5 Compute dC_i/dP_i from (11.4.13) $(i = 1, 2, \ldots, N)$.

Step 6 Determine updated values of unit output P_i $(i = 1, 2, \ldots, N)$. Read P_i from each incremental operating cost table. If P_i exceeds a limit value, set P_i to the limit value.

Step 7 Compare P_i determined in Step 6 with the previous value $(i = 1, 2, \ldots, N)$. If the change in each unit output is less than a specified tolerance ε_1, go to Step 8. Otherwise, return to Step 4.

Step 8 Compute P_L from (11.4.14).

Step 9 If $\left| \left(\sum_{i=1}^{N} P_i \right) - P_L - P_T \right|$ is less than a specified tolerance ε_2, stop. Otherwise set $m = m + 1$ and return to Step 2.

Instead of having their values stored in tabular form for this procedure, the incremental cost curves could instead be represented by nonlinear functions such as polynomials. Then, in Step 3 and Step 5 each unit output P_i would be computed from the nonlinear functions instead of being read from a table. Note that this procedure assumes that the total load demand P_T is constant. In practice, this economic dispatch program is executed every few minutes with updated values of P_T.

Coordination of Economic Dispatch with LFC

Both the load-frequency control (LFC) and economic dispatch objectives are achieved by adjusting the reference power settings of turbine-governors on control. Figure 11.9 shows an *automatic generation control* strategy for achieving both objectives in a coordinated manner. As shown, the area control error (ACE) is first computed, and a share K_{1i} ACE is allocated to each unit. Second, the deviation of total actual generation from total desired generation is computed, and a share $K_{2i} \sum (P_{iD} - P_i)$ is allocated to unit i. Third, the deviation of actual generation from desired generation of unit i is computed, and $(P_{iD} - P_i)$ is allocated to unit i. An error signal formed from these three components and multiplied by a control gain K_{3i} determines the

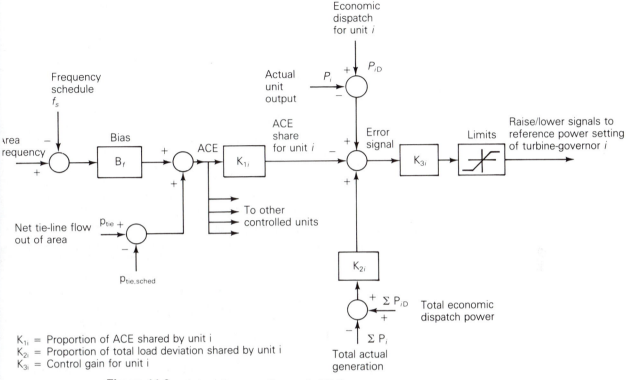

K_{1i} = Proportion of ACE shared by unit i
K_{2i} = Proportion of total load deviation shared by unit i
K_{3i} = Control gain for unit i

Figure 11.9 Automatic generation control [11]

raise or lower signals that are sent to the turbine-governor of each unit i on control.

In practice, raise or lower signals are dispatched to the units at discrete time intervals of 2 to 10 seconds. The desired outputs P_{iD} of units on control, determined from the economic dispatch program, are updated at slower intervals, typically every 2 to 10 minutes.

Other Types of Units

The economic dispatch criterion has been derived for a power-system area consisting of fossil-fuel generating units. In practice, however, an area has a mix of different types of units including fossil-fuel, nuclear, pumped-storage hydro, hydro, and other types.

Although the fixed costs of a nuclear unit may be high, their operating costs are low due to inexpensive nuclear fuel. As such, nuclear units are normally base-loaded at their rated outputs. That is, the reference power settings of turbine-governors for nuclear units are held constant at rated output; therefore, these units do not participate in LFC or economic dispatch.

Pumped-storage hydro is a form of energy storage. During off-peak hours these units are operated as synchronous motors to pump water to a higher elevation. Then during peak-load hours the water is released and the units are operated as synchronous generators to supply power. As such, pumped-storage hydro units are used for light-load build-up and peak-load shaving. Economic operation of the area is improved by pumping during off-peak hours when the area λ is low, and by generating during peak-load hours when λ is high. Techniques are available for incorporating pumped-storage hydro units into economic dispatch of fossil-fuel units [7].

In an area consisting of hydro plants located along a river, the objective is to maximize the energy generated over the yearly water cycle rather than to minimize total operating costs. Reservoirs are used to store water during high-water or light-load periods, although some water may have to be released through spillways. Also, there are constraints on water levels due to river transportation, irrigation, or fishing requirements. Optimal strategies are available for coordinating outputs of plants along a river [8]. Economic dispatch strategies for mixed fossil-fuel/hydro systems are also available [9, 10, 11].

Techniques are also available for including reactive power flows in the economic dispatch formultion, whereby both active and reactive powers are selected to minimize total operating costs. In particular, reactive injections from generators, switched capacitor banks, and static var systems, along with transformer tap settings, can be selected to minimize transmission-line losses [11]. However, electric utility companies usually control reactive power locally. That is, the reactive power output of each generator is selected to control the generator terminal voltage, and the reactive power output of each capacitor bank or static var system located at a power system bus is selected to control the voltage magnitude at that bus. In this way, the reactive power flows on transmission lines are low, and the need for central dispatch of reactive power is eliminated.

PROBLEMS

Section 11.2

11.1 An area of an interconnected 60-Hz power system has three turbine-generator units rated 100, 200, and 600 MVA. The regulation constants of the units are 0.04, 0.05, and 0.06 per unit, respectively, based on their ratings. Each unit is initially operating at one-half its own rating when the load suddenly decreases by 100 MW. Determine (a) the unit area frequency response characteristic β on a 100-MVA base, (b) the steady-state increase in area frequency, and (c) the MW decrease in mechanical power output of each turbine. Assume that the reference power setting of each turbine-governor remains constant. Neglect losses and the dependence of load on frequency.

11.2 Each unit in Problem 11.1 is initially operating at one-half its own rating when the load suddenly increases by 75 MW. Determine (a) the steady-state decrease in area frequency, and (b) the MW increase in mechanical power output of each turbine. Assume that the reference power setting of each turbine-generator remains constant. Neglect losses and the dependence of load on frequency.

11.3 Each unit in Problem 11.1 is initially operating at one-half its own rating when the frequency increases by 0.003 per unit. Determine the MW decrease of each unit. The reference power setting of each turbine-governor is fixed. Neglect losses and the dependence of load on frequency.

11.4 Repeat Problem 11.3 if the frequency decreases by 0.005 per unit. Determine the MW increase of each unit.

Section 11.3

11.5 A 60-Hz power system consists of two interconnected areas. Area 1 has 1000 MW of generation and an area frequency response characteristic $\beta_1 = 500$ MW/Hz. Area 2 has 2000 MW of generation and $\beta_2 = 800$ MW/Hz. Each area is initially operating at one-half its total generation, at $\Delta p_{tie1} = \Delta p_{tie2} = 0$ and at 60 Hz, when the load in area 1 suddenly increases by 400 MW. Determine the steady-state frequency error and the steady-state tie-line error Δp_{tie} of each area. Assume that the reference power settings of all turbine-governors are fixed. That is, LFC is not employed in any area. Neglect losses and the dependence of load on frequency.

11.6 Repeat Problem 11.5 if LFC is employed in area 2 alone. The area 2 frequency bias coefficient is set at $B_{f2} = \beta_2 = 800$ MW/Hz. Assume that LFC in area 1 is inoperative due to a computer failure.

11.7 Repeat Problem 11.5 if LFC is employed in both areas. The frequency bias coefficients are $B_{f1} = \beta_1 = 500$ MW/Hz and $B_{f2} = \beta_2 = 800$ MW/Hz.

11.8 Repeat Problem 11.6 if there is a third area, with 4000 MW of generation and $\beta_3 = 1500$ MW/Hz. The load in area 1 increases by 400 MW. LFC is employed in area 2 alone. All three areas are interconnected.

11.9 Rework Problems 11.5 through 11.7 when the load in area 2 suddenly decreases by 400 MW. The load in area 1 does not change.

Section 11.4

11.10 An area of an interconnected power system has two fossil-fuel units operating on economic dispatch. The variable operating costs of these units are given by

$$C_1 = \begin{cases} 2P_1 + 0.02P_1^2 & \text{for } 0 < P_1 \leqslant 100 \quad \text{MW} \\ 6P_1 \; \dfrac{\$}{hr} & \text{for } P_1 > 100 \quad \text{MW} \end{cases}$$

$$C_2 = 0.03P_2^2 \; \frac{\$}{hr}$$

where P_1 and P_2 are in megawatts. Determine the power output of each unit, the incremental operating cost, and the total cost C_T that minimizes C_T as the load demand P_T varies from 200 to 700 MW. Generating-unit inequalities and transmission losses are neglected.

11.11 Rework Problem 11.10 if the units are subject to the following inequality constraints:

$$100 \leqslant P_1 \leqslant 500 \quad \text{MW}$$

$$50 \leqslant P_2 \leqslant 300 \quad \text{MW}$$

11.12 Rework Problem 11.11 if total transmission losses for the power system are given by

$$P_L = 2 \times 10^{-4}P_1^2 + 1 \times 10^{-4}P_2^2 \quad \text{MW}$$

11.13 Rework Problems 11.10 through 11.12 if the operating cost of unit 2 is changed to:

$$C_2 = 0.04P_2^2 \quad \$/hr$$

Compare the results with those of Problems 11.10 through 11.12.

11.14   Expand the summations in (11.4.14) for $N = 2$, and verify the formula for $\partial P_L/\partial P_i$ given by (11.4.15). Assume $B_{ij} = B_{ji}$.

CASE STUDY QUESTIONS

A.   What are the differences in automatic generation control (AGC) of an electric utility with predominantly hydroelectric generation versus a utility with predominantly fossil-fuel generation?

B.   Should energy management systems (EMS) at power-system control centers be totally automated? During emergency conditions, should operating decisions be made by the computer or by experienced dispatch personnel?

References

1.   IEEE Committee Report, "Computer Representation of Excitation Systems," *IEEE Transactions PAS*, vol. PAS-87 (June 1968), pp. 1460–1464.

2.   M. S. Sarma, *Electric Machines* (Dubuque, IA: Brown, 1985).

3.   IEEE Committee Report, "Dynamic Models for Steam and Hydro Turbines in Power System Studies," *IEEE Transactions PAS*, vol. PAS-92, no. 6 (November/December 1973), pp. 1904–1915.

4.   N. Cohn, *Control of Generation and Power Flow on Interconnected Systems* (New York: Wiley, 1971).

5.   L. K. Kirchmayer, *Economic Operation of Power Systems* (New York: Wiley, 1958).

6.   L. K. Kirchmayer and G. W. Stagg, "Evaluation of Methods of Coordinating Incremental Fuel Costs and Incremental Transmission Losses," *Transactions AIEE*, vol. 71, part III (1952), pp. 513–520.

7.   G. H. McDaniel and A. F. Gabrielle, "Dispatching Pumped Storage Hydro," *IEEE Transmission PAS*, vol. PAS-85 (May 1966), pp. 465–471.

8.   E. B. Dahlin and E. Kindingstad, "Adaptive Digital River Flow Predictor for Power Dispatch," *IEEE Transactions PAS*, vol. PAS-83 (April 1964), pp. 320–327.

9.   L. K. Kirchmayer, *Economic Control of Interconnected Systems* (New York: Wiley, 1959).

10.   J. H. Drake et al., "Optimum Operation of a Hydrothermal System," *Transactions AIEE* (*Power Apparatus and Systems*), vol. 62 (August 1962), pp. 242–250.

11.   A. J. Wood and B. F. Wollenberg, *Power Generation, Operation, and Control* (New York: Wiley, 1989).

12.   J. Kimbriel, L. Douglas, and J. Bodmann, "Scada Upgraded with AGC Using Different Vendor's Computer," *Electrical World*, *205*, 11 (November 1991), pp. 32–34.

13.   J. Reason, "Full Graphics Boosts Dispatcher's Capacity," *Electrical World*, *205*, 11 (November 1991), pp. 34–35.

CHAPTER 12

TRANSMISSION LINES: TRANSIENT OPERATION

Three-phase 1500-MVA, 525-kV/241-kV substation transformer bank
(Courtesy of Florida Power and Light Company)

Transient overvoltages caused by lightning strikes to transmission lines and by switching operations are of fundamental importance in selecting equipment insulation levels and surge-protection devices. We must, therefore, understand the nature of transmission-line transients.

During our study of the steady-state performance of transmission lines in

Chapter 6, the line constants R, L, G, and C were recognized as distributed rather than lumped constants. When a line with distributed constants is subjected to a disturbance such as a lightning strike or a switching operation, voltage and current waves arise and travel along the line at a velocity near the speed of light. When these waves arrive at the line terminals, reflected voltage and current waves arise and travel back down the line, superimposed on the initial waves.

Because of line losses, traveling waves are attenuated and essentially die out after a few reflections. Also, the series inductances of transformer windings effectively block the disturbances, thereby preventing them from entering generator windings. However, due to the reinforcing action of several reflected waves, it is possible for voltage to build up to a level that could cause transformer insulation or line insulation to arc over and suffer damage.

Circuit breakers, which can operate within 50 ms, are too slow to protect against lightning or switching surges. Lightning surges can rise to peak levels within a few microseconds and switching surges within a few hundred microseconds—fast enough to destroy insulation before a circuit breaker could open. However, protective devices are available. Called surge arresters, these can be used to protect equipment insulation against transient overvoltages. These devices limit voltage to a ceiling level and absorb the energy from lightning and switching surges.

We begin this chapter with a discussion of traveling waves on single-phase lossless lines (Section 12.1). Boundary conditions are given in Section 12.2, and the Bewley lattice diagram for organizing reflections is presented in Section 12.3. Discrete-time models of single-phase lines and of lumped RLC elements are derived in Section 12.4, and the effects of line losses and multiconductor lines are discussed in Sections 12.5 and 12.6. In Section 12.7 power-system overvoltages including lightning surges, switching surges, and power-frequency overvoltages are discussed, followed by an introduction to insulation coordination in Section 12.8. Finally, the personal computer program TRANS-MISSION-LINE TRANSIENTS, suitable for power-system analysis and design with respect to insulation coordination, is described in Section 12.9.

CASE STUDY Transients occurring on electric utility transmission or distribution lines as well as on communication lines (telephone, TV cables) can reach vulnerable customer facilities, such as computer and solid-state equipment. Undesirable effects of transients includes safety hazards, equipment damage, and business losses. Two reports on transients are presented here. The first describes methods customers use to protect against transients and other power problems including the use of surge suppressors, "ground windows," and uninterruptible power supplies [22]. The second describes metal oxide varistor (MOV) arresters used by electric utilities to protect distribution equipment against transients [23].

Protecting Computer Systems Against Power Transients

FRANÇOIS MARTZLOFF

For the third time in less than three weeks, the sky darkened and thunder rumbled in the distance. With that ominous warning, the appointed "thunder scout" decided it was time to pull the plugs of the central unit and remote terminals of his CAD/CAM computer system. Better to shut down the operation than risk damage to the systems as had occurred in the two previous storms.

But pulling the plugs did not help. When the storm was over and the system was restarted, permanent damage had been done to it—to the chagrin of both the operator and the service contractor.

In this common example, the damage was caused not by power-line surges but by differences in ground potential at various terminals of the system. The oversimplified assumption that power-line surge problems could be eliminated had led the uninformed operator to attempt a simple prevention step. Not only did it not work, but it created a safety hazard plugging the line cords removed the safety ground, leaving the equipment still connected to the data lines where the surges were occurring.

Understanding the general causes of, and remedies for, power transients can help users of small computer systems, especially stand-alones, protect their systems with do-it-yourself methods. More complex systems may need the attention of a specialist. Systems designers should also be aware of the way users hook up their systems, the potential damage that could be caused by power transients, and side effects of incorrectly applied measures.

Growing concern among computer users that power-line surges may damage equipment or cause loss of data has created a market for surge suppressors. But clear performance standards in the industry are lacking, and several standards-writing groups are striving to develop adequate ones. To make a difficult choice among these devices, the consumer should learn some basic rules about selecting and installing a suitable surge suppressor. Even the best surge suppressor, if incorrectly applied, might not work and could cause adverse side effects.

Transient origins

While the term "transient" is often understood as a transient overvoltage, it is also more broadly interpreted as the occurrence of any disturbance, either on the power line or the computer system's data line.

The most obvious source of an electrical disturbance is a lightning strike, but the lightning bolt need not hit power lines to cause damage. Because the electromagnetic field radiated by the lightning current couples into the conductors of power lines or data lines, it induces transient voltages along these conductors. Also, as the lightning current spreads into the

Copyright © 1990 IEEE. Reprinted, with permission, from IEEE Spectrum, 27, 4 (April 1990), pp. 37–40.

ground, it produces differences in potential at points that are normally at "ground" potential. Conductors spanning some distance between their ends in the area where the lightning current is spreading will be exposed to these differences of potential, or to a transient overvoltage.

Though the direct effects of lightning can be dramatic, their relatively low rate of occurrence can lull one into complacency, and most of their widespread indirect effects can be overcome through sound protection practices. On the other hand, electrostatic discharges, which could be considered miniature lightning discharges, require only the fingertips of mortals rather than an Olympian fistful of lightning bolts to have very serious effects ["How to defeat electrostatic discharge," *IEEE Spectrum*, August 1989, pp. 36–40].

A less obvious but more frequent source of transients is switching sequences in the power system. Switching can be a normal, recurrent operation such as turning a local load on and off, or it could entail occasionally clearing an overload or short circuit.

These switching transients cover a wide range of frequencies and amplitudes. Some have a brief duration (nanoseconds) and involve little energy (millijoules). While they present little risk of damage, their high-frequency spectrum makes them likely sources of interference. Others have a longer duration (microseconds or even milliseconds) and involve greater energy (up to hundreds of joules) with lower frequencies. They have the opposite trait of low risk of interference because of the relatively low frequencies, but because of the longer duration and increased energy, they have a higher damage risk.

Another source of disturbance to computer systems is the occurrence of an undervoltage that could be caused by a nearby startup of heavy loads or by distant faults, such as lightning-induced line flashover, falling trees, or utility lines downed by runaway vehicles. While transient overvoltages can be easily suppressed—a more correct description would be "mitigated"—by a simple added device that diverts the excess energy, the reduced energy associated with an undervoltage cannot be supplemented by a

simple device. Different methods are needed for a solution of that problem.

Over the years, the need to learn more about the characteristics of these transients has led to various projects aimed at monitoring power-line disturbances. One result of these projects—which are performed by isolated researchers, sometimes with equipment designed by the researchers rather than commercial equipment—is that their reports are based on different assumptions and definitions of disturbances. Comparing results can thus become difficult and confusing [see Fig. 1].

Leaving the problems of monitoring transients and designing protective devices to the specialists, an informed user can take several steps toward buttressing the reliability and integrity of the system. The first step is to distinguish between mere temporary upset and permanent damage, each of which has a different impact on the user, depending on the relative importance of the operation. For a commercial setup, disrupting the operation can be more expensive than repairing the damage so that protecting data integrity ranks high. For an engineer working at his home computer, however, damage protection may be more important than some data loss. In this case, limiting the protection expense to avoiding damage and accepting interruptions may be preferred.

Vulnerable stand-alones

Small computer "systems" can be categorized as stand-alone systems or distributed systems. A stand-alone system is typically a one-operator setup consisting of a desktop computer coupled with a printer, or any microcomputer not linked to a network. Distributed systems range from a simple stand-alone augmented by a telephone or other network link to multiple-station systems or process control systems with remote sensors and actuators.

Found in offices, laboratories, and homes, stand-alone systems can be disrupted or damaged by two possible causes. First, transients with low amplitudes (less than 1000 volts) are buffered by the computer's power supply but might still couple into circuits and

	Typical system configurations	Threat	Solution
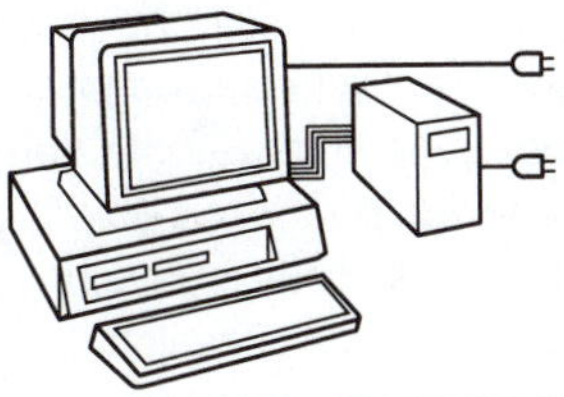	Stand-alone with peripherals		
	in same outlet	Line transient	Spike suppressor
	in different outlets	Ground potential differences	Local ground window
	Power-line and data-line interfaces (such as a facsimile or answering machine)	Differences in ground potential during surges, even with individual line protection	Local ground window
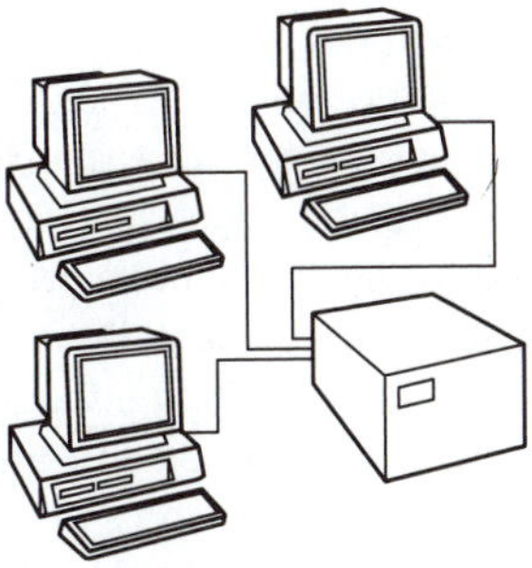	Distributed system with remote terminals (such as three PCs connected to a printer or three dumb terminals linked to a central processing unit)	Line transients in individual cords; operation of built-in suppression raises "ground" potential	Local ground window at each terminal; an optical-fiber link is an alternative
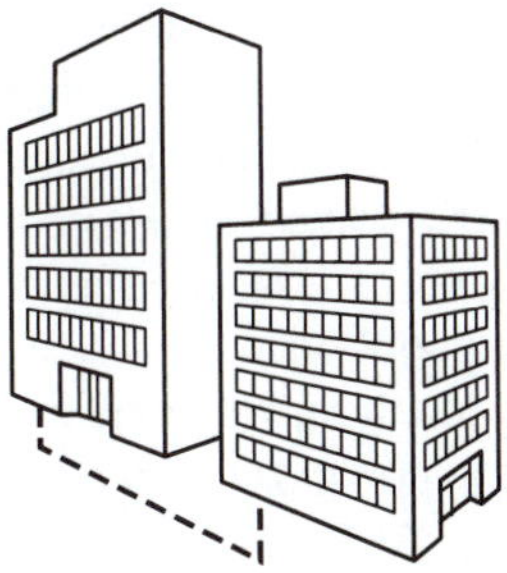	Systems in separate buildings	Line transients in individual cords; ground potential differences	Special grounding (a specialist's task) because of problems due to ground grid design, National Electrical Code issues, installation, and so on; an optical-fiber link is an alternative

Figure 1 *Common troublesome scenarios*

cause glitches. Transients of high amplitudes (over 1000 V) may at worst damage the power input components and are likely to cause glitches at best. Second, power interruptions (sags or outages) can cause a momentary shutdown.

Transient damage protection for these systems is simple to achieve and is probably built in to some degree. However, until the day arrives when equipment has its transient capability stated on the nameplate (which may be sooner than expected because the Europeans are increasingly concerned with electromagnetic compatibility issues), the user has no way to know the extent of that protection. The European approach, motivated by a 1989 Directive on Electromagnetic Compatibility promulgated by the European Community Council, requires that equipment must operate satisfactorily in a specified environment without introducing intolerable disturbances into that environment. Thus, this ability is likely to be stated explicitly, in addition to the usual voltage, frequency, and current ratings now required.

So far, the approach has been one of purchasing additional peace of mind by inserting a separate surge suppressor (also called spike protector and transient voltage suppressor) on the power cord. Prices for these devices range over an order of magnitude, and claims of performance may include the fastest response (an irrelevant issue) and the lowest clamping voltage (a reliability risk because the transient protector may fail under abnormal power fluctuations).

Though its basic technology does not change rapidly, details of the rating and packaging of a surge suppressor are driven by market competition. Ideally, its rating should reflect three basic requirements: the nominal line voltage, the surge current capability, and the clamping voltage during the surge. All of these should be stated by the maker of the device with due consideration to the user's needs for protection and long-term reliability.

At this time, there is only one performance standard in the United States for transient voltage surge suppressors, UL 1449, which was developed by the Underwriters Laboratories. This standard specifies primarily the safety aspects of the product, but does contain some performance specifications. The UL label on a surge suppressor means that a test has been applied to the device, reflecting industry consensus standards on the severity of the environment. In the UL test, a specified surge current is applied to the device and the maximum resulting voltage is measured; this is then indicated on the product.

Product literature for some devices, however, makes claims of response time in nanoseconds—even picoseconds—a feature that is not important in a power system. Nanosecond pulses do not propagate very far in power systems, and measuring a picosecond response time in support of the claim would be a technical challenge. Likewise, emphasis on achieving the lowest clamping voltage only demonstrates imbalance in the design goals: the object of a surge suppressor is to lessen the surge level from the thousands of volts that can occur occasionally; it is not to shave off the last tens of volts from the protection level in a "lower is better" bid for ranking in the purchaser's choice. An excessively low clamping voltage introduces the risk of premature aging, even failure, of the device when the power line goes through repeated momentary overvoltages, or "swells."

The second type of disturbance, a sag or outage, cannot be corrected by a surge suppressor. The computer operation is interrupted when the sag or outage exceeds the capability of the internal dc supply to power the logic and memory circuits. Most computers have a built-in capability to maintain operation for a short time when this power is lost, but that supply is drained out if the interruption is long enough. If the computer is using a disk drive when the sag occurs, a shutdown is likely; in an office using several identical machines, some ride through a disturbance while others, especially those reading from a disk, shut down and have to be restarted. Protection against such sags and outages requires an uninterruptible power supply (UPS), which is now readily available. In fact, the volume of UPS production as well as competition has brought prices down so low that purchasing one becomes a viable solution and, for users dependent on the continuity of their operation, a must.

Unexpected problems

A power outage or sag on distributed systems has the same effect as for stand-alone systems. A more subtle problem, however, has crept in for some sophisticated systems that include automatic restart, or rebooting, after a power interruption. Anecdotes have circulated of damage caused by repeated sags during the automatic rebooting sequence, typically occurring because of multiple lightning strokes or during fault clearing with automatic reclosing by the utility system.

In the case of surges, as soon as a simple stand-alone system is augmented by peripherals, additional remote terminals, networking, and sensors that require a data link, the threat that the system will be affected increases. Even what may appear as a stand-alone system, such as a simple desktop pair of a PC linked with a printer, might be at risk if the two units are plugged into different power receptacles fed by separate branch circuits from the breakers.

In addition to the risk of interference or damage from surges on the power line, the data-line input and output ports are also vulnerable. Several mechanisms can inject interfering or damaging transients into the data lines of distributed systems. First, a problem could result because data lines act as antennas that can collect energy from electromagnetic fields and feed it, as noise or surges, to the data port's input or output, the driver or the receiver of the computer, or its peripherals.

The problem's severity increases with the length of the data link. Within the same room, the risk of damage is minimal. But as the communication link reaches farther out, the risk increases that a surge would not only interfere with a system but could damage it. Though there may be an unknown (to the user) built-in protection or inherent capability of the data port components to withstand these surges, little is known about the occurence of surges on data lines, compared to that on ac power lines, which makes the task of designing protection difficult.

For users of computer systems in the same room or the same corner of a building, the built-in capability probably suffices. For systems with longer reach, the ultimate protection is an optical-fiber link with no metallic jacket, which provides immunity against noise collection as well as possible surge damage. For these complex systems, however, the do-it-yourself approach should be replaced by one that has been designed by a specialist.

Another mechanism that could cause trouble is the difference in the potential of objects at nominal "ground" potential occurring during surge events. Most data links operate with the signal reference conductor (shield or one wire of a group) connected to the chassis of the equipment. This chassis is in turn connected to the grounding conductor of the power cord supplying the equipment, a requirement of the National Electrical Code. Thus, if lightning or power system faults inject a high current in the site's ground conductors, the potential of the "grounded" points at the two ends of the data link differs. This potential difference causes a current to flow into the data link, possibly exceeding the capability of the input or output components.

The user can stay with conductors for the data link or convert (or initially design) it to an optical-fiber link, an approach that is becoming increasingly popular as hardware costs fall with economies of scale. However, if the conversion electronics at the ends of the fiber link are disturbed by electrical noise, that noise will be faithfully transmitted, not blocked.

If a conductive data link is to remain, the remedy is to insert protective devices that are complementary for the power line and data line. These devices typically operate by limiting the overvoltage or attenuating the higher frequencies by filtering, which works effectively on the power line but not on the data link. Here, filtering is not possible because it would affect the signals; limiting the overvoltages will eliminate that damage risk, but might still let through a spurious signal. Thus, data integrity may be more difficult to achieve unless the software includes inherent immunity or fault tolerance.

Side effects

Avoiding damage with protective devices may then seem to require only the insertion of a power-line surge suppressor at the wall receptacle and a data-line surge suppressor at the input to the computer. This apparent simplicity, however, is deceptive because the very operation of this device, if incorrectly installed, can have a side effect that would put the data link components at risk, a mechanism that is only beginning to be fully recognized.

Still another mechanism can be demonstrated by a scenario that can occur in any building with power and telephone service. The incoming telephone line is provided with surge suppressors (carbon blocks or gas tubes) that divert surges to the nearest grounded conductor, generally a nearby water pipe. The manufacturer of the computer or modem used for the computer–telephone-line linkup may have provided a protective device within the equipment. Alternately, the surge-conscious user may have inserted a protective device in the power cord. But should a surge occur on either the data line or the power line, the corresponding protective device will dutifully divert that surge to the nearest ground. Since the "nearest ground" may not be the same for the connection of the two suppressors, the surge current in the ground connection raises the potential of one side with respect to the other, placing the data input at risk.

The solution is a miniature setup of the "ground window" concept developed by telephone companies in protecting their central station switches: all cables entering a room or a complete floor in a building are routed through a single "window" where grounding conductors, shields, and ground connections of protective devices are bonded together. In this manner, there cannot be any potential difference between the various ground reference points within the room or floor. [See Figure 2.]

Some surge suppressor manufacturers have adapted that concept to a portable version of the ground window, a device consisting of a suppressor for the power line and one for the data line, but packaged in a single box most likely sharing the same ground connection. This local ground window is now found in computer or hobby stores and is easily recognizable because it features both a power connection (male plug for connection to a wall receptacle and female receptacles for powering the loads) and a pair of data link connectors (input and output). Depending on what is needed, these connectors can be a standard telephone jack, a multipin RS232, or a cable television coaxial connector. The device is then

inserted near the computer, with the power cord and data link routed through its connectors.

Another protection scheme is always available: disconnect the system when not in use! In fact, some of the consumer guidance folders inserted by the utilities with their monthly bills mention that approach. That option may not be practical for commercial operations, where some link could be left connected, creating the risk of ungrounded equipment. Thus, if applied, every link to the outside world must be disconnected.

To probe further

A good source of information on the basics of lightning is the book *Understanding Lightning* by Martin A. Uman, available from Academic Press, New York, 1971. Solutions to noise problems are given a general treatment in the second edition of *Noise Reduction Techniques in Electronic Systems* by Henry W. Ott, available from John Wiley & Sons, New York, 1989. Fundamentals of surge protection techniques are treated in *Protection of Electronic Circuits from Overvoltages* by Ronald B. Standler, also available from John Wiley & Sons, 1989. Another useful reference is *Uninterruptible Power Supplies* by David C. Griffith, published by Marcel Dekker, New York, 1989.

Guidance on the nature and severity of transients (not specifications for protective devices) is given in the *IEEE Guide for Surge Voltages in Low Voltage AC Power Circuits*, American National Standards Institute, C62.41-1980, available from the IEEE Service Center, 445 Hoes Lane, Box 1331, Piscataway, N.J.; 800-678-IEEE.

A paper by François Martzloff and Thomas Gruzs titled "Power Quality Site Surveys: Facts, Fiction and Fallacies" in the November 1988 IEEE *Industry Applications Society Transactions* presents a review of recording, analyzing, and reporting transient disturbances on power lines. Another paper, "Coupling, Propagation, and Side Effects of Surges in an Industrial Wiring System," by Martzloff in the same *Transactions* is in press. The journal is available from the IEEE Service Center.

Waveform	Type of disturbance	Type of equipment affected	Effective protection equipment
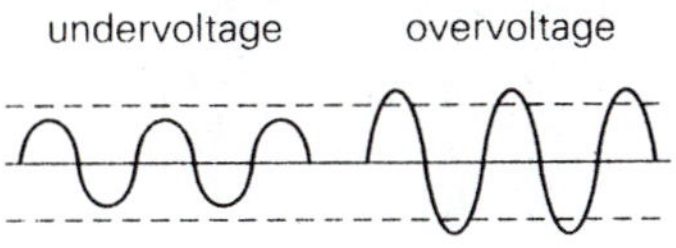 undervoltage overvoltage	Undervoltage or over-voltage are conditions of abnormally low or high voltages lasting more than a few seconds and caused by circuit over- loads, poor voltage regula- tion, and intentional reductions by the utility (brownout)	All equipment is affected, although most equipment is designed to tolerate 120 volts ± 10 percent	Voltage regulator, line conditioner, or uninterruptible power supply (UPS)
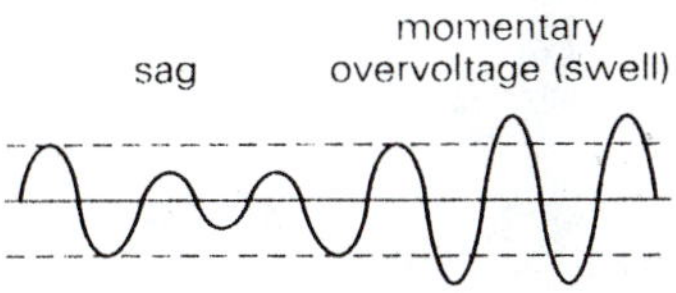 sag momentary overvoltage (swell)	Voltage sags (voltage decreases outside normal tolerance lasting less than a few seconds) that are often caused when heavy loads are started, by lightning, and by power system faults; swells are brief voltage increases often caused by sudden load decreases or turn-off of heavy equipment	Sags affect power-downsensing circuitry on computers and large controllers and can cause equipment to shut down; swells can damage equipment (including spike suppressors) that have insufficient tolerance	Voltage regulator, ferro-resonant transformer, line conditioner, or UPS
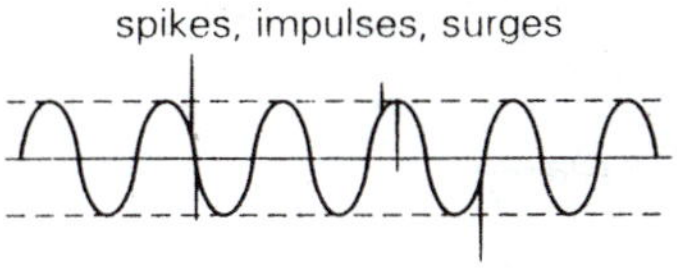 spikes, impulses, surges	Spikes (impulses, switching surges, or lightning surges) are microsecond- to millisecond-long voltage increases, ranging in amplitude from 200 to 6000 volts and caused by light-ning, switching of heavy loads, and short circuits or power system faults	Electronic loads can be destroyed and transformer or motor insulation broken down	Spike suppressor (also called surge suppressor), or some line conditioners
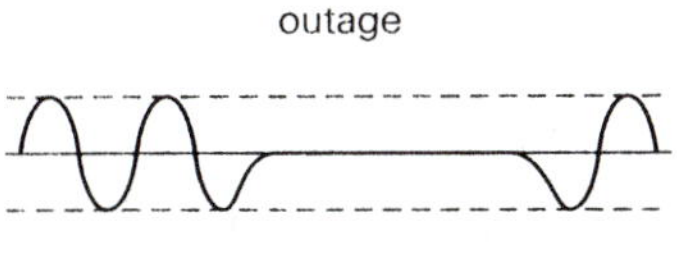 outage	An outage is the complete loss of power for several milliseconds to several hours and may be caused by power system faults, accidents involving power lines, transformer failures, and generator failures. Some sensitive equipment may be disrupted by outages as short as 15 ms	All equipment is affected	UPS or standby power supply
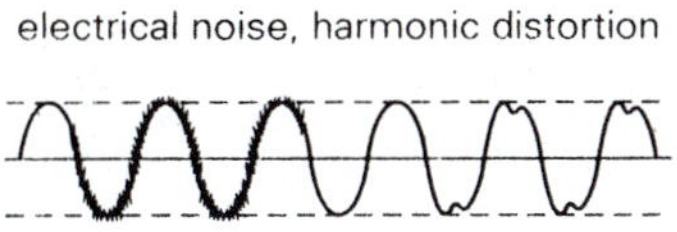 electrical noise, harmonic distortion	Electrical noise is a distortion of the normal sinewave power and can be caused by radar and radio transmitters, flourescent lights, power electronic control circuits, arcing utility and industrial equipment, loads with solid-state rectifiers, and switching power supplies typically used in computer systems	Electrical noise disturbs microprocessor-based equipment, such as microcomputers and programmable controllers; harmonic distortion causes motor loads, such as compressors, pumps, and disk drives, to overheat	Filter, isolation transformer, UPS, or some line conditioners

Figure 2 *Types of disturbances and effective protection equipment*

How to Get Top Performance from Polymer MOV Arresters

Introduction of new materials in arresters over the past few years has provided dramatic improvements in performance over the old construction using gapped silicon carbide in porcelain. But at the same time, these new materials demand extra care when specifying arresters and have introduced some unknowns. Polymer housings cannot be evaluated by the same standards as porcelain housing; metal oxide varistors cannot be specified as though they were silicon carbide blocks. In the following articles, two authors offer valuable guidelines for the selection of modern arresters.

Not All Polymer Housings Are the Same Quality

JEFFRY MACKEVICH

Danger of fragmentation of porcelain surge arresters is the principal reason why utilities are rapidly converting to polymer-housed arresters (Fig. 1). Most engineers know that this is not the only advantage offered by polymer arresters, but few have fully considered the design aspects that make a top-quality polymer arrester. Even fewer have access to laboratories where they can conduct their own comprehensive evaluation testing. Industry standards do not yet address the unique performance aspects of these products.

For a start, don't assume that all polymer arresters are inherently safe. This is a potentially dangerous assumption. Testing has shown that in some failure modes, internal parts can be expelled.

Low price can be false economy

Since the arrester is a protective device, it is often assumed that if it fails it has sacrificed itself to protect

Reprinted with permission from Electrical World.

Figure 1 *Polymer arresters are shatterproof, but they're not all the same.*

another piece of equipment. A nonviolent failure that occurs on a sunny day may be considered no differently from a failure that occurs during a lightning storm. Even a utility that keeps accurate failure statistics usually does not record the likely cause of failure or the nameplate information, from which an arrester performance data base could be developed.

Without performance records, utilities buy arresters on a low-bid price. But an arrester that fails for any reason other than excessive energy input is not providing value to the user. Buying on price represents false economy if the total operating or ownership costs are higher than they would be with a more reliable unit.

Check multiple failure modes

Porcelain arresters fail because either excessive energy input or moisture ingress produces an internal flashover. That is why US arrester standards require a pressure-relief/fault-current withstand test. In this test, a fuse wire is assembled inside the arrester to produce the flashover and determine how the arrester would perform under actual failure conditions.

For a polymer arrester, the fuse wire simulates moisture ingress or tracking of the fiberglass casing. But neither of these should occur in a well-designed polymer arrester and the user should not base the evaluation solely on the fuse-wire test. Two other tests are available to evaluate different modes of failure: One tests the arrester with failed metal-oxide varistors; the other checks for thermal runaway.

Failed varistors can be simulated by applying excessive energy to the disks and failing them before assembly into the arrester body. Alternatively, a small hole can be drilled through the disks and a fuse wire inserted.

Fast thermal runaway testing takes a functioning arrester and drives it into failure by overvoltage from a short-circuit generator. A fast-thermal-runaway test simulates a number of likely field failures and should always be required by the purchaser. This is because metal-oxide arresters fail from excessive energy if the varistor temperature exceeds the arrester's point of thermal instability. Thermal instability means that the power dissipation (heat generation) within the varistors, which is a function of varistor temperature, exceeds the cooling capacity of the arrester.

Slow thermal runaway is an alternative test that is the subject of some controversy. In this test, a constant heating current is applied by a low power source to raise the arrester's internal temperature and reduce its mechanical strength. At the point of varistor failure, the short-circuit current is applied and the weakened arrester subjected to the fault energy and related mechanical forces. The test simulates a condition that might occur when one phase of a floating wye/delta transformer bank is open. The unenergized phase can experience overvoltage, which produces a low-magnitude heating current. This is followed by available system fault current when the open phase is re-energized. The same situation may also be produced by a ferroresonance condition.

Check for internal air space

Polymer arresters can experience moisture vapor transmission (MVT) if there is an internal air space. MVT is the passage of moisture molecules through the polymer material. It occurs when there are relative humidity, pressure, and temperature gradients across the moisture-permeable polymer housing. Moisture collects on the inside of the arrester and causes failure. If there is no internal air space, MVT cannot occur.

Also, an arrester designed without internal air space has better heat-transfer capability because of the intimate contact between the arrester body and the varistors. The better the heat transfer, the less possibility the varistor temperature will rise to the point of thermal runaway.

One measure of arrester reliability is the temperature at which thermal instability occurs. The higher the temperature, the more durable and reliable the arrester.

Check arrester-body design

One major unknown and cause for concern for many users is the polymer material. While there are perceptions and opinions about one base polymer over another, numerous other aspects affect overall performance. Carefully consider the housing design—creepage and strike distances, rain-shed configuration, and terminal electrode design. Check also the use of the polymer material in similar applications.

The product type that most closely approximates an arrester is a shielded power cable termination. By contrast, insulators are significantly different from arresters in both construction and service conditions. They must withstand a different mechanical load and electric field distribution, and are not the best proxy for judging polymer performance in an arrester.

Ask the arrester supplier to document materials testing conducted, including tracking wheel and salt-fog chamber, accelerated ultraviolet exposure from natural and synthetic light, and thermal aging. The basis for testing should be explained, in addition to actual experience with that polymer material in similar applications. The supplier should also be able to furnish independent materials studies to verify the performance of the polymer material. Consider the length of industry experience with similar applications, the breadth of the supplier's product line using out-door, non-tracking, weather-resistant polymer ma-terials, and the technical support provided by the supplier to help the user understand polymer arrester performance.

Moisture-seal systems must be of proven design. The housing should have primary end seals (Fig. 2) to prevent moisture from reaching the housing inter-face, and a robust sealing material that fills the inter-face. This not only prevents moisture vapor trans-mission, but also stops axial migration of moisture if an end seal is compromised. Greases do not provide the necessary long-term life. Permanent, non-migra-ting sealants should be preferred.

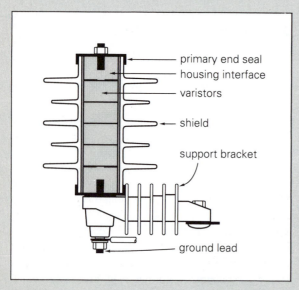

Figure 2 Housing design must ensure complete absence of internal airspace and positive seal against moisture ingress

Why Higher MOV Ratings May Be Necessary

RAMON MANCAO, JAMES BURKE, AND ANDREW MYERS

Because MOV arresters do not incorporate a gap in their design, they are considered to be more prone to damage from steady-state overvoltages. Particular attention should be paid to accurate calculation of line-to-neutral overvoltages on unfaulted phases of a multigrounded-neutral circuit during a fault to ground on one phase.

Two methods of calculation are compared below. The first, CALC1, is the typical utility approach using symmetrical components, which considers the neutral conductor and ground to be perfectly coupled. The second method, CALC2, considers an equivalent electrical network that takes into account the substa-tion ground-mat impedance and the coupling of ground footing resistances on the neutral wire. Re-sults of this second method in conjunction with field measurements suggest that the currently recommen-ded overvoltage factor of 1.25 p.u. may be inadequate and a factor of 1.35 p.u. should be considered.

Fig. 3 shows the equivalent circuit used for CALC2. The parameters of each line section between ground footing resistances consist of the self and mutual impedances of the phase and neutral wires. The sharing of fault currents between the neutral wire and earth determines the voltage drop across each section. The maximum rise of the line-to-neutral volt-age on the unfaulted phases is the vectorial sum of the voltage drops of each neutral-wire section from the fault to the substation and the phase-to-neutral voltage of the source.

If ground footing resistance is assumed to be 25 ohms, the new calculation method predicts that the highest overvoltage (1.313 p.u.) occurs when the fault is two miles from the substation. The older method predicts that the highest overvoltage (1.290 p.u.) occurs when the fault is at the end of the line. If ground footing resistance is 100 ohms, the over-voltage reaches 1.339 p.u. when the fault is 3 miles from the substation. Fig. 4 shows graphically how

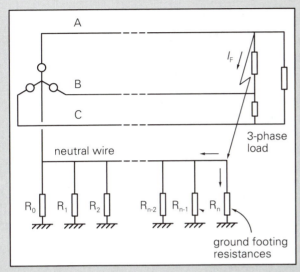

Figure 3 *Equivalent circuit shows how ground-mat imped-ance and ground-footing resistance on the neutral are taken into account*

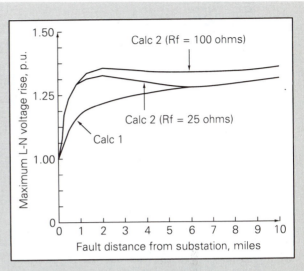

Figure 4 *New calculation method predicts considerably higher overvoltages when fault is close to substation*

these results differ considerably from typical industry calculations.

A 1971 working group of the IEEE surge protec-tion committee suggested that an overvoltage factor of 1.25 p.u. be used for arrester selection. This factor was based on a 1.2 p.u. overvoltage component that included the effect of transformer saturation, as well as a 5% regulation factor. This overvoltage was con-sidered conservative because the system parameters were thought to be pessimistic. But random field measurements show per-unit overvoltages in excess of 25%, indicating that transformer saturation cannot always be relied upon to limit temporary overvoltages. Voltage regulation on a distribution feeder often ex-

ceeds 1.05 p.u. and has been measured in excess of 1.10 p.u.

The analysis performed in this study shows that it is highly probable that voltages up to or exceeding 1.30 p.u. will occur on typical four-wire multigrounded circuits during line-to-ground faults if transformer saturation is neglected. Even if ground footing resis-tance as low as 25 ohms is obtained, lower overvolt-ages are not assured. Small-diameter neutral wires can significantly increase overvoltage. Broken neu-trals can raise overvoltages to 1.5 p.u. if transformers do not saturate.

Based on this study, a new overvoltage factor of 1.35 p.u. is recommended for utilities using MOV arresters on multigrounded systems.

SECTION 12.1

TRAVELING WAVES ON SINGLE-PHASE LOSSLESS LINES

We first consider a single-phase two-wire lossless transmission line. Figure 12.1 shows a line section of length Δx meters. If the line has a loop inductance L H/m and a line-to-line capacitance C F/m, then the line section has a series inductance $L\Delta x$ H and shunt capacitance $C\Delta x$ F, as shown. In Chapter 6, the direction of line position x was selected to be from the receiving end ($x = 0$) to the sending end ($x = l$); this selection was unimportant, since the variable x was subsequently eliminated when relating the steady-state sending-end quantities V_s and I_s to the receiving-end quantities V_R and I_R. Here, however, we are interested in voltages and current waveforms traveling along the line.

Figure 12.1

Single-phase two-wire loss-less line section of length Δx

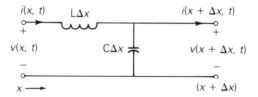

Therefore, we select the direction of increasing x as being from the sending end ($x = 0$) toward the receiving end ($x = l$).

Writing a KVL and KCL equation for the circuit in Figure 12.1,

$$v(x + \Delta x, t) - v(x, t) = -L\Delta x \frac{\partial i(x, t)}{\partial t} \tag{12.1.1}$$

$$i(x + \Delta x, t) - i(x, t) = -C\Delta x \frac{\partial v(x, t)}{\partial t} \tag{12.1.2}$$

Dividing (12.1.1) and (12.1.2) by Δx and taking the limit as $\Delta x \to 0$, we obtain

$$\frac{\partial v(x, t)}{\partial x} = -L \frac{\partial i(x, t)}{\partial t} \tag{12.1.3}$$

$$\frac{\partial i(x, t)}{\partial x} = -C \frac{\partial v(x, t)}{\partial t} \tag{12.1.4}$$

We use partial derivatives here because $v(x, t)$ and $i(x, t)$ are differentiated with respect to both position x and time t. Also, the negative signs in (12.1.3) and (12.1.4) are due to the reference direction for x. For example, with a positive value of $\partial i/\partial t$ in Figure 12.1, $v(x, t)$ decreases as x increases.

Taking the Laplace transform of (12.1.3) and (12.1.4),

$$\frac{dV(x, s)}{dx} = -sLI(x, s) \tag{12.1.5}$$

$$\frac{dI(x, s)}{dx} = -sCV(x, s) \tag{12.1.6}$$

where zero initial conditions are assumed. $V(x, s)$ and $I(x, s)$ are the Laplace transforms of $v(x, t)$ and $i(x, t)$. Also, ordinary rather than partial derivatives are used since the derivatives are now with respect to only one variable, x.

Next we differentiate (12.1.5) with respect to x and use (12.1.6), in order to eliminate $I(x, s)$:

$$\frac{d^2V(x, s)}{dx^2} = -sL\,\frac{dI(x, s)}{dx} = s^2LCV(x, s)$$

or

$$\frac{d^2V(x, s)}{dx^2} - s^2LCV(x, s) = 0 \qquad (12.1.7)$$

Similarly, (12.1.6) can be differentiated in order to obtain

$$\frac{d^2I(x, s)}{dx^2} - s^2LCI(x, s) = 0 \qquad (12.1.8)$$

Equation (12.1.7) is a linear, second-order homogeneous differential equation. By inspection, its solution is

$$V(x, s) = V^+(s)e^{-sx/v} + V^-(s)e^{+sx/v} \qquad (12.1.9)$$

where

$$v = \frac{1}{\sqrt{LC}} \quad \text{m/s} \qquad (12.1.10)$$

Similarly, the solution to (12.1.8) is

$$I(x, s) = I^+(s)e^{-sx/v} + I^-(s)e^{+sx/v} \qquad (12.1.11)$$

The reader can quickly verify that these solutions satisfy (12.1.7) and (12.1.8). The "constants" $V^+(s)$, $V^-(s)$, $I^+(s)$, and $I^-(s)$, which in general are functions of s but are independent of x, can be evaluated from the boundary conditions at the sending and receiving ends of the line. The superscripts $+$ and $-$ refer to waves traveling in the positive x and negative x directions, soon to be explained.

Taking the inverse Laplace transform of (12.1.9) and (12.1.11), and recalling the time shift properly, $\mathscr{L}[f(t - \tau)] = F(s)e^{-s\tau}$, we obtain

$$v(x, t) = v^+\left(t - \frac{x}{v}\right) + v^-\left(t + \frac{x}{v}\right) \qquad (12.1.12)$$

$$i(x, t) = i^+\left(t - \frac{x}{v}\right) + i^-\left(t + \frac{x}{v}\right) \qquad (12.1.13)$$

where the functions $v^+(\ )$, $v^-(\ )$, $i^+(\ )$, and $i^-(\ )$, can be evaluated from the boundary conditions.

We now show that $v^+(t - x/v)$ represents a voltage wave traveling in the positive x direction with velocity $v = 1/\sqrt{LC}$ m/s. Consider any wave

$f^+(u)$, where $u = t - x/v$. Suppose that this wave begins at $u = u_0$, as shown in Figure 12.2(a). At time $t = t_1$, the wavefront is at $u_0 = (t_1 - x_1/v)$, or at $x_1 = v(t_1 - u_0)$. At a later time, t_2, the wavefront is at $u_0 = (t_2 - x_2/v)$ or at $x_2 = v(t_2 - u_0)$. As shown in Figure 12.2(b), the wavefront has moved in the positive x direction a distance $(x_2 - x_1) = v(t_2 - t_1)$ during time $(t_2 - t_1)$. The velocity is therefore $(x_2 - x_1)/(t_2 - t_1) = v$.

Figure 12.2

The function $f^+(u)$, where $u = \left(t - \dfrac{x}{v}\right)$

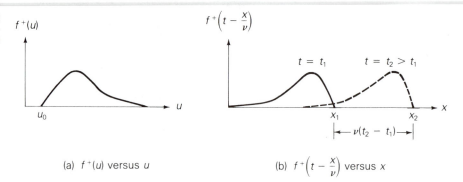

(a) $f^+(u)$ versus u

(b) $f^+\left(t - \dfrac{x}{v}\right)$ versus x

Similarly, $i^+(t - x/v)$ represents a current wave traveling in the positive x direction with velocity v. We call $v^+(t - x/v)$ and $i^+(t - x/v)$ the *forward* traveling voltage and current waves. It can be shown analogously that $v^-(t + x/v)$ and $i^-(t + x/v)$ travel in the negative x direction with velocity v. We call $v^-(t + x/v)$ and $i^-(t + x/v)$ the *backward* traveling voltage and current waves.

Recall from (6.4.16) that for a lossless line $f\lambda = 1/\sqrt{LC}$. It is now evident that the term $1/\sqrt{LC}$ in this equation is v, the velocity of propagation of voltage and current waves along the lossless line. Also, recall from Chapter 5 that L is proportional to μ and C is proportional to ε. For overhead lines, $v = 1/\sqrt{LC}$ is approximately equal to $1/\sqrt{\mu\varepsilon} = 1/\sqrt{\mu_0\varepsilon_0} = 3 \times 10^8$ m/s, the speed of light in free space. For cables, the relative permittivity $\varepsilon/\varepsilon_0$ may be 3 to 5 or even higher, resulting in a value of v lower than that for overhead lines.

We next evaluate the terms $I^+(s)$ and $I^-(s)$. Using (12.1.9) and (12.1.10) in (12.1.6),

$$\frac{s}{v}[-I^+(s)e^{-sx/v} + I^-(s)e^{+sx/v}] = -sC[V^+(s)e^{-sx/v} + V^-(s)e^{+sx/v}]$$

Equating the coefficients of $e^{-sx/v}$ on both sides of this equation,

$$I^+(s) = (vC)V^+(s) = \frac{V^+(s)}{\sqrt{\dfrac{L}{C}}} = \frac{V^+(s)}{Z_c} \tag{12.1.14}$$

where

$$Z_c = \sqrt{\frac{L}{C}} \quad \Omega \tag{12.1.15}$$

Similarly, equating the coefficients of $e^{+sx/v}$,

$$I^-(s) = \frac{-V^-(s)}{Z_c} \tag{12.1.16}$$

Thus, we can rewrite (12.1.11) and (12.1.13) as

$$I(x, s) = \frac{1}{Z_c}[V^+(s)e^{-sx/v} - V^-(s)e^{+sx/v}] \tag{12.1.17}$$

$$i(x, t) = \frac{1}{Z_c}\left[v^+\left(t - \frac{x}{v}\right) - v^-\left(t + \frac{x}{v}\right)\right] \tag{12.1.18}$$

Recall from (6.4.3) that $Z_c = \sqrt{L/C}$ is the characteristic impedance (also called surge impedance) of a lossless line.

SECTION 12.2

BOUNDARY CONDITIONS FOR SINGLE-PHASE LOSSLESS LINES

Figure 12.3 shows a single-phase two-wire lossless line terminated by an impedance $Z_R(s)$ at the receiving end and a source with Thévenin voltage $E_G(s)$ and with Thévenin impedance $Z_G(s)$ at the sending end. $V(x, s)$ and $I(x, s)$ are the Laplace transforms of the voltage and current at position x. The line has length l, surge impedance $Z_c = \sqrt{L/C}$, and velocity $v = 1/\sqrt{LC}$. We assume that the line is initially unenergized.

Figure 12.3

Single-phase two-wire loss-less line with source and load terminations

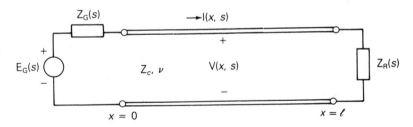

From Figure 12.3, the boundary condition at the receiving end is

$$V(l, s) = Z_R(s)I(l, s) \tag{12.2.1}$$

Using (12.1.9) and (12.1.17) in (12.2.1),

$$V^+(s)e^{-sl/v} + V^-(s)e^{+sl/v} = \frac{Z_R(s)}{Z_c}[V^+(s)e^{-sl/v} - V^-(s)e^{+sl/v}]$$

Solving for $V^-(l, s)$

$$V^-(l, s) = \Gamma_R(s)V^+(s)e^{-2s\tau} \tag{12.2.2}$$

where

$$\Gamma_R(s) = \frac{\dfrac{Z_R(s)}{Z_c} - 1}{\dfrac{Z_R(s)}{Z_c} + 1} \quad \text{per unit} \tag{12.2.3}$$

$$\tau = \frac{l}{v} \quad \text{seconds} \tag{12.2.4}$$

$\Gamma_R(s)$ is called the *receiving-end voltage reflection coefficient*. Also, τ, called the *transit time* of the line, is the time it takes a wave to travel the length of the line.

Using (12.2.2) in (12.1.9) and (12.1.17),

$$V(x, s) = V^+(s)[e^{-sx/v} + \Gamma_R(s)e^{s[(x/v)-2\tau]}] \tag{12.2.5}$$

$$I(x, s) = \frac{V^+(s)}{Z_c}[e^{-sx/v} - \Gamma_R(s)e^{s[(x/v)-2\tau]}] \tag{12.2.6}$$

From Figure 12.3 the boundary condition at the sending end is

$$V(0, s) = E_G(s) - Z_G(s)I(0, s) \tag{12.2.7}$$

Using (12.2.5) and (12.2.6) in (12.2.7),

$$V^+(s)[1 + \Gamma_R(s)e^{-2s\tau}] = E_G(s) - \left[\frac{Z_G(s)}{Z_c}\right]V^+(s)[1 - \Gamma_R(s)e^{-2s\tau}]$$

Solving for $V^+(s)$,

$$V^+(s)\left\{\left[\frac{Z_G(s)}{Z_c} + 1\right] - \Gamma_R(s)e^{-2s\tau}\left[\frac{Z_G(s)}{Z_c} - 1\right]\right\} = E_G(s)$$

$$V^+(s)\left[\frac{Z_G(s)}{Z_c} + 1\right]\left\{1 - \Gamma_R(s)\Gamma_s(s)e^{-2s\tau}\right\} = E_G(s)$$

or

$$V^+(s) = E_G(s)\left[\frac{Z_c}{Z_G(s) + Z_c}\right]\left[\frac{1}{1 - \Gamma_R(s)\Gamma_s(s)e^{-2s\tau}}\right] \tag{12.2.8}$$

where

$$\Gamma_s(s) = \frac{\dfrac{Z_G(s)}{Z_c} - 1}{\dfrac{Z_G(s)}{Z_c} + 1} \tag{12.2.9}$$

$\Gamma_s(s)$ is called the *sending-end voltage reflection coefficient*. Using (12.2.9) in

(12.2.5) and (12.2.6), the complete solution is

$$V(x, s) = E_G(s)\left[\frac{Z_c}{Z_G(s) + Z_c}\right]\left[\frac{e^{-sx/v} + \Gamma_R(s)e^{s[(x/v) - 2\tau]}}{1 - \Gamma_R(s)\Gamma_S(s)e^{-2s\tau}}\right] \tag{12.2.10}$$

$$I(x, s) = \left[\frac{E_G(s)}{Z_G(s) + Z_c}\right]\left[\frac{e^{-sx/v} - \Gamma_R(s)e^{s[(x/v) - 2\tau]}}{1 - \Gamma_R(s)\Gamma_S(s)e^{-2s\tau}}\right] \tag{12.2.11}$$

where

$$\Gamma_R(s) = \frac{\dfrac{Z_R(s)}{Z_c} - 1}{\dfrac{Z_R(s)}{Z_c} + 1} \quad \text{per unit} \qquad \Gamma_S(s) = \frac{\dfrac{Z_G(s)}{Z_c} - 1}{\dfrac{Z_G(s)}{Z_c} + 1} \quad \text{per unit} \tag{12.2.12}$$

$$Z_c = \sqrt{\frac{L}{C}} \;\; \Omega \qquad v = \frac{1}{\sqrt{LC}} \;\; \text{m/s} \qquad \tau = \frac{l}{v} \;\; \text{s} \tag{12.2.13}$$

The following four examples illustrate this general solution. All four examples refer to the line shown in Figure 12.3, which has length l, velocity v, characteristic impedance Z_c, and is initially unenergized.

| **EXAMPLE 12.1** | **Single-phase lossless-line transients: step-voltage source at sending end, matched load at receiving end** |

Let $Z_R = Z_c$ and $Z_G = 0$. The source voltage is a step, $e_G(t) = E u_{-1}(t)$. (a) Determine $v(x, t)$ and $i(x, t)$. Plot the voltage and current versus time t at the center of the line and at the receiving end.

Solution **a.** From (12.2.12) with $Z_R = Z_c$ and $Z_G = 0$,

$$\Gamma_R(s) = \frac{1 - 1}{1 + 1} = 0 \qquad \Gamma_S(s) = \frac{0 - 1}{0 + 1} = -1$$

The Laplace transform of the source voltage is $E_G(s) = E/s$. Then, from (12.2.10) and (12.2.11),

$$V(x, s) = \left(\frac{E}{s}\right)(1)(e^{-sx/v}) = \frac{E e^{-sx/v}}{s}$$

$$I(x, s) = \frac{(E/Z_c)}{s} e^{-sx/v}$$

Taking the inverse Laplace transform,

$$v(x, t) = E u_{-1}\left(t - \frac{x}{v}\right)$$

$$i(x, t) = \frac{E}{Z_c} u_{-1}\left(t - \frac{x}{v}\right)$$

b. At the center of the line, where $x = l/2$,

$$v\left(\frac{l}{2}, t\right) = Eu_{-1}\left(t - \frac{\tau}{2}\right) \qquad i\left(\frac{l}{2}, t\right) = \frac{E}{Z_c} u_{-1}\left(t - \frac{\tau}{2}\right)$$

At the receiving end, where $x = l$,

$$v(l, t) = Eu_{-1}(t - \tau) \qquad i(l, t) = \frac{E}{Z_c} u_{-1}(t - \tau)$$

These waves, plotted in Figure 12.4, can be explained as follows. At $t = 0$ the ideal step voltage of E volts, applied to the sending end, encounters Z_c, the characteristic impedance of the line. Therefore, a forward traveling step voltage wave of E volts is initiated at the sending end. Also, since the ratio of the forward traveling voltage to current is Z_c, a forward traveling step current wave of (E/Z_c) amperes is initiated. These waves travel in the positive x direction, arriving at the center of the line at $t = \tau/2$, and at the end of the line at $t = \tau$. The receiving-end load is *matched* to the line; that is, $Z_R = Z_c$. For a matched load, $\Gamma_R = 0$, and therefore no backward traveling waves are initiated. In steady-state, the line with matched load is energized at E volts with current E/Z_c amperes.

Figure 12.4

Voltage and current wave-forms for Example 12.1

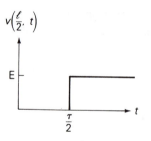

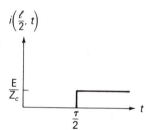

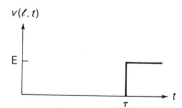

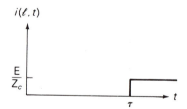

■

| **EXAMPLE 12.2** | **Single-phase lossless-line transients: step-voltage source matched at sending end, open receiving end** |

The receiving end is open. The source voltage at the sending end is a step $e_G(t) = Eu_{-1}(t)$, with $Z_G(s) = Z_c$. (a) Determine $v(x, t)$ and $i(x, t)$. (b) Plot the voltage and current versus time t at the center of the line.

Solution   **a.** From (12.2.12),

$$\Gamma_R(s) = \lim_{Z_R \to \infty} \frac{\dfrac{Z_R}{Z_c} - 1}{\dfrac{Z_R}{Z_c} + 1} = 1 \qquad \Gamma_S(s) = \frac{1 - 1}{1 + 1} = 0$$

The Laplace transform of the source voltage is $E_G(s) = E/s$. Then, from (12.2.10) and (12.2.11),

$$V(x, s) = \frac{E}{s}\left(\frac{1}{2}\right)[e^{-sx/v} + e^{s[(x/v) - 2\tau]}]$$

$$I(x, s) = \frac{E}{s}\left(\frac{1}{2Z_c}\right)[e^{-sx/v} - e^{s[(x/v) - 2\tau]}]$$

Taking the inverse Laplace transform,

$$v(x, t) = \frac{E}{2} u_{-1}\left(t - \frac{x}{v}\right) + \frac{E}{2} u_{-1}\left(t + \frac{x}{v} - 2\tau\right)$$

$$i(x, t) = \frac{E}{2Z_c} u_{-1}\left(t - \frac{x}{v}\right) - \frac{E}{2Z_c} u_{-1}\left(t + \frac{x}{v} - 2\tau\right)$$

b. At the center of the line, where $x = l/2$,

$$v\left(\frac{l}{2}, t\right) = \frac{E}{2} u_{-1}\left(t - \frac{\tau}{2}\right) + \frac{E}{2} u_{-1}\left(t - \frac{3\tau}{2}\right)$$

$$i\left(\frac{l}{2}, t\right) = \frac{E}{2Z_c} u_{-1}\left(t - \frac{\tau}{2}\right) - \frac{E}{2Z_c} u_{-1}\left(t - \frac{3\tau}{2}\right)$$

These waves are plotted in Figure 12.5. At $t = 0$ the step voltage source of E

Figure 12.5

Voltage and current wave-
forms for Example 12.2

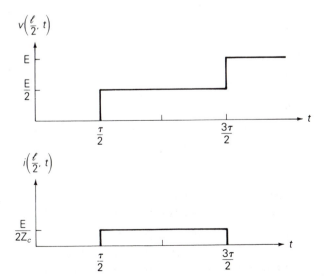

volts encounters the source impedance $Z_G = Z_c$ in series with the characteristic impedance of the line, Z_c. Using voltage division, the sending-end voltage at $t = 0$ is E/2. Therefore, a forward traveling step voltage wave of E/2 volts and a forward traveling step current wave of $E/(2Z_c)$ amperes are initiated at the sending end. These waves arrive at the center of the line at $t = \tau/2$. Also, with $\Gamma_R = 1$, the backward traveling voltage wave equals the forward traveling voltage wave, and the backward traveling current wave is the negative of the forward traveling current wave. These backward traveling waves, which are initiated at the receiving end at $t = \tau$ when the forward traveling waves arrive there, arrive at the center of the line at $t = 3\tau/2$ and are superimposed on the forward traveling waves. No additional forward or backward traveling waves are initiated because the source impedance is matched to the line; that is, $\Gamma_S(s) = 0$. In steady-state the line, which is open at the receiving end, is energized at E volts with zero current.   ∎

| EXAMPLE 12.3 | **Single-phase lossless-line transients: step-voltage source matched at sending end, capacitive load at receiving end** |

The receiving end is terminated by a capacitor with C_R farads, which is initially unenergized. The source voltage at the sending end is a unit step $e_G(t) = Eu_{-1}(t)$, with $Z_G = Z_c$. Determine and plot $v(x, t)$ versus time t at the sending end of the line.

Solution   From (12.2.12) with $Z_R = \dfrac{1}{sC_R}$ and $Z_G = Z_c$,

$$\Gamma_R(s) = \frac{\dfrac{1}{sC_R Z_c} - 1}{\dfrac{1}{sC_R Z_c} + 1} = \frac{-s + \dfrac{1}{Z_c C_R}}{s + \dfrac{1}{Z_c C_R}}$$

$$\Gamma_S(s) = \frac{1 - 1}{1 + 1} = 0$$

Then, from (12.2.10), with $E_G(s) = E/s$,

$$V(x, s) = \frac{E}{s}\left(\frac{1}{2}\right)\left[e^{-sx/v} + \left(\frac{-s + \dfrac{1}{Z_c C_R}}{s + \dfrac{1}{Z_c C_R}}\right) e^{s[(x/v) - 2\tau]} \right]$$

$$= \frac{E}{2}\left[\frac{e^{-sx/v}}{s} + \frac{1}{s}\left(\frac{-s + \dfrac{1}{Z_c C_R}}{s + \dfrac{1}{Z_c C_R}}\right) e^{s[(x/v) - 2\tau]} \right]$$

Using partial fraction expansion of the second term above,

$$V(x, s) = \frac{E}{2}\left[\frac{e^{-sx/v}}{s} + \left(\frac{1}{s} - \frac{2}{s + \frac{1}{Z_c C_R}}\right)e^{s[(x/v) - 2\tau]}\right]$$

The inverse Laplace transform is

$$v(x, t) = \frac{E}{2} u_{-1}\left(t - \frac{x}{v}\right) + \frac{E}{2}[1 - 2e^{(-1/Z_c C_R)(t + x/v - 2\tau)}]u_{-1}\left(t + \frac{x}{v} - 2\tau\right)$$

At the sending end, where $x = 0$,

$$v(0, t) = \frac{E}{2} u_{-1}(t) + \frac{E}{2}[1 - 2e^{(-1/Z_c C_R)(t - 2\tau)}]u_{-1}(t - 2\tau)$$

$v(0, t)$ is plotted in Figure 12.6. As in Example 12.2, a forward traveling step voltage wave of E/2 volts is initiated at the sending end at $t = 0$. At $t = \tau$, when the forward traveling wave arrives at the receiving end, a backward traveling wave is initiated. The backward traveling voltage wave, an exponential with initial value $-E/2$, steady-state value $+E/2$, and time constant $Z_c C_R$, arrives at the sending end at $t = 2\tau$, where it is superimposed on the forward traveling wave. No additional waves are initiated, since the source impedance is matched to the line. In steady-state, the line and the capacitor at the receiving end are energized at E volts with zero current.

Figure 12.6

Voltage waveform for Example 12.3

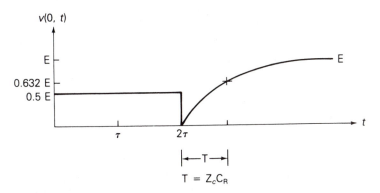

The capacitor at the receiving end can also be viewed as a short circuit at the instant $t = \tau$, when the forward traveling wave arrives at the receiving end. For a short circuit at the receiving end, $\Gamma_R = -1$, and therefore the backward traveling voltage wavefront is $-E/2$, the negative of the forward traveling wave. However, in steady-state the capacitor is an open circuit, for which $\Gamma_R = +1$, and the steady-state backward traveling voltage wave equals the forward traveling voltage wave. ∎

| EXAMPLE 12.4 | **Single-phase lossless-line transients: step-voltage source with unmatched source resistance at sending end, unmatched resistive load at receiving end** |

At the receiving end, $Z_R = Z_c/3$. At the sending end, $e_G(t) = Eu_{-1}(t)$ and $Z_G = 2Z_c$. Determine and plot the voltage versus time at the center of the line.

Solution   From (12.2.12),

$$\Gamma_R = \frac{\frac{1}{3} - 1}{\frac{1}{3} + 1} = -\frac{1}{2} \qquad \Gamma_S = \frac{2 - 1}{2 + 1} = \frac{1}{3}$$

From (12.2.10), with $E_G(s) = E/s$,

$$V(x, s) = \frac{E}{s}\left(\frac{1}{3}\right)\frac{[e^{-sx/v} - \frac{1}{2}e^{s[(x/v) - 2\tau]}]}{1 + (\frac{1}{6}e^{-2s\tau})}$$

The preceding equation can be rewritten using the following geometric series:

$$\frac{1}{1 + y} = 1 - y + y^2 - y^3 + y^4 - \cdots$$

with $y = \frac{1}{6}e^{-2s\tau}$,

$$V(x, s) = \frac{E}{3s}\left[e^{-sx/v} - \frac{1}{2}e^{s[(x/v) - 2\tau]}\right]$$

$$\left[1 - \frac{1}{6}e^{-2s\tau} + \frac{1}{36}e^{-4s\tau} - \frac{1}{216}e^{-6s\tau} + \cdots\right]$$

Multiplying the terms within the brackets,

$$V(x, s) = \frac{E}{3s}\left[e^{-sx/v} - \frac{1}{2}e^{s[(x/v) - 2\tau]} - \frac{1}{6}e^{-s[(x/v) + 2\tau]} + \frac{1}{12}e^{s[(x/v) - 4\tau]}\right.$$

$$\left. + \frac{1}{36}e^{-s[(x/v) + 4\tau]} - \frac{1}{72}e^{s[(x/v) - 6\tau]} + \cdots\right]$$

Taking the inverse Laplace transform,

$$v(x, t) = \frac{E}{3}\left[u_{-1}\left(t - \frac{x}{v}\right) - \frac{1}{2}u_{-1}\left(t + \frac{x}{v} - 2\tau\right) - \frac{1}{6}u_{-1}\left(t - \frac{x}{v} - 2\tau\right)\right.$$

$$+ \frac{1}{12}u_{-1}\left(t + \frac{x}{v} - 4\tau\right) + \frac{1}{36}u_{-1}\left(t - \frac{x}{v} - 4\tau\right)$$

$$\left. - \frac{1}{72}u_{-1}\left(t + \frac{x}{v} - 6\tau\right)\cdots\right]$$

At the center of the line, where $x = l/2$,

$$v\left(\frac{l}{2}, t\right) = \frac{E}{3}\left[u_{-1}\left(t - \frac{\tau}{2}\right) - \frac{1}{2}u_{-1}\left(t - \frac{3\tau}{2}\right) - \frac{1}{6}u_{-1}\left(t - \frac{5\tau}{2}\right)\right.$$

$$+ \frac{1}{12}u_{-1}\left(t - \frac{7\tau}{2}\right) + \frac{1}{36}u_{-1}\left(t - \frac{9\tau}{2}\right)$$

$$\left. - \frac{1}{72}u_{-1}\left(t - \frac{11\tau}{2}\right)\cdots\right]$$

Figure 12.7

Example 12.4

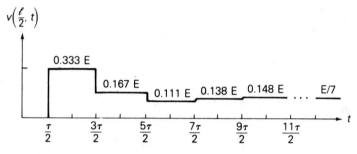

(a) Voltage waveform

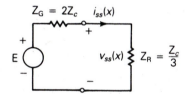

(b) Steady-state solution

$v(l/2, t)$ is plotted in Figure 12.7(a). Since neither the source nor the load is matched to the line, the voltage at any point along the line consists of an infinite series of forward and backward traveling waves. At the center of the line, the first forward traveling wave arrives at $t = \tau/2$; then a backward traveling wave arrives at $3\tau/2$, another forward traveling wave arrives at $5\tau/2$, another backward traveling wave at $7\tau/2$, and so on.

The steady-state voltage can be evaluated from the final value theorem. That is,

$$v_{ss}(x) = \lim_{t \to \infty} v(x, t) = \lim_{s \to 0} sV(x, s)$$

$$= \lim_{s \to 0} \left\{ s\left(\frac{E}{s}\right)\left(\frac{1}{3}\right) \frac{[e^{-sx/v} - \frac{1}{2}e^{s[(x/v) - 2\tau]}]}{1 + \frac{1}{6}e^{-2s\tau}} \right\}$$

$$= E\left(\frac{1}{3}\right)\left(\frac{1 - \frac{1}{2}}{1 + \frac{1}{6}}\right) = \frac{E}{7}$$

The steady-state solution can also be evaluated from the circuit in Figure 12.7(b). Since there is no steady-state voltage drop across the lossless line when a dc source is applied, the line can be eliminated, leaving only the source and load. The steady-state voltage is then, by voltage division,

$$v_{ss}(x) = E\left(\frac{Z_R}{Z_R + Z_G}\right) = E\left(\frac{\frac{1}{3}}{\frac{1}{3} + 2}\right) = \frac{E}{7}$$

∎

SECTION 12.3

BEWLEY LATTICE DIAGRAM

A lattice diagram developed by L. V. Bewley [2] conveniently organizes the reflections that occur during transmission-line transients. For the Bewley lattice diagram, shown in Figure 12.8, the vertical scale represents time and is scaled in units of τ, the transient time of the line. The horizontal scale

Figure 12.8

Bewley lattice diagram for Example 12.5

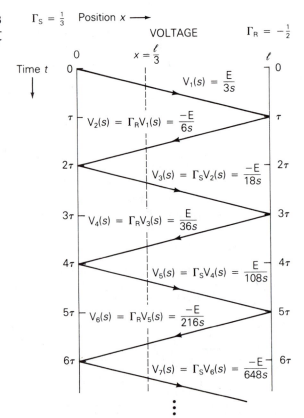

represents line position x, and the diagonal lines represent traveling waves. Each reflection is determined by multiplying the incident wave arriving at an end by the reflection coefficient at that end. The voltage $v(x, t)$ at any point x and t on the diagram is determined by adding all the terms directly above that point.

EXAMPLE 12.5

Lattice diagram: single-phase lossless line

For the line and terminations given in Example 12.4, draw the lattice diagram and plot $v(l/3, t)$ versus time t.

Solution

The lattice diagram is shown in Figure 12.8. At $t = 0$, the source voltage encounters the source impedance and the line characteristic impedance, and the first forward traveling wave is determined by voltage division:

$$V_1(s) = E_G(s) \left[\frac{Z_c}{Z_c + Z_G} \right] = \frac{E}{s} \left[\frac{1}{1 + 2} \right] = \frac{E}{3sx}$$

which is a step with magnitude $(E/3)$ volts. The next traveling wave, a backward one, is $V_2(s) = \Gamma_R(s)V_1(s) = (-\frac{1}{2})V_1(s) = -E/(6s)$, and the next wave, a forward one, is $V_3(s) = \Gamma_S(s)V_2(s) = (\frac{1}{3})V_2(s) = -E/(18s)$. Subsequent waves are calculated in a similar manner.

The voltage at $x = l/3$ is determined by drawing a vertical line at $x = l/3$ on the lattice diagram, shown dashed in Figure 12.8. Starting at the top of the dashed line, where $t = 0$, and moving down, each voltage wave is added at the time it intersects the dashed line. The first wave v_1 arrives at $t = \tau/3$, the second v_2 arrives at $5\tau/3$, v_3 at $7\tau/3$, and so on. $v(l/3, t)$ is plotted in Figure 12.9.

Figure 12.9

Voltage waveform for Example 12.5

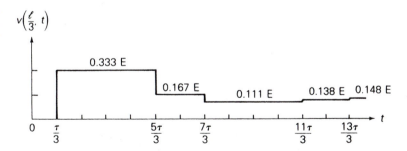

Figure 12.10 shows a forward traveling voltage wave V_A^+ arriving at the junction of two lossless lines A and B with characteristic impedances Z_A and

Figure 12.10

Junction of two single-phase lossless lines

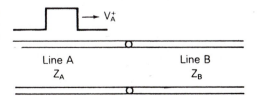

Z_B, respectively. This could be, for example, the junction of an overhead line and a cable. When V_A^+ arrives at the junction, both a reflection V_A^- on line A and a refraction V_B^+ on line B will occur. Writing a KVL and KCL equation at the junction,

$$V_A^+ + V_A^- = V_B^+ \tag{12.3.1}$$

$$I_A^+ + I_A^- = I_B^+ \tag{12.3.2}$$

Recall that $I_A^+ = V_A^+/Z_A$, $I_A^- = -V_A^-/Z_A$, and $I_B^+ = V_B^+/Z_B$. Using these relations in (12.3.2),

$$\frac{V_A^+}{Z_A} - \frac{V_A^-}{Z_A} = \frac{V_B^+}{Z_B} \tag{12.3.3}$$

Solving (12.3.1) and (12.3.3) for V_A^- and V_B^+ in terms of V_A^+ yields

$$V_A^- = \Gamma_{AA} V_A^+ \tag{12.3.4}$$

where

$$\Gamma_{AA} = \frac{\dfrac{Z_B}{Z_A} - 1}{\dfrac{Z_B}{Z_A} + 1} \tag{12.3.5}$$

and

$$V_B^+ = \Gamma_{BA} V_A \tag{12.3.6}$$

where

$$\Gamma_{BA} = \frac{2\left(\dfrac{Z_B}{Z_A}\right)}{\dfrac{Z_B}{Z_A} + 1} \tag{12.3.7}$$

Note that Γ_{AA}, given by (12.3.5), is similar to Γ_R, given by (12.2.12), except that Z_B replaces Z_R. Thus for waves arriving at the junction from line A, the "load" at the receiving end of line A is the characteristic impedance of line B.

EXAMPLE 12.6   **Lattice diagram: overhead line connected to a cable, single phase lossless lines**

As shown in Figure 12.10, a single-phase lossless overhead line with $Z_A = 400\,\Omega$, $v_A = 3 \times 10^8$ m/s, and $l_A = 30$ km is connected to a single phase lossless cable with $Z_B = 100\,\Omega$, $v_B = 2 \times 10^8$ m/s, and $l_B = 20$ km. At the sending end of line A, $e_g(t) = E u_{-1}(t)$ and $Z_G = Z_A$. At the receiving end of line B, $Z_R = 2Z_B = 200\,\Omega$. Draw the lattice diagram for $0 \le t \le 0.6$ ms and plot the voltage at the junction versus time. The line and cable are initially unenergized.

Solution From (12.2.13),

$$\tau_A = \frac{30 \times 10^3}{3 \times 10^8} = 0.1 \times 10^{-3} \quad \text{s} \qquad \tau_B = \frac{20 \times 10^3}{2 \times 10^8} = 0.1 \times 10^{-3} \quad \text{s}$$

From (12.2.12), with $Z_G = Z_A$ and $Z_R = 2Z_B$,

$$\Gamma_S = \frac{1-1}{1+1} = 0 \qquad \Gamma_R = \frac{2-1}{2+1} = \frac{1}{3}$$

From (12.3.5) and (12.3.6), the reflection and refraction coefficients for waves arriving at the junction from line A are

$$\left. \Gamma_{AA} = \frac{\dfrac{100}{400} - 1}{\dfrac{100}{400} + 1} = \frac{-3}{5} \qquad \Gamma_{BA} = \frac{2 \dfrac{100}{400}}{\dfrac{100}{400} + 1} = \frac{2}{5} \right\} \text{ from line A}$$

Reversing A and B, the reflection and refraction coefficients for waves returning to the junction from line B are

$$\left. \Gamma_{BB} = \frac{\dfrac{400}{100} - 1}{\dfrac{400}{100} + 1} = \frac{3}{5} \qquad \Gamma_{AB} = \frac{2 \dfrac{400}{100}}{\dfrac{400}{100} + 1} = \frac{8}{5} \right\} \text{ from line B}$$

The lattice diagram is shown in Figure 12.11. Using voltage division, the first forward traveling voltage wave is

$$V_1(s) = E_G(s) \left(\frac{Z_A}{Z_A + Z_G} \right) = \frac{E}{s} \left(\frac{1}{2} \right) = \frac{E}{2s}$$

When v_1 arrives at the junction, a reflected wave v_2 and refracted wave v_3 are initiated. Using the reflection and refraction coefficients for line A,

$$V_2(s) = \Gamma_{AA} V_1(s) = \left(\frac{-3}{5} \right) \left(\frac{E}{2s} \right) = \frac{-3E}{10s}$$

$$V_3(s) = \Gamma_{BA} V_1(s) = \left(\frac{2}{5} \right) \left(\frac{E}{2s} \right) = \frac{E}{5s}$$

When v_2 arrives at the receiving end of line B, a reflected wave $V_4(s) = \Gamma_R V_3(s) = \frac{1}{3}(E/5s) = (E/15s)$ is initiated. When v_4 arrives at the junction, reflected wave v_5 and refracted wave v_6 are initiated. Using the reflection and refraction coefficients for line B,

$$V_5(s) = \Gamma_{BB} V_4(s) = \left(\frac{3}{5} \right) \left(\frac{E}{15s} \right) = \frac{E}{25s}$$

$$V_6(s) = \Gamma_{AB} V_4(s) = \left(\frac{8}{5} \right) \left(\frac{E}{15s} \right) = \frac{8E}{75s}$$

Subsequent reflections and refractions are calculated in a similar manner.

Figure 12.11

Lattice diagram for Example 12.6

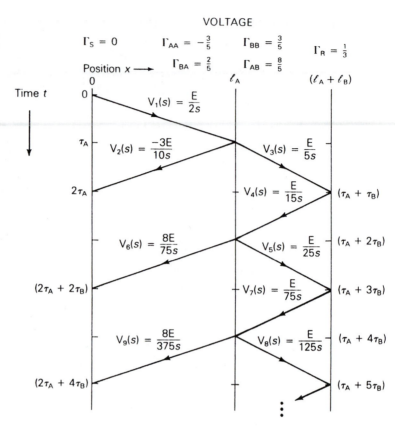

The voltage at the junction is determined by starting at $x = l_A$ at the top of the lattice diagram, where $t = 0$. Then, moving down the lattice diagram, voltage waves either just to the left or just to the right of the junction are added when they occur. For example, looking just to the right of the junction at $x = l_A^+$, the voltage wave v_3, a step of magnitude E/5 volts occurs at $t = \tau_A$. Then at $t = (\tau_A + 2\tau_B)$, two waves v_4 and v_5, which are steps of magnitude E/15 and E/25, are added to v_3. $v(l_A, t)$ is plotted in Figure 12.12.

Figure 12.12

Junction voltage for Example 12.6

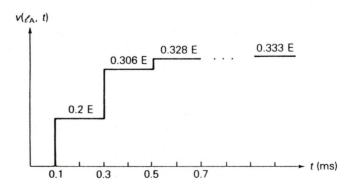

The steady-state voltage is determined by removing the lossless lines and calculating the steady-state voltage across the receiving-end load:

$$v_{ss}(x) = E \left(\frac{Z_R}{Z_R + Z_G} \right) = E \left(\frac{200}{200 + 400} \right) = \frac{E}{3}$$ ∎

Figure 12.13

Junction of lossless lines A, B, C, D, and so on

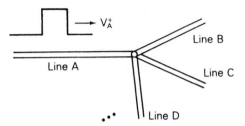

The preceding analysis can be extended to the junction of more than two lossless lines, as shown in Figure 12.13. Writing a KVL and KCL equation at the junction for a voltage V_A^+ arriving at the junction from line A,

$$V_A^+ + V_A^- = V_B^+ = V_C^+ = V_D^+ = \cdots \tag{12.3.8}$$

$$I_A^+ + I_A^- = I_B^+ + I_C^+ + I_D^+ + \cdots \tag{12.3.9}$$

Using $I_A^+ = V_A^+/Z_A$, $I_A^- = -V_A^-/Z_A$, $I_B^+ = V_B^+/Z_B$, and so on in (12.3.9),

$$\frac{V_A^+}{Z_A} - \frac{V_A^-}{Z_A} = \frac{V_B^+}{Z_B} + \frac{V_C^+}{Z_C} + \frac{V_D^+}{Z_D} + \cdots \tag{12.3.10}$$

Equations (12.3.8) and (12.3.10) can be solved for V_A^-, V_B^+, V_C^+, V_D^+, and so on in terms of V_A^+. (See Problem 12.11.)

SECTION 12.4

DISCRETE-TIME MODELS OF SINGLE-PHASE LOSSLESS LINES AND LUMPED RLC ELEMENTS

Our objective in this section is to develop discrete-time models of single-phase lossless lines and of lumped RLC elements suitable for computer calculation of transmission-line transients at discrete-time intervals $t = \Delta t, 2\Delta t, 3\Delta t$, and so on. The discrete-time models are presented as equivalent circuits consisting of lumped resistors and current sources. The current sources in the models represent the past history of the circuit—that is, the history at times $t - \Delta t$, $t - 2\Delta t$, and so on. After interconnecting the equivalent circuits of all the components in any given circuit, nodal equations can then be written for each discrete time. Discrete-time models, first developed by L. Bergeron [3], are presented first.

Single-phase Lossless Line

From the general solution of a single-phase lossless line, given by (12.1.12) and (12.1.18), we obtain

$$v(x, t) + Z_c i(x, t) = 2v^+ \left(t - \frac{x}{v} \right) \tag{12.4.1}$$

$$v(x, t) - Z_c i(x, t) = 2v^- \left(t + \frac{x}{v} \right) \tag{12.4.2}$$

In (12.4.1), the left side $(v + Z_c i)$ remains constant when the argument $(t - x/v)$ is constant. Therefore, to a fictitious observer traveling at velocity v in the positive x direction along the line, $(v + Z_c i)$ remains constant. If τ is the transit time from terminal k to terminal m of the line, the value of $(v + Z_c i)$ observed at time $(t - \tau)$ at terminal k must equal the value at time t at terminal m. That is,

$$v_k(t - \tau) + Z_c i_k(t - \tau) = v_m(t) + Z_c i_m(t) \tag{12.4.3}$$

where k and m denote terminals k and m, as shown in Figure 12.14(a).

Figure 12.14

Single-phase two-wire loss-less line

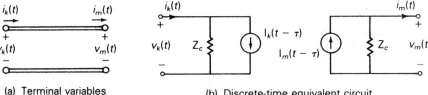

(a) Terminal variables   (b) Discrete-time equivalent circuit

Similarly, $(v - Z_c i)$ in (12.4.2) remains constant when $(t + x/v)$ is constant. To a fictitious observer traveling at velocity v in the negative x direction, $(v - Z_c i)$ remains constant. Therefore, the value of $(v - Z_c i)$ at time $(t - \tau)$ at terminal m must equal the value at time t at terminal k. That is,

$$v_m(t - \tau) - Z_c i_m(t - \tau) = v_k(t) - Z_c i_k(t) \tag{12.4.4}$$

Equation (12.4.3) is rewritten as

$$i_m(t) = I_m(t - \tau) - \frac{1}{Z_c} v_m(t) \tag{12.4.5}$$

where

$$I_m(t - \tau) = i_k(t - \tau) + \frac{1}{Z_c} v_k(t - \tau) \tag{12.4.6}$$

Similarly, (12.4.4) is rewritten as

$$i_k(t) = I_k(t - \tau) + \frac{1}{Z_c} v_k(t) \tag{12.4.7}$$

where

$$I_k(t - \tau) = i_m(t - \tau) - \frac{1}{Z_c} v_m(t - \tau) \tag{12.4.8}$$

Also, using (12.4.7) in (12.4.6),

$$I_m(t - \tau) = I_k(t - 2\tau) + \frac{2}{Z_c} v_k(t - \tau) \tag{12.4.9}$$

and using (12.4.5) in (12.4.8),

$$I_k(t - \tau) = I_m(t - 2\tau) - \frac{2}{Z_c} v_m(t - \tau) \tag{12.4.10}$$

Equations (12.4.5) and (12.4.7) are represented by the circuit shown in Figure 12.14(b). The current sources $I_m(t - \tau)$ and $I_k(t - \tau)$ shown in this figure, which are given by (12.4.9) and (12.4.10), represent the past history of the transmission line.

Note that in Figure 12.14(b) terminals k and m are not directly connected. The conditions at one terminal are "felt" indirectly at the other terminal after a delay of τ seconds.

Lumped Inductance

As shown in Figure 12.15(a) for a constant lumped inductance L,

$$v(t) = L \frac{di(t)}{dt} \tag{12.4.11}$$

Integrating this equation from time $(t - \Delta t)$ to t,

$$\int_{t-\Delta t}^{t} di(t) = \frac{1}{L} \int_{t-\Delta t}^{t} v(t)\, dt \tag{12.4.12}$$

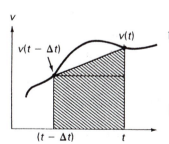

Trapezoidal Integration Rule

Using the trapezoidal rule of integration,

$$i(t) - i(t - \Delta t) = \left(\frac{1}{L}\right)\left(\frac{\Delta t}{2}\right) [v(t) + v(t - \Delta t)]$$

Rearranging:

$$i(t) = \frac{v(t)}{(2L/\Delta t)} + \left[i(t - \Delta t) + \frac{v(t - \Delta t)}{(2L/\Delta t)} \right]$$

Figure 12.15

Lumped inductance

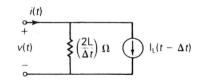

(a) Continuous time circuit   (b) Discrete-time circuit

or

$$i(t) = \frac{v(t)}{(2L/\Delta t)} + I_L(t - \Delta t) \tag{12.4.13}$$

where

$$I_L(t - \Delta t) = i(t - \Delta t) + \frac{v(t - \Delta t)}{(2L/\Delta t)} = I_L(t - 2\Delta t) + \frac{v(t - \Delta t)}{(L/\Delta t)} \tag{12.4.14}$$

Equations (12.4.13) and (12.4.14) are represented by the circuit shown in Figure 12.15(b). As shown, the inductor is replaced by a resistor with resistance $(2L/\Delta t)\,\Omega$. A current source $I_L(t - \Delta t)$ given by (12.4.14) is also included. $I_L(t - \Delta t)$ represents the past history of the inductor. Note that the trapezoidal rule introduces an error of the order $(\Delta t)^3$.

Lumped Capacitance

As shown in Figure 12.16(a) for a constant lumped capacitance C,

$$i(t) = C\frac{dv(t)}{dt} \tag{12.4.15}$$

Integrating from time $(t - \Delta t)$ to t,

$$\int_{t-\Delta t}^{t} dv(t) = \frac{1}{C} \int_{t-\Delta t}^{t} i(t)\, dt \tag{12.4.16}$$

Using the trapezoidal rule of integration,

$$v(t) - v(t - \Delta t) = \frac{1}{C}\left(\frac{\Delta t}{2}\right)[i(t) + i(t - \Delta t)]$$

Rearranging,

$$i(t) = \frac{v(t)}{(\Delta t/2C)} - I_C(t - \Delta t) \tag{12.4.17}$$

where

$$I_C(t - \Delta t) = i(t - \Delta t) + \frac{v(t - \Delta t)}{(\Delta t/2C)} = -I_C(t - 2\Delta t) + \frac{v(t - \Delta t)}{(\Delta t/4C)} \tag{12.4.18}$$

Equations (12.4.17) and (12.4.18) are represented by the circuit in Figure 12.16(b). The capacitor is replaced by a resistor with resistance $(\Delta t/2C)\,\Omega$. A

Figure 12.16

Lumped capacitance

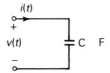

(a) Continuous time circuit

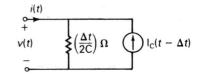

(b) Discrete-time circuit

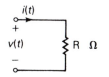

(a) Continuous time circuit

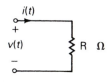

(b) Discrete-time circuit

Figure 12.17

Lumped resistance

current source $I_C(t - \Delta t)$, which represents the capacitor's past history, is also included.

Lumped Resistance

The discrete model of a constant lumped resistance R, as shown in Figure 12.17, is the same as the continuous model. That is,

$$v(t) = Ri(t) \tag{12.4.19}$$

Nodal Equations

A circuit consisting of single-phase lossless transmission lines and constant lumped RLC elements can be replaced by the equivalent circuits given in Figures 12.14(b), 12.15(b), 12.16(b), and 12.17(b). Then, writing nodal equations, the result is a set of linear algebraic equations that determine the bus voltages at each instant t.

EXAMPLE 12.7   **Discrete-time equivalent circuit, single-phase lossless line transients, computer solution**

For the circuit given in Example 12.3, replace the circuit elements by their discrete-time equivalent circuits and write the nodal equations that determine the sending-end and receiving-end voltages. Then, using a digital computer, compute the sending-end and receiving-end voltages for $0 \leq t \leq 9$ ms. For numerical calculations, assume $E = 100$ V, $Z_c = 400\,\Omega$, $C_R = 5\,\mu F$, $\tau = 1.0$ ms, and $\Delta t = 0.1$ ms.

Solution   The discrete model is shown in Figure 12.18, where $v_k(t)$ represents the sending-end voltage $v(0, t)$ and $v_m(t)$ represents the receiving-end voltage $v(l, t)$.

Figure 12.18

Discrete-time equivalent circuit for Example 12.7

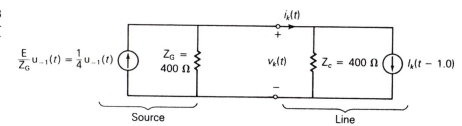

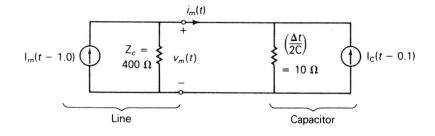

Also, the sending-end voltage source $e_G(t)$ in series with Z_G is converted to an equivalent current source in parallel with Z_G. Writing nodal equations for this circuit,

$$\begin{bmatrix} \left(\dfrac{1}{400} + \dfrac{1}{400}\right) & 0 \\ 0 & \left(\dfrac{1}{400} + \dfrac{1}{10}\right) \end{bmatrix} \begin{bmatrix} v_k(t) \\ v_m(t) \end{bmatrix} = \begin{bmatrix} \frac{1}{4} - I_k(t - 1.0) \\ I_m(t - 1.0) + I_c(t - 0.1) \end{bmatrix}$$

Solving,

$$v_k(t) = 200[\tfrac{1}{4} - I_k(t - 1.0)] \tag{a}$$

$$v_m(t) = 9.75610[I_m(t - 1.0) + I_c(t - 0.1)] \tag{b}$$

The current sources in these equations are, from (12.4.9), (12.4.10), and (12.4.18), with the argument $(t - \tau)$ replaced by t,

$$I_m(t) = I_k(t - 1.0) + \tfrac{2}{400}v_k(t) \tag{c}$$

$$I_k(t) = I_m(t - 1.0) - \tfrac{2}{400}v_m(t) \tag{d}$$

$$I_c(t) = -I_c(t - 0.1) + \tfrac{1}{5}v_m(t) \tag{e}$$

Equations (a) through (e) above are in a form suitable for digital computer solution. A scheme for iteratively computing v_k and v_m is as follows, starting at $t = 0$:

1. Compute $v_k(t)$ and $v_m(t)$ from equations (a) and (b).
2. Compute $I_m(t)$, $I_k(t)$, and $I_c(t)$ from equations (c), (d), and (e). Store $I_m(t)$ and $I_k(t)$.
3. Change t to $(t + \Delta t) = (t + 0.1)$ and return to (1) above.

Note that since the transmission line and capacitor are unenergized for time t less than zero, the current sources $I_m(\ )$, $I_k(\ )$, and $I_c(\ )$ are zero whenever their arguments $(\ )$ are negative. Note also from equations (a) through (e) that it is necessary to store the past ten values of $I_m(\ )$ and $I_k(\ )$.

A personal computer program written in BASIC that executes the above scheme and the computational results are shown in Figure 12.19. The plotted sending-end voltage $v_k(t)$ can be compared with the results of Example 12.3.

Figure 12.19

Example 12.7

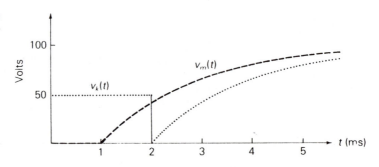

OUTPUT			COMPUTER PROGRAM LISTING
Time ms	VK Volts	VM Volts	
0.00	50.00	0.00	10 REM EXAMPLE 12.7
0.20	50.00	0.00	20 LPRINT " TIME VK VM "
0.40	50.00	0.00	30 LPRINT " ms Volts Volts "
0.60	50.00	0.00	40 IC = 0
0.80	50.00	0.00	50 T = 0
1.00	50.00	2.44	60 KPRINT = 2
1.20	50.00	11.73	65 REM T IS TIME. KPRINT DETERMINES THE PRINTOUT INTERVAL.
1.40	50.00	20.13	70 REM LINES 110 to 210 COMPUTE EQS(a)–(e) FOR THE FIRST
1.60	50.00	27.73	80 REM TEN TIME STEPS (A TOTAL OF ONE ms) DURING WHICH
1.80	50.00	34.61	90 REM THE CURRENT SOURCES ON THE RIGHT HAND SIDE
2.00	2.44	40.83	100 REM OF THE EQUATIONS ARE ZERO. TEN VALUES OF
2.20	11.73	46.46	105 REM CURRENT SOURCES IK(J) AND IM(J) ARE STORED.
2.40	20.13	51.56	110 FOR J = 1 TO 10
2.60	27.73	56.17	120 VK = 200/4
2.80	34.61	60.34	130 VM = 9.7561*IC
3.00	40.83	64.12	140 IM(J) = (2/400)*VK
3.20	46.46	67.53	150 IK(J) = (−2/400)*VM
3.40	51.56	70.62	160 IC = −IC + (1/5)*VM
3.60	56.17	73.42	170 Z = (J − 1)/KPRINT
3.80	60.34	75.95	180 M = INT(Z)
4.00	64.12	78.24	190 IF M = Z THEN LPRINT USING "*** ** "; T,VK,VM
4.20	67.53	80.31	200 T = T + .1
4.40	70.62	82.18	210 NEXT J
4.60	73.42	83.88	220 REM LINES 250 to 420 COMPUTE EQS(a)–(e) FOR TIME T
4.80	75.95	85.41	230 REM EQUAL TO AND GREATER THAN 1.0 ms. THE PAST TEN
5.00	78.24	86.80	240 REM VALUES OF IK(J) AND IM(J) ARE STORED
5.20	80.31	88.06	250 FOR J = 1 TO 10
5.40	82.18	89.20	260 REM LINE 270 IS EQ(a).
5.60	83.88	90.22	270 VK = 200*((1/4) − IK(J))
5.80	85.41	91.15	280 REM LINE 290 IS EQ(b).
6.00	86.80	92.00	290 VM = 9.7561*(IM(J) + IC)
6.20	88.06	92.76	300 REM LINE 310 IS EQ(e).
6.40	89.20	93.45	310 IC = −IC + (1/5)*VM
6.60	90.22	94.07	320 REM LINES 330–360 ARE EQS (c) and (d).
6.80	91.15	94.64	330 C1 = IK(J) + (2/400)*VK
7.00	92.00	95.15	340 C2 = IM(J) − (2/400)*VM
7.20	92.76	95.61	350 IM(J) = C1
7.40	93.45	96.03	360 IK(J) = C2
7.60	94.07	96.40	370 Z = (J − 1)/KPRINT
7.80	94.64	96.75	380 M = INT(Z)
8.00	95.15	97.06	390 IF M = Z THEN LPRINT USING "*** ** "; T,VK,VM
8.20	95.61	97.34	400 T = T + .1
8.40	96.03	97.59	410 NEXT J
8.60	96.40	97.82	420 IF T < 9.0 THEN GOTO 250
8.80	96.75	98.03	430 STOP
9.00	97.06	98.22	

Example 12.7 can be generalized to compute bus voltages at discrete-time intervals for an arbitrary number of buses, single-phase lossless lines, and lumped RLC elements. When current sources instead of voltage sources are employed, the unknowns are all bus voltages, for which nodal equations $\mathbf{YV} = \mathbf{I}$ can be written at each discrete-time instant. Also, the dependent current sources in $\mathbf{I}$ are written in terms of bus voltages and current sources at prior times. For computational convenience, the time interval Δt can be chosen constant so that the bus admittance matrix $\mathbf{Y}$ is a constant real symmetric matrix as long as the RLC elements are constant.

SECTION 12.5

LOSSY LINES

Transmission-line series resistance or shunt conductance causes the following:

1. Attenuation
2. Distortion
3. Power losses

These effects are briefly discussed as follows.

Attenuation

When constant series resistance R Ω/m and shunt conductance G S/m are included in the circuit of Figure 12.1 for a single-phase two-wire line, (12.1.3) and (12.1.4) become

$$\frac{\partial v(x,\ t)}{\partial x} = -R i(x,\ t) - L\,\frac{\partial i(x,\ t)}{\partial t} \tag{12.5.1}$$

$$\frac{\partial i(x,\ t)}{\partial x} = -G v(x,\ t) - C\,\frac{\partial v(x,\ t)}{\partial t} \tag{12.5.2}$$

Taking the Laplace transform of these, equations analogous to (12.1.7) and (12.1.8) are

$$\frac{d^2 V(x,\ s)}{dx^2} - \gamma^2(s)V(x,\ s) = 0 \tag{12.5.3}$$

$$\frac{d^2 I(x,\ s)}{dx^2} - \gamma^2(s)I(x,\ s) = 0 \tag{12.5.4}$$

where

$$\gamma(s) = \sqrt{(R + sL)(G + sC)} \tag{12.5.5}$$

The solution to these equations is

$$V(x, s) = V^+(s)e^{-\gamma(s)x} + V^-(s)e^{+\gamma(s)x} \tag{12.5.6}$$

$$I(x, s) = I^+(s)e^{-\gamma(s)x} + I^-(s)e^{+\gamma(s)x} \tag{12.5.7}$$

In general, it is impossible to obtain a closed form expression for $v(x, t)$ and $i(x, t)$, which are the inverse Laplace transforms of these equations. However, for the special case of a *distortionless* line, which has the property $R/L = G/C$, the inverse Laplace transform can be obtained as follows. Rewrite (12.5.5) as

$$\gamma(s) = \sqrt{LC[(s + \delta)^2 - \sigma^2]} \tag{12.5.8}$$

where

$$\delta = \frac{1}{2}\left(\frac{R}{L} + \frac{G}{C}\right) \tag{12.5.9}$$

$$\sigma = \frac{1}{2}\left(\frac{R}{L} - \frac{G}{C}\right) \tag{12.5.10}$$

For a distortionless line, $\sigma = 0$, $\delta = R/L$, and (12.5.6) and (12.5.7) become

$$V(x, s) = V^+(s)e^{-\sqrt{LC}[s+(R/L)]x} + V^-(s)e^{+\sqrt{LC}[s+(R/L)]x} \tag{12.5.11}$$

$$I(x, s) = I^+(s)e^{-\sqrt{LC}[s+(R/L)]x} + I^-(s)e^{+\sqrt{LC}[s+(R/L)]x} \tag{12.5.12}$$

Using $v = 1/\sqrt{LC}$ and $\sqrt{LC}(R/L) = \sqrt{RG} = \alpha$ for the distortionless line, the inverse transform of these equations is

$$v(x, t) = e^{-\alpha x}v^+\left(t - \frac{x}{v}\right) + e^{+\alpha x}v^-\left(t + \frac{x}{v}\right) \tag{12.5.13}$$

$$i(x, t) = e^{-\alpha x}i^+\left(t - \frac{x}{v}\right) + e^{\alpha x}i^-\left(t + \frac{x}{v}\right) \tag{12.5.14}$$

These voltage and current waves consist of forward and backward traveling waves similar to (12.1.12) and (12.1.13) for a lossless line. However, for the lossy distortionless line, the waves are attenuated versus x due to the $e^{\pm\alpha x}$ terms. Note that the attenuation term $\alpha = \sqrt{RG}$ is constant. Also, the attenuated waves travel at constant velocity $v = 1/\sqrt{LC}$. Therefore, waves traveling along the distortionless line do not change their shape; only their magnitudes are attenuated.

Distortion

For sinusoidal steady-state waves, the propagation constant $\gamma(j\omega)$ is, from (12.5.5), with $s = j\omega$

$$\gamma(j\omega) = \sqrt{(R + j\omega L)(G + j\omega C)} = \alpha + j\beta \tag{12.5.15}$$

For a lossless line, $R = G = 0$; therefore, $\alpha = 0$, $\beta = \omega\sqrt{LC}$, and the phase velocity $v = \omega/\beta = 1/\sqrt{LC}$ is constant. Thus, sinusoidal waves of all frequencies travel at constant velocity v without attenuation along a lossless line.

For a distortionless line $(R/L) = (G/C)$, and $\gamma(j\omega)$ can be rewritten, using (12.5.8)–(12.5.10), as

$$\gamma(j\omega) = \sqrt{LC\left(j\omega + \frac{R}{L}\right)^2} = \sqrt{LC}\left(j\omega + \frac{R}{L}\right)$$

$$= \sqrt{RG} + j\frac{\omega}{v} = \alpha + j\beta \tag{12.5.16}$$

Since $\alpha = \sqrt{RG}$ and $v = 1/\sqrt{LC}$ are constant, sinusoidal waves of all frequencies travel along the distortionless line at constant velocity with constant attenuation—that is, without distortion.

It can also be shown that for frequencies above 1 MHz, practical transmission lines with typical constants R, L, G, and C tend to be distortionless. Above 1 MHz, α and β can be approximated by

$$\alpha \simeq \frac{R}{2}\sqrt{\frac{C}{L}} + \frac{G}{2}\sqrt{\frac{L}{C}} \tag{12.5.17}$$

$$\beta \simeq \omega\sqrt{LC} = \frac{\omega}{v} \tag{12.5.18}$$

Therefore, sinusoidal waves with frequencies above 1 MHz travel along a practical line undistorted at constant velocity $v = 1/\sqrt{LC}$, with attenuation α given by (12.5.17).

At frequencies below 1 MHz these approximations do not hold, and lines are generally not distortionless. For typical transmission and distribution lines, (R/L) is much greater than (G/C) by a factor of 1000 or so. Therefore, the condition $(R/L) = (G/C)$ for a distortionless line does not hold.

Figure 12.20

Distortion and attenuation of surges on a 132-kV overhead line [4]

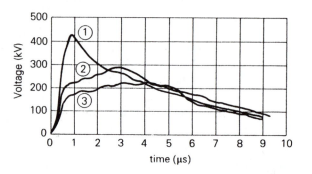

① Start (0. miles)

② Tower 150 (1.449 miles)

③ Tower 130 (4.97 miles)

Figure 12.20 shows the effect of distortion and attenuation of voltage surges based on experiments on a 132-kV overhead transmission line [4]. The shapes of the surges at three points along the line are shown. Note how distortion reduces the front of the wave and builds up the tail as it travels along the line.

Power Losses

Power losses are associated with series resistance R and shunt conductance G. When a current I flows along a line, I^2R losses occur, and when a voltage V appears across the conductors, V^2G losses occur. V^2G losses are primarily due to insulator leakage and corona for overhead lines, and to dielectric losses for cables. For practical lines operating at rated voltage and rated current, I^2R losses are much greater than V^2G losses.

As discussed above, the analysis of transients on single-phase two-wire lossy lines with constant parameters R, L, G, and C is complicated. The analysis becomes more complicated when skin effect is included, which means that R is not constant but frequency-dependent. Additional complications arise for a single-phase line consisting of one conductor with earth return, where Carson [5] has shown that both series resistance and inductance are frequency-dependent.

In view of these complications, the solution of transients on lossy lines is best handled via digital computation techniques. A single-phase line of length *l* can be approximated by a lossless line with half the total resistance $(Rl/2)\,\Omega$ lumped in series with the line at both ends. For improved accuracy, the line can be divided into various line sections, and each section can be approximated by a lossless line section, with a series resistance lumped at both ends. Simulations have shown that accuracy does not significantly improve with more than two line sections.

Discrete-time equivalent circuits of a single-phase lossless line, Figure 12.14, together with a constant lumped resistance, Figure 12.17, can be used to approximate a lossy line section, as shown in Figure 12.21. Also, digital techniques for modeling frequency-dependent line parameters [6, 7] are available but are not discussed here.

Figure 12.21

Approximate model of a lossy line segment

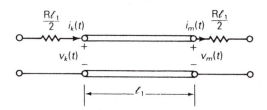

(a) Lossless line segment of length ℓ_1 with lumped line resistance

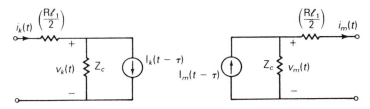

(b) Discrete-time model

SECTION 12.6

MULTICONDUCTOR LINES

Up to now we have considered transients on single-phase two-wire lines. For a transmission line with n conductors above a ground plane, waves travel in n "modes," where each mode has its own wave velocity and its own surge impedance. In this section we illustrate "modal analysis" for a relatively simple three-phase line [8].

Given a three-phase, lossless, completely transposed line consisting of three conductors above a perfectly conducting ground plane, the transmission-line equations are

$$\frac{d\mathbf{V}(x,\ s)}{dx} = -s\mathbf{LI}(x,\ s) \tag{12.6.1}$$

$$\frac{d\mathbf{I}(x,\ s)}{dx} = -s\mathbf{CV}(x,\ s) \tag{12.6.2}$$

where

$$\mathbf{V}(x,\ s) = \begin{bmatrix} V_{ag}(x,\ s) \\ V_{bg}(x,\ s) \\ V_{cg}(x,\ s) \end{bmatrix} \qquad \mathbf{I}(x,\ s) = \begin{bmatrix} I_a(x,\ s) \\ I_b(x,\ s) \\ I_c(x,\ s) \end{bmatrix} \tag{12.6.3}$$

Equations (12.6.1) and (12.6.2) are identical to (12.1.5) and (12.1.6) except that scalar quantities are replaced by vector quantities. $\mathbf{V}(x,s)$ is the vector of line-to-ground voltages and $\mathbf{I}(x,s)$ is the vector of line currents. For a completely transposed line, the 3×3 inductance matrix $\mathbf{L}$ and capacitance matrix $\mathbf{C}$ are given by

$$\mathbf{L} = \begin{bmatrix} L_s & L_m & L_m \\ L_m & L_s & L_m \\ L_m & L_m & L_s \end{bmatrix} \quad \text{H/m} \tag{12.6.4}$$

$$\mathbf{C} = \begin{bmatrix} C_s & C_m & C_m \\ C_m & C_s & C_m \\ C_m & C_m & C_s \end{bmatrix} \quad \text{F/m} \tag{12.6.5}$$

For any given line configuration, $\mathbf{L}$ and $\mathbf{C}$ can be computed from the equations given in Sections 5.7 and 5.11. Note that L_s, L_m, and C_s are positive, whereas C_m is negative.

We now transform the phase quantities to modal quantities. First we define

$$\begin{bmatrix} V_{ag}(x, s) \\ V_{bg}(x, s) \\ V_{cg}(x, s) \end{bmatrix} = \mathbf{T_V} \begin{bmatrix} V^0(x, s) \\ V^+(x, s) \\ V^-(x, s) \end{bmatrix} \qquad (12.6.6)$$

$$\begin{bmatrix} I_a(x, s) \\ I_b(x, s) \\ I_c(x, s) \end{bmatrix} = \mathbf{T_I} \begin{bmatrix} I^0(x, s) \\ I^+(x, s) \\ I^-(x, s) \end{bmatrix} \qquad (12.6.7)$$

$V^0(x, s)$, $V^+(x, s)$, and $V^-(x, s)$ are denoted *zero-mode*, *positive-mode*, and *negative-mode* voltages, respectively. Similarly, $I^0(x, s)$, $I^+(x, s)$, and $I^-(x, s)$ are *zero-*, *positive-*, and *negative-mode* currents. $\mathbf{T_V}$ and $\mathbf{T_I}$ are 3×3 constant transformation matrices, soon to be specified. Denoting $\mathbf{V}_m(x, s)$ and $\mathbf{I}_m(x, s)$ as the modal voltage and modal current vectors,

$$\mathbf{V}(x, s) = \mathbf{T_V}\mathbf{V}_m(x, s) \qquad (12.6.8)$$

$$\mathbf{I}(x, s) = \mathbf{T_I}\mathbf{I}_m(x, s) \qquad (12.6.9)$$

Using (12.6.8) and (12.6.9) in (12.6.1),

$$\mathbf{T_V} \frac{d\mathbf{V}_m(x, s)}{dx} = -s\mathbf{L}\mathbf{T_I}\mathbf{I}_m(x, s)$$

or

$$\frac{d\mathbf{V}_m(x, s)}{dx} = -s(\mathbf{T}_v^{-1}\mathbf{L}\mathbf{T_I})\mathbf{I}_m(x, s) \qquad (12.6.10)$$

Similarly, using (12.6.8) and (12.6.9) in (12.6.2),

$$\frac{d\mathbf{I}_m(x, s)}{dx} = -s(\mathbf{T_I}^{-1}\mathbf{C}\mathbf{T}_v)\mathbf{V}_m(x, s) \qquad (12.6.11)$$

The objective of the modal transformation is to diagonalize the matrix products within the parentheses of (12.6.10) and (12.6.11), thereby decoupling these vector equations. For a three-phase completely transposed line, $\mathbf{T_V}$ and $\mathbf{T_I}$ are given by

$$\mathbf{T_V} = \mathbf{T_I} = \begin{bmatrix} 1 & 1 & 1 \\ 1 & -2 & 1 \\ 1 & 1 & -2 \end{bmatrix} \qquad (12.6.12)$$

Also, the inverse transformation matrices are

$$\mathbf{T_V}^{-1} = \mathbf{T_I}^{-1} = \tfrac{1}{3}\begin{bmatrix} 1 & 1 & 1 \\ 1 & -1 & 0 \\ 1 & 0 & -1 \end{bmatrix} \qquad (12.6.13)$$

Substituting (12.6.12), (12.6.13), (12.6.4), and (12.6.5) into (12.6.10) and (12.6.11) yields

$$\frac{d}{dx}\begin{bmatrix} V^0(x,\ s) \\ V^+(x,\ s) \\ V^-(x,\ s) \end{bmatrix} = \begin{bmatrix} -s(L_s + 2L_m) & 0 & 0 \\ 0 & -s(L_s - L_m) & 0 \\ 0 & 0 & -s(L_s - L_m) \end{bmatrix}\begin{bmatrix} I^0(x,\ s) \\ I^+(x,\ s) \\ I^-(x,\ s) \end{bmatrix}$$

(12.6.14)

$$\frac{d}{dx}\begin{bmatrix} I^0(x,\ s) \\ I^+(x,\ s) \\ I^-(x,\ s) \end{bmatrix} = \begin{bmatrix} -s(C_s + 2C_m) & 0 & 0 \\ 0 & -s(C_s - C_m) & 0 \\ 0 & 0 & -s(C_s - C_m) \end{bmatrix}\begin{bmatrix} V^0(x,\ s) \\ V^+(x,\ s) \\ V^-(x,\ s) \end{bmatrix}$$

(12.6.15)

From (12.6.14) and (12.6.15), the zero-mode equations are

$$\frac{dV^0(x,\ s)}{dx} = -s(L_s + 2L_m)I^0(x,\ s) \tag{12.6.16}$$

$$\frac{dI^0(x,\ s)}{dx} = -s(C_s + 2C_m)V^0(x,\ s) \tag{12.6.17}$$

These equations are identical in form to those of a two-wire lossless line, (12.1.5) and (12.1.6). By analogy, the zero-mode waves travel at velocity

$$v^0 = \frac{1}{\sqrt{(L_s + 2L_m)(C_s + 2C_m)}} \quad \text{m/s} \tag{12.6.18}$$

and the zero-mode surge impedance is

$$Z_c^0 = \sqrt{\frac{L_s + 2L_m}{C_s + 2C_m}} \quad \Omega \tag{12.6.19}$$

Similarly, the positive- and negative-mode velocities and surge impedances are

$$v^+ = v^- = \frac{1}{\sqrt{(L_s - L_m)(C_s - C_m)}} \quad \text{m/s} \tag{12.6.20}$$

$$Z_c^+ = Z_c^- = \sqrt{\frac{L_s - L_m}{C_s - C_m}} \quad \Omega \tag{12.6.21}$$

These equations can be extended to more than three conductors—for example, to a three-phase line with shield wires or to a double-circuit three-phase line. Although the details are more complicated, the modal transformation is straightforward. There is one mode for each conductor, and each mode has its own wave velocity and its own surge impedance.

The solution of transients on multiconductor lines is best handled via digital computer methods, and such programs are available [9, 10]. Digital techniques are also available to model the following effects:

1. Nonlinear and time-varying RLC elements [8]
2. Lossy lines with frequency-dependent line parameters [6, 7, 12]

SECTION 12.7

POWER-SYSTEM OVERVOLTAGES

Overvoltages encountered by power-system equipment are of three types:

1. Lightning surges
2. Switching surges
3. Power frequency (50 or 60 Hz) overvoltages

Lightning

Lightning is the greatest single cause of overhead transmission and distribution line outages. Data obtained over a 14-year period from electric utility companies in the United States and Canada and covering 25,000 miles of transmission show that lightning accounted for about 26% of outages on 230-kV circuits and about 65% of outages on 345-kV circuits [13]. A similar study in Britain, also over a 14-year period, covering 50,000 faults on distribution lines shows that lightning accounted for 47% of outages on circuits up to and including 33 kV [14].

The electrical phenomena that occur within clouds leading to a lightning strike are complex and not totally understood. Several theories [15, 16, 17] generally agree, however, that charge separation occurs within clouds. Wilson [15] postulates that falling raindrops attract negative charges and therefore leave behind masses of positively charged air. The falling raindrops bring the negative charge to the bottom of the cloud, and upward air drafts carry the positively charged air and ice crystals to the top of the cloud, as shown in Figure 12.22. Negative charges at the bottom of the cloud induce a positively charged region, or "shadow," on the earth directly below the cloud. The electric field lines shown in Figure 12.22 originate from the positive charges and terminate at the negative charges.

When voltage gradients reach the breakdown strength of the humid air within the cloud, typically 5 to 15 kV/cm, an ionized path or downward *leader* moves from the cloud toward the earth. The leader progresses somewhat randomly along an irregular path, in steps. These leader steps, about 50 m long, move at a velocity of about 10^5 m/s. As a result of the opposite charge distribution under the cloud, another upward leader may rise to meet the downward leader. When the two leaders meet, a lightning discharge occurs, which neutralizes the charges.

Figure 12.22

Postulation of charge separation within clouds [16]

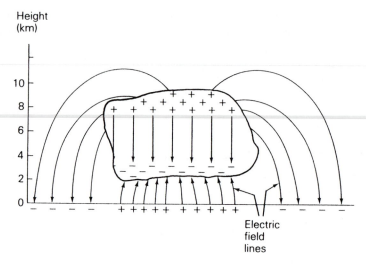

The current involved in a lightning stroke typically rises to a peak value within 1 to 10 μs, and then diminishes to one-half the peak within 20 to 100 μs. The distribution of peak currents is shown in Figure 12.23 [20]. This curve represents the percentage of strokes that exceed a given peak current.

Figure 12.23

Frequency of occurrence of lightning currents that exceed a given peak value [20]

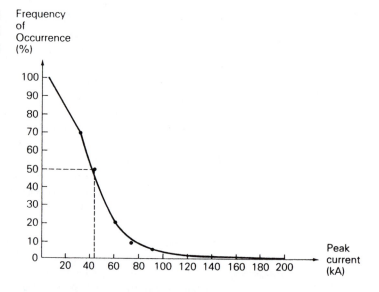

For example, 50% of all strokes have a peak current greater than 45 kA. In extreme cases, the peak current can exceed 200 kA. Also, test results indicate that approximately 90% of all strokes are negative.

It has also been shown that what appears to the eye as a single flash of lightning is often the cumulative effect of many strokes. A typical flash consists of typically 3 to 5, and occasionally as many as 40, strokes, at intervals of 50 ms.

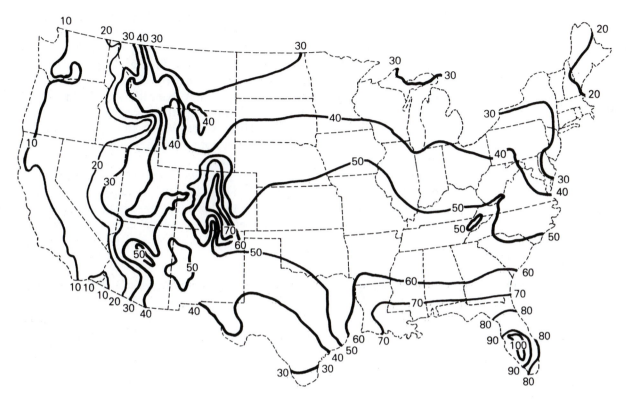

Figure 12.24   Isokeraunic map showing average thunderstorm days per year within the continental United States [18]

The frequency of occurrence of thunderstorms in a given geographical location is of interest to transmission-line designers. Figure 12.24 is an isokeraunic map prepared by the U.S. Weather Bureau, which shows the average number of thunderstorm days in the contiguous United States [18]. As shown, the greatest thunderstorm activity occurs in Florida, with 70 to 100 storm days per year, and in New Mexico, with a maximum of 70. On the West Coast, the average number of thunderstorm days is as low as 3 to 10 per year.

A typical transmission-line design goal is to have an average of less than 0.50 lightning outages per year per 100 miles of transmission. For a given overhead line with a specified voltage rating, the following factors affect this design goal:

1. Tower height
2. Number and location of shield wires
3. Number of standard insulator discs per phase wire
4. Tower impedance and tower-to-ground impedance

It is well known that lightning strikes tall objects. Thus, shorter, H-frame structures are less susceptible to lightning strokes than taller, lattice towers.

Also, shorter span lengths with more towers per kilometer can reduce the number of strikes.

Shield wires installed above phase conductors can effectively shield the phase conductors from direct lightning strokes. Figure 12.25 illustrates the effect of shield wires. Experience has shown that the chance of a direct hit to phase conductors located within $\pm 30°$ arcs beneath the shield wires is reduced by a factor of 1000 [18]. Some lightning strokes are, therefore, expected to hit these overhead shield wires. When this occurs, traveling voltage and current waves propagate in both directions along the shield wire that is hit. When a wave arrives at a tower, a reflected wave returns toward the point where the lightning hit, and two refracted waves occur. One refracted wave moves along the shield wire into the next span. And since the shield wire is electrically connected to the tower, the other refracted wave moves down the tower, its energy being harmlessly diverted to ground.

Figure 12.25

Effect of shield wires

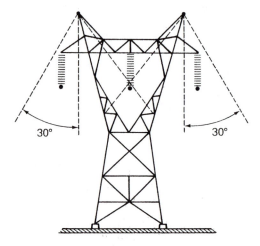

However, if the tower impedance or tower-to-ground impedance is too high, IZ voltages that are produced could exceed the breakdown strength of the insulator discs that hold the phase wires. The number of insulator discs per string (see Table 5.1) is selected to avoid insulator flashover. Also, tower impedances and tower footing resistances are designed to be as low as possible. If the inherent tower construction does not give a naturally low resistance to ground, driven ground rods can be employed. Sometimes buried conductors running under the line (called *counterpoise*) are employed.

Switching Surges

The magnitudes of overvoltages due to lightning surges are not significantly affected by the power-system voltage. On the other hand, overvoltages due to switching surges are directly proportional to system voltage. Consequently, lightning surges are less important for EHV transmission above 345 kV and for UHV transmission, which has improved insulation. Switching surges become the limiting factor in insulation coordination for system voltages above 345 kV.

One of the simplest and largest overvoltages can occur when an open-circuited line is energized, as shown in Figure 12.26. Assume that the circuit breaker closes at the instant the sinusoidal source voltage has a peak value $\sqrt{2}\,V$. Assuming zero source impedance, a forward traveling voltage wave of magnitude $\sqrt{2}\,V$ occurs. When this wave arrives at the open-circuited receiving end, where $\Gamma_R = +1$, the reflected voltage wave superimposed on the forward wave results in a maximum voltage of $2\sqrt{2}\,V = 2.83\,V$. Even higher voltages can occur when a line is reclosed after momentary interruption.

Figure 12.26

Energizing an open-circuited line

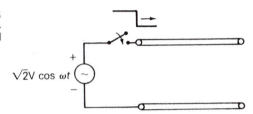

In order to reduce overvoltages due to line energizing or reclosing, resistors are almost always preinserted in circuit breakers at 345 kV and above. Resistors ranging from 200 to 800 Ω are preinserted when EHV circuit breakers are closed, and subsequently bypassed. When a circuit breaker closes, the source voltage divides across the preinserted resistors and the line, thereby reducing the initial line voltage. When the resistors are shorted out, a new transient is initiated, but the maximum line voltage can be substantially reduced by careful design.

Dangerous overvoltages can also occur during a single line-to-ground fault on one phase of a transmission line. When such a fault occurs, a voltage equal and opposite to that on the faulted phase occurs at the instant of fault inception. Traveling waves are initiated on both the faulted phase and, due to capacitive coupling, the unfaulted phases. At the line ends, reflections are produced and are superimposed on the normal operating voltages of the unfaulted phases. Kimbark and Legate [19] show that a line-to-ground fault can create an overvoltage on an unfaulted phase as high as 2.1 times the peak line-to-neutral voltage of the three-phase line.

Power Frequency Overvoltages

Sustained overvoltages at the fundamental power frequency (60 Hz in the United States) or at higher harmonic frequencies (such as 120 Hz, 180 Hz, and so on) occur due to load rejection, to ferroresonance, or to permanent faults. These overvoltages are normally of long duration, seconds to minutes, and are weakly damped.

SECTION 12.8

INSULATION COORDINATION

Insulation coordination is the process of correlating electric equipment insulation strength with protective device characteristics so that the equipment is protected against expected overvoltages. The selection of equipment insulation strength and the protected voltage level provided by protective devices depends on engineering judgment and cost.

Figure 12.27

Equipment insulation strength

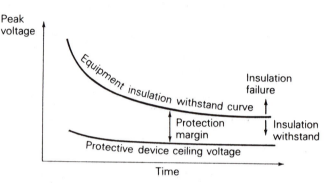

As shown by the top curve in Figure 12.27, equipment insulation strength is a function of time. Equipment insulation can generally withstand high transient overvoltages only if they are of sufficiently short duration. However, determination of insulation strength is somewhat complicated. During repeated tests with identical voltage waveforms under identical conditions, equipment insulation may fail one test and withstand another.

Figure 12.28

Standard impulse voltage waveform

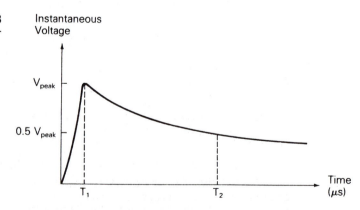

For purposes of insulation testing, a standard impulse voltage wave, as shown in Figure 12.28, is defined. The impulse wave shape is specified by giving the time T_1 in microseconds for the voltage to reach its peak value and the time T_2 for the voltage to decay to one-half its peak. One standard wave is a 1.2×50 wave, which rises to a peak value at $T_1 = 1.2\,\mu s$ and decays to one-half its peak at $T_2 = 50\,\mu s$.

Basic insulation level or BIL is defined as the peak value of the standard impulse voltage wave in Figure 12.28. Standard BILs adopted by the IEEE are shown in Table 12.1. Equipment conforming to these BILs must be capable of withstanding repeated applications of the standard waveform of positive or negative polarity without insulation failure. Also, these standard BILs apply to equipment regardless of how it is grounded. For nominal system voltages 115 kV and above, solidly grounded equipment with the reduced BILs shown in the table have been used.

Table 12.1

Standard and reduced basic insulation levels [18]

NOMINAL SYSTEM VOLTAGE	STANDARD BIL	REDUCED BIL*
kVrms	kV	kV
1.2	45	
2.5	60	
5.0	75	
8.7	95	
15	110	
23	150	
34.5	200	
46	250	
69	350	
92	450	
115	550	450
138	650	550
161	750	650
196	900	750
230	1050	825– 900
287	1300	1000–1100
345	1550	1175–1300
500		1300–1800
765		1675–2300

*For solidly grounded systems
These BILs are based on $1.2 \times 50\,\mu s$ voltage waveforms. They apply to internal (or non-self-restoring) insulation such as transformer insulation, as well as to external (or self-restoring) insulation, such as transmission-line insulation, on a statistical basis.

BILs are often expressed in per-unit, where the base voltage is the maximum value of nominal line-to-ground system voltage. Consider for example a 345-kV system, for which the maximum value of nominal line-to-ground voltage is $\sqrt{2}(345/\sqrt{3}) = 281.7\,kV$. The 1550-kV standard BIL shown in Table 12.1 is then $(1550/281.7) = 5.5$ per unit.

Note that overhead-transmission-line insulation, which is external insulation, is usually self-restoring. When a transmission-line insulator string flashes over, a short circuit occurs. After circuit breakers open to deenergize the line, the insulation of the string usually recovers, and the line can be rapidly reenergized. However, transformer insulation, which is internal, is not self-restoring. When transformer insulation fails, the transformer must be removed for repair or replaced.

To protect equipment such as a transformer against overvoltages higher than its BIL, a protective device, such as that shown in Figure 12.29, is employed. Such protective devices are generally connected in parallel with the equipment from each phase to ground. As shown in Figure 12.27, the function of the protective device is to maintain its voltage at a ceiling voltage below the BIL of the equipment it protects. The difference between the equipment breakdown voltage and the protective device ceiling voltage is the *protection margin.*

Figure 12.29

Single-line diagram of equipment and protective device

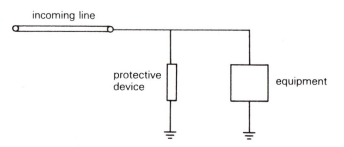

Protective devices should satisfy the following four criteria:

1. Provide a high or infinite impedance during normal system voltages, to minimize steady-state losses

2. Provide a low impedance during surges, to limit voltage

3. Dissipate or store the energy in the surge without damage to itself

4. Return to open-circuit conditions after the passage of a surge

One of the simplest protective devices is the rod gap, two metal rods with a fixed air gap, which is designed to spark over at specified overvoltages. Although it satisfies the first two protective device criteria, it dissipates very little energy and it cannot clear itself after arcing over.

A surge arrester, consisting of an air gap in series with a nonlinear silicon carbide resistor, satisfies all four criteria. The gap eliminates losses at normal voltages and arcs over during overvoltages. The resistor has the property that its resistance decreases sharply as the current through it increases, thereby limiting the voltage across the resistor to a specified ceiling. The resistor also dissipates the energy in the surge. Finally, following the passage of a surge, various forms of arc control quench the arc within the gap and restore the surge arrester to normal open-circuit conditions.

The "gapless" surge arrester, consisting of a nonlinear metal oxide resistor with no air gap, also satisfies all four criteria. At normal voltages the resistance is extremely high, limiting steady-state currents to microamperes and steady-state losses to a few watts. During surges, the resistance sharply decreases, thereby limiting overvoltage while dissipating surge energy. After the surge passes, the resistance naturally returns to its original high value. One advantage of the gapless arrester is that its ceiling voltage is closer to its normal operating voltage than is the conventional arrester, thus permitting reduced BILs and potential savings in the capital cost of equipment insulation.

There are four classes of surge arresters: station, intermediate, distribution, and secondary. Station arresters, which have the heaviest construction, are designed for the greatest range of ratings and have the best protective characteristics. Intermediate arresters, which have moderate construction, are designed for systems with nominal voltages 138 kV and below. Distribution arresters are employed with lower-voltage transformers and lines, where there is a need for economy. Secondary arresters are used for nominal system voltages below 1000 V. A summary of the protective characteristics of station arresters is given in Table 12.2 [20]. This summary is based on manufacturers' catalog information.

Note that arrester currents due to lightning surges are generally less than the lightning currents shown in Figure 12.23. In the case of direct strokes to transmission-line phase conductors, traveling waves are set up in two directions from the point of the stroke. Flashover of line insulation diverts part of the lightning current from the arrester. Only in the case of a direct stroke to a phase conductor very near an arrester, where no line flashover occurs, does the arrester discharge the full lightning current. The probability of this occurrence can be significantly reduced by using overhead shield wires to shield transmission lines and substations. Recommended practice for substations with unshielded lines is to select an arrester discharge current of at least 20 kA (even higher if the isokeraunic level is above 40 thunderstorm days per year). For substations with shielded lines, lower arrester discharge currents, from 5 to 20 kA, have been found satisfactory in most situations [20].

| EXAMPLE 12.8 | **Surge arrester: protective margin and discharge voltage** |

Determine the protective margin for a 120-kV station arrester employed in a 115-kV three-phase system with a 450-kV BIL, based on a 1.2×50 standard impulse voltage wave. Also determine the maximum discharge voltage across the arrester for a 20-kA arrester discharge current.

Solution From Table 12.2, for a 120-kV surge arrester, the maximum discharge voltage for a $1.2 \times 50 \, \mu s$ impulse voltage wave ranges from 272 to 300 kV, depending on arrester manufacturer. Therefore, the protective margin ranges from $(450 - 300) = 150$ kV to $(450 - 272) = 178$ kV. Also from Table 12.2, for a 20-kA discharge current, the maximum discharge voltage across the 120-kV arrester ranges from 300 to 316 kV. ∎

VOLTAGE RATING OF ARRESTER	IMPULSE SPARKOVER VOLTAGE			SWITCHING SURGE SPARKOVER VOLTAGE
	FRONT-OF-WAVE		$1.2 \times 50\text{-}\mu s$	
	RATE OF RISE OF TEST VOLTAGE ($kV/\mu s$)	kV CREST (RANGE OF MAXIMA)	kV CREST (RANGE OF MAXIMA)	kV CREST (RANGE OF MAXIMA)
3	25	10–18	10–14	—
6	50	19–28	16–23	—
9	75	28.5–38	24–32	—
12	100	36–48	32–41	—
15	125	45–57	40–51	—
21	175	63–76	54–68	—
24	200	71–86	62–77	—
30	250	89–103	77–93	—
36	300	107–118	92–108	—
39	325	115–125	100–114	—
48	400	143–148	122–132	—
60	500	170–190	141–165	136–153
72	600	204–226	169–190	163–178
90	750	254–275	210–235	203–215
96	800	270–295	218–245	218–225
108	900	304–325	245–270	245–250
120	1000	338–360	272–300	272–275
144	1200	400–430	326–346	325–326
168	1400	460–525	380–404	380–381
180	1500	490–565	400–430	400–410
192	1600	520–600	427–460	426–435
240	2000	620–735	535–577	533–545
258	2000	766–790	575–620	573–585
276	2000	820–840	615–664	612–630
294	2000	875–885	653–675	653–675
312	2000	924–935	690–750	693–710
372	2000	1078–1100	870–890	790–830
396	2000	1140–1176	925–950	840–885
420	2000	1200–1250	980–1005	890–940
444	2000	1265–1320	1035–1055	940–990
468	2000	1326–1390	1090–1110	992–1045
492	2000	1385–1425	1160–1165	1045–1090
540	2000	1515–1555	1274–1280	1145–1200
576	2000	1616–1665	1359–1380	1225–1285
612	2000	1700–1765	1440–1480	1300–1370
648	2000	1790–1865	1525–1570	1380–1445
684	2000	1880–1960	1610–1680	1455–1525

Table 12.2 Characteristics of station-type arresters [20]

DISCHARGE VOLTAGE FOR 8 × 20-μs DISCHARGE CURRENT WAVE

kV CREST FOR 1500 A (RANGE OF MAXIMA)	kV CREST FOR 3000 A (RANGE OF MAXIMA)	kV CREST FOR 5000 A (RANGE OF MAXIMA)	kV CREST FOR 10000 A (RANGE OF MAXIMA)	kV CREST FOR 20000 A (RANGE OF MAXIMA)	kV CREST FOR 40000 A (RANGE OF MAXIMA)
4.7–6	5.3–6.5	6–7	6.7–7.5	7.7–8.3	—
9.3–11	10–12	12–13	13.4–14.3	15.3–16.3	—
13.9–17	16–18	18–19	20–21.5	22.9–24.3	—
18.5–22	21.3–24	23.5–25.5	26.7–28.5	30.1–32.1	—
23.1–27.5	26.6–30	29.5–32	33.4–36	38.2–40	—
32.3–38.5	37.2–42	41–45	46.8–50	53.4–55.5	—
36.9–44	42.5–48	47–51	53.4–57	61–63.5	—
46.1–55	53.1–60	59–64	66.9–72	76.3–79	—
55.3–66	63.7–72	70.5–76	80–85	91.5–94.5	—
60–71.5	69–78	76.5–82.5	86.5–92	99.1–102	—
73.8–88	84.9–96	94–102	106–114	122–126	—
95–109	110–120	118–130	132–143	150–158	—
114–131	130–144	141–155	159–170	180–189	—
142–163	162–180	176–194	199–213	225–237	—
151–174	173–192	188–218	212–227	240–253	—
170–196	194–216	212–245	238–256	270–284	—
188–218	216–240	235–272	265–285	300–316	—
226–262	260–288	282–311	318–342	360–379	—
263–305	303–336	329–362	371–399	420–442	—
281–327	324–360	353–388	397–455	450–505	—
300–348	346–384	376–414	424–427	480–495	—
374–436	432–480	470–518	530–570	605–630	—
402–438	465–474	505–515	569–575	650–666	—
429–468	496–507	540–570	609–615	690–714	—
458–472	528–532	576–595	653–653	735–758	—
485–530	562–574	611–620	688–693	780–805	874–961
562–610	655–680	726–738	809–826	932–955	1136–1145
599–672	697–726	772–785	861–880	990–1015	1109–1226
634–713	739–770	819–830	913–930	1050–1070	1176–1294
670–753	781–814	866–875	965–977	1110–1130	1243–1358
707–794	823–860	913–930	1018–1040	1170–1200	1310–1441
742–830	865–925	958–1000	1070–1115	1232–1290	1500–1515
814–890	949–990	1052–1070	1173–1195	1350–1390	1646–1663
868–950	1012–1060	1122–1150	1251–1285	1440–1480	1755–1780
924–1010	1076–1130	1193–1220	1330–1360	1531–1580	1865–1885
977–1070	1138–1190	1261–1290	1407–1440	1619–1670	1974–1996
1031–1130	1153–1260	1331–1360	1489–1520	1709–1765	2063–2107

SECTION 12.9

PERSONAL COMPUTER PROGRAM: TRANSMISSION-LINE TRANSIENTS

It is evident from the discussion in this chapter that transmission-line transients problems are complicated and that closed-form solutions are impossible except for simple idealized cases. Accordingly, both analog and digital computation techniques have been developed.

Analog simulators for solving transmission-line transients are known in the industry by trade names such as the "Transient Network Analyzer" (TNA) and "ANACOM." The TNA is composed of electric circuit components including lumped RLC elements and voltage sources connected to represent the various components of the power system under study [21]. Three-phase transmission lines are represented by T or π circuits. For sufficient accuracy, each line is divided into sections—for example, into 15- to 20-km lengths—and a π circuit for each section is connected in cascade. Also, electric circuit models of transformers, shunt reactors, surge arresters, and other components have been developed to match the characteristics of the actual devices. The TNA is energized by a three-phase generator with impedances added to represent source impedance. The TNA is scaled in both voltage and current; for example, 1 A in the model represents 1 kA in the system and 1 V represents 1 kV. Electrical transients caused by line-switching operations or faults are initiated in the TNA via solid-state switches. Switching operations can be performed many times per second, with the responses observed on oscilloscopes that are synchronously triggered with switching events. Many factors influence transients, such as the instant during a 60-Hz cycle at which a switching event occurs. Accordingly, one of the great advantages of the TNA is that it can perform many simulations in a short time and collect data quickly and economically.

Digital computer programs are also available for solving transmission-line transients [9, 10]. Discrete-time models of three-phase transmission lines, transformers, shunt reactors, surge arresters, and other components can be interconnected to represent a power system under study. The program sets up and solves a set of nodal equations that determine the bus voltages at each discrete-time instant. Digital techniques for modeling nonlinear and time-varying RLC elements [11] as well as lossy lines with frequency-dependent line parameters including earth resistance [6, 7, 12] are also available. One advantage of the digital techniques is solution accuracy, especially for high-frequency transient effects above 5 kHZ.

The personal computer software package that accompanies this text includes the program entitled "TRANSMISSION-LINE TRANSIENTS," which computes voltage transients for the single-phase circuit shown in Figure 12.30. This circuit consists of 3 single-phase lossless line sections, 10 buses, 4 independent current sources, and 10 lumped parallel RLC elements. The line sections and RLC elements are represented by the discrete-time equivalent circuits given in Figures 12.14–12.17. A lossy line section can be

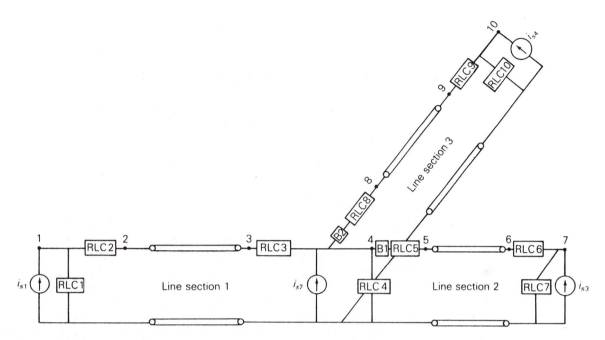

Figure 12.30   Circuit diagram for the program TRANSMISSION-LINE TRANSIENTS

represented by lumping half the line section series resistance at each end of the line section. BASIC functions can be used to represent the independent current sources as square waves, ramps, exponentials, or other wave shapes. The circuit in Figure 12.30 is assumed to be initially unenergized.

Input data for the computer program consist of: (1) surge impedance Z_c and transit time τ for each line section; (2) RLC values for each of the lumped elements; (3) function representation for each independent current source; (4) calculation time interval Δt and final time $t_F = N \Delta t$. The program user can also change the input data in order to simulate piecewise constant RLC elements.

The program replaces the circuit elements in Figure 12.30 by their discrete-time equivalent circuits and solves the resulting nodal equations at each discrete-time instant Δt, $2\Delta t$, ..., $N \Delta t$. Dependent current sources arising from the models of line sections and of lumped L and C elements are rewritten in terms of bus voltages and current sources at prior times.

The program output data consist of the ten bus voltages at each discrete-time instant 0, Δt, $2 \Delta t$, ..., $N \Delta t$.

EXAMPLE 12.9   **TRANSMISSION-LINE TRANSIENTS program: lightning strike**

Lightning strikes the center of a single-phase 20-kV, 10-km line that is terminated at one end by a 100-MVA, 20-kV/8-kV transformer and at the other end by an open circuit. The line is initially unenergized. Run the TRANSMISSION-LINE TRANSIENTS program to compute the voltage at each end of the line and at the line center versus time. Use the following data for computations:

line: $20\,kV$, $l = 10\,km$, $Z_c = 300\,\Omega$, $v = 3.0 \times 10^8\,m/s$,
                    $R = 0.05\,\Omega/km$

transformer $10\,MVA$, $20\,kV/8\,kV$, 200-kV BIL, winding
                    capacitance-to-ground $C_{TR} = 6 \times 10^{-9}\,F$

lightning surge: An ideal square wave current source with
                    magnitude $20\,kA$ and duration $20\,\mu s$

calculation time interval: $\Delta t = 0.1\,\mu s$

final time: $t_F = 150\,\mu s$

Solution   The discrete-time equivalent circuit is shown in Figure 12.31. The circuit for each 5-km line section is taken from Figure 12.14, with $Z_c = 300\,\Omega$ and $\tau = l/(2v) = 10 \times 10^3/(2 \times 3 \times 10^8) = 16.67 \times 10^{-6}\,s = 16.67\,\mu s$. The total line resistance of $(0.05\,\Omega/km)(10\,km) = 0.50\,\Omega$ is divided by 4 to give $0.125\,\Omega$ lumped at each end of both line sections. The discrete-time equivalent circuit of the transformer capacitance is taken from Figure 12.16.

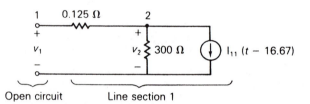

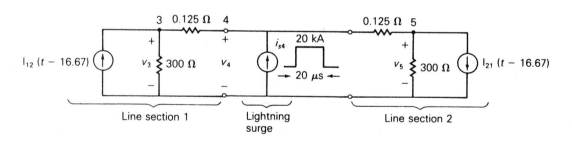

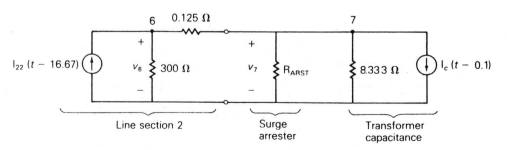

Figure 12.31   Discrete-time equivalent circuit for Examples 12.9 and 12.10

The program input data are given in Figure 12.32. Output data at 3 microsecond printout intervals are shown in Table 12.3 and Figure 12.33. Note that the transformer voltage exceeds its BIL rating. Also, the high transient overvoltage across the center and ends of the line indicate flashover of the line insulation. ■

Figure 12.32

Input data for Example 12.9

LINE DATA FOR EXAMPLE 12.9

LINE SECTION	CHARACTERISTIC IMPEDANCE	TRANSIT TIME
	OHMS	MICRO-SECONDS
1	300	16.67
2	300	16.67

CIRCUIT BREAKER ONE IS CLOSED
CIRCUIT BREAKER TWO IS OPEN

RLC DATA FOR EXAMPLE 12.9

RLC ELEMENT	1/R			1/L			C		
	FIRST VALUE	SECOND VALUE	CHANGE VOLTAGE	FIRST VALUE	SECOND VALUE	CHANGE VOLTAGE	FIRST VALUE	SECOND VALUE	CHANGE VOLTAGE
	1/OHMS	1/OHMS	kV	1/HENRIES	1/HENRIES	kV	FARADS	FARADS	kV
1	0	0	0	0	0	0	0	0	0
2	8	8	0	0	0	0	0	0	0
3	8	8	0	0	0	0	0	0	0
4	0	0	0	0	0	0	0	0	0
5	8	8	0	0	0	0	0	0	0
6	8	8	0	0	0	0	0	0	0
7	0	0	55	0	0	0	6.0D–09	6.0D–09	0

INDEPENDENT CURRENT SOURCE DATA FOR EXAMPLE 12.9

SOURCE	TYPE	I	T1	T2
		kA or kA/S	MICRO-SECONDS	MICRO-SECONDS
1	S	0	0	0
2	S	20	0	20
3	S	0	0	0

Table 12.3　Output data for Example 12.9

| TIME | BUS VOLTAGES | | |
| | V1 | V4 | V7 |
MICRO-SECONDS	kV	kV	kV
0	0.0	3001.3	0.0
3	0.0	3001.3	0.0
6	0.0	3001.3	0.0
9	0.0	3001.3	0.0
12	0.0	3001.3	0.0
15	0.0	3001.3	0.0
18	6000.0	3001.3	3317.6
21	6000.0	0.0	5493.2
24	6000.0	0.0	5904.3
27	6000.0	0.0	5981.9
30	6000.0	0.0	5996.6
33	6000.0	0.0	5999.4
36	6000.0	4768.0	5999.9
39	0.0	5767.2	1627.1
42	0.0	5956.0	307.4
45	0.0	5991.7	58.1
48	0.0	5998.4	11.0
51	9.8	5999.7	3002.7
54	4868.3	3955.5	5433.0
57	5786.2	747.3	5892.7
60	5959.6	141.2	5979.7
63	5992.4	26.7	5996.2
66	5998.6	5.0	5999.3
69	5999.7	3246.9	5999.9
72	3633.5	5479.5	1820.1
75	686.5	5901.6	344.3
78	129.7	5981.4	65.1
81	24.5	5996.5	12.3
84	−692.7	5999.3	−1248.6
87	4733.8	8838.0	2522.0
90	5760.5	1670.9	4944.6
93	5954.7	315.9	5725.3
96	5991.4	59.7	5933.9
99	5998.4	11.3	5984.8
102	5999.7	276.7	5996.6
105	4064.1	3949.9	6428.5
108	768.9	5429.6	2493.5
111	145.5	5857.7	712.7
114	27.5	5966.6	180.3
117	−1057.3	5992.5	−1228.6
120	−42.0	6855.5	2275.5
123	3968.9	4405.8	4849.5
126	5448.2	1420.0	5698.0
129	5864.0	379.3	5926.9
132	5968.3	92.6	5983.2
135	5992.9	−2568.8	5996.2
138	8817.8	2215.5	6681.2
141	4522.1	4875.3	2695.2
144	1393.9	5710.0	780.7
147	365.3	5930.5	198.9
150	55.9	5984.1	771.9

Table 12.4　Output data for Example 12.10

| TIME | BUS VOLTAGES | | |
| | V1 | V4 | V7 |
MICRO-SECONDS	kV	kV	kV
0	0.0	3001.3	0.0
3	0.0	3001.3	0.0
6	0.0	3001.3	0.0
9	0.0	3001.3	0.0
12	0.0	3001.3	0.0
15	0.0	3001.3	0.0
18	6000.0	3001.3	88.6
21	6000.0	0.0	88.6
24	6000.0	0.0	88.6
27	6000.0	0.0	88.6
30	6000.0	0.0	88.6
33	6000.0	0.0	88.6
36	6000.0	91.1	88.6
39	0.0	91.1	−7.8
42	0.0	91.1	−1.5
45	0.0	91.1	−0.3
48	0.0	91.1	−0.1
51	−5812.9	91.1	88.6
54	−5812.9	−18.9	88.6
57	−5812.9	−3.6	88.6
60	−5812.9	−0.7	88.6
63	−5812.9	−0.1	88.6
66	−5812.9	−0.0	88.6
69	−5812.9	−5813.0	88.6
72	−17.3	−5813.0	−8.7
75	−3.3	−5813.0	−1.6
78	−0.6	−5813.0	−0.3
81	−0.1	−5813.0	−0.1
84	−5813.0	−5813.0	−85.9
87	−5813.0	−42.1	−85.9
90	−5813.0	−8.0	−85.9
93	−5813.0	−1.5	−85.9
96	−5813.0	−0.3	−85.9
99	−5813.0	−0.1	−85.9
102	−5813.0	−88.3	−85.9
105	−19.4	−88.3	−10.1
108	−3.7	−88.3	−8.0
111	−0.7	−88.3	−2.7
114	−0.1	−88.3	−0.7
117	5631.7	−88.3	−85.8
120	5631.6	17.4	−85.8
123	5631.6	−11.6	−85.8
126	5631.6	−5.0	−85.8
129	5631.6	−1.5	−85.8
132	5631.6	−0.4	−85.8
135	5631.6	5631.8	−85.8
138	3.9	5631.8	−8.8
141	−12.9	5631.8	−8.5
144	−5.0	5631.8	−2.9
147	−1.4	5631.8	−0.8
150	5632.8	5631.8	82.9

Figure 12.33

Transformer voltage plots for Examples 12.9 and 12.10

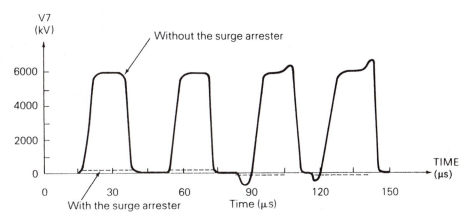

| EXAMPLE 12.10 |

TRANSMISSION-LINE TRANSIENTS program: lightning strike with surge-arrester protection

Repeat Example 12.9 with a 21-kV surge arrester installed adjacent to the transformer. The arrester is represented by a piecewise constant resistance given by $R_{ARST} = 2 \times 10^6 \, \Omega$ for $V_{ARST} < 55 \, kV$ and $R_{ARST} = 4.5 \, \Omega$ for $V_{ARST} \geqq 55 \, kV$.

Solution   The surge-arrester resistor is shown in Figure 12.31. Logic in the computer program is used to change the arrester resistance when its voltage exceeds 55 kV. The output is shown in Table 12.4 and Figure 12.33. Note that the transformer voltage is held by the arrester to 88.6 kV, well below the transformer BIL. However, flashover of the line insulation is still expected to occur.   ∎

The transformer in Examples 12.9 and 12.10 is represented by its winding capacitance-to-ground. Lightning or switching surges are effectively blocked by transformer leakage inductance, which acts as an open circuit at high frequencies. As such, generators are protected against transmission-line transient overvoltages by generator step-up transformers.

Figure 12.34 shows typical ranges of transformer winding capacitance-

Figure 12.34

Transformer winding capacitance-to-ground versus transformer size [1]

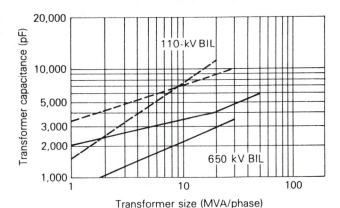

to-ground versus transformer size, with BIL of the highest voltage winding as a parameter [1]. This figure gives the total capacitance-to-ground of the highest voltage winding on a per-phase basis. For a Y-connected transformer bank, 0.30 to 0.41 of the capacitance values from Figure 12.34 are the approximate values for each winding. For a Δ-connected bank, the full value from the figure is used for each winding.

PROBLEMS

Section 12.2

12.1 From the results of Example 12.2, plot the voltage and current profiles along the line at times $\tau/2$, τ, and 2τ. That is, plot $v(x, \tau/2)$ and $i(x, \tau/2)$ versus x for $0 \leqslant x \leqslant l$; then plot $v(x, \tau)$, $i(x, \tau)$, $v(x, 2\tau)$, and $i(x, 2\tau)$ versus x.

12.2 Rework Example 12.2 if the source voltage at the sending end is a ramp, $e_G(t) = Eu_{-2}(t) = Etu_{-1}(t)$, with $Z_G = Z_c$.

12.3 Referring to the single-phase two-wire lossless line shown in Figure 12.3, the receiving end is terminated by an inductor with L_R henries. The source voltage at the sending end is a step, $e_G(t) = Eu_{-1}(t)$ with $Z_G = Z_c$. Both the line and inductor are initially unenergized. Determine and plot the voltage at the center of the line $v(l/2, t)$ versus time t.

12.4 Rework Problem 12.3 if $Z_R = Z_c$ at the receiving end and the source voltage at the sending end is $e_G(t) = Eu_{-1}(t)$, with an inductive source impedance $Z_G(s) = sL_G$. Both the line and source inductor are initially unenergized.

12.5 Rework Example 12.4 with $Z_R = 4Z_c$ and $Z_G = Z_c/3$.

12.6 The single-phase, two-wire lossless line in Figure 12.3 has a series inductance $L = (1/3) \times 10^{-6}$ H/m, a shunt capacitance $C = (1/3) \times 10^{-10}$ F/m, and a 30-km line length. The source voltage at the sending end is a step $e_G(t) = 100u_{-1}(t)$ kV with $Z_G(s) = 100\,\Omega$. The receiving-end load consists of a 100-Ω resistor in parallel with a 2-mH inductor. The line and load are initially unenergized. Determine (a) the characteristic impedance in ohms, the wave velocity in m/s, and the transit time in ms for this line; (b) the sending-and receiving-end voltage reflection coefficients in per-unit; (c) the Laplace transform of the receiving-end current, $I_R(s)$; and (d) the receiving-end current $i_R(t)$ as a function of time.

12.7 The single-phase, two-wire lossless line in Figure 12.3 has a series inductance $L = 2 \times 10^{-6}$ H/m, a shunt capacitance $C = 1.25 \times 10^{-11}$ F/m, and a 100-km line length. The source voltage at the sending end is a step $e_G(t) = 100u_{-1}(t)$ kV with a source impedance equal to the characteristic impedance of the line. The receiving-end load consists of a 100-mH inductor in series with a 1-μF capacitor. The line and load are initially unenergized. Determine (a) the characteristic impedance in Ω, the wave velocity in m/s, and the transit time in ms for this line; (b) the sending- and receiving-end voltage reflection coefficients in per-unit; (c) the receiving-end voltage $v_R(t)$ as a function of time; and (d) the steady-state receiving-end voltage.

12.8 The single-phase, two-wire lossless line in Figure 12.3 has a series inductance $L = 0.999 \times 10^{-6}$ H/m, a shunt capacitance $C = 1.112 \times 10^{-11}$ F/m, and a 60-km line length. The source voltage at the sending end is a ramp $e_G(t) = Etu_{-1}(t) = Eu_{-2}(t)$ kV with a source impedance equal to the characteristic impedance of the line. The receiving-end load consists of a 150-Ω resistor in parallel with a 1-μF capacitor. The line and load are initially unenergized. Determine (a) the characteristic impedance in Ω, the wave velocity in m/s, and the transit time in ms for this line; (b) the sending- and receiving-end voltage reflection coefficients in per-unit; (c) the Laplace transform of the sending-end voltage, $V_S(s)$; and (d) the sending-end voltage $v_S(t)$ as a function of time.

Section 12.3

12.9   Draw the Bewley lattice diagram for Problem 12.5, and plot $v(l/3, t)$ versus time t for $0 \leqslant t \leqslant 5\tau$. Also plot $v(x, 3\tau)$ versus x for $0 \leqslant x \leqslant l$.

12.10   Rework Problem 12.9 if the source voltage is a pulse of magnitude E and duration $\tau/10$; that is, $e_G(t) = E[u_{-1}(t) - u_{-1}(t - \tau/10)]$. $Z_R = 4Z_c$ and $Z_G = Z_c/3$ are the same as in Problem 12.9.

12.11   Rework Example 12.6 if the source impedance at the sending end of line A is $Z_G = Z_A/4 = 100\,\Omega$, and the receiving end of line B is short-circuited, $Z_R = 0$.

12.12   Rework Example 12.6 if the overhead line and cable are interchanged. That is, $Z_A = 100\,\Omega$, $v_A = 2 \times 10^8\,\text{m/s}$, $l_A = 20\,\text{km}$, $Z_B = 400\,\Omega$, $v_B = 3 \times 10^8\,\text{m/s}$, and $l_B = 30\,\text{km}$. The step voltage source $e_G(t) = Eu_{-1}(t)$ is applied to the sending end of line A with $Z_G = Z_A = 100\,\Omega$, and $Z_R = 2Z_B = 800\,\Omega$ at the receiving end. Draw the lattice diagram for $0 \leqslant t \leqslant 0.6\,\text{ms}$ and plot the junction voltage versus time t.

12.13   As shown in Figure 12.35, a single-phase two-wire lossless line with $Z_c = 400\,\Omega$, $v = 3 \times 10^8\,\text{m/s}$, and $l = 100\,\text{km}$ has a 400-Ω resistor, denoted R_J, installed across the center of the line, thereby dividing the line into two sections, A and B. The source

Figure 12.35

Circuit for Problem 12.13

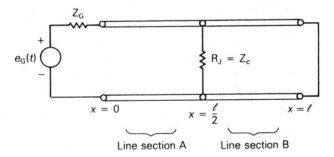

voltage at the sending end is a pulse of magnitude 100 V and duration 0.1 ms. The source impedance is $Z_G = Z_c = 400\,\Omega$, and the receiving end of the line is short-circuited. (a) Show that for an incident voltage wave arriving at the center of the line from either line section, the voltage reflection and refraction coefficients are given by

$$\Gamma_{BB} = \Gamma_{AA} = \frac{\left(\dfrac{Z_{eq}}{Z_c}\right) - 1}{\left(\dfrac{Z_{eq}}{Z_c}\right) + 1} \qquad \Gamma_{AB} = \Gamma_{BA} = \frac{2\left(\dfrac{Z_{eq}}{Z_c}\right)}{\left(\dfrac{Z_{eq}}{Z_c}\right) + 1}$$

where

$$Z_{eq} = \frac{R_J Z_c}{R_J + Z_c}$$

(b) Draw the Bewley lattice diagram for $0 \leqslant t \leqslant 6\tau$. (c) Plot $v(l/2, t)$ versus time t for $0 \leqslant t \leqslant 6\tau$ and plot $v(x, 6\tau)$ versus x for $0 \leqslant x \leqslant l$.

12.14 The junction of four single-phase two-wire lossless lines, denoted A, B, C, and D, is shown in Figure 12.13. Consider a voltage wave v_A^+ arriving at the junction from line A. Using (12.3.8) and (12.3.9), determine the voltage reflection coefficient Γ_{AA} and the voltage refraction coefficients Γ_{BA}, Γ_{CA}, and Γ_{DA}.

12.15 Referring to Figure 12.3, the source voltage at the sending end is a step $e_G(t) = Eu_{-1}(t)$ with an inductive source impedance $Z_G(s) = sL_G$, where $L_G/Z_c = \tau/3$. At the receiving end, $Z_R = Z_c/4$. The line and source inductance are initially unenergized. (a) Draw the Bewley lattice diagram for $0 \leqslant t \leqslant 5\tau$. (b) Plot $v(l, t)$ versus time t for $0 \leqslant t \leqslant 5\tau$.

12.16 As shown in Figure 12.36, two identical, single-phase, two-wire, lossless lines are connected in parallel at both the sending and receiving ends. Each line has a 400-Ω characteristic impedance, 3×10^8 m/s velocity of propagation, and 100-km line length. The source voltage at the sending end is a 100-kV step with source impedance $Z_G = 100\,\Omega$. The receiving end is shorted ($Z_R = 0$). Both lines are initially unenergized.

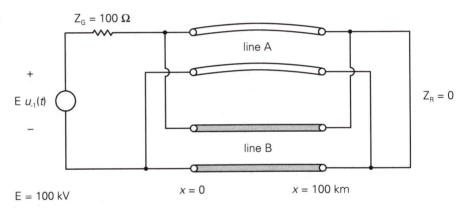

Figure 12.36

Circuit for Problem 12.16

(a) Determine the first forward traveling voltage waves that start at time $t = 0$ and travel on each line toward the receiving end. (b) Determine the sending- and receiving-end voltage reflection coefficients in per-unit. (c) Draw the Bewley lattice diagram for $0 < t < 2.0$ ms. (d) Plot the voltage at the center of one line versus time t for $0 < t < 2.0$ ms.

12.17 As shown in Figure 12.37, an ideal current source consisting of a 10-kA pulse with 50-μs duration is applied to the junction of a single-phase, lossless cable and a single-phase, lossless overhead line. The cable has a 200-Ω characteristic impedance, 2×10^8 m/s velocity of propagation, and 20-km length. The overhead line has a 300-Ω characteristic impedance, 3×10^8 m/s velocity of propagation, and 60-km length. The sending end of the cable is terminated by a 400-Ω resistor, and the receiving end of the overhead line is terminated by a 100-Ω resistor. Both the line and cable are initially unenergized. (a) Determine the voltage reflection coefficients Γ_S, Γ_R, Γ_{AA}, Γ_{AB}, Γ_{BA}, and Γ_{BB}. (b) Draw the Bewley lattice diagram for $0 < t < 0.8$ ms. (c) Determine and plot the voltage $v(0, t)$ at $x = 0$ versus time t for $0 < t < 0.8$ ms.

Figure 12.37

Circuit for Problem 12.17

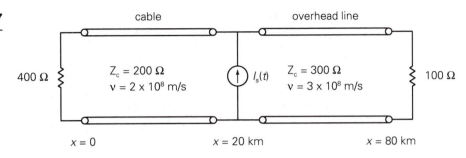

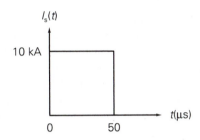

Section 12.4

12.18   For the circuit given in Problem 12.3, replace the circuit elements by their discrete-time equivalent circuits and write nodal equations in a form suitable for computer solution of the sending-end and receiving-end voltages. Give equations for all dependent sources. Assume $E = 1000\,V$, $L_R = 10\,mH$, $Z_c = 100\,\Omega$, $v = 2 \times 10^8\,m/s$, $l = 40\,km$, and $\Delta t = 0.02\,ms$.

12.19   Repeat Problem 12.18 for the circuit given in Problem 12.13. Assume $\Delta t = 0.03333\,ms$.

12.20   For the circuit given in Problem 12.7, replace the circuit elements by their discrete-time equivalent circuits. Use $\Delta t = 100\,\mu s = 1 \times 10^{-4}\,s$. Determine and show all resistance values on the discrete-time circuit. Write nodal equations for the discrete-time circuit, giving equations for all dependent sources. Then solve the nodal equations and determine the sending- and receiving-end voltages at the following times: $t = 100, 200, 300, 400, 500,$ and $600\,\mu s$.

12.21   For the circuit given in Problem 12.8, replace the circuit elements by their discrete-time equivalent circuits. Use $\Delta t = 50\,\mu s = 5 \times 10^{-5}\,s$ and $E = 100\,kV$. Determine and show all resistance values on the discrete-time circuit. Write nodal equations for the discrete-time circuit, giving equations for all dependent sources. Then solve the nodal equations and determine the sending- and receiving-end voltages at the following times: $t = 50, 100, 150, 200, 250,$ and $300\,\mu s$.

Section 12.5

12.22   Rework Problem 12.18 for a lossy line with a constant series resistance $R = 0.3\,\Omega/km$. Lump half of the total resistance at each end of the line.

Section 12.8

12.23   Repeat Example 12.8 for a 372-kV station arrester employed in a 345-kV system with a 1550-kV BIL.

12.24 Select a surge arrester from Table 12.2 for the high-voltage side of a three-phase 600-MVA, 500-kV Y/13.8-kV Δ transformer. The high-voltage windings of the transformer have a 1550-kV BIL and a solidly grounded neutral. A minimum protective margin of 1.4 per unit based on a $1.2 \times 50 \, \mu s$ voltage waveform is required. Lightning discharge currents are limited to 40 kA.

12.25 The high-voltage windings of a transformer are connected to a 138-kV, three-phase transmission line. The BIL of these windings is 650 kV. (a) Determine the BIL of the windings in per-unit. (b) Select a surge arrester from Table 12.2 that protects the windings against $1.2 \times 50 \, \mu s$ surges with a minimum protection margin of 3.0 per unit. The selected surge arrester should remain in an open-circuit condition for all $1.2 \times 50 \, \mu s$ surges with peak voltages up to 2.0 per unit.

Section 12.9

12.26 Solve the nodal equations for the sending- and receiving-end voltages at the times given in Problem 12.20 using the TRANSMISSION-LINE TRANSIENTS program.

12.27 Solve the nodal equations for the sending- and receiving-end voltages at the times given in Problem 12.21 using the TRANSMISSION-LINE TRANSIENTS program.

12.28 Run the TRANSMISSION-LINE TRANSIENTS program for Examples 12.9 and 12.10 if the lightning surge strikes the line 3 km from the transformer.

12.29 Investigate the effect of different transformer winding capacitances-to-ground on the transformer voltage in Example 12.9. Run two additional cases of the TRANSMISSION-LINE TRANSIENTS program for Example 12.9: case (1), $C_{TR} = 3 \times 10^{-9}$ F; case (2), $C_{TR} = 9 \times 10^{-9}$ F. Which case gives the largest transformer voltage?

12.30 Recommended practice is to install surge arresters as close as possible to the equipment they protect. As shown in Figure 12.38, a single-phase unenergized line is terminated by a transformer at the receiving end and is matched at the sending end.

Figure 12.38

Circuit for Problem 12.30

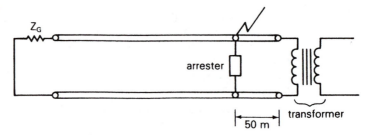

A surge arrester is located 50 m from the transformer. Lightning strikes the surge arrester. Run the TRANSMISSION-LINE TRANSIENTS program to compute the transformer voltage and arrester voltage. Does the arrester protect the transformer? Use the following data:

line:	20 kV. $l = 10$ km, $Z_c = 300 \, \Omega$, $v = 3 \times 10^8$ m/s, $R = 0.05 \, \Omega/$km
transformer:	100 MVA, 20 kV/8 kV, high-voltage winding BIL = 200 kV; high-voltage winding capacitance-to-ground, three cases: $C_{TR} = 1.5 \times 10^{-10}$, 1.5×10^{-9}, and 1.5×10^{-8} F
arrester:	$R_{ARST} = 2 \times 10^6 \, \Omega$ for $V_{ARST} < 55$ kV $R_{ARST} = 4.5 \, \Omega$ for $V_{ARST} \geq 55$ kV

lightning surge: a 30 kA, $1.5 \times 50\ \mu s$ current source which is approximated
                       by a double exponential

calculation interval: $\Delta t = 0.1\ \mu s$

final time: $t_{\mathrm{F}} = 5\ \mu s$

CASE STUDY QUESTIONS

A. What are the differences between surge arresters used to protect electric utility equipment and surge suppressors used to protect utility customers' computers and electronic equipment?

B. Should electric utility companies be held responsible for lightning strikes that propagate through utility transmission and distribution networks to customers and damage customers' computers and electronic facilities?

References

1. A. Greenwood, *Electrical Transients in Power Systems* (New York: Wiley Interscience, 1971).

2. L. V. Bewley, *Travelling Waves on Transmission Systems*, 2d ed. (New York: Wiley, 1951).

3. L. Bergeron, *Water Hammer in Hydraulics and Wave Surges in Electricity* (New York: Wiley, 1961).

4. H. M. Lacey, "The Lightning Protection of High-Voltage Overhead Transmission and Distribution Systems," *Proc. IEE, 96* (1949), p. 287.

5. J. R. Carson, "Wave Propagation in Overhead Wires with Ground Return," *Bell System Technical Journal 5* (1926), pp. 539–554.

6. W. S. Meyer and H. W. Dommel, "Numerical Modelling of Frequency-Dependent Transmission Line Parameters in an Electromagnetic Transients Program," *IEEE Transactions PAS,* vol. PAS-99 (September/October 1974), pp. 1401–1409.

7. A. Budner, "Introduction of Frequency-Dependent Line Parameters into an Electromagnetics Transients Program," *IEEE Transactions PAS*, vol. PAS-89 (January 1970), pp. 88–97.

8. D. E. Hedman, "Propagation on Overhead Transmission Lines: I—Theory of Modal Analysis and II—Earth Conduction Effects and Practical Results," *IEEE Transactions PAS* (March 1965), pp. 200–211.

9. H. W. Dommel, "A Method for Solving Transient Phenomena in Multiphase Systems," *Proceedings 2nd Power Systems Computation Conference,* Stockholm, 1966.

10. H. W. Dommel, "Digital Computer Solution of Electromagnetic Transients in Single- and Multiphase Networks," *IEEE Transactions PAS*, vol. PAS-88 (1969), pp. 388–399.

11. H. W. Dommel, "Nonlinear and Time-Varying Elements in Digital Simulation of Electromagnetic Transients," *IEEE Transactions PAS*, vol. PAS-90 (November/December 1971), pp. 2561–2567.

12. S. R. Naidu, *Transitorios Electromagnéticos em Sistemas de Potência*, Eletrobras/UFPb, Brazil, 1985.

13. "Report of Joint IEEE-EEI Committee on EHV Line Outages," *IEEE Transactions PAS, 86* (1967), p. 547.

14.   R. A. W. Connor and R. A. Parkins, "Operations Statistics in the Management of Large Distribution System," *Proc. IEE, 113* (1966), p. 1823.

15.   C. T. R. Wilson, "Investigations on Lightning Discharges and on the Electrical Field of Thunderstorms," *Phil. Trans. Royal Soc.*, Series A, *221* (1920), p. 73.

16.   G. B. Simpson and F. J. Scrase, "The Distribution of Electricity in Thunderclouds," *Proc. Royal Soc.*, Series A, *161* (1937), p. 309.

17.   B. F. J. Schonland and H. Collens, "Progressive Lightning," *Proc. Royal Soc.*, Series A, *143* (1934), p. 654.

18.   Westinghouse Electric Corporation, *Electrical Transmission and Distribution Reference Book*, 4th ed. (East Pittsburgh, PA, 1964).

19.   E. W. Kimbark and A. C. Legate, "Fault Surge Versus Switching Surge, A Study of Transient Voltages Caused by Line to Ground Faults," *IEEE Transactions PAS, 87* (1968), p. 1762.

20.   *Guide for the Application of Valve-Type Surge Arresters for Alternating-Current Systems*, ANSI C62.2-1981 (New York: American National Standards Institute, 1981).

21.   C. Concordia, "The Transient Network Analyzer for Electric Power System Problems," Supplement to *CIGRE Committee No. 13 Report*, 1956.

22.   F. Martzloff, "Protecting Computer Systems Against Power Transients," *IEEE Spectrum, 27*, 4 (April 1990), pp. 37–40.

23.   J. Mackevich, "How to Get Top Performance from Polymer MOV Arresters, *Electrical World, 206*, 2 (February 1992), pp. 54–56.

24.   W. R. Newcott, "Lightning, Nature's High-Voltage Spectacle," *National Geographic*, 184, 1 (July 1993), pp. 83–103.

CHAPTER 13

TRANSIENT STABILITY

1300 MW generating unit consisting of a cross-compound steam turbine and
two 722-MVA synchronous generators
(Courtesy of American Electric Power)

Power-system stability refers to the ability of synchronous machines to move from one steady-state operating point following a disturbance to another steady-state operating point, without losing synchronism [1]. There are three types of power-system stability: steady-state, transient, and dynamic.

Steady-state stability, discussed in Chapter 6, involves slow or gradual changes in operating points. Steady-state stability studies, which are usually performed with a power-flow computer program (Chapter 7), ensure that phase angles across transmission lines are not too large, that bus voltages are close to nominal values, and that generators, transmission lines, transformers, and other equipment are not overloaded.

Transient stability, the main focus of this chapter, involves major disturbances such as loss of generation, line-switching operations, faults, and sudden load changes. Following a disturbance, synchronous machine frequencies undergo transient deviations from synchronous frequency (60 Hz), and machine power angles change. The objective of a transient stability study is to determine whether or not the machines will return to synchronous frequency with new steady-state power angles. Changes in power flows and bus voltages are also of concern.

Elgerd [2] gives an interesting mechanical analogy to the power-system transient stability program. As shown in Figure 13.1, a number of masses representing synchronous machines are interconnected by a network of elastic strings representing transmission lines. Assume that this network is initially at rest in steady-state, with the net force on each string below its break point, when one of the strings is cut, representing the loss of a transmission line. As a result, the masses undergo transient oscillations and the forces on the strings fluctuate. The system will then either settle down to a new steady-state operating point with a new set of string forces, or additional strings will break, resulting in an even weaker network and eventual system collapse. That is, for a given disturbance, the system is either transiently stable or unstable.

Figure 13.1

Mechanical analog of power-system transient stability [2]

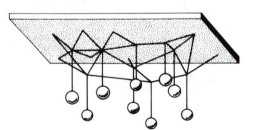

In today's large-scale power systems with many synchronous machines interconnected by complicated transmission networks, transient stability studies are best performed with a digital computer program. For a specified disturbance, the program alternately solves, step by step, algebraic power-flow equations representing a network and nonlinear differential equations representing synchronous machines. Both predisturbance, disturbance, and postdisturbance computations are performed. The program output includes power angles and frequencies of synchronous machines, bus voltages, and power flows versus time.

In many cases, transient stability is determined during the first swing of machine power angles following a disturbance. During the first swing, which typically lasts about 1 second, the mechanical output power and the internal voltage of a generating unit are often assumed constant. However, where multiswings lasting several seconds are of concern, models of turbine-governors and excitation systems (for example, see Figures 11.3 and 11.5) as well as more detailed machine models can be employed to obtain accurate transient stability results over the longer time period.

Dynamic stability involves an even longer time period, typically several minutes. It is possible for controls to affect dynamic stability even though transient stability is maintained. The action of turbine-governors, excitation systems, tap-changing transformers, and controls from a power-system dispatch center can interact to stabilize or destabilize a power system several minutes after a disturbance has occurred.

In order to simplify transient stability studies, the following assumptions are made:

1. Only balanced three-phase systems and balanced disturbances are considered. Therefore, only positive-sequence networks are employed.

2. Deviations of machine frequencies from synchronous frequency (60 Hz) are small, and dc offset currents and harmonics are neglected. Therefore, the network of transmission lines, transformers, and impedance loads is essentially in steady-state; and voltages, currents, and powers can be computed from algebraic power-flow equations.

In Section 13.1 we introduce the swing equation, which determines synchronous machine rotor dynamics. In Section 13.2 a simplified model of a synchronous machine and a Thévenin equivalent of a system consisting of lines, transformers, loads, and other machines are given. Then in Section 13.3 the equal-area criterion is presented; this gives a direct method for determining the transient stability of one machine connected to a system equivalent. Numerical integration techniques for solving swing equations step by step are discussed in Section 13.4 and used in Section 13.5 to determine multimachine stability. The personal computer program TRANSIENT STABILITY, which can be used for the design and analysis of power systems with regard to transient stability, is described in Section 13.6. Finally, Section 13.7 discusses design methods for improving power-system transient stability.

| **CASE STUDY** | In the United States, electric utilities grew first as isolated systems. Gradually, however, neighboring utilities began to interconnect, which allowed utility companies to draw upon each others' generation reserves during time of need and to schedule power transfers that take advantage of energy-cost differences. |

ferences. Although overall system reliability and economy have improved dramatically through interconnection, there is a remote possibility that an initial disturbance may lead to instability and a regional blackout. The following article reviews the most historic power system transient stability event: the 1965 Northeast blackout [9].

The Great Blackout

BOB McCAW

As we go to press, the immediate cause of the Great East Coast Blackout has been announced. Most of the dramatic and sensational aspects have been covered completely by the general press. And we don't have any intention of trying to explain the entire sequence of events, or whether the complete collapse of the system was predictable. We do feel, however, that certain comments are in order at this time; also, a few basic facts should be brought to the attention of the general public.

Reasons for interconnections are sound

First, the complete philosophy of interconnection, or power-pooling, is good, sound business. (Some critics asked why a loss of power in one region should affect another area.) Next, there is no lack of generating capacity *reserve* in the regions affected. FPC reports show that there is about 15% reserve capacity—*over expected peak*—for the area affected.

Next, the utility industry—in both public and private segments—cannot be accused of lack of planning. In fact, utilities are constantly predicting their load demands for many years into the future. And the technique of estimating load-growth employs every means that our present technology offers. Computers have been doing this work for some time, and

Reprinted with permission from Power Engineering

historical records have a double check on the predictions.

National power survey made long-term recommendations

On the national scene, we must recall the recently completed National Power Survey. This brought together many of the nation's top power authorities, who recommended a long-range pattern of interconnections and power-pooling. This was originated, and coordinated, by the FPC.

It appears to us that we must follow this general course for the immediate and long-range future:

1. Take every possible precaution against hysteria, and any *concerted* drive to inject more controls by the government. Some additional controls may be needed over the number and types of interconnections, but this should be carefully worked out with industry representatives.

2. We must not rush into any so-called *preventive-measure* projects, until all facts are known, and until a careful study has been made of all governing factors.

3. After the experts have accurately determined the overall reasons for this recent catastrophe. we must lend all possible support in providing additional precautionary measures to prevent a recurrence.

The following pages present a round-up of opinion, excerpts from the National Power Survey, and other data pertinent to the Blackout. We believe that facts, not flights of fancy, are needed to assess the situation as it now stands.

Faulty relay in Ontario blacks out New York & New England

Blackout hits large area

Worst power failure in history occurred Tuesday evening, Nov. 9, 1965. At the height of the rush hour, complete darkness descended, and practically everything came to a standstill, over an area of 83,000 square miles. About 30 million people were affected in eight states and part of Canada. The system involved was the Canadian-United States eastern power complex (CANUSE).

Full inquiry ordered by President

An immediate investigation was launched by the Federal Power Commission under orders from President Johnson. The Federal Bureau of Investigation, Office of Emergency Planning, Defense Department, Department of the Interior and other agencies were also ordered to cooperate in the investigation. In addition, power industry representatives and Canadian authorities took part.

Faulty relay found to be source of trouble

At first, little hope was held out for pinpointing the exact reason for the failure in a short time. However, on November 15th it was announced that the cause had been found. The failure was attributed to a faulty relay at Sir Adam Beck Plant No. 2 at Queenston, Ontario. [See Fig. 1.] This plant is part of the Ontario Hydro-Electric System on the Niagara River. Six lines run into Ontario from the Beck plant, and the line controlled by the faulty relay was carrying 300 Mw. Failure of the relay dumped this load onto the other five lines. Even though these lines were not overloaded, all five tripped out.

Here's the chain reaction

Total power flow on these lines had been 1600 Mw into Ontario, including 500 Mw being imported from the Power Authority of the State of New York. All this power was suddenly dumped on the New York system. The resultant surge knocked out the Power Authority's main east-west transmission line and shut down seven units that had been feeding the Northeast grid. The resulting drain on systems to the south and east caused the whole system to collapse.

New York City, for example, had been drawing about 300 Mw from the network just before the failure. Loss of the upstate plants caused a sudden reversal of flow and placed a heavy drain on the City generators. The load was much greater than the plants still in service could supply, and the result was a complete collapse. Automatic equipment shut down the units to protect them from damage.

After the total failure the individual systems started up in sections, and most power was restored by a little after midnight. However, Manhattan, with the greatest concentration of load, was not fully restored until after six o'clock in the morning.

First guesses blamed ground or short circuit

Many guesses were made as to the cause of the blackout, but most were far from right. A common opinion was that the failure had occurred in upstate New York because the first known tripout was N.Y. Power Authority's line.

In the early hours, the only thing sure was that the fault had occurred somewhere in the Lake Ontario region. This was evident from the times recorded as the blackout advanced across the entire region. Detailed studies of oscillographs and power flow charts showed that the fault lay in the Ontario Hydro system or the N.Y. Power Authority's main line. Careful checking then disclosed the faulty relay. It was estimated that at least two more weeks would be needed to make a complete analysis of the blackout. Following this, government and industry leaders will make a study of the reliability of power grids.

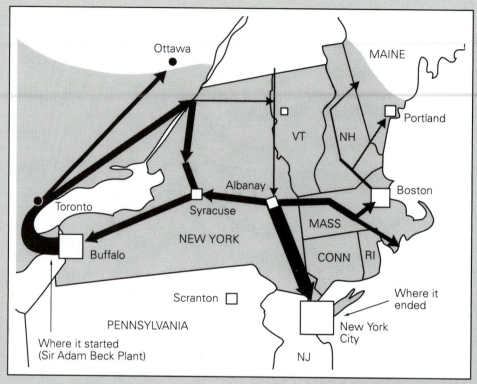

Figure 1 Area affected by blackout

National power policy at stake

Of one thing there is no doubt: the utilities are on trial before the public. By-and-large the people accepted the blackout in good spirits, and disorders and damage were surprisingly scarce. However, we might not be so lucky a second time. If the blackout had occurred during adverse weather conditions, the results could have been disastrous.

Stronger grids are planned

Our present policy is directed toward a strong national grid, as spelled out in the National Power Survey. Our 1980 transmission pattern, as projected by the Federal Power Commission, includes the possibility of three major east-west EHV interties and a northwest-middle south connection. Lengths range up to nearly 2000 miles, and power flows as high as 4000 Mw are predicted.

According to the report there are no insurmountable obstacles to successful operation of large interconnected systems. Techniques developed during the last decade are claimed to demonstrate that large networks can be operated in parallel with a high degree of operating stability, and with increasing dependence on automated controls.

Lighter inter-regional connections suggested

After this experience with cascading tripouts, much reconsideration might be given to the plans for a strong national grid. At least one utility company official expressed the belief that the entire country could have been blacked out if a strong grid had been in operation at the time of failure. Philip Sporn of American Electric Power Co. suggested that it might be better to have strong intra-regional systems and rather light inter-regional connections. On the other hand, Interior Secretary Udall claimed that what was needed was better interconnections. He emphasized that interconnections are the best preventives of power failures, but the system must be improved.

Careful deliberation needed to set future program

Clearly, we can expect investigations, studies and reexaminations of where we are headed. Nothing in recent years has given this nation such cause for concern as the massive blackout that occurred in a region served by some of our strongest utilities. Our whole economy is now almost completely dependent on electric power.

However, we should certainly avoid pushing the panic button. In any complex system, any one of hundreds of faults can cause an outage. Only rare combinations of faults, however, result in a cascading of tripouts and a complete shutdown over an entire region.

Even in this case, not all protective equipment failed to operate. For example, look at what happened to the Pennsylvania-New Jersey-Maryland interconnection going into New York State. When the trouble started, all six lines dropped out simultaneously. This caused a jolt to the P-J-M system, but it was then cleared from the major problem afflicting the northeast.

Pumped storage plants can help

In the future, the Cornwall pumped-storage plant could be a big factor in preventing the recurrence of such a failure. The plant will represent a large reserve that can be put into operation on short notice. If, in spite of all precautions, another blackout should occur, the pumped-storage plant will make it easier to put the system back on the line. One reason that Consolidated Edison took so long to get started up in some sections was the lack of auxiliary power. Some plants had no auxiliary electric power and no steam auxiliaries; so they couldn't be started up without outside help.

This has been a sobering experience and possibly a needed warning to reexamine some of our policies and plans. The warning will not have been wasted if we learn from it how to build a reliable grid to supply all of the power that will be needed in the years ahead.

SECTION 13.1

THE SWING EQUATION

Consider a generating unit consisting of a three-phase synchronous generator and its prime mover. The rotor motion is determined by Newton's second law, given by

$$J\alpha_m(t) = T_m(t) - T_e(t) = T_a(t) \tag{13.1.1}$$

where J = total moment of inertia of the rotating masses, kgm^2

α_m = rotor angular acceleration, rad/s^2

T_m = mechanical torque supplied by the prime mover minus the retarding torque due to mechanical losses, Nm

T_e = electrical torque that accounts for the total three-phase electrical power output of the generator, plus electrical losses, Nm

T_a = net accelerating torque, Nm

Also, the rotor angular acceleration is given by

$$\alpha_m(t) = \frac{d\omega_m(t)}{dt} = \frac{d^2\theta_m(t)}{dt^2} \tag{13.1.2}$$

$$\omega_m(t) = \frac{d\theta_m(t)}{dt} \tag{13.1.3}$$

where ω_m = rotor angular velocity, rad/s

θ_m = rotor angular position with respect to a stationary axis, rad

T_m and T_e are positive for generator operation. In steady-state T_m equals T_e, the accelerating torque T_a is zero, and, from (13.1.1), the rotor acceleration α_m is zero, resulting in a constant rotor velocity called *synchronous speed*. When T_m is greater than T_e, T_a is positive and α_m is therefore positive, resulting in increasing rotor speed. Similarly, when T_m is less than T_e, the rotor speed is decreasing.

It is convenient to measure the rotor angular position with respect to a synchronously rotating reference axis instead of a stationary axis. Accordingly, we define

$$\theta_m(t) = \omega_{msyn}t + \delta_m(t) \tag{13.1.4}$$

where ω_{msyn} = synchronous angular velocity of the rotor, rad/s

δ_m = rotor angular position with respect to a synchronously rotating reference, rad

Using (13.1.2) and (13.1.4), (13.1.1) becomes

$$J\frac{d^2\theta_m(t)}{dt^2} = J\frac{d^2\delta_m(t)}{dt^2} = T_m(t) - T_e(t) = T_a(t) \tag{13.1.5}$$

It is also convenient to work with power rather than torque, and to work in per-unit rather than in actual units. Accordingly, we multiply (13.1.5) by $\omega_m(t)$ and divide by S_{rated}, the three-phase voltampere rating of the generator:

$$\frac{J\omega_m(t)}{S_{rated}}\frac{d^2\delta_m(t)}{dt^2} = \frac{\omega_m(t)T_m(t) - \omega_m(t)T_e(t)}{S_{rated}}$$

$$= \frac{p_m(t) - p_e(t)}{S_{rated}} = p_{mp.u.}(t) - p_{ep.u.}(t) = p_{ap.u.}(t) \tag{13.1.6}$$

where $p_{mp.u.}$ = mechanical power supplied by the prime mover minus mechanical losses, per unit

$p_{ep.u.}$ = electrical power output of the generator plus electrical losses, per unit

Finally, it is convenient to work with a normalized inertia constant, called the H constant, which is defined as

$$H = \frac{\text{stored kinetic energy at synchronous speed}}{\text{generator voltampere rating}}$$

$$= \frac{\frac{1}{2}J\omega_{msyn}^2}{S_{rated}} \quad \text{joules/VA or per unit-seconds} \tag{13.1.7}$$

The H constant has the advantage that it falls within a fairly narrow range, normally between 1 and 10 p.u.-s, whereas J varies widely, depending on generating unit size and type. Solving (13.1.7) for J and using in (13.1.6),

$$2H \frac{\omega_m(t)}{\omega_{msyn}^2} \frac{d^2\delta_m(t)}{dt^2} = p_{mp.u.}(t) - p_{ep.u.}(t) = p_{ap.u.}(t) \tag{13.1.8}$$

Defining per-unit rotor angular velocity,

$$\omega_{p.u.}(t) = \frac{\omega_m(t)}{\omega_{msyn}} \tag{13.1.9}$$

Equation (13.1.8) becomes

$$\frac{2H}{\omega_{msyn}} \omega_{p.u.}(t) \frac{d^2\delta_m(t)}{dt^2} = p_{mp.u.}(t) - p_{ep.u.}(t) = p_{ap.u.}(t) \tag{13.1.10}$$

For a synchronous generator with P poles, the electrical angular acceleration α, electrical radian frequency ω, and power angle δ are

$$\alpha(t) = \frac{P}{2} \alpha_m(t) \tag{13.1.11}$$

$$\omega(t) = \frac{P}{2} \omega_m(t) \tag{13.1.12}$$

$$\delta(t) = \frac{P}{2} \delta_m(t) \tag{13.1.13}$$

Similarly, the synchronous electrical radian frequency is

$$\omega_{syn} = \frac{P}{2} \omega_{msyn} \tag{13.1.14}$$

The per-unit electrical frequency is

$$\omega_{p.u.}(t) = \frac{\omega(t)}{\omega_{syn}} = \frac{\frac{2}{P}\omega(t)}{\frac{2}{P}\omega_{syn}} = \frac{\omega_m(t)}{\omega_{msyn}} \tag{13.1.15}$$

Therefore, using (13.1.13–13.1.15), (13.1.10) can be written as

$$\frac{2H}{\omega_{syn}} \omega_{p.u.}(t) \frac{d^2\delta(t)}{dt^2} = p_{mp.u.}(t) - p_{ep.u.}(t) = p_{ap.u.}(t) \tag{13.1.16}$$

Equation (13.1.16), called the per-unit *swing equation*, is the fundamental equation that determines rotor dynamics in transient stability studies. Note that it is nonlinear due to $p_{ep.u.}(t)$, which is shown in Section 13.2 to be a nonlinear function of δ. Equation (13.1.16) is also nonlinear due to the $\omega_{p.u.}(t)$ term. However, in practice the rotor speed does not vary significantly from synchronous speed during transients. That is, $\omega_{p.u.}(t) \simeq 1.0$, which is often assumed in (13.1.16) for hand calculations.

Equation (13.1.16) is a second-order differential equation that can be rewritten as two first-order differential equations. Differentiating (13.1.4), and then using (13.1.3) and (13.1.12)–(13.1.14), we obtain

$$\frac{d\delta(t)}{dt} = \omega(t) - \omega_{\text{syn}} \tag{13.1.17}$$

Using (13.1.17) in (13.1.16),

$$\frac{2H}{\omega_{\text{syn}}} \omega_{p.u.}(t) \frac{d\omega(t)}{dt} = p_{mp.u.}(t) - p_{ep.u.}(t) = p_{ap.u.}(t) \tag{13.1.18}$$

Equations (13.1.17) and (13.1.18) are two first-order differential equations.

EXAMPLE 13.1

Generator per-unit swing equation and power angle during a short circuit

A three-phase, 60-Hz, 500-MVA, 15-kV, 32-pole hydroelectric generating unit has an H constant of 2.0 p.u.-s. (a) Determine ω_{syn} and ω_{msyn}. (b) Give the per-unit swing equation for this unit. (c) The unit is initially operating at $p_{mp.u.} = p_{ep.u.} = 1.0$, $\omega = \omega_{\text{syn}}$, and $\delta = 10°$ when a three-phase-to-ground bolted short circuit at the generator terminals causes $p_{ep.u.}$ to drop to zero for $t \geqslant 0$. Determine the power angle 3 cycles after the short circuit commences. Assume $p_{mp.u.}$ remains constant at 1.0 per unit. Also assume $\omega_{p.u.}(t) = 1.0$ in the swing equation.

Solution   **a.** For a 60-Hz generator,

$$\omega_{\text{syn}} = 2\pi 60 = 377 \text{ rad/s}$$

and, from (13.1.14), with P = 32 poles,

$$\omega_{msyn} = \frac{2}{P} \omega_{\text{syn}} = \left(\frac{2}{32}\right) 377 = 23.56 \quad \text{rad/s}$$

b. From (13.1.16), with H = 2.0 p.u.-s,

$$\frac{4}{2\pi 60} \omega_{p.u.}(t) \frac{d^2\delta(t)}{dt^2} = p_{mp.u.}(t) - p_{ep.u.}(t)$$

c. The initial power angle is

$$\delta(0) = 10° = 0.1745 \quad \text{radian}$$

Also, from (13.1.17), at $t = 0$,

$$\frac{d\delta(0)}{dt} = 0$$

Using $p_{mp.u.}(t) = 1.0$, $p_{ep.u.} = 0$, and $\omega_{p.u.}(t) = 1.0$, the swing equation from (b) is

$$\left(\frac{4}{2\pi60}\right)\frac{d^2\delta(t)}{dt^2} = 1.0 \quad t \geqslant 0$$

Integrating twice and using the above initial conditions,

$$\frac{d\delta(t)}{dt} = \left(\frac{2\pi60}{4}\right)t + 0$$

$$\delta(t) = \left(\frac{2\pi60}{8}\right)t^2 + 0.1745$$

At $t = 3$ cycles $= \dfrac{3 \text{ cycles}}{60 \text{ cycles/second}} = 0.05$ second,

$$\delta(0.05) = \left(\frac{2\pi60}{8}\right)(0.05)^2 + 0.1745$$

$$= 0.2923 \text{ radian} = 16.75° \qquad\qquad ■$$

EXAMPLE 13.2   **Equivalent swing equation: two generating units**

A power plant has two three-phase, 60-Hz generating units with the following ratings:

       Unit 1: 500 MVA, 15 kV, 0.85 power factor, 32 poles, $H_1 = 2.0$ p.u.-s
       Unit 2: 300 MVA, 15 kV, 0.90 power factor, 16 poles, $H_2 = 2.5$ p.u.-s

(a) Give the per-unit swing equation of each unit on a 100-MVA system base. (b) If the units are assumed to "swing together," that is, $\delta_1(t) = \delta_2(t)$, combine the two swing equations into one equivalent swing equation.

Solution   a. If the per-unit powers on the right-hand side of the swing equation are converted to the system base, then the H constant on the left-hand side must also be converted. That is,

$$H_{new} = H_{old}\frac{S_{old}}{S_{new}} \quad \text{per unit}$$

Converting H_1 from its 500-MVA rating to the 100-MVA system base,

$$H_{1new} = H_{1old}\frac{S_{old}}{S_{new}} = (2.0)\left(\frac{500}{100}\right) = 10 \quad \text{p.u.-s}$$

Similarly, converting H_2,

$$H_{2new} = (2.5)\left(\frac{300}{100}\right) = 7.5 \quad \text{p.u.-s}$$

The per-unit swing equations on the system base are then

$$\frac{2H_{1new}}{\omega_{syn}}\omega_{1p.u.}(t)\frac{d^2\delta_1(t)}{dt^2} = \frac{20.0}{2\pi 60}\omega_{1p.u.}(t)\frac{d^2\delta_1(t)}{dt^2} = p_{m1p.u.}(t) - p_{e1p.u.}(t)$$

$$\frac{2H_{2new}}{\omega_{syn}}\omega_{2p.u.}(t)\frac{d^2\delta_2(t)}{dt^2} = \frac{15.0}{2\pi 60}\omega_{2p.u.}(t)\frac{d^2\delta_2(t)}{dt^2} = p_{m2p.u.}(t) - p_{e2p.u.}(t)$$

b. Letting:

$$\delta(t) = \delta_1(t) = \delta_2(t)$$

$$\omega_{p.u.}(t) = \omega_{1p.u.}(t) = \omega_{2p.u.}(t)$$

$$p_{mp.u.}(t) = p_{m1p.u.}(t) + p_{m2p.u.}(t)$$

$$p_{ep.u.}(t) = p_{e1p.u.}(t) + p_{e2p.u.}(t)$$

and adding the above swing equations

$$\frac{2(H_{1new} + H_{2new})}{\omega_{syn}}\omega_{p.u.}(t)\frac{d^2\delta(t)}{dt^2} = \frac{35.0}{2\pi 60}\omega_{p.u.}(t)\frac{d^2\delta(t)}{dt^2} = p_{mp.u.}(t) - p_{ep.u.}(t)$$

When transient stability studies involving large-scale power systems with many generating units are performed with a digital computer, computation time can be reduced by combining the swing equations of those units that swing together. Such units, which are called *coherent machines*, usually are connected to the same bus or are electrically close, and they are usually remote from network disturbances under study. ∎

SECTION 13.2

SIMPLIFIED SYNCHRONOUS MACHINE MODEL AND SYSTEM EQUIVALENTS

Figure 13.2 shows a simplified model of a synchronous machine, called the classical model, that can be used in transient stability studies. As shown, the synchronous machine is represented by a constant internal voltage E' behind its direct axis transient reactance X_d'. This model is based on the following assumptions:

1. The machine is operating under balanced three-phase positive-sequence conditions.

Figure 13.2

Simplified synchronous machine model for transient stability studies

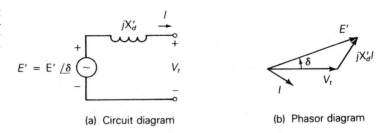

(a) Circuit diagram

(b) Phasor diagram

2. Machine excitation is constant.

3. Machine losses, saturation, and saliency are neglected.

In transient stability programs, more detailed models can be used to represent exciters, losses, saturation, and saliency. However, the simplified model reduces model complexity while maintaining reasonable accuracy in stability calculations.

Each generator in the model is connected to a system consisting of transmission lines, transformers, loads, and other machines. To a first approximation the system can be represented by an "infinite bus" behind a system reactance. An infinite bus is an ideal voltage source that maintains constant voltage magnitude, constant phase, and constant frequency.

Figure 13.3 shows a synchronous generator connected to a system equivalent. The voltage magnitude V_{bus} and $0°$ phase of the infinite bus are constant. The phase angle δ of the internal machine voltage is the machine power angle with respect to the infinite bus.

The equivalent reactance between the machine internal voltage and the infinite bus is $X_{eq} = (X'_d + X)$. From (7.10.3), the real power delivered by the synchronous generator to the infinite bus is

$$p_e = \frac{E'V_{bus}}{X_{eq}} \sin \delta \tag{13.2.1}$$

During transient disturbances both E' and V_{bus} are considered constant in (13.2.1). Thus p_e is a sinusoidal function of the machine power angle δ.

Figure 13.3

Synchronous generator connected to a system equivalent

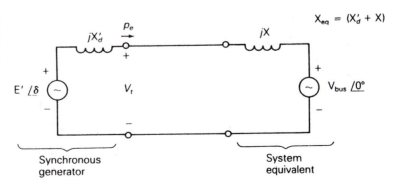

Synchronous generator

System equivalent

| **EXAMPLE 13.3** | **Generator internal voltage and real power output versus power angle** |

Figure 13.4 shows a single-line diagram of a three-phase, 60-Hz synchronous generator, connected through a transformer and parallel transmission lines to an infinite bus. All reactances are given in per-unit on a common system base. If the infinite bus receives 1.0 per unit real power at 0.95 p.f. lagging, determine (a) the internal voltage of the generator and (b) the equation for the electrical power delivered by the generator versus its power angle δ.

Figure 13.4

Single-line diagram for Example 13.3

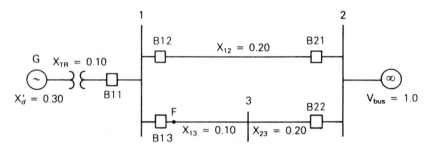

Solution   a. The equivalent circuit is shown in Figure 13.5, from which the equivalent reactance between the machine internal voltage and infinite bus is

$$X_{eq} = X'_d + X_{TR} + X_{12}\|(X_{13} + X_{23})$$

$$= 0.30 + 0.10 + 0.20\|(0.10 + 0.20)$$

$$= 0.520 \quad \text{per unit}$$

Figure 13.5

Equivalent circuit for Example 13.3

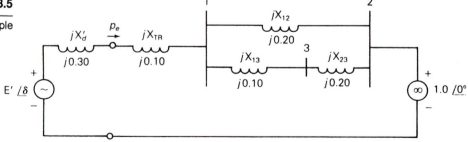

The current into the infinite bus is

$$I = \frac{P}{V_{bus}(\text{p.f.})} \big/\!\!-\cos^{-1}(\text{p.f.}) = \frac{(1.0)}{(1.0)(0.95)} \big/\!\!-\cos^{-1}0.95$$

$$= 1.05263\big/\!-18.195°$$

and the machine internal voltage is

$$E' = E'\underline{/\delta} = V_{bus} + jX_{eq}I$$
$$= 1.0\underline{/0°} + (j0.520)(1.05263\underline{/-18.195°})$$
$$= 1.0\underline{/0°} + 0.54737\underline{/71.805°}$$
$$= 1.1709 + j0.5200$$
$$= 1.2812\underline{/23.946°} \quad \text{per unit}$$

b. From (13.2.1),

$$p_e = \frac{(1.2812)(1.0)}{0.520} \sin \delta = 2.4638 \sin \delta \quad \text{per unit}$$ ∎

SECTION 13.3

THE EQUAL-AREA CRITERION

Consider a synchronous generating unit connected through a reactance to an infinite bus. Plots of electrical power p_e and mechanical power p_m versus power angle δ are shown in Figure 13.6. p_e is a sinusoidal function of δ, as given by (13.2.1).

Figure 13.6

p_e and p_m versus δ

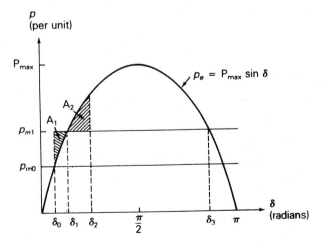

Suppose the unit is initially operating in steady-state at $p_e = p_m = p_{m0}$ and $\delta = \delta_0$, when a step change in p_m from p_{m0} to p_{m1} occurs at $t = 0$. Due to rotor inertia, the rotor position cannot change instantaneously. That is, $\delta_m(0^+) = \delta_m(0^-)$; therefore, $\delta(0^+) = \delta(0^-) = \delta_0$ and $p_e(0^+) = p_e(0^-)$. Since $p_m(0^+) = p_{m1}$ is greater than $p_e(0^+)$, the acceleration power $p_a(0^+)$ is positive and, from (13.1.16), $(d^2\delta)/(dt^2)(0^+)$ is positive. The rotor accelerates and δ

increases. When δ reaches δ_1, $p_e = p_{m1}$ and $(d^2\delta)/(dt^2)$ becomes zero. However, $d\delta/dt$ is still positive and δ continues to increase, overshooting its final steady-state operating point. When δ is greater than δ_1, p_m is less than p_e, p_a is negative, and the rotor decelerates. Eventually, δ reaches a maximum value δ_2 and then swings back toward δ_1. Using (13.1.16), which has no damping, δ would continually oscillate around δ_1. However, damping due to mechanical and electrical losses causes δ to stabilize at its final steady-state operating point δ_1. Note that if the power angle exceeded δ_3, then p_m would exceed p_e and the rotor would accelerate again, causing a further increase in δ and loss of stability.

One method for determining stability and maximum power angle is to solve the nonlinear swing equation via numerical integration techniques using a digital computer. This method, which is applicable to multimachine systems, is described in Section 13.4. However, there is also a direct method for determining stability that does not involve solving the swing equation; this method is applicable for one machine connected to an infinite bus or for two machines. The method, called the *equal-area criterion*, is described in this section.

In Figure 13.6, p_m is greater than p_e during the interval $\delta_0 < \delta < \delta_1$, and the rotor is accelerating. The shaded area A_1 between the p_m and p_e curves is called the accelerating area. During the interval $\delta_1 < \delta < \delta_2$, p_m is less than p_e, the rotor is decelerating, and the shaded area A_2 is the decelerating area. At both the initial value $\delta = \delta_0$ and the maximum value $\delta = \delta_2$, $d\delta/dt = 0$. The equal-area criterion states that $A_1 = A_2$.

In order to derive the equal-area criterion for one machine connected to an infinite bus, assume $\omega_{\text{p.u.}}(t) = 1$ in (13.1.16), giving

$$\frac{2H}{\omega_{\text{syn}}} \frac{d^2\delta}{dt^2} = p_{m\text{p.u.}} - p_{e\text{p.u.}} \tag{13.3.1}$$

Multiplying by $d\delta/dt$ and using

$$\frac{d}{dt}\left[\frac{d\delta}{dt}\right]^2 = 2\left(\frac{d\delta}{dt}\right)\left(\frac{d^2\delta}{dt^2}\right)$$

(13.3.1) becomes

$$\frac{2H}{\omega_{\text{syn}}}\left(\frac{d^2\delta}{dt^2}\right)\left(\frac{d\delta}{dt}\right) = \frac{H}{\omega_{\text{syn}}}\frac{d}{dt}\left[\frac{d\delta}{dt}\right]^2 = (p_{m\text{p.u.}} - p_{e\text{p.u.}})\frac{d\delta}{dt} \tag{13.3.2}$$

Multiplying (13.3.2) by dt and integrating from δ_0 to δ,

$$\frac{H}{\omega_{\text{syn}}} \int_{\delta_0}^{\delta} d\left[\frac{d\delta}{dt}\right]^2 = \int_{\delta_0}^{\delta} (p_{m\text{p.u.}} - p_{e\text{p.u.}})\,d\delta$$

or

$$\frac{H}{\omega_{\text{syn}}}\left[\frac{d\delta}{dt}\right]^2\bigg|_{\delta_0}^{\delta} = \int_{\delta_0}^{\delta} (p_{m\text{p.u.}} - p_{e\text{p.u.}})\,d\delta \tag{13.3.3}$$

The above integration begins at δ_0 where $d\delta/dt = 0$, and continues to an arbitrary δ. When δ reaches its maximum value, denoted δ_2, $d\delta/dt$ again equals zero. Therefore, the left-hand side of (13.3.3) equals zero for $\delta = \delta_2$ and

$$\int_{\delta_0}^{\delta_2} (p_{mp.u.} - p_{ep.u.}) \, d\delta = 0 \tag{13.3.4}$$

Separating this integral into positive (accelerating) and negative (decelerating) areas, we arrive at the equal-area criterion

$$\int_{\delta_0}^{\delta_1} (p_{mp.u.} - p_{ep.u.}) \, d\delta + \int_{\delta_1}^{\delta_2} (p_{mp.u.} - p_{ep.u.}) \, d\delta = 0$$

or

$$\underbrace{\int_{\delta_0}^{\delta_1} (p_{mp.u.} - p_{ep.u.}) \, d\delta}_{A_1} = \underbrace{\int_{\delta_1}^{\delta_2} (p_{ep.u.} - p_{mp.u.}) \, d\delta}_{A_2} \tag{13.3.5}$$

In practice, sudden changes in mechanical power usually do not occur, since the time constants associated with prime mover dynamics are on the order of seconds. However, stability phenomena similar to that described above can also occur from sudden changes in electrical power, due to system faults and line switching. The following three examples are illustrative.

EXAMPLE 13.4

Equal-area criterion: transient stability during a three-phase fault

The synchronous generator shown in Figure 13.4 is initially operating in the steady-state condition given in Example 13.3, when a temporary three-phase-to-ground bolted short circuit occurs on line 1–3 at bus 1, shown as point F in Figure 13.4. Three cycles later the fault extinguishes by itself. Due to a relay misoperation, all circuit breakers remain closed. Determine whether stability is or is not maintained and determine the maximum power angle. The inertia constant of the generating unit is 3.0 per unit-seconds on the system base. Assume p_m remains constant throughout the disturbance. Also assume $\omega_{p.u.}(t) = 1.0$ in the swing equation.

Solution Plots of p_e and p_m versus δ are shown in Figure 13.7. From Example 13.3 the

Figure 13.7

p–δ plot for Example 13.4

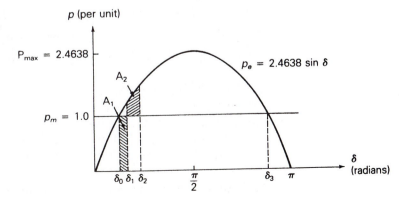

initial operating point is $p_e(0^-) = p_m = 1.0$ per unit and $\delta(0^+) = \delta(0^-) = \delta_0 = 23.95° = 0.4179$ radian. At $t = 0$, when the short circuit occurs, p_e instantaneously drops to zero and remains at zero during the fault since power cannot be transferred past faulted bus 1. From (13.1.16), with $\omega_{p.u.}(t) = 1.0$,

$$\frac{2H}{\omega_{syn}} \frac{d^2\delta(t)}{dt^2} = p_{mp.u.} \qquad 0 \leqslant t \leqslant 0.05 \quad s$$

Integrating twice with initial condition $\delta(0) = \delta_0$ and $\dfrac{d\delta(0)}{dt} = 0$,

$$\frac{d\delta(t)}{dt} = \frac{\omega_{syn} p_{mp.u.}}{2H} t + 0$$

$$\delta(t) = \frac{\omega_{syn} p_{mp.u.}}{4H} t^2 + \delta_0$$

At $t = 3$ cycles $= 0.05$ second,

$$\delta_1 = \delta(0.05 \text{ s}) = \frac{2\pi 60}{12}(0.05)^2 + 0.4179$$

$$= 0.4964 \text{ radian} = 28.44°$$

The accelerating area A_1, shaded in Figure 13.7, is

$$A_1 = \int_{\delta_0}^{\delta_1} p_m d\delta = \int_{\delta_0}^{\delta_1} 1.0 d\delta = (\delta_1 - \delta_2) = 0.4964 - 0.4179 = 0.0785$$

At $t = 0.05$ s the fault extinguishes and p_e instantaneously increases from zero to the sinusoidal curve in Figure 13.7. δ continues to increase until the decelerating area A_2 equals A_1. That is,

$$A_2 = \int_{\delta_1}^{\delta_2} (p_{max} \sin \delta - p_m) d\delta$$

$$= \int_{0.4964}^{\delta_2} (2.4638 \sin \delta - 1.0) d\delta = A_1 = 0.0785$$

Integrating,

$$2.4638[\cos(0.4964) - \cos \delta_2] - (\delta_2 - 0.4964) = 0.0785$$

$$2.4638 \cos \delta_2 + \delta_2 = 2.5843$$

The above nonlinear algebraic equation can be solved iteratively to obtain

$$\delta_2 = 0.7003 \text{ radian} = 40.12°$$

Since the maximum angle δ_2 does not exceed $\delta_3 = (180° - \delta_0) = 156.05°$, stability is maintained. In steady-state, the generator returns to its initial operating point $p_{ess} = p_m = 1.0$ per unit and $\delta_{ss} = \delta_0 = 23.95°$.

Note that as the fault duration increases, the risk of instability also increases. The *critical clearing time*, denoted t_{cr}, is the longest fault duration allowable for stability.   ∎

| EXAMPLE 13.5 | **Equal-area criterion: critical clearing time for a temporary three-phase fault** |

Assuming the temporary short circuit in Example 13.4 lasts longer than 3 cycles, calculate the critical clearing time.

Figure 13.8

p–δ plot for Example 13.5

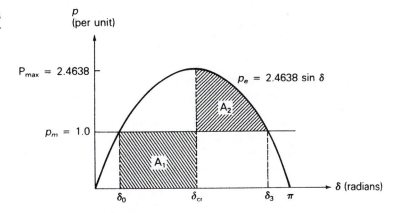

Solution   The p–δ plot is shown in Figure 13.8. At the critical clearing angle, denoted δ_{cr}, the fault is extinguished. The power angle then increases to a maximum value $\delta_3 = 180° - \delta_0 = 156.05° = 2.7236$ radians, which gives the maximum decelerating area. Equating the accelerating and decelerating areas,

$$A_1 = \int_{\delta_0}^{\delta_{cr}} p_m\,d\delta = A_2 = \int_{\delta_{cr}}^{\delta_3} (P_{max} \sin \delta - p_m)\,d\delta$$

$$\int_{0.4179}^{\delta_{cr}} 1.0\,d\delta = \int_{\delta_{cr}}^{2.7236} (2.4638 \sin\,\delta - 1.0)\,d\delta$$

Solving for δ_{cr},

$$(\delta_{cr} - 0.4179) = 2.4638[\cos\,\delta_{cr} - \cos(2.7236)] - (2.7236 - \delta_{cr})$$

$$2.4638 \cos \delta_{cr} = +0.05402$$

$$\delta_{cr} = 1.5489 \text{ radians} = 88.74°$$

From the solution to the swing equation given in Example 13.4,

$$\delta(t) = \frac{\omega_{syn}P_{mp.u.}}{4H}\,t^2 + \delta_0$$

Solving

$$t = \sqrt{\frac{4H}{\omega_{syn}P_{mp.u.}}\,(\delta(t) - \delta_0)}$$

Using $\delta(t_{cr}) = \delta_{cr} = 1.5489$ and $\delta_0 = 0.4179$ radian,

$$t_{cr} = \sqrt{\frac{12}{(2\pi60)(1.0)} (1.5489 - 0.4179)}$$

$$= 0.1897 \text{ s} = 11.38 \text{ cycles}$$

If the fault is cleared before $t = t_{cr} = 11.38$ cycles, stability is maintained. Otherwise, the generator goes out of synchronism with the infinite bus. That is, stability is lost. ∎

| **EXAMPLE 13.6** | **Equal-area criterion: critical clearing angle for a cleared three-phase fault** |

The synchronous generator in Figure 13.4 is initially operating in the steady-state condition given in Example 13.3 when a permanent three-phase-to-ground bolted short circuit occurs on line 1–3 at bus 3. The fault is cleared by opening the circuit breakers at the ends of line 1–3 and line 2–3. These circuit breakers then remain open. Calculate the critical clearing angle. As in previous examples, H = 3.0 p.u.-s, $p_m = 1.0$ per unit and $\omega_{p.u.} = 1.0$ in the swing equation.

Solution   From Example 13.3, the equation for the prefault electrical power, denoted p_{e1} here, is $p_{e1} = 2.4638 \sin \delta$ per unit. The faulted network is shown in Figure 13.9(a), and the Thévenin equivalent of the faulted network, as viewed from the generator internal voltage source, is shown in Figure 13.9(b). The Thévenin reactance is

$$X_{Th} = 0.40 + 0.20\|0.10 = 0.46666 \quad \text{per unit}$$

and the Thévenin voltage source is

$$V_{Th} = 1.0\underline{/0^\circ} \left[\frac{X_{13}}{X_{13} + X_{12}} \right] = 1.0\underline{/0^\circ} \frac{0.10}{0.30}$$

$$= 0.33333\underline{/0^\circ} \quad \text{per unit}$$

From Figure 13.9(b), the equation for the electrical power delivered by the generator to the infinite bus during the fault, denoted p_{e2}, is

$$p_{e2} = \frac{E'V_{Th}}{X_{Th}} \sin \delta = \frac{(1.2812)(0.3333)}{0.46666} \sin \delta = 0.9152 \sin \delta \quad \text{per unit}$$

The postfault network is shown in Figure 13.9(c), where circuit breakers have opened and removed lines 1–3 and 2–3. From this figure, the postfault electrical power delivered, denoted p_{e3}, is

$$p_{e3} = \frac{(1.2812)(1.0)}{0.60} \sin \delta = 2.1353 \sin \delta \quad \text{per unit}$$

The p–δ curves as well as the accelerating area A_1 and decelerating area A_2 corresponding to critical clearing are shown in Figure 13.9(d). Equating A_1

Figure 13.9

Example 13.6

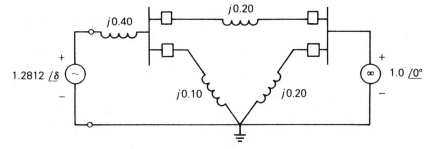

(a) Faulted network

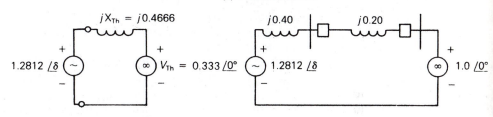

(b) Thévenin equivalent
of the faulted network

(c) Postfault conditions

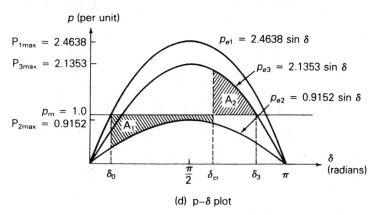

(d) p–δ plot

and A_2,

$$A_1 = \int_{\delta_0}^{\delta_{cr}} (p_m - P_{2max} \sin \delta)\, d\delta = A_2 = \int_{\delta_{cr}}^{\delta_3} (P_{3max} \sin \delta - p_m)\, d\delta$$

$$\int_{0.4179}^{\delta_{cr}} (1.0 - 0.9152 \sin \delta)\, d\delta = \int_{\delta_{cr}}^{2.6542} (2.1353 \sin \delta - 1.0)\, d\delta$$

Solving for δ_{cr},

$$(\delta_{cr} - 0.4179) + 0.9152(\cos \delta_{cr} - \cos 0.4179)$$

$$= 2.1353(\cos \delta_{cr} - \cos 2.6542) - (2.6542 - \delta_{cr})$$

$$-1.2201 \cos \delta_{cr} = 0.4868$$

$$\delta_{cr} = 1.9812 \text{ radians} = 113.5°$$

If the fault is cleared before $\delta = \delta_{cr} = 113.5°$, stability is maintained. Otherwise, stability is lost.   ■

NUMERICAL INTEGRATION OF THE SWING EQUATION

The equal-area criterion is applicable to one machine and an infinite bus or to two machines. For multimachine stability problems, however, numerical integration techniques can be employed to solve the swing equation for each machine.

Given a first-order differential equation

$$\frac{dx}{dt} = f(x) \tag{13.4.1}$$

one relatively simple integration technique is Euler's method [1], illustrated in Figure 13.10. The integration step size is denoted Δt. Calculating the slope at the beginning of the integration interval, from (13.4.1),

$$\frac{dx_t}{dt} = f(x_t) \tag{13.4.2}$$

Figure 13.10

Euler's method

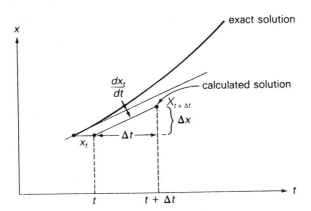

The new value $x_{t+\Delta t}$ is calculated from the old valued x_t by adding the increment Δx,

$$x_{t+\Delta t} = x_t + \Delta x = x_t + \left(\frac{dx_t}{dt}\right)\Delta t \tag{13.4.3}$$

As shown in the figure, Euler's method assumes that the slope is constant over the entire interval Δt. An improvement can be obtained by calculating the slope at both the beginning and end of the interval, and then averaging these slopes. The modified Euler's method is illustrated in Figure 13.11. First, the slope at the beginning of the interval is calculated from

Figure 13.11

Modified Euler's method

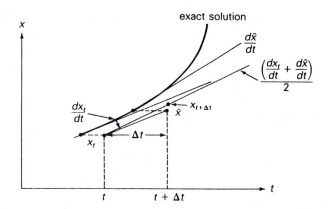

(13.4.1) and used to calculate a preliminary value $\tilde{x}$ given by

$$\tilde{x} = x_t + \left(\frac{dx_t}{dt}\right)\Delta t \tag{13.4.4}$$

Next the slope at $\tilde{x}$ is calculated:

$$\frac{d\tilde{x}}{dt} = f(\tilde{x}) \tag{13.4.5}$$

Then the new value is calculated using the average slope:

$$x_{t+\Delta t} = x_t + \frac{\left(\dfrac{dx_t}{dt} + \dfrac{d\tilde{x}}{dt}\right)}{2}\Delta t \tag{13.4.6}$$

We now apply the modified Euler's method to calculate machine frequency ω and power angle δ. Letting x be either δ or ω, the old values at the beginning of the interval are denoted δ_t and ω_t. From (13.1.17) and (13.1.18), the slopes at the beginning of the interval are

$$\frac{d\delta_t}{dt} = \omega_t - \omega_{\text{syn}} \tag{13.4.7}$$

$$\frac{d\omega_t}{dt} = \frac{p_{\text{ap.u.}t}\,\omega_{\text{syn}}}{2H\omega_{\text{p.u.}t}} \tag{13.4.8}$$

where $p_{\text{ap.u.}t}$ is the per-unit accelerating power calculated at $\delta = \delta_t$, and $\omega_{\text{p.u.}t} = \omega_t/\omega_{\text{syn}}$. Applying (13.4.4), preliminary values are

$$\tilde{\delta} = \delta_t + \left(\frac{d\delta_t}{dt}\right)\Delta t \tag{13.4.9}$$

$$\tilde{\omega} = \omega_t + \left(\frac{d\omega_t}{dt}\right)\Delta t \tag{13.4.10}$$

Next, the slopes at $\tilde{\delta}$ and $\tilde{\omega}$ are calculated, again using (13.1.17) and (13.1.18):

$$\frac{d\tilde{\delta}}{dt} = \tilde{\omega} - \omega_{\text{syn}} \tag{13.4.11}$$

$$\frac{d\tilde{\omega}}{dt} = \frac{\tilde{p}_{ap.u.}\omega_{\text{syn}}}{2H\tilde{\omega}_{\text{p.u.}}} \tag{13.4.12}$$

where $\tilde{p}_{ap.u.}$ is the per-unit accelerating power calculated at $\delta = \tilde{\delta}$, and $\tilde{\omega}_{\text{p.u.}} = \tilde{\omega}/\omega_{\text{syn}}$. Applying (13.4.6), the new values at the end of the interval are

$$\delta_{t+\Delta t} = \delta_t + \frac{\left(\dfrac{d\delta_t}{dt} + \dfrac{d\tilde{\delta}}{dt}\right)}{2}\Delta t \tag{13.4.13}$$

$$\omega_{t+\Delta t} = \omega_t + \frac{\left(\dfrac{d\omega_t}{dt} + \dfrac{d\tilde{\omega}}{dt}\right)}{2}\Delta t \tag{13.4.14}$$

This procedure, given by (13.4.7)–(13.4.13), begins at $t = 0$ with specified initial values δ_0 and ω_0, and continues iteratively until $t = T$, a specified final time. Calculations are best performed using a digital computer.

EXAMPLE 13.7	**Euler's method: computer solution to swing equation and critical clearing time**

Verify the critical clearing angle determined in Example 13.6, and calculate the critical clearing time by applying the modified Euler's method to solve the swing equation for the following two cases:

> **Case 1** The fault is cleared at $\delta = 1.95$ radians $= 112°$
> (which is less than δ_{cr})
>
> **Case 2** The fault is cleared at $\delta = 2.09$ radians $= 120°$
> (which is greater than δ_{cr})

For calculations, use a step size $\Delta t = 0.01$ s, and solve the swing equation from $t = 0$ to $t = T = 0.85$ s.

Solution   Equations (13.4.7)–(13.4.14) are solved by a digital computer program written in BASIC. From Example 13.6, the initial conditions at $t = 0$ are

$$\delta_0 = 0.4179 \quad \text{rad}$$

$$\omega_0 = \omega_{\text{syn}} = 2\pi 60 \quad \text{rad/s}$$

Also, the H constant is 3.0 p.u.-s, and the faulted accelerating power is

$$p_{ap.u.} = 1.0 - 0.9152 \sin\delta$$

The postfault accelerating power is

$$p_{ap.u.} = 1.0 - 2.1353 \sin\delta \quad \text{per unit}$$

The computer program and results at 0.02 s printout intervals are listed in Table 13.1. As shown, these results agree with Example 13.6, since the system

Table 13.1 Computer calculation of swing curves for Example 13.7

CASE 1 STABLE			CASE 2 UNSTABLE			PROGRAM LISTING
TIME s	DELTA rad	OMEGA rad/s	TIME s	DELTA rad	OMEGA rad/s	
0.000	0.418	376.991	0.000	0.418	376.991	
0.020	0.426	377.778	0.020	0.426	377.778	10 REM EXAMPLE 13.7
0.040	0.449	378.547	0.040	0.449	378.547	20 REM SOLUTION TO SWING EQUATION
0.060	0.488	379.283	0.060	0.488	379.283	30 REM THE STEP SIZE IS DELTA
0.080	0.541	379.970	0.080	0.541	379.970	40 REM THE CLEARING ANGLE IS DLTCLR
0.100	0.607	380.599	0.100	0.607	380.599	50 DELTA + .01
0.120	0.685	381.159	0.120	0.685	381.159	60 DLTCLR = 1.95
0.140	0.773	381.646	0.140	0.773	381.646	70 J = 1
0.160	0.870	382.056	0.160	0.870	382.056	80 PMAX = .9152
0.180	0.975	382.392	0.180	0.975	382.392	90 PI = 3.1415927 #
0.200	1.086	382.660	0.200	1.086	382.660	100 T = 0
0.220	1.202	382.868	0.220	1.202	382.868	110 X1 = .4179
0.240	1.321	383.027	0.240	1.321	383.027	120 X2 = 2*PI*60
0.260	1.443	383.153	0.260	1.443	383.153	130 LPRINT " TIME DELTA OMEGA"
0.280	1.567	383.262	0.280	1.567	383.262	140 LPRINT " s rad rad/s"
0.300	1.694	383.370	0.300	1.694	383.370	150 LPRINT USING "# # # # #.# # #"; T;X1;X2
0.320	1.823	383.495	0.320	1.823	383.495	160 FOR K = 1 TO 86
0.340	1.954	383.658	0.340	1.954	383.658	170 REM LINE 180 IS EQ(13.4.7)
FAULT CLEARED			0.360	2.090	383.876	180 X3 = X2 − (2*PI*60)
0.360	2.076	382.516				190 IF J = 2 THEN GOTO 240
0.380	2.176	381.510	FAULT CLEARED			200 IF X1 > DLTCLR OR X1 = DLTCLR THEN PMAX = 2.1353
0.400	2.257	380.638	0.380	2.217	382.915	
0.420	2.322	379.886	0.400	2.327	382.138	210 IF X1 > DLTCLR OR X1 = DLTCLR THEN LPRINT" FAULT CLEARED"
0.440	2.373	379.237	0.420	2.424	381.546	
0.460	2.413	378.674	0.440	2.511	381.135	220 IF X1 > DLTCLR OR X1 = DLTCLR THEN J = 2
0.480	2.441	378.176	0.460	2.591	380.902	
0.500	2.460	377.726	0.480	2.668	380.844	230 REM LINES 240 AND 250 ARE EQ(13.4.8)
0.520	2.471	377.307	0.500	2.746	380.969	240 X4 = 1 − PMAX*SIN(X1)
0.540	2.473	376.900	0.520	2.828	381.288	250 X5 = X4*(2*PI*60)*(2*PI*60)/(6*X2)
0.560	2.467	376.488	0.540	2.919	381.824	260 REM LINE 270 IS EQ(13.4.9)
0.580	2.453	376.056	0.560	3.022	382.609	270 X6 = X1 + X3*DELTA
0.600	2.429	375.583	0.580	3.145	383.686	280 REM LINE 290 IS EQ(13.4.10)
0.620	2.396	375.053	0.600	3.292	385.111	290 X7 = X2 + X5*DELTA
0.640	2.351	374.446	0.620	3.472	386.949	300 REM LINE 310 IS EQ(13.4.11)
0.660	2.294	373.740	0.640	3.693	389.265	310 X8 = X7 − 2*PI*60
0.680	2.221	372.917	0.660	3.965	392.099	320 REM LINES 330 AND 340 ARE EQ(13.4.12)
0.700	2.130	371.960	0.680	4.300	395.426	330 X9 = 1 − PMAX*SIN(X6)
0.720	2.019	370.855	0.700	4.704	399.079	340 X10 = X9*(2*PI*60)*(2*PI*60)/(6*X7)
0.740	1.884	369.604	0.720	5.183	402.689	350 REM LINE 360 IS EQ(13.4.13)
0.760	1.723	368.226	0.740	5.729	405.683	360 X1 = X1 + (X3 + X8)*(DELTA/2)
0.780	1.533	366.773	0.760·	6.325	407.477	370 REM LINE 380 IS EQ(13.4.14)
0.800	1.314	365.341	0.780	6.941	407.812	380 X2 = X2 + (X5 + X10)*(DELTA/2)
0.820	1.068	364.070	0.800	7.551	406.981	390 T = K*DELTA
0.840	0.799	363.143	0.820	8.139	405.711	400 Z = K/2
0.860	0.516	362.750	0.840	8.702	404.819	410 M = INT(Z)
			0.860	9.257	404.934	420 IF M = Z THEN LPRINT USING "# # # # #.# # #";T;X1;X2
						430 NEXT K
						440 END

is stable for Case 1 and unstable for Case 2. Also from Table 13.1, the critical clearing time is between 0.34 and 0.36 s.   ∎

In addition to Euler's method, there are many other numerical integration techniques, such as Runge–Kutta, Picard's method, and Milne's predictor-corrector method [1]. Comparison of the methods shows a trade-off of accuracy versus computation complexity. The Euler method is a relatively simple method to compute, but requires a small step size Δt for accuracy. Some of the other methods can use a larger step size for comparable accuracy, but the computations are more complex.

SECTION 13.5

MULTIMACHINE STABILITY

The numerical integration methods discussed in Section 13.4 can be used to solve the swing equations for a multimachine stability problem. However, a method is required for computing machine output powers for a general network. Figure 13.12 shows a general N-bus power system with M synchronous machines. Each machine is the same as that represented by the simplified model of Figure 13.2, and the internal machine voltages are denoted $E'_1, E'_2, \ldots, E'_M$. The M machine terminals are connected to system buses denoted G1, G2, ..., GM in Figure 13.12. All loads are modeled here as constant admittances. Writing nodal equations for this network,

$$\begin{bmatrix} Y_{11} & Y_{12} \\ Y_{12}^{T} & Y_{22} \end{bmatrix} \begin{bmatrix} V \\ E \end{bmatrix} = \begin{bmatrix} 0 \\ I \end{bmatrix} \tag{13.5.1}$$

Figure 13.12

N-bus power-system representation for transient stability studies

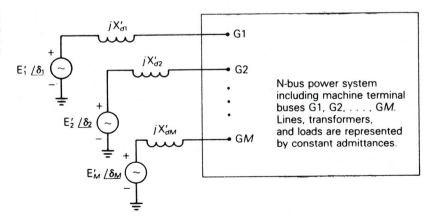

jX'_{d1}

$E'_1 \underline{/\delta_1}$

jX'_{d2}

$E'_2 \underline{/\delta_2}$

jX'_{dM}

$E'_M \underline{/\delta_M}$

G1

G2

GM

N-bus power system including machine terminal buses G1, G2, . . . , GM. Lines, transformers, and loads are represented by constant admittances.

where

$$V = \begin{bmatrix} V_1 \\ V_2 \\ \vdots \\ V_N \end{bmatrix} \quad \text{is the } N \text{ vector of bus voltages} \qquad (13.5.2)$$

$$E = \begin{bmatrix} E'_1 \\ E'_2 \\ \vdots \\ E'_M \end{bmatrix} \quad \text{is the } M \text{ vector of machine voltages} \qquad (13.5.3)$$

$$I = \begin{bmatrix} I_1 \\ I_2 \\ \vdots \\ I_M \end{bmatrix} \quad \begin{array}{l} \text{is the } M \text{ vector of machine currents} \\ \text{(these are current sources)} \end{array} \qquad (13.5.4)$$

$$\begin{bmatrix} Y_{11} & Y_{12} \\ \hline Y_{12}^{\mathrm{T}} & Y_{22} \end{bmatrix} \quad \text{is an } (N + M) \times (N + M) \text{ admittance matrix} \qquad (13.5.5)$$

The admittance matrix in (13.5.5) is partitioned in accordance with the N system buses and M internal machine buses, as follows:

Y_{11} is $N \times N$

Y_{12} is $N \times M$

Y_{22} is $M \times M$

Y_{11} is similar to the bus admittance matrix used for power flows in Chapter 7, except that load admittances and inverted generator impedances are included. That is, if a load is connected to bus n, then that load admittance is added to the diagonal element Y_{11nn}. Also, $(1/jX'_{dn})$ is added to the diagonal element Y_{11GnGn}.

Y_{22} is a diagonal matrix of inverted generator impedances. That is,

$$Y_{22} = \begin{bmatrix} \dfrac{1}{jX'_{d1}} & & & 0 \\ & \dfrac{1}{jX'_{d2}} & & \\ & & \ddots & \\ 0 & & & \dfrac{1}{jX'_{dM}} \end{bmatrix} \qquad (13.5.6)$$

Also, the *km*th element of Y_{12} is

$$Y_{12km} = \begin{cases} \dfrac{-1}{jX'_{dn}} & \text{if } k = Gn \text{ and } m = n \\ 0 & \text{otherwise} \end{cases} \tag{13.5.7}$$

Writing (13.5.1) as two separate equations,

$$Y_{11}V + Y_{12}E = 0 \tag{13.5.8}$$

$$Y_{12}^{\mathrm{T}}V + Y_{22}E = I \tag{13.5.9}$$

Assuming E is known, (13.5.8) is a linear equation in V that can be solved either iteratively or by Gauss elimination. Using the Gauss–Seidel iterative method given by (7.2.9), the kth component of V is

$$V_k(i+1) = \frac{1}{Y_{11kk}}\left[-\sum_{n=1}^{M} Y_{12kn}E_n - \sum_{n=1}^{k-1} Y_{11kn}V_n(i+1) - \sum_{n=k+1}^{N} Y_{11kn}V_n(i) \right] \tag{13.5.10}$$

After V is computed, the machine currents can be obtained from (13.5.9). That is,

$$I = \begin{bmatrix} I_1 \\ I_2 \\ \vdots \\ I_M \end{bmatrix} = Y_{12}^{\mathrm{T}}V + Y_{22}E \tag{13.5.11}$$

The (real) electrical power output of machine n is then

$$p_{en} = \mathrm{Re}[E_n I_n^*] \qquad n = 1, 2, \ldots, M \tag{13.5.12}$$

We are now ready to outline a computation procedure for solving a transient stability problem. The procedure alternately solves the swing equations representing the machines and the above algebraic power-flow equations representing the network. We use the modified Euler method of Section 13.4 to solve the swing equations and the Gauss–Seidel iterative method to solve the power-flow equations. The procedure is now given in the following 11 steps.

Transient stability computation procedure

Step 1   Run a prefault power-flow program to compute initial bus voltages V_k, $k = 1, 2, \ldots, N$, initial machine currents I_n, and initial machine electrical power outputs p_{en}, $n = 1, 2, \ldots, M$. Set machine mechanical power outputs, $p_{mn} = p_{en}$. Set initial machine frequencies, $\omega_n = \omega_{\mathrm{syn}}$. Compute the load admittances.

Step 2   Compute the internal machine voltages:

$$E_n = \mathrm{E}_n\underline{/\delta_n} = V_{Gn} + (jX'_{dn})I_n \quad n = 1, 2, \ldots, M$$

where V_{Gn} and I_n are computed in Step 1. The magnitudes E_n will remain constant throughout the study. The angles δ_n are the initial power angles.

Step 3 Compute Y_{11}. Modify the $(N \times N)$ power-flow bus admittance matrix by including the load admittances and inverted generator impedances.

Step 4 Compute Y_{22} from (13.5.6) and Y_{12} from (13.5.7).

Step 5 Set time $t = 0$.

Step 6 Is there a switching operation, change in load, short circuit, or change in data? For a switching operation or change in load, modify the bus admittance matrix. For a short circuit, set the faulted bus voltage [in (13.5.10)] to zero.

Step 7 Using the internal machine voltages $E_n = E_n \underline{/\delta_n}$, $n = 1$, $2, \ldots, M$, with the values of δ_n at time t, compute the machine electrical powers p_{en} at time t from (13.5.10)–(13.5.12).

Step 8 Using p_{en} computed in Step 7 and the values of δ_n and ω_n at time t, compute the preliminary estimates of power angles $\tilde{\delta}_n$ and machine speeds $\tilde{\omega}_n$ at time $(t + \Delta t)$ from (13.4.7)–(13.4.10).

Step 9 Using $E_n = E_n \underline{/\tilde{\delta}_n}$, $n = 1, 2, \ldots, M$, compute the preliminary estimates of the machine electrical powers $\tilde{p}_{en}$ at time $(t + \Delta t)$ from (13.5.10)–(13.5.12).

Step 10 Using $\tilde{p}_{en}$ computed in Step 9, as well as $\tilde{\delta}_n$ and $\tilde{\omega}_n$ computed in Step 8, compute the final estimates of power angles δ_n and machine speeds ω_n at time $(t + \Delta t)$ from (13.4.11)–(13.4.14).

Step 11 Set time $t = t + \Delta t$. Stop if $t \geq T$. Otherwise, return to Step 6.

EXAMPLE 13.8 **Modifying power-flow Y_{bus} for application to multimachine stability**

Consider a transient stability study for the power system given in Example 7.9, with the 200-Mvar shunt capacitor of Example 7.13 installed at bus 2. Machine transient reactances are $X'_{d1} = 0.20$ and $X'_{d2} = 0.10$ per unit on the system base. Determine the admittance matrices Y_{11}, Y_{22}, and Y_{12}.

Solution From Example 7.9, the power system has $N = 5$ buses and $M = 2$ machines. The second row of the 5×5 bus admittance matrix used for power flows is calculated in Example 7.9. Calculating the other rows in the same manner, we obtain

$$Y_{bus} = \begin{bmatrix} (0.932 - j12.43) & 0 & 0 & 0 & (-0.932 + j12.43) \\ 0 & (0.670 - j7.115) & 0 & (-0.223 + j2.480) & (-0.446 + j4.96) \\ 0 & 0 & (1.865 - j24.86) & (-1.865 + j24.86) & 0 \\ 0 & (-0.223 + j2.480) & (-1.865 + j24.86) & (2.980 - j37.0) & (-0.893 + j9.92) \\ (-0.932 + j12.43) & (-0.446 + j4.960) & 0 & (-0.893 + j9.920) & (2.271 - j27.15) \end{bmatrix} \begin{matrix} \text{per} \\ \text{unit} \end{matrix}$$

To obtain Y_{11}, Y_{bus} is modified by including load admittances and inverted generator impedances. From Table 7.1, the load at bus 3 is $P_{L3} + jQ_{L3} = 0.2 + j0.1$ per unit and the voltage at bus 3 is $V_3 = 1.05$ per unit. Representing this load as a constant admittance,

$$Y_{load3} = \frac{P_{L3} - jQ_{L3}}{V_3^2} = \frac{0.2 - j0.1}{(1.05)^2} = 0.1814 - j0.0907 \quad \text{per unit}$$

Similarly, the load admittance at bus 2 is

$$Y_{load2} = \frac{P_{L2} - jQ_{L2}}{V_2^2} = \frac{2 - j0.7 + j0.5}{(0.968)^2} = 2.134 - j0.213$$

where V_2 is obtained from Example 7.13 and the 200-Mvar (0.5 per unit) shunt capacitor bank is included in the bus 2 load.

The inverted generator impedances are: for machine 1 connected to bus 1,

$$\frac{1}{jX'_{d1}} = \frac{1}{j0.20} = -j5.0 \text{ per unit}$$

and for machine 2 connected to bus 3,

$$\frac{1}{jX'_{d2}} = \frac{1}{j0.10} = -j10.0 \text{ per unit}$$

To obtain Y_{11}, add $(1/jX'_{d1})$ to the first diagonal element of Y_{bus}, add Y_{load2} to the second diagonal element, and add $Y_{load3} + (1/jX'_{d2})$ to the third diagonal element. The 5×5 matrix Y_{11} is then

$$Y_{11} = \begin{bmatrix} (0.932 - j17.43) & 0 & 0 & 0 & (-0.932 + j12.43) \\ 0 & (2.804 - j7.328) & 0 & (-0.223 + j2.480) & (-0.446 + j4.96) \\ 0 & 0 & (2.0465 - j34.951) & (-1.865 + j24.86) & 0 \\ 0 & (-0.223 + j2.480) & (-1.865 + j24.86) & (2.980 - j37.0) & (-0.893 + j9.92) \\ (-0.932 + j12.43) & (-0.446 + j4.960) & 0 & (-0.893 + j9.920) & (2.271 - j27.15) \end{bmatrix} \text{per unit}$$

From (13.5.6), the 2×2 matrix Y_{22} is

$$Y_{22} = \begin{bmatrix} \dfrac{1}{jX'_{d1}} & 0 \\ 0 & \dfrac{1}{jX'_{d2}} \end{bmatrix} = \begin{bmatrix} -j5.0 & 0 \\ 0 & -j10.0 \end{bmatrix} \text{per unit}$$

From Figure 7.2, generator 1 is connected to bus 1 (therefore, bus G1 = 1 and generator 2 is connected to bus 3 (therefore G2 = 3). From (13.5.7), the 5×2 matrix Y_{12} is

$$Y_{12} = \begin{bmatrix} j5.0 & 0 \\ 0 & 0 \\ 0 & j10.0 \\ 0 & 0 \\ 0 & 0 \end{bmatrix} \text{per unit}$$

■

SECTION 13.6

PERSONAL COMPUTER PROGRAM: TRANSIENT STABILITY

The personal computer software package that accompanies this text includes the program entitled "TRANSIENT STABILITY," which computes machine power angles and frequencies in a balanced three-phase power system subjected to disturbances. This program also computes machine angular accelerations, machine electrical power outputs, and bus voltage magnitudes.

Input data for the program include: (1) the bus admittance matrix, initial bus voltages, initial machine currents, and initial machine electrical power outputs, all obtained from the program POWER FLOW described in Chapter 7; and (2) the per-unit inertia constant and direct axis transient reactance of each synchronous machine.

The program executes the transient stability computation procedure given in Section 13.5. The program user selects the type of each disturbance and the time at which each disturbance begins. Disturbance types include: switching operations (opening or closing circuit breakers selected by the program user), three-phase short circuits, changes in loads, and changes in input data. The program user also selects the integration step size Δt and the final time T.

Output data consist of the power angle, frequency, and electrical power output of each machine versus time, as well as bus voltages versus time. The program user selects the outputs to be printed and the print time interval.

EXAMPLE 13.9	**TRANSIENT STABILITY program:** **fault clearing with high-speed reclosure**

Use the program TRANSIENT STABILITY to study a temporary three-phase short circuit at bus 5 of the power system given in Example 7.9. For prefault conditions, use the power-flow output given in Example 7.13, where a 200-Mvar shunt capacitor bank is installed at bus 2. Machine transient reactances are $X'_{d1} = 0.20$ and $X'_{d2} = 0.10$ per unit, as given in Example 13.8, and machine inertia constants are $H_1 = 5.0$ and $H_2 = 50$ p.u.-s (where machine 2 represents a large system). The short circuit is cleared by opening circuit breakers B1, B51, and B52 at $t = 0.05\,s$ (3 cycles), followed by reclosing these circuit breakers. Assume that the temporary fault has already self-extinguished when reclosure occurs.

Run the following two cases:

Case 1 Reclosure at $t = 0.27\,s$ (13 cycles after fault clearing)

Case 2 Reclosure at $t = 0.30\,s$ (15 cycles after fault clearing)

For computation purposes, select an integration step size $\Delta t = 0.01\,s$ and final time $T = 0.75\,s$.

Solution   The power-flow output given in Example 7.13 is selected as the input to the TRANSIENT STABILITY program, along with the machine reactances and inertia constants given above. The first disturbance selected is a short circuit at bus 5 (that is, $V_5 = 0$) at time $t = 0$. The second disturbance selected is the opening of circuit breakers B1, B51, and B52 to clear the fault at $t = 0.05\,\text{s}$. The third disturbance selected is the reclosure of these breakers at $t = 0.27\,\text{s}$ for Case 1 and at $t = 0.30\,\text{s}$ for Case 2. The condition $V_5 = 0$ is also removed.

Machine power angle outputs are listed in Table 13.2 and plotted in Figure 13.13 versus time for both cases. The printout interval is 0.03 s. As shown for Case 1, power angle δ_1 reaches a maximum value and then swings back. With damping included, δ_1 and δ_2 would settle down to new steady-state values.

Case 2 is unstable. The power angle δ_1 of machine 1 exceeds 180 degrees and diverges away from δ_2. That is, machine 1 pulls out of synchronism with machine 2 on the first swing.

Table 13.2   Transient stability output for Example 13.9

CASE (1)					CASE (2)				
TIME	DELTA1	OMEGA1	DELTA2	OMEGA2	TIME	DELTA1	OMEGA1	DELTA2	OMEGA2
SECONDS	DEGREES	RAD/SEC	DEGREES	RAD/SEC	SECONDS	DEGREES	RAD/SEC	DEGREES	RAD/SEC
FAULT AT BUS 5					FAULT AT BUS 5				
0.00	11.16	376.99	6.10	376.99	0.00	11.16	376.99	6.10	376.99
0.03	12.02	377.99	6.19	377.10	0.03	12.02	377.99	6.19	377.10
FAULT CLEARED					FAULT CLEARED				
0.06	14.61	379.02	6.46	377.17	0.06	14.61	379.02	6.46	377.17
0.09	19.03	380.10	6.75	377.15	0.09	19.03	380.10	6.75	377.15
0.12	25.31	381.18	7.02	377.14	0.12	25.31	381.18	7.02	377.14
0.15	33.43	382.26	7.27	377.13	0.15	33.43	382.26	7.27	377.13
0.18	43.41	383.33	7.50	377.12	0.18	43.41	383.33	7.50	377.12
0.21	55.22	384.40	7.72	377.11	0.21	55.22	384.40	7.72	377.11
0.24	68.87	385.47	7.91	377.10	0.24	68.87	385.47	7.91	377.10
RECLOSURE					0.27	84.36	386.53	8.09	377.09
0.27	84.36	386.53	8.09	377.09	RECLOSURE				
0.30	99.39	384.95	8.47	377.34	0.30	101.66	387.59	8.25	377.08
0.33	111.78	383.48	9.29	377.60	0.33	118.65	386.21	8.63	377.34
0.36	121.78	382.17	10.56	377.86	0.36	133.51	385.13	9.44	377.59
0.39	129.66	381.01	12.27	378.11	0.39	146.83	384.43	10.68	377.83
0.42	135.66	379.98	14.41	378.36	0.42	159.27	384.11	12.30	378.03
0.45	139.97	379.02	16.99	378.61	0.45	171.49	384.17	14.25	378.21
0.48	142.66	378.09	19.98	378.86	0.48	184.17	384.64	16.48	378.36
0.51	143.72	377.11	23.39	379.10	0.51	198.01	385.54	18.93	378.46
0.54	143.02	376.04	27.24	379.35	0.54	213.78	386.90	21.51	378.52
0.57	140.37	374.82	31.52	379.61	0.57	232.34	388.78	24.15	378.52
0.60	135.47	373.43	36.24	379.87	0.60	254.56	391.15	26.73	378.45
0.63	128.04	371.87	41.41	380.13	0.63	281.23	393.92	29.14	378.31
0.66	117.82	370.22	47.00	380.37	0.66	312.80	396.75	31.25	378.12
0.69	104.82	368.67	52.99	380.57	0.69	348.90	399.05	33.02	377.93
0.72	89.39	367.48	59.27	380.71	0.72	388.05	400.20	34.52	377.82
0.75	72.47	366.98	65.73	380.76	0.75	427.95	399.94	35.95	377.85

Figure 13.13

Machine power angle swing
curves for Example 13.9

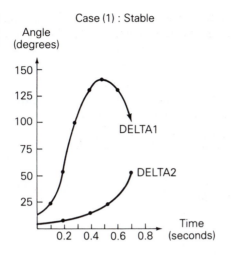

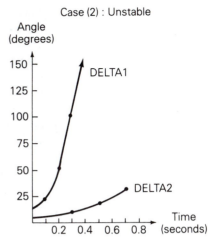

SECTION 13.7

DESIGN METHODS FOR IMPROVING TRANSIENT STABILITY

Design methods for improving power-system transient stability include the following:

1. Improved steady-state stability
   a. Higher system voltage levels
   b. Additional transmission lines
   c. Smaller transmission-line series reactances
   d. Smaller transformer leakage reactances
   e. Series capacitive transmission-line compensation
2. High-speed fault clearing
3. High-speed reclosure of circuit breakers
4. Single-pole switching
5. Larger machine inertia, lower transient rectance
6. Fast responding, high-gain exciters
7. Fast valving
8. Braking resistors

We discuss these design methods in the following paragraphs.

1. Increasing the maximum power transfer in steady-state can also improve transient stability, allowing for increased power transfer through the unfaulted portion of a network during disturbances. Upgrading voltage on

existing transmission or opting for higher voltages on new transmission increases line loadability (6.5.6). Additional parallel lines increase power-transfer capability. Reducing system reactances also increases power-transfer capability. Lines with bundled phase conductors have lower series reactances than lines that are not bundled. Oversized transformers with lower leakage reactances also help. Series capacitors reduce the total series reactances of a line by compensating for the series line inductance.

2. High-speed fault clearing is fundamental to transient stability. Standard practice for EHV systems is 1-cycle relaying and 2-cycle circuit breakers, allowing for fault clearing within 3 cycles (0.05 s). Ongoing research is presently aimed at reducing these to one-half cycle relaying and 1-cycle circuit breakers.

3. The majority of transmission-line short circuits are temporary, with the fault arc self-extinguishing within 5 to 40 cycles (depending on system voltage) after the line is deenergized. High-speed reclosure of circuit breakers can increase postfault transfer power, thereby improving transient stability. Conservative practice for EHV systems is to employ high-speed reclosure only if stability is maintained when reclosing into a permanent fault with subsequent reopening and lockout of breakers.

4. Since the majority of short circuits are single line-to-ground, relaying schemes and independent-pole circuit breakers can be used to clear a faulted phase while keeping the unfaulted phases of a line operating, thereby maintaining some power transfer across the faulted line. Studies have shown that single line-to-ground faults are self-clearing even when only the faulted phase is deenergized. Capacitive coupling between the energized unfaulted phases and the deenergized faulted phase is, in most cases, not strong enough to maintain an arcing short circuit [5].

5. Inspection of the swing equation, (13.1.16), shows that increasing the per-unit inertia constant H of a synchronous machine reduces angular acceleration, thereby slowing down angular swings and increasing critical clearing times. Stability is also improved by reducing machine transient reactances, which increases power-transfer capability during fault or postfault periods [*see* (13.2.1)]. Unfortunately, present-day generator manufacturing trends are toward lower H constants and higher machine reactances, which are a detriment to stability.

6. Modern machine excitation systems with fast thyristor controls and high amplifier gains (to overcome generator saturation) can rapidly increase generator field excitation after sensing low terminal voltage during faults. The effect is to rapidly increase internal machine voltages during faults, thereby increasing generator output power during fault and postfault periods. Critical clearing times are also increased [6].

7. Some steam turbines are equipped with fast valving to divert steam flows and rapidly reduce turbine mechanical power outputs. During faults near the generator, when electrical power output is reduced, fast valving action acts to balance mechanical and electrical power, providing reduced accel-

eration and longer critical clearing times. The turbines are designed to withstand thermal stresses due to fast valving [7].

8. In power systems with generation areas that can be temporarily separated from load areas, braking resistors can improve stability. When separation occurs, the braking resistor is inserted into the generation area for a second or two, preventing or slowing acceleration in the generation area. Shelton et al. [8] describe a 3-GW-s braking resistor.

PROBLEMS

Section 13.1

13.1 A three-phase, 60-Hz, 400-MVA, 13.8-kV, 4-pole steam turbine-generating unit has an H constant of 5.0 p.u.-s. Determine: (a) ω_{syn} and ω_{msyn}; (b) the kinetic energy in joules stored in the rotating masses at synchronous speed; (c) the mechanical angular acceleration α_m and electrical angular acceleration α if the unit is operating at synchronous speed with an accelerating power of 400 MW.

13.2 Calculate J in kg m^2 for the generating unit given in Problem 13.1.

13.3 Generator manufacturers often use the term WR^2, which is the weight in pounds of all the rotating parts of a generating unit (including the prime mover) multiplied by the square of the radius of gyration in feet. WR^2/32.2 is then the total moment of inertia of the rotating parts in slug-ft^2. (a) Determine a formula for the stored kinetic energy in ft-lb of a generating unit in terms of WR^2 and rotor angular velocity ω_m. (b) Show that

$$H = \frac{2.31 \times 10^{-4} WR^2 (rpm)^2}{S_{rated}} \quad \text{per unit-seconds}$$

where S_{rated} is the voltampere rating of the generator and rpm is the synchronous speed in r/min. Noted that 1 ft-lb = 746/550 = 1.356 joules. (c) Evaluate H for a three-phase generating unit rated 800 MVA, 3600 r/min, with WR^2 = 4,000,000 lb-ft^2.

13.4 The generating unit in Problem 13.1 is initially operating at $p_{mp.u.} = p_{ep.u.} = 0.7$ per unit, $\omega = \omega_{syn}$, and $\delta = 12°$ when a fault reduces the generator electrical power output by 70%. Determine the power angle δ five cycles after the fault commences. Assume that the accelerating power remains constant during the fault. Also assume that $\omega_{p.u.}(t) = 1.0$ in the swing equation.

13.5 Repeat Problem 13.4 for a bolted three-phase fault at the generator terminals that reduces the electrical power output to zero. Compare the power angle with that determined in Problem 13.4.

13.6 A third generating unit rated 400 MVA, 15 kV, 0.90 power factor, 16 poles, with $H_3 = 3.5$ p.u.-s is added to the power plant in Example 13.2. Assuming all three units swing together, determine an equivalent swing equation for the three units.

Section 13.2

13.7 The synchronous generator in Figure 13.4 delivers 0.9 per-unit real power at 1.08 per-unit terminal voltage. Determine: (a) the reactive power output of the generator; (b) the generator internal voltage; and (c) an equation for the electrical power delivered by the generator versus power angle δ.

13.8   The generator in Figure 13.4 is initially operating in the steady-state condition given in Problem 13.7 when a three-phase-to-ground bolted short circuit occurs at bus 3. Determine an equation for the electrical power delivered by the generator versus power angle δ during the fault.

Section 13.3

13.9   The generator in Figure 13.4 is initially operating in the steady-state condition given in Example 13.3 when circuit breaker B12 inadvertently opens. Use the equal-area criterion to calculate the maximum value of the generator power angle δ. Assume $\omega_{p.u.}(t) = 1.0$ in the swing equation.

13.10   The generator in Figure 13.4 is initially operating in the steady-state condition given in Example 13.3 when a temporary three-phase-to-ground short circuit occurs at point F. Three cycles later, circuit breakers B13 and B22 permanently open to clear the fault. Use the equal-area criterion to determine the maximum value of the power angle δ.

13.11   If breakers B13 and B22 in Problem 13.10 open later than 3 cycles after the fault commences, determine the critical clearing time.

13.12   Rework Problem 13.10 if circuit breakers B13 and B22 open after 3 cycles and then reclose when the power angle reaches 35°. Assume that the temporary fault has already self-extinguished when the breakers reclose.

13.13   The generator in Figure 13.4 is initially operating in the steady-state condition given in Problem 13.7 when circuit breaker B12 inadvertently opens. Use the equal-area criterion to calculate the maximum value of the generator power angle δ. Assume $\omega_{p.u.}(t) = 1.0$ in the swing equation.

13.14   The generator in Figure 13.4 is initially operating in the steady-state condition given in Problem 13.7 when a temporary three-phase-to-ground short circuit occurs at point F. Three cycles later, circuit breakers B13 and B22 permanently open to clear the fault. Use the equal-area criterion to calculate the maximum value of the generator power angle δ. Assume $\omega_{p.u.}(t) = 1.0$ in the swing equation.

13.15   If breakers B13 and B22 in Problem 13.14 open later than three cycles after the fault commences, determine the critical clearing time.

Section 13.4

13.16   Verify the maximum power angle determined in Problem 13.9 by applying the modified Euler's method to numerically integrate the swing equation. Write and run a computer program.

13.17   Investigate the effect of generating-unit damping torque on the maximum power angle in Problem 13.16. Damping, which is caused by friction and windage, can be represented by subtracting from the per-unit accelerating power $p_{ap.u.}(t)$ [used in (13.4.8) and (13.4.12)] the term $B\omega_{p.u.}(t)$, where B is a per-unit damping coefficient. Compare the maximum power angle using $B = 0.01$ per unit with that computed in Problem 13.16. Discuss the effect of generating-unit damping torques on stability.

13.18   Verify the critical clearing time determined in Problem 13.11 by applying the modified Euler's method. Write and run a computer program.

13.19   In Problem 13.12, assume that the circuit breakers open at $t = 3$ cycles and then reclose at $t = 24$ cycles (instead of when δ reaches 35°). Determine the maximum power angle by applying the modified Euler method. Write and run a computer program.

Section 13.5

13.20   Consider the six-bus power system shown in Figure 13.14, where all data is given in per-unit on a common system base. All resistances as well as transmission-line capacitances are neglected. (a) Determine the 6 × 6 per-unit bus admittance matrix Y_{bus} suitable for a power-flow computer program. (b) Determine the per-unit admittance matrices Y_{11}, Y_{12}, and Y_{22} given in (13.5.5), which are suitable for a transient stability study.

Figure 13.14

Single-line diagram of a six-bus power system (per-unit values are shown)

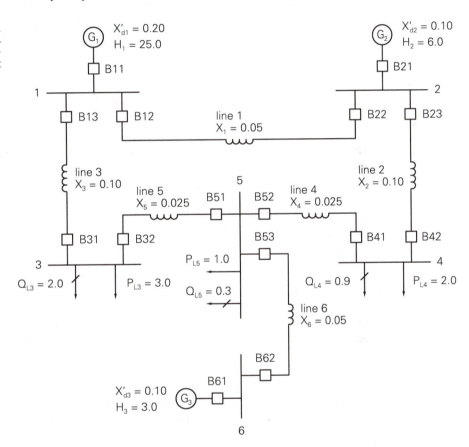

13.21   Modify the matrices Y_{11}, Y_{12}, and Y_{22} determined in Problem 13.20 for (a) the case when circuit breakers B12 and B22 open to remove line 1–2; and (b) the case when the load $P_{L4} + jQ_{L4}$ is removed.

Sections 13.6 and 13.7

13.22   Run the program TRANSIENT STABILITY for Example 13.9. Verify that Case 1 is stable and Case 2 is unstable.

13.23   Investigate the effect of varying the transient reactance X'_{d1} of machine 1 in Example 13.9. Run the program TRANSIENT STABILITY for (a) $X'_{d1} = 0.1$ and (b) $X'_{d1} = 0.4$ per unit. Discuss the effect of X'_{d1} on stability.

13.24   Investigate the effect of varying the inertia constant H in Example 13.9. Run the program TRANSIENT STABILITY for (a) $H_1 = 3.0$ and (b) $H_1 = 8.0$ p.u.-s. Discuss the effect of H_1 on stability.

13.25   Investigate the effect of varying the fault-clearing time in Example 13.9. Run the program TRANSIENT STABILITY, with the circuit breakers opening at (a) $t = 0.025$ s (1.5 cycles) and (b) $t = 0.075$ s (4.5 cycles), instead of at $t = 0.05$ s. In both cases, assume that reclosure occurs 13 cycles after fault clearing. Discuss the effect of fault-clearing time on stability.

13.26   Rework Example 13.9 with an additional line installed between buses 4 and 5. The parameters of the added line equal those of existing line 4–5. First run the program POWER FLOW with the additional line to determine prefault conditions. Then, using the POWER FLOW output as prefault conditions, run the program TRANSIENT STABILITY for both cases given in Example 13.9. Discuss the effect of the additional line on stability.

13.27   Figure 13.15 shows a single-line diagram of the same circuit given in Example 13.3, except that bus numbers 1 and 2 are interchanged and a new bus 4 is defined for the generator. All reactances are given in per-unit on a common system base. The infinite bus voltage is $1.0\underline{/0°}$ per unit, and the generator initially delivers 1.0 per unit real power at 1.05 per unit terminal voltage, as in Problem 13.7. Using the POWER FLOW program, create the bus input data with bus 1 as the swing bus, buses 2 and 3 as load buses, and bus 4 as a constant voltage magnitude bus. Also create the line input and transformer input data files. Then run the POWER FLOW program and verify the generator reactive power output determined in Problem 13.7.

Figure 13.15

Circuit for Problems 13.27 and 13.28

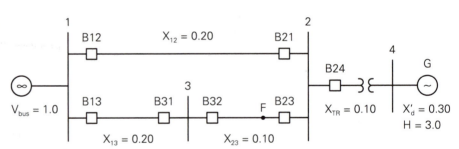

13.28   Using the POWER FLOW input/output data from Problem 13.27, run the TRANSIENT STABILITY program to rework Problem 13.14. First create the machine input data with $H = 100$ per unit-s and $X'_d = 0.01$ per unit, approximating the infinite bus. For the generator, $H = 3.0$ per unit-s and $X'_d = 0.30$ per unit. Then run the TRANSIENT STABILITY program. Select an integration time interval $\Delta t = 0.01$ seconds and a final time $T = 1.5$ s. Also select two disturbance times, one at $t = 0$ and one at $t = 0.05$ s (3 cycles). At the first disturbance time, a three-phase fault occurs at bus 2 (point F in Figure 13.13). At the second disturbance time, the breakers for line 2–3 (at bus 2 and bus 3) permanently open to clear the fault. Verify the maximum generator power angle determined in Problem 13.14. Discuss any difference between the result from the TRANSIENT STABILITY program and that from the equal-area criterion.

13.29   Run the TRANSIENT STABILITY program to verify the critical clearing time determined in Problem 13.15. As in Problem 13.28, select two disturbance times, with the first at $t = 0$. Select the following two cases for the second disturbance time: Case (1) 0.01 s before the critical clearing time determined in Problem 13.15; Case (2) 0.01 s after the critical clearing time. Plot the infinite bus and machine power angles δ_1 and δ_2 versus time for both cases. Show that Case (1) is stable and Case (2) is unstable, thereby verifying the critical clearing time within ± 0.01 s. Discuss any difference between the critical clearing time determined from the TRANSIENT STABILITY program and that from the equal-area criterion.

13.30 Use the program TRANSIENT STABILITY to study a three-phase short circuit at bus 2 of the power system given in Figure 13.13. First run the program POWER FLOW with bus 1 as the swing bus, where $V_1 = 1.0$ per unit and $\delta_1 = 0°$, and with buses 2 and 6 as constant voltage magnitude buses, where $V_6 = V_2 = 1.0$ per unit, $P_{G2} = 2.0$ per unit, and $P_{G3} = 3.0$ per unit. Using the POWER FLOW output as prefault conditions, run the program TRANSIENT STABILITY. Assume that the short circuit at bus 2 is cleared by permanently opening breakers B21, B22, and B23. Determine the critical clearing time.

CASE STUDY QUESTIONS

A. How have electric utility companies changed their power-system planning practices since the 1965 Northeast blackout?

B. Since 1965, has the U.S. electric power system become less prone to blackouts or more prone?

C. Will there be another blackout as large in scale as the 1965 Northeast blackout?

References

1. G. W. Stagg and A. H. El-Abiad, *Computer Methods in Power Systems* (New York: McGraw-Hill, 1968).

2. O. I. Elgerd, *Electric Energy Systems Theory*, 2d ed. (New York: McGraw-Hill, 1982).

3. C. A. Gross, *Power System Analysis* (New York: Wiley, 1979).

4. W. D. Stevenson, Jr., *Elements of Power System Analysis*, 4th ed. (New York: McGraw-Hill, 1982).

5. E. W. Kimbark, "Suppression of Ground-Fault Arcs on Single-Pole Switched EHV Lines by Shunt Reactors," *IEEE Trans PAS, 83* March 1964, pp. 285–290.

6. K. R. McClymont et al., "Experience with High-Speed Rectifier Excitation Systems," *IEEE Trans PAS*, vol. PAS-87 (June 1986), pp. 1464–1470.

7. E. W. Cushing et al., "Fast Valving as an Aid to Power System Transient Stability and Prompt Resynchronization and Rapid Reload after Full Load Rejection," *IEEE Trans PAS*, vol. PAS-90 (November/December 1971), pp. 2517–2527.

8. M. L. Shelton et al., "Bonneville Power Administration 1400 MW Braking Resistor," *IEEE Trans PAS*, vol. PAS-94 (March/April 1975), pp. 602–611.

9. R. McCaw, "The Great Blackout," *Power Engineering* (December 1965), pp. 36A–36C.

APPENDIX

CONSTANT (UNITS)	TYPE	SYMBOL	TURBO-GENERATOR (SOLID ROTOR)	WATER-WHEEL GENERATOR (WITH DAMPERS)	SYNCHRO-NOUS CONDENSER	SYNCHRO-NOUS MOTOR
Reactances (per unit)	Synchronous	X_d	1.1	1.15	1.80	1.20
		X_q	1.08	0.75	1.15	0.90
	Transient	X_d'	0.23	0.37	0.40	0.35
		X_q'	0.23	0.75	1.15	0.90
	Subtransient	X_d''	0.12	0.24	0.25	0.30
		X_q''	0.15	0.34	0.30	0.40
	Negative-sequence	X_2	0.13	0.29	0.27	0.35
	Zero-sequence	X_0	0.05	0.11	0.09	0.16
Resistances (per unit)	Positive-sequence	R (dc)	0.003	0.012	0.008	0.01
		R (ac)	0.005	0.012	0.008	0.01
	Negative-sequence	R_2	0.035	0.10	0.05	0.06
Time constants (seconds)	Transient	T_{d0}'	5.6	5.6	9.0	6.0
		T_d'	1.1	1.8	2.0	1.4
	Subtransient	$T_d'' = T_q''$	0.035	0.035	0.035	0.036
	Armature	T_a	0.16	0.15	0.17	0.15

Adapted from E. W. Kimbark, *Power-System Stability: Synchronous Machines* (New York: Dover Publications, 1956/1968), Chap. 12.

Table A.1   Typical average values of synchronous-machine constants

Table A.2

Typical transformer leakage reactances

RATING OF HIGHEST VOLTAGE WINDING	BIL OF HIGHEST VOLTAGE WINDING	LEAKAGE REACTANCE	
kV	kV	per unit*	
DISTRIBUTION TRANSFORMERS			
2.4	30	0.023–0.049	
4.8	60	0.023–0.049	
7.2	75	0.026–0.051	
12	95	0.026–0.051	
23	150	0.052–0.055	
34.5	200	0.052–0.055	
46	250	0.057–0.063	
69	350	0.065–0.067	
POWER TRANSFORMERS 10 MVA AND BELOW			
8.7	110	0.050–0.058	
25	150	0.055–0.058	
34.5	200	0.060–0.065	
46	250	0.060–0.070	
69	350	0.070–0.075	
92	450	0.070–0.085	
115	550	0.075–0.100	
138	650	0.080–0.105	
161	750	0.085–0.011	
POWER TRANSFORMERS ABOVE 10 MVA			
		SELF-COOLED OR FORCED-AIR-COOLED	FORCED-OIL-COOLED
8.7	110	0.050–0.063	0.082–0.105
34.5	200	0.055–0.075	0.090–0.128
46	250	0.057–0.085	0.095–0.143
69	350	0.063–0.095	0.103–0.158
92	450	0.060–0.118	0.105–0.180
115	550	0.065–0.135	0.107–0.195
138	650	0.070–0.140	0.117–0.245
161	750	0.075–0.150	0.125–0.250
230	900	0.070–0.160	0.120–0.270
345	1300	0.080–0.170	0.130–0.280
500	1550	0.100–0.200	0.160–0.340
765		0.110–0.210	0.190–0.350

*Per-unit reactances are based on the transformer rating

Table A.3 Characteristics of copper conductors, hard drawn, 97.3% conductivity

Column groups — r_a Resistance (Ohms per Conductor per Mile) at 25°C (77°F) and 50°C (122°F); x_a Inductive Reactance (Ohms per Conductor per Mile at 1 Ft Spacing); x'_a Shunt Capacitive Reactance (Megohms per Conductor per Mile at 1 Ft Spacing).

Circular Mils	A.W.G. or B.&S.	No. of Strands	Diam. of Individual Strands (in)	Outside Diam. (in)	Breaking Strength (lb)	Weight (lb per Mile)	Approx. Current Capacity* (A)	GMR at 60 Hz (ft)	r 25°C dc	r 25°C 25Hz	r 25°C 50Hz	r 25°C 60Hz	r 50°C dc	r 50°C 25Hz	r 50°C 50Hz	r 50°C 60Hz	x_a 25Hz	x_a 50Hz	x_a 60Hz	x'_a 25Hz	x'_a 50Hz	x'_a 60Hz
1 000 000	....	37	0.1644	1.151	43 830	16 300	1 300	0.0368	0.0585	0.0594	0.0620	0.0634	0.0640	0.0648	0.0672	0.0685	0.1666	0.333	0.400	0.216	0.1081	0.0901
900 000	....	37	0.1560	1.092	39 510	14 670	1 220	0.0349	0.0650	0.0658	0.0682	0.0695	0.0711	0.0718	0.0740	0.0752	0.1693	0.339	0.406	0.220	0.1100	0.0916
800 000	....	37	0.1470	1.029	35 120	13 040	1 130	0.0329	0.0731	0.0739	0.0760	0.0772	0.0800	0.0806	0.0826	0.0837	0.1722	0.344	0.413	0.224	0.1121	0.0934
750 000	....	37	0.1424	0.997	33 400	12 230	1 090	0.0319	0.0780	0.0787	0.0807	0.0818	0.0853	0.0859	0.0878	0.0888	0.1739	0.348	0.417	0.226	0.1132	0.0943
700 000	....	37	0.1375	0.963	31 170	11 410	1 040	0.0308	0.0836	0.0842	0.0861	0.0871	0.0914	0.0920	0.0937	0.0947	0.1759	0.352	0.422	0.229	0.1145	0.0954
600 000	....	37	0.1273	0.891	27 020	9 781	940	0.0285	0.0975	0.0981	0.0997	0.1006	0.1066	0.1071	0.1086	0.1095	0.1799	0.360	0.432	0.235	0.1173	0.0977
500 000	....	37	0.1162	0.814	22 510	8 151	840	0.0260	0.1170	0.1175	0.1188	0.1196	0.1280	0.1283	0.1296	0.1303	0.1845	0.369	0.443	0.241	0.1205	0.1004
500 000	....	19	0.1622	0.811	21 590	8 151	840	0.0256	0.1170	0.1175	0.1188	0.1196	0.1280	0.1283	0.1296	0.1303	0.1853	0.371	0.445	0.241	0.1206	0.1005
450 000	....	19	0.1539	0.770	19 750	7 336	780	0.0243	0.1300	0.1304	0.1316	0.1323	0.1422	0.1426	0.1437	0.1443	0.1879	0.376	0.451	0.245	0.1224	0.1020
400 000	....	19	0.1451	0.726	17 560	6 521	730	0.0229	0.1462	0.1466	0.1477	0.1484	0.1600	0.1603	0.1613	0.1619	0.1909	0.382	0.458	0.249	0.1245	0.1038
350 000	....	19	0.1357	0.679	15 590	5 706	670	0.0214	0.1671	0.1675	0.1684	0.1690	0.1828	0.1831	0.1840	0.1845	0.1943	0.388	0.466	0.254	0.1269	0.1058
350 000	....	12	0.1708	0.710	15 140	5 706	670	0.0225	0.1671	0.1675	0.1684	0.1690	0.1828	0.1831	0.1840	0.1845	0.1918	0.384	0.460	0.251	0.1253	0.1044
300 000	....	19	0.1257	0.629	13 510	4 891	610	0.01987	0.1950	0.1953	0.1961	0.1966	0.213	0.214	0.214	0.215	0.1982	0.396	0.476	0.259	0.1296	0.1080
300 000	....	12	0.1581	0.657	13 170	4 891	610	0.0208	0.1950	0.1953	0.1961	0.1966	0.213	0.214	0.214	0.215	0.1957	0.392	0.470	0.256	0.1281	0.1068
250 000	....	19	0.1147	0.574	11 360	4 076	540	0.01813	0.234	0.234	0.235	0.235	0.256	0.256	0.257	0.257	0.203	0.406	0.487	0.266	0.1329	0.1108
250 000	....	12	0.1443	0.600	11 130	4 076	540	0.01902	0.234	0.234	0.235	0.235	0.256	0.256	0.257	0.257	0.200	0.401	0.481	0.263	0.1313	0.1094
211 600	4/0	19	0.1055	0.528	9 617	3 450	480	0.01668	0.276	0.277	0.277	0.278	0.302	0.303	0.303	0.303	0.207	0.414	0.497	0.272	0.1359	0.1132
211 600	4/0	12	0.1328	0.552	9 483	3 450	490	0.01750	0.276	0.277	0.277	0.278	0.302	0.303	0.303	0.303	0.205	0.409	0.491	0.269	0.1343	0.1119
211 600	4/0	7	0.1739	0.522	9 154	3 450	480	0.01579	0.276	0.277	0.277	0.278	0.302	0.303	0.303	0.303	0.210	0.420	0.503	0.273	0.1363	0.1136
167 800	3/0	12	0.1183	0.492	7 556	2 736	420	0.01559	0.349	0.349	0.349	0.350	0.381	0.381	0.382	0.382	0.210	0.421	0.505	0.277	0.1384	0.1153
167 800	3/0	7	0.1548	0.464	7 366	2 736	420	0.01404	0.349	0.349	0.349	0.350	0.381	0.381	0.382	0.382	0.216	0.431	0.518	0.281	0.1405	0.1171
133 100	2/0	7	0.1379	0.414	5 926	2 170	360	0.01252	0.440	0.440	0.440	0.440	0.481	0.481	0.481	0.481	0.222	0.443	0.532	0.289	0.1445	0.1205
105 500	1/0	7	0.1228	0.368	4 752	1 720	310	0.01113	0.555	0.555	0.555	0.555	0.606	0.607	0.607	0.607	0.227	0.455	0.546	0.298	0.1488	0.1240
83 690	1	7	0.1093	0.328	3 804	1 364	270	0.00992	0.699	0.699	0.699	0.699	0.765	Same as dc	Same as dc	Same as dc	0.233	0.467	0.560	0.306	0.1528	0.1274
83 690	1	3	0.1670	0.360	3 620	1 351	270	0.01016	0.692	0.692	0.692	0.692	0.757	Same as dc	Same as dc	Same as dc	0.232	0.464	0.557	0.299	0.1495	0.1246
66 370	2	7	0.0974	0.292	3 045	1 082	230	0.00883	0.881	0.882	0.882	0.882	0.964	Same as dc	Same as dc	Same as dc	0.239	0.478	0.574	0.314	0.1570	0.1308
66 370	2	3	0.1487	0.320	2 913	1 071	240	0.00903	0.873	Same as dc	Same as dc	Same as dc	0.955	Same as dc	Same as dc	Same as dc	0.238	0.476	0.571	0.307	0.1537	0.1281
66 370	2	1	....	0.258	3 003	1 061	220	0.00836	0.864	Same as dc	Same as dc	Same as dc	0.945	Same as dc	Same as dc	Same as dc	0.242	0.484	0.581	0.323	0.1614	0.1345
52 630	3	7	0.0867	0.260	2 433	858	200	0.00787	1.112	Same as dc	Same as dc	Same as dc	1.216	Same as dc	Same as dc	Same as dc	0.245	0.490	0.588	0.322	0.1611	0.1343
52 630	3	3	0.1325	0.285	2 359	850	200	0.00805	1.101	Same as dc	Same as dc	Same as dc	1.204	Same as dc	Same as dc	Same as dc	0.244	0.488	0.585	0.316	0.1578	0.1315
52 630	3	1	....	0.229	2 439	841	190	0.00745	1.090	Same as dc	Same as dc	Same as dc	1.192	Same as dc	Same as dc	Same as dc	0.248	0.496	0.595	0.331	0.1656	0.1380
41 740	4	3	0.1180	0.254	1 879	674	180	0.00717	1.388	Same as dc	Same as dc	Same as dc	1.518	Same as dc	Same as dc	Same as dc	0.250	0.499	0.599	0.324	0.1619	0.1349
41 740	4	1	....	0.204	1 970	667	170	0.00663	1.374	Same as dc	Same as dc	Same as dc	1.503	Same as dc	Same as dc	Same as dc	0.254	0.507	0.609	0.339	0.1697	0.1415
33 100	5	3	0.1050	0.226	1 505	534	150	0.00638	1.750	Same as dc	Same as dc	Same as dc	1.914	Same as dc	Same as dc	Same as dc	0.256	0.511	0.613	0.332	0.1661	0.1384
33 100	5	1	....	0.1819	1 591	529	140	0.00590	1.733	Same as dc	Same as dc	Same as dc	1.895	Same as dc	Same as dc	Same as dc	0.260	0.519	0.623	0.348	0.1738	0.1449
26 250	6	3	0.0935	0.201	1 205	424	130	0.00568	2.21	Same as dc	Same as dc	Same as dc	2.41	Same as dc	Same as dc	Same as dc	0.262	0.523	0.628	0.341	0.1703	0.1419
26 250	6	1	....	0.1620	1 280	420	120	0.00526	2.18	Same as dc	Same as dc	Same as dc	2.39	Same as dc	Same as dc	Same as dc	0.265	0.531	0.637	0.356	0.1779	0.1483
20 820	7	1	....	0.1443	1 030	333	110	0.00468	2.75	Same as dc	Same as dc	Same as dc	3.01	Same as dc	Same as dc	Same as dc	0.271	0.542	0.651	0.364	0.1821	0.1517
16 510	8	1	....	0.1285	826	264	90	0.00417	3.47	Same as dc	Same as dc	Same as dc	3.80	Same as dc	Same as dc	Same as dc	0.277	0.554	0.665	0.372	0.1862	0.1552

*For conductor at 75°C, air at 25°C, wind 1.4 miles per hour (2 ft/sec), frequency = 60Hz.

Table A.4 Characteristics of aluminum cable, steel, reinforced (Aluminum Company of America)—ACSR

Code Word	Circular Mils Aluminum	Al Strands	Al Strand Diameter (Inches)	Steel Strands	Steel Strand Diameter (Inches)	Outside Diameter (Inches)	Copper Equivalent* Circular Mils or A.W.G.	Ultimate Strength (Pounds)	Weight (Pounds per Mile)	Geometric Mean Radius at 60 Hz (Feet)	Approx. Current Carrying Capacity (Amps)	r_a 25°C dc	r_a 25°C 25 Hz	r_a 25°C 50 Hz	r_a 25°C 60 Hz	r_a 50°C† dc	r_a 50°C† 25 Hz	r_a 50°C† 50 Hz	r_a 50°C† 60 Hz	x_a 60 Hz	x'_a 60 Hz
Joree	2 515 000	76	0.1819	19	0.0849	1.880		61 700		0.0621									0.0450	0.337	0.0755
Thrasher	2 312 000	76	0.1744	19	0.0814	1.802		57 300		0.0595									0.0482	0.342	0.0767
Kiwi	2 167 000	72	0.1735	7	0.1157	1.735		49 800		0.0570									0.0511	0.348	0.0778
Bluebird	2 156 000	84	0.1602	19	0.0961	1.762		60 300		0.0588									0.0505	0.344	0.0774
Chukar	1 781 000	84	0.1456	19	0.0874	1.602		51 000		0.0534									0.0598	0.355	0.0802
Falcon	1 590 000	54	0.1716	19	0.1030	1.545	1 000 000	56 000	10 777	0.0520	1 380	0.0587	0.0588	0.0590	0.0591	0.0646	0.0656	0.0675	0.0684	0.359	0.0814
Parrot	1 510 500	54	0.1673	19	0.1004	1.506	950 000	53 200	10 237	0.0507	1 340	0.0618	0.0619	0.0621	0.0622	0.0680	0.0690	0.0710	0.0720	0.362	0.0821
Plover	1 431 000	54	0.1628	19	0.0977	1.465	900 000	50 400	9 699	0.0493	1 300	0.0652	0.0653	0.0655	0.0656	0.0718	0.0729	0.0749	0.0760	0.365	0.0830
Martin	1 351 000	54	0.1582	19	0.0949	1.424	850 000	47 600	9 160	0.0479	1 250	0.0691	0.0692	0.0694	0.0695	0.0761	0.0771	0.0792	0.0803	0.369	0.0838
Pheasant	1 272 000	54	0.1535	19	0.0921	1.382	800 000	44 800	8 621	0.0465	1 200	0.0734	0.0735	0.0737	0.0738	0.0808	0.0819	0.0840	0.0851	0.372	0.0847
Grackle	1 192 500	54	0.1486	19	0.0892	1.338	750 000	43 100	8 082	0.0450	1 160	0.0783	0.0784	0.0786	0.0788	0.0862	0.0872	0.0894	0.0906	0.376	0.0857
Finch	1 113 000	54	0.1436	19	0.0862	1.293	700 000	40 200	7 544	0.0435	1 110	0.0839	0.0840	0.0842	0.0844	0.0924	0.0935	0.0957	0.0969	0.380	0.0867
Curlew	1 033 500	54	0.1384	7	0.1384	1.246	650 000	37 100	7 019	0.0420	1 060	0.0903	0.0905	0.0907	0.0909	0.0994	0.1005	0.1025	0.1035	0.385	0.0878
Cardinal	954 000	54	0.1329	7	0.1329	1.196	600 000	34 200	6 479	0.0403	1 010	0.0979	0.0980	0.0981	0.0982	0.1078	0.1088	0.1118	0.1128	0.390	0.0890
Canary	900 000	54	0.1291	7	0.1291	1.162	566 000	32 300	6 112	0.0391	970	0.104	0.104	0.104	0.104	0.1145	0.1155	0.1175	0.1185	0.393	0.0898
Crane	874 500	54	0.1273	7	0.1273	1.146	550 000	31 400	5 940	0.0386	950	0.107	0.107	0.107	0.108	0.1178	0.1188	0.1218	0.1228	0.395	0.0903
Condor	795 000	54	0.1214	7	0.1214	1.093	500 000	28 500	5 399	0.0368	900	0.117	0.117	0.118	0.119	0.1288	0.1308	0.1358	0.1378	0.401	0.0917
Drake	795 000	26	0.1749	7	0.1360	1.108	500 000	31 200	5 770	0.0375	900	0.117	0.117	0.117	0.117	0.1288	0.1288	0.1288	0.1288	0.399	0.0912
Mallard	795 000	30	0.1628	19	0.0977	1.140	500 000	38 400	6 517	0.0393	910	0.117	0.117	0.117	0.117	0.1288	0.1288	0.1288	0.1288	0.393	0.0904
Crow	715 500	54	0.1151	7	0.1151	1.036	450 000	26 300	4 859	0.0349	830	0.131	0.131	0.131	0.132	0.1442	0.1452	0.1472	0.1482	0.407	0.0932
Starling	715 500	26	0.1659	7	0.1290	1.051	450 000	28 100	5 193	0.0355	840	0.131	0.131	0.131	0.131	0.1442	0.1442	0.1442	0.1442	0.405	0.0928
Redwing	715 500	30	0.1544	19	0.0926	1.081	450 000	34 600	5 865	0.0372	840	0.131	0.131	0.131	0.131	0.1442	0.1442	0.1442	0.1442	0.399	0.0920
Flamingo	666 600	54	0.1111	7	0.1111	1.000	419 000	24 500	4 527	0.0337	800	0.140	0.140	0.141	0.141	0.1541	0.1571	0.1591	0.1601	0.412	0.0943
Rook	636 000	54	0.1085	7	0.1085	0.977	400 000	23 600	4 319	0.0329	770	0.147	0.147	0.148	0.148	0.1618	0.1638	0.1678	0.1688	0.414	0.0950
Grosbeak	636 000	26	0.1564	7	0.1216	0.990	400 000	25 000	4 616	0.0335	780	0.147	0.147	0.147	0.147	0.1618	0.1618	0.1618	0.1618	0.412	0.0946
Egret	636 000	30	0.1456	19	0.0874	1.019	400 000	31 500	5 213	0.0351	780	0.147	0.147	0.147	0.147	0.1618	0.1618	0.1618	0.1618	0.406	0.0937
Peacock	605 000	54	0.1059	7	0.1059	0.953	380 500	22 500	4 109	0.0321	750	0.154	0.155	0.155	0.155	0.1695	0.1715	0.1755	0.1775	0.417	0.0957
Squab	605 000	26	0.1525	7	0.1186	0.966	380 500	24 100	4 391	0.0327	760	0.154	0.154	0.154	0.154	0.1700	0.1720	0.1720	0.1720	0.415	0.0953
Dove	556 500	26	0.1463	7	0.1138	0.927	350 000	22 400	4 039	0.0313	730	0.168	0.168	0.168	0.168	0.1849	0.1859	0.1859	0.1859	0.420	0.0965
Eagle	556 500	30	0.1362	7	0.1362	0.953	350 000	27 200	4 588	0.0328	730	0.168	0.168	0.168	0.168	0.1849	0.1849	0.1849	0.1849	0.415	0.0957
Hawk	477 000	26	0.1355	7	0.1054	0.858	300 000	19 430	3 462	0.0290	670	0.196	0.196	0.196	0.196	0.216	0.216	0.216	0.216	0.430	0.0988
Hen	477 000	30	0.1261	7	0.1261	0.883	300 000	23 300	3 933	0.0304	670	0.196	0.196	0.196	0.196	0.216	0.216	0.216	0.216	0.424	0.0980
Ibis	397 500	26	0.1236	7	0.0961	0.783	250 000	16 190	2 885	0.0265	590	0.235	0.235	0.235	0.235	0.259	0.259	0.259	0.259	0.441	0.1015
Lark	397 500	30	0.1151	7	0.1151	0.806	250 000	19 980	3 277	0.0278	600	0.235	0.235	0.235	0.235	0.259	0.259	0.259	0.259	0.435	0.1006
Linnet	336 400	26	0.1138	7	0.0855	0.721	4/0	14 050	2 442	0.0244	530	0.278	0.278	0.278	0.278	0.306	0.306	0.306	0.306	0.451	0.1039
Oriole	336 400	30	0.1059	7	0.1059	0.741	4/0	17 040	2 774	0.0255	530	0.278	0.278	0.278	0.278	0.306	0.306	0.306	0.306	0.445	0.1032
Ostrich	300 000	26	0.1074	7	0.0835	0.680	188 700	12 650	2 178	0.0230	490	0.311	0.311	0.311	0.311	0.342	0.342	0.342	0.342	0.458	0.1057
Piper	300 000	30	0.1000	7	0.1000	0.700	188 700	15 430	2 473	0.0241	500	0.311	0.311	0.311	0.311	0.342	0.342	0.342	0.342	0.462	0.1049
Partridge	266 800	26	0.1013	7	0.0788	0.642	3/0	11 250	1 936	0.0217	460	0.350	0.350	0.350	0.350	0.385	0.385	0.385	0.385	0.465	0.1074

Table headings (as printed): r_a Resistance (Ohms per Conductor per Mile): 25°C (77°F) Small Currents; 50°C (122°F) Current Approx. 75% Capacity‡. For the Eagle through Partridge rows the AC resistance values are "Same as dc." x_a Inductive Reactance (Ohms per Conductor per Mile at 1 Ft Spacing, All Currents). x'_a Shunt Capacitive Reactance (Megohms per Conductor per Mile at 1 Ft Spacing).

*Based on copper 97%, aluminum 61% conductivity.

†For conductor at 75°C, air at 25°C, wind 1.4 miles per hour (2 ft/sec), frequency = 60 Hz.

‡ "Current Approx. 75% Capacity" is 75% of the "Approx. Current Carrying Capacity in Amps" and is approximately the current which will produce 50°C conductor temp. (25°C rise) with 25°C air temp., wind 1.4 miles per hour.

INDEX